Few-Body Systems

Suppl. 8

Few-Body Systems

Editor: H. Mitter, Graz | Managing Editor: W. Plessas, Graz

Supplementum 8

Few-Body Problems in Physics '95

In memoriam
Professor Paul Urban (* 15. 6. 1905, † 13. 2. 1995)

Proceedings of the XVth European Conference on Few-Body Problems in Physics, Peñiscola (Castellón), Spain, June 5—9, 1995

Edited by
R. Guardiola

Springer-Verlag Wien GmbH

Dr. R. Guardiola
Departamento di Física Nuclear
Universidad de Valencia
Burjasot, Valencia, Spain

Printing was supported by the Bundesministerium für Wissenschaft, Forschung und Kunst, Wien, Austria

Originally published by Springer-Verlag Wien New York in 1996
Softcover reprint of the hardcover 1st edition 1996

Typesetting: Camera ready by the editor

Printed on acid-free and chlorine-free bleached paper

With 213 Figures and 1 Frontispiece

Die Deutsche Bibliothek – CIP-Einheitsaufnahme

Few body problems in physics '95: in memoriam Professor Paul Urban (15. 6. 1905 – 13. 2. 1995); proceedings of the XVth European Conference on Few Body Problems in Physics, Peñiscola (Castellón), Spain, June 5–9, 1995 / ed. by R. Guardiola. – Wien ; New York : Springer, 1995
(Few body systems / Supplementum ; 8)
ISBN 978-3-7091-9429-4 ISBN 978-3-7091-9427-0 (eBook)
DOI 10.1007/ 978-3-7091-9427-0
NE: Guardiola, Rafael [Hrsg.]; Urban, Paul: Festschrift; European Conference on Few Body Problems in Physics (15, 1995, Peñiscola); GT

ISSN 0177-8811
ISBN 978-3-7091-9429-4

Organizing Committee

R. Guardiola, Departamento Física Nuclear Universidad de Valencia	Chairman
J. Navarro, Inst. Física Corpuscular (IFIC) CSIC and Universidad de Valencia	
E. Oset, Departamento Física Teórica Universidad de Valencia	
J. Ros, Departamento Física Teórica Universidad de Valencia	
V. Vento, Departamento Física Teórica Universidad de Valencia	Scientific secretary

International Advisory Committee

Y. Akaishi (Tokyo)
J. Arvieux (LNS)
V. Belyaev (Dubna)
L. Bogdanova (ITEP)
A. Deprit (NBS)
J. Deutsch (Louvaine)
T.E.O. Ericson (CERN)
S. Fantoni (SISSA)
B. Frois (DAPNIA)
D. von Harrach (Mainz)
K. Killian (Jülich)
D. Measday (TRIUMF)
W. Plessas (Graz)
D.O. Riska (Helsinki)
F. Santos (Lisbonne)
D. Skopik (SAL)
J.A. Tjon (Utrecht)
J.A. Alonso (Valladolid)
B.L.G. Bakker (Amsterdam)
R.F. Bishop (Manchester)
C. Ciofi degli Atti (Perugia)
J.J. de Swart (Nijmegen)
P. de Witt-Huberts (NIKHEF)
M. Fabre de la Ripelle (Orsay)
F. Fernández (Salamanca)
F. Gross (CEBAF and Williamsburg)
U. Kaldor (Tel Aviv)
S. Kowalski (MIT)
G. Orlandini (Trento)
M. Rho (Saclay)
W. Sandhas (Bonn)
B. Schoch (Bonn)
I. Slaus (Zagreb)

Invited speakers

G. Anton (Bonn)
V. Belyaev (Dubna)
J. van den Brand (NIKHEF)
C. Ciofi degli Atti (Perugia)
A. Faessler (Tübingen)
W. Gloeckle (Bochum)
U. Kaldor (Tel Aviv)
L. Kiseleva (Cambridge)
S. Kowalski (MIT)
F.M. Lev (Dubna)
H. Oberhummer (Wien)
W. Plessas (Graz)
S. Rosati (Pisa)
J.A. Tjon (Utrecht)
J. Arvieux (Saclay)
A. Boudard (Saclay)
J.A. Carlson (Los Alamos)
J. de Swart (Nijmegen)
J.L. Friar (Los Alamos)
K. Holinde (Jülich)
O. Kartavtsev (Dubna)
E. Klempt (Bonn)
E. Krotscheck (Texas A& M and Linz)
I. Lindgren (Gothenburg)
E. Oset (Valencia)
C. Petitjean (PSI)
G. Smith (TRIUMF)
T. Walcher (Mainz)

Sponsoring Institutions

Dirección General de Investigación Científica y Técnica
Institució Valenciana d'Estudis i Investigació
Instituto de Física Corpuscular, Centro mixto CSIC–Universidad de Valencia
Vicerrectorado de Investigación de la Universidad de Valencia
Direcció General d'Ensenyaments Universitaris i Investigació, Generalitat Valenciana
Banco Bilbao Vizcaya

Foreword

The fifteenth *European Conference on Few–Body Problems in Physics* has taken place during the week of June 5th to 9th, in the lovely village of Peniscola, approximately midway between Barcelona and Valencia on the Mediterranean coast. This conference continues the tradition initiated in 1972 at Budapest, where the first conference took place, and followed in Graz (1973), Tübingen (1975), Vlieland (1976), Uppsala (1977), Dubna (1979), Sesimbra (1980), Ferrara (1981), Tbilisi (1984), Fontevraud (1987), Uzhgorod (1990), Elba (1991) and Amsterdam (1993).

During this week, a total of one hundred and fifty one scientist were exchanging their knowledge and initiatives in this broad field of Few–Body Physics. Even if the name of the conference restricts its domain to Europe, there has been an important participation of scientists from non-European countries.

A conference with more than twenty years of tradition is already an autonomous being, with a noticeable inertia. Nevertheless, it is a reasonable thought to bend this inertia trying to introduce some innovation, of course, without any damage to the basic structure and objectives of the conference. The way we have found, the reasonable way in our minds, to bring some innovation to this field has been to configure an advisory committee integrated both by experts of the field as well as by renowned experts of other areas. Below there is the full list of our advisors, to whom we are pleased to acknowledge the flux of ideas, subjects and names who led us to configure the skeleton of the program, i.e., the basic structure of the sessions and the invited lecturers and topics. Both their advice and their criticism have been very positive.

The main topics in which the conference has focussed were the theoretical methodology, the experimental measurements and future plans, the exotic atomic systems, the subnuclear degrees of freedom and the role of relativistic effects in few-body systems. All these main topics, as well as complementary subjects, were developed along seventeen sessions, thirteen of them plenary and four parallel sessions. The preparation of the invited talks, as well as the presentation of the selected contributions, has been excellent. Certainly, there is not an *excellence-meter* which permits to give a measurement, but there is an indirect measure: being at a physical distance of fifty meters to the shore, with an excellent weather and a calm sea, the average attendance to the sessions was over ninety per cent. In a large measure, the smooth development of the sessions has been the result of an appropriate control of them by the chairpersons.

We want specially to mention the two panel sessions, one devoted to experimental few-body physics and the other to relativistic effects. These sessions were basically preceded by almost all talks related with their subject, with the aim of serving both as a resume of the present status of the subject and as an advance of the lines of future development. Panel sessions are risky. Their success depends of the interaction between the collectivity and the coordinators of the panel. We believe that in our case we had a strong coupling, the merit being consequence of the excellent work of the coordinators and the high receptivity of the audience. Their names appear below in these proceedings, but we are pleased to mention them explicitly here: Professors Arvieux, Lev, Tjon and van den Brand.

Also an special mention has to be awarded to the speakers of the outlying subjects, Professors Lindgren, Kaldor and Krotscheck. Their communication ability largely compensated the fact of dealing with an audience basically far away of their field of expertise.

One of the aspects which has required a significant effort for the organizers was getting financial help to cover, at least partially, the residence and/or travel expenses of scientists coming from institutions or countries with a delicate economical situation. Our efforts, however, would have been unsuccessful if we had not had the help of several institutions and the dedicated work of some persons. In this respect we must deeply acknowledge the help of the University of Valencia and his Vice-rector of Research, Professor Juli Peretó, and the effort and patience of Professor Willibald Plessas, in his role of coordinator of the Few-Body Physics network of the European program INTAS. Many scientists have transmitted to us their acknowledgement for having found funds for their attendance. We must divert these acknowledgements to the above mentioned persons and institutions.

The explicitly session of the conference does not appear specifically in this book. Better to say, this last session *is this book*. All talks but two of them are included here. The authors have made a good work and also in an incredible short time. I acknowledge the patience of many authors who have attended immediately my requirements directed to a better presentation of the proceedings, as well as the careful job of many other authors whose contributions have not required any further interchange of messages. Again it is my duty to acknowledge the patience of Professor Plessas, this time in his role of Managing Editor of *Few-Body Systems Supplementum*, whose collaboration has significantly simplified the preparation of the final version of this book.

The conference has been satisfactorily developed because of the financial help received from official and private institutions. We want explicitly mention the help of the DGICyT, IVEI, IFIC, Universidad de Valencia, Generalitat Valenciana and the Bilbao-Vizcaya Bank.

At the middle of February the physics community was shocked by the death of Professor Paul Urban. In addition to his role as scientist, Professor Urban had strong ties with the few–body physics community. For that reason, the organizing committee decided to devote this conference to him, as a last tribute to his effort in favour of the scientific knowledge. This decision was formally comunicated in the opening session of the conference by the chairman of the organizing committee.

Valencia, September 24th 1995

Rafael Guardiola
for the Organizing Committee

Contents

Session 1: Methods in the Few-Body Problem

Chair: *M. Fabre de la Ripelle*

Session 2: Methods in the Few-Body Problem

Chair: *S.A. Sofianos*

Session 3: Atomic and Molecular Systems

Chair: *E.O. Alt*

Session 4: Methods in the Few-Body Problem

Chair: *R. Adam*

Session 5: Strong and Electromagnetic Interactions

Chair: *W. von Witsch*

Session 6: Strong and Electromagnetic Interactions

Chair: *E. Jans*

Session 7: Strong and Electromagnetic Interactions

Chair: *R. van Dantzig*

Panel Session 8: Few–Body Systems Probed in Several Ways

Session 9: Atomic Clusters and Exotic Atoms

Chair: *A. Fabrocini*

Session 10: Nuclear Forces, Chirality and Relativistic Effects

Chair: *P. U. Sauer*

Session 11: Antiprotonic and Strange Systems

Chair: *D.W. Sprung*

Session 12: Few–Body Problems in Astrophysics

Chair: *P. Dortmans*

Session 13a: Quark Structure of Hadrons

Chair: *F. Fernández*

Session 13b: Nuclear Structure and Reactions

Chair: *F.D. Santos*

Session 14a: Physics with Strange Particles and Antiprotons

Chair: *M.A. Nagarajan*

Session 14b: Relativistic Effects and Antiprotonic Atoms

Chair: *G. Orlandini*

Session 15: The Baryon–Baryon Interaction

Chair: *T.E.O. Ericson*

Panel Session 16: Relativistic Effects in the Few-Body Problem

Session 17: Closing session

Chair: *B.L.G. Bakker*

Few-Body Systems Suppl. 8, 1–8 (1995)

Few-Body Systems

Paul Urban: Promoter of Modern Theoretical Physics*

W. Plessas

Institute for Theoretical Physics, University of Graz, Universitätsplatz 5, A-8010 Graz, Austria

Ladies and Gentlemen, Dear Colleagues:

Some of you might have been surprised when learning about the devotion of this Conference to the memory of Paul Urban. Some of you, especially the younger colleagues, might not know the person or even the name *Paul Urban*. Others will well understand why the Conference Organizers have spontaneously decided to devote this European Conference on Few-Body Problems in Physics to Paul Urban, once they had received the message about his decease in the middle of February of this year. Of course, in few-body physics we have no Urban equation or Urban potential, like we have a Faddeev equation or a Yukawa or a Yamaguchi potential. Nevertheless, as I hope to make clear by the following presentation, the decision of the Conference Organizers appears highly justified, in view of the rôle that Paul Urban had played in the development of the whole modern nuclear and particle theory, especially also of subatomic few-body physics, from about the middle of this century until quite recently.

While he had been very active in physics till the late seventies, Paul Urban may be considered as one of the big old men who had pioneered the origin of modern physics during the turbulent times of the first half of this century. Being born in Vienna on June 15th, 1905 - in a few days from now he would have celebrated his 90th birthday - he was one of the last eye-witnesses of imperial Austria, the First World War, the exciting times - both in politics and physics - of the following decades, the Second World War, and the reconstruction of Austria and Europe after that catastrophe. Paul Urban spent his childhood and youth in Vienna. He was educated in the comprehensive spirit of the cultural blossom of the late Austrian Empire. This influenced his spirit and thinking thoroughly, as it became evident from his attitude towards science, art, society, and also every-day life. He often told us nice stories. For example, when he usually observed, on the day of Corpus Christi (the notable Christian holiday

*Memorial Speech Given at the Beginning of the 15th European Conference on Few-Body Problems in Physics, Peñiscola, 1995

in spring), the Austrian Emperor Franz Josef to follow, walking, the traditional procession to the service in the church of Hietzing, the quarter of Vienna where also the castle of Schönbrunn is located and where Urban's family lived.

Paul Urban attended a humanistic high school, with Greek, Latin, French, English,... He also got an almost professional education in musics as a pianist (both of his parents had graduated from the Konservatorium Vienna). However, at an instance, he decided to better become a technician. So he went to the Technische Hochschule, now Technical University, of Vienna and enrolled in electrical and mechanical engineering. After he had graduated with a diploma as an engineer in 1928, Paul Urban accepted a job at Siemens in Berlin. After two years, however, he returned back to Austria and joined the Austrian Federal Railways.

At that time he did not only earn a license as a driver of electric locomotives (in order to be able to perform test runs with newly designed steering systems and engines) but, most importantly for his further career, he began to study physics at the University of Vienna. He came under the guidance of Hans Thirring and profitted a lot from the world-open atmosphere at that physics institute. Among his teachers was also the experimental physicist Felix Ehrenhaft, who is known as discoverer of electrophoresis and for the measurement of the charge of the electron. Paul Urban wrote a thesis, in which he dealt with wave mechanics, in particular the calculation of eigenvalues of the Schrödinger equation within the Padé approximation. He received his Ph.D. in 1935 and continued to work in theoretical physics during his free time, keeping close contacts with the Viennese physicists.

Figure 1. Paul Urban. *15.6.1905 in Purkersdorf, Lower Austria. †13.2.1995 in Graz, Styria.

In March 1938 the Nazis took power in Austria. Shortly afterwards Paul Urban was fired by the - now - German Railways (Deutsche Reichsbahnen) because of "lacking political loyalty".

For one year he was without a job but he used the time to publish (mostly together with Theodor Sexl) several papers in physics, about radioactivity and classical diffraction theory. After the breakout of WW II in 1939, Paul Urban was drafted by the German Army, and served as a sergeant in a sapper company, where he was mainly concerned with constructing bridges. A year later, however, he was dismissed from the Army because of an injury and subsequent health problems. Then, in 1940, Paul Urban's professional career as a physicist started when he was appointed assistant at the Physics Institute of the University of Vienna.

Paul Urban passed his habilitation for theoretical physics in 1942 but did not get, as it would have been usual, the "venia docendi", i.e. the right to teach; he was still considered "politically unreliable". During that time there were several Nazi physicists installed on influential positions in Austria.

During the last days of WW II, the Viennese Physics Institute was rescued to the countryside. In this course, Urban came to Innsbruck (Tyrol) where he was offered a position by Arthur March, a former collaborator of Erwin Schrödinger. In 1947 Paul Urban was called to Graz. The position was then vacant, as the professor who had been installed by the Nazis had fled after the end of the war.

Before the beginning of WW II the Physics Institute of the University of Graz (the building of which was constructed still during the regency of emperor Franz Josef, see Fig. 2) enjoyed an excellent reputation. This stemmed from the early times when among other notable scientists Ernst Mach (1866 - 1867) as well as Ludwig Boltzmann (1869 - 1873 and 1876 - 1890) and later on Victor Franz Hess (1920 - 1931 and 1937 - 1938) as well as Erwin Schrödinger (1936 - 1938) were professors there (see Figs. 3 - 7). The latter two, both of them Nobel Laureates, had worked at the Institute until 1938. Despite their certainly worldwide recognition they were dismissed within a few days after the Nazis took power. They had to flee the country - a big physics tradition was abruptly cut.

Paul Urban, who had known the whole physics scene of Austria before WW II, felt the responsibility to restore physics research to the best quality possible. When he came to Graz, there was almost nothing left to build upon. Even the physics building was partly ruined due to bombardment in the last period of the war. He started to run the institute with one assistant and a half-day secretary. When in 1975, after 28 years, he gave over the institute to his successor and former doctoral student Heinrich Mitter, the institute had a staff of 14 permanent positions (among that 6 professors), 4 temporary positions and 2 secretaries, and 1 librarian! Urban certainly knew what were the most important prerequisites to restore the international reputation of the institute: i) to care for good teaching and ii) to perform forefront research. Soon after he had assumed the directorate of his institute, he succeeded to expand his "troop".

Figure 2. The Physics Institute in Graz as it looked around the beginning of the 20th century. The classicistic façade was restored after WW II and is preserved until today.

Figure 3. Ernst Mach, 1838 - 1916. Professor at the University of Graz during 1866 - 1867.

Figure 4. Ludwig Boltzmann, 1844 - 1906. Professor at the University of Graz during 1869 - 1873 and 1876 - 1890.

Figure 5. Photograph of physicists in Graz taken in 1885. Sitting from left to right: E. Aulinger, A. von Ettingshausen (then assistant of Boltzmann), L. Boltzmann, I. Klemencic, V. Hausmanninger. Standing from left to right: W. Nernst (then student of Boltzmann; Nobel Prize in chemistry 1920), F. Streintz, S. Arrhenius (then student of Boltzmann; Nobel Prize in chemistry 1903), R. Hiecke.

The range of physics problems attacked by Paul Urban and his collaborators is unusually wide, at least by present standards. He profitted from the good education he had received in Vienna, the broad studies he had performed both in engineering and physics, and specifically his good knowledge of mathematical techniques. It went always in the same manner: He located the problems, analyzed them by studying the mathematical machinery needed, and had them worked out by his assistants or doctoral students under his permanent and very strict control. In this way many contributions to particle, nuclear, atomic, and condensed-matter physics were achieved. In the centre of interest were electromagnetic processes (electron scattering, bremsstrahlung, radiative corrections,...); some of these results also got summarized in Urban's book *Topics in Applied Quantum Electrodynamics* (Springer, Vienna, 1970). Furthermore Paul Urban worked on general quantum field theory, in particular, symmetries and structural problems as well as weak interactions. He, together with his students and assistants, also attacked the meson-nucleon (pion-nucleon) and nucleon-nucleon interactions. Latter work, together with Harald Zingl (who deceased much too early in 1986 [1]), led directly to few-nucleon problems. This area of research started in Graz in the late sixties/early seventies and resulted among others in the construction of the so-called Graz potentials and in

Figure 6. Erwin Schrödinger, 1887 - 1961. Professor at the University of Graz during 1936 - 1938. Nobel Prize in physics 1933 (together with Paul A.M. Dirac) for the discovery of new and productive forms of atomic theory.

Figure 7. Victor Franz Hess, 1883 - 1964. Professor at the University of Graz during 1920 - 1931 and 1937 - 1938. Nobel Prize in physics 1936 (together with Carl D. Anderson) for the discovery of cosmic radiation.

the solution of manifold problems in few-nucleon physics. Our group originates from those early attempts.

Paul Urban always sought for a vivid atmosphere at his Institute. He eagerly cared for international contacts in order to embed especially the doctoral students and younger assistants into an open environment and to confront them with worldwide competition. He succeeded in inviting many renowned physicists at all times for shorter or longer visits to the Institute. Among them were, e.g., Otto Hahn, Lise Meitner, Werner Heisenberg, Karl Wildermuth (pupil and assistant of the latter; he became Dr.rer.nat.h.c. of the University of Graz in 1989). Urban also went often abroad to visit and keep close contacts with his colleagues, e.g., Arnold Sommerfeld in Munich, Louis de Broglie in Paris, Mauro Piccone in Rome, Eugen Guth in Oak Ridge, Victor Weisskopf in Cambridge (USA), Karl Wildermuth in Tübingen, Fritz Rohrlich in Syracuse,...

Resulting from the attempts of establishing and maintaining good scientific (and personal) contacts with physicists abroad was also the foundation of the Schladming Winter School in 1962. This institution over the years developed into a highly recognized one. Many important phycisists of the recent decades lectured in Schladming, among them 7 Nobel Laureates so far. For few-body physics certainly the Schools in 1972 (on "Multiparticle Aspects" [2] with lecturers M. Gell-Mann, E. Predazzi, W. Sandhas, W. Thirring, J. Wess, K. Wildermuth and others), in 1974 (on "Progress in Particle Physics" [3] with lecturers A. Bassetto, J.S. Bell, H. Pietschmann, F. Rohrlich, W. Sandhas and others), in 1985 (on "Nucleon-Nucleon and Nucleon-Antinucleon Interactions" [4] with lecturers A.M. Green, N. Isgur, K. Kilian, I. Šlaus, R. Vinh Mau and others), and in 1993 (on "Substructures of Matter" [5] with lecturers G. Altarelli, H. Arenhövel, E. Reya, B. Schoch, W.J. Stirling, R.E. Taylor, G. Wolf and others) were most important.

Among the lasting initiatives of Paul Urban is also the foundation of the Austrian journal *Acta Physica Austriaca* (together with Fritz Kohlrausch and Hans Thirring) in 1948 and likewise the foundation of the Austrian Physical Society in 1950. He edited *Acta Physica Austriaca* for 37 years (until 1985). Then the journal got transformed into *Few-Body Systems*, the meanwhile established forum of the few-body-physics community, and Paul Urban became its Honorary Editor.

Paul Urban retired in 1975. The last Schladming School he directed was the one in 1976. Heinrich Mitter, who was called from Tübingen to Graz, became his successor both as chair and director of the Schladming Schools. As Professor Emeritus, Urban spent an agreeable life in Graz. Only, over the recent years he became more and more handicapped by a continuously weakening eyesight, what could not be halted by any medical means. Nevertheless, he visited *his* Institute almost daily. He discussed with the people around and tried to keep up with the more modern development of physical theory. He also did not get tired with always encouraging especially the students and younger colleagues to work hard and perform original research.

A long time after his retirement, on the occasion of his 85th birthday in 1990, Paul Urban donated a considerable amount of his private savings to a foundation at our University in order to promote scientific contacts. The Paul-Urban Foundation grants scholarships for sending abroad young physicists graduated from our Institute and for inviting scientists to the Institute for research collaborations.

Paul Urban was a member of several academies and physical societies. He also received many medals of honour in Austria and foreign countries. He himself was proudest of the Erwin-Schrödinger Award by the Austrian Academy of Sciences with the citation "Dem Begründer der Österreichischen Schule der modernen theoretischen Physik" (to the founder of the Austrian School of modern theoretical physics) and of the honorary doctorate awarded to him by the University of Tübingen "in Anerkennung seiner Verdienste als Wissenschaftler auf weiten Gebieten der theoretischen Physik, als akademischer Lehrer einer Generation von Physikern und als aktiver und unermüdlicher Förderer der internationalen wissenschaftlichen Zusammenarbeit" (in recognition of his merits as scientist in broad areas of theoretical physics, as academic teacher of a generation of physicists, and as active and tireless promoter of international scientific collaboration).

For more background information (especially on the physics scene after WW II until the late seventies), a list of publications of Paul Urban, a summary of awards he received etc. we refer to his biography [6].

As an academic teacher and scientist, Paul Urban has impressed on several generations of physicists. Also, and in particular, few-body physics has profitted a lot from his initiatives. I am sure that many colleagues will appreciate his works and deeds, those who knew Paul Urban personally will always keep a fond memory of him.

Acknowledgement. I want to express my gratitude to Prof. A. Hohenester for providing some of the photographs and for giving valuable hints about the history of our Institute. I am also grateful to V. Christian, C. Neubacher, and W. Schweiger for help and advice in processing the photographs.

References

1. W. Plessas: Few-Body Systems **2**, 53 (1987)
2. P. Urban (ed.): *Elementary Particle Physics: Multiparticle Aspects* (Acta Physica Austriaca Suppl. 9). Wien: Springer 1972
3. P. Urban (ed.): *Progress in Particle Physics* (Acta Physica Austriaca Suppl. 13). Wien: Springer 1974
4. H. Mitter and W. Plessas (eds.): *Nucleon-Nucleon and Nucleon-Antinucleon Interactions* (Acta Physica Austriaca Suppl. 27). Wien: Springer 1985
5. L. Mathelitsch and W. Plessas (eds.): *Substructures of Matter as Revealed with Electroweak Probes* (Lecture Notes in Physics, vol. 426). Berlin: Springer 1994
6. P. Urban: *Erinnerungen.* Graz: Institute for Theoretical Physics 1985

Few-Body Systems Suppl. 8, 9–20 (1995)

Few-Body Systems

Nuclear Three- and Four-Body Systems

W. Glöckle[1], H. Kamada[1], H. Witała[2], D. Hüber[1], J. Golak[2], K. Miyagawa[3], S. Ishikawa[4]

[1] Institut für Theoretische Physik II, Ruhr Universität Bochum, 44780 Bochum, Germany

[2] Institute of Physics, Jagellonian University, Cracow, Poland

[3] Faculty of Liberal Arts and Science, Okayama University of Science, Ridai-cho Okayama 700, Japan

[4] Department of Physics, Hosei University, 2-17-1 Fujimi, Chiyoda-ku, Tokyo 102, Japan

Abstract. Recent insight in 3- and 4-nucleon bound state properties, the hypertriton and $3N$ scattering are reviewed. New results for inclusive electron scattering on ^{3}He (^{3}H) including exactly the final state interaction are given.

In this presentation we restrict ourselves to the Faddeev-Yakubovsky approach for the three- and four-body systems. Other techniques and results are addressed elsewhere in this volume.

1 Three- and Four-Nucleon Bound States

Binding energies are the first spot to check the nuclear Hamiltonian, but wave-function properties are equally important. We shall address both. We work in momentum space and thus use the integral forms of the Faddeev-Yakubovsky equations. For three nucleons the bound state is decomposed as

$$\Psi = (1 + P)\ \psi \tag{1}$$

and the Faddeev component ψ obeys the Faddeev equation

$$\psi = G_0\ t\ P\ \psi \tag{2}$$

Here G_0 is the free $3N$ propagator, t the NN t-matrix between particles 2 and 3 and $P = P_{12}P_{23} + P_{13}P_{23}$. In case of four nucleons the bound state is decomposed into 18 Faddeev-Yakubovsky components

$$\Psi = (1 + P - P_{34}\ P + \tilde{P})(1 - P_{34})\ \psi_1 + (1 + P - P_{34}\ P + \tilde{P})\ \psi_2 \tag{3}$$

Table 1. Our ^{3}H and ^{4}He binding energies based on $j \leq 4$ NN forces.

Potential	p_d	$E_{^3\mathrm{H}}$	Potential	p_d	$E_{^3\mathrm{H}}$	$E_{^4\mathrm{He}}$
Bonn B (cd)	4.99	7.92	Bonn B	4.99	8.14	27.07
Nijm II (cd)	5.65	7.64	Nijm78	5.39	7.63	25.10
Nijm I (cd)	5.68	7.73	Paris	5.77	7.46	24.32
Nijm93 (cd)	5.76	7.66	Ruhrpot	5.85	7.64	24.60
AV18 (cd)	5.78	7.65	AV14	6.08	7.68	24.73
			AV8	6.08	7.79	25.31

with $\tilde{P} \equiv P_{13}\, P_{24}$. The two components ψ_1 and ψ_2 stand for the 3+1 and 2+2 partitions, respectively, and obey the set of Yakubovsky equations

$$\begin{aligned} \psi_1 &= -G_0\, T\, P\, P_{34}\, \psi_1 + G_0\, T\, P\, \psi_2 \\ \psi_2 &= G_0\, \tilde{T}\, \tilde{P}\, (1 - P_{34})\, \psi_1 \end{aligned} \tag{4}$$

The kernels in Eq. (4) contain T-matrices to the underlying subcluster problems (3 particles and noninteracting pairs) and the free $4N$ propagator G_0. They are defined via

$$\begin{aligned} T &= t + T\, P\, G_0\, t \\ \tilde{T} &= t + \tilde{T}\tilde{P}\, G_0\, t \end{aligned} \tag{5}$$

We solve the integral equations (2), (4) - (5) in a partial wave momentum space basis leading to amplitudes in 2 and 3 variables and refer the reader to [1]. Any NN force can be used and one achieves essentially fully converged results. Some of them have been already reported in [2] and we update them in Table 1. We agree fairly well with the results of the Pisa group [3], 24.78 and 25.34 MeV for AV14 and AV8, resp., but disagree slightly with the GFMC results [4], 24.95 and 25.78 MeV. In Table 1 (cd) denotes charge dependent calculations. The large binding energies for Bonn B are likely due to its inherent nonlocality. Namely using a local potential generated by inverse scattering theory [5] out of the Bonn B NN phaseshifts leads to 7.76 MeV in comparison to 8.10 MeV for the original Bonn B (18 channels). See also [6] for further arguments.

With the present day NN forces there is underbinding and 3NF mechanisms are needed in addition. The missing binding energies, however, are a few percent of the total potential energies provided by the NN forces. See [7] for a recent study on the inclusion of the π-π, π-ρ and ρ-ρ exchange TM 3NF [8].

Nucleon momentum distributions $n(p)$ and probabilities $C(r)$ to find 2 nucleons at a distance r in d, ^{3}H and ^{4}He are shown in Fig. 1 for AV14. Detailed behaviours for large p and small r depend on the choice of the NN force.

Of interest are also state-dependent correlation functions

$$C_{lsjt}(\boldsymbol{r}) \equiv \langle \Psi | \delta(\boldsymbol{r} - \boldsymbol{r}_{ij})\; P_{lsjt} | \Psi \rangle \tag{6}$$

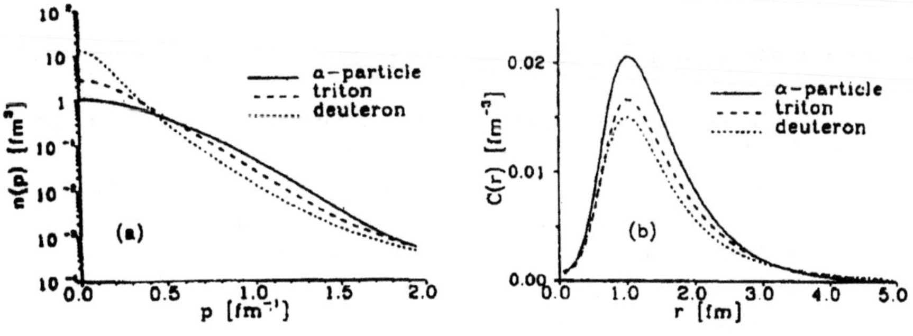

Figure 1. a) Nucleon momentum distributions, b) NN correlation functions.

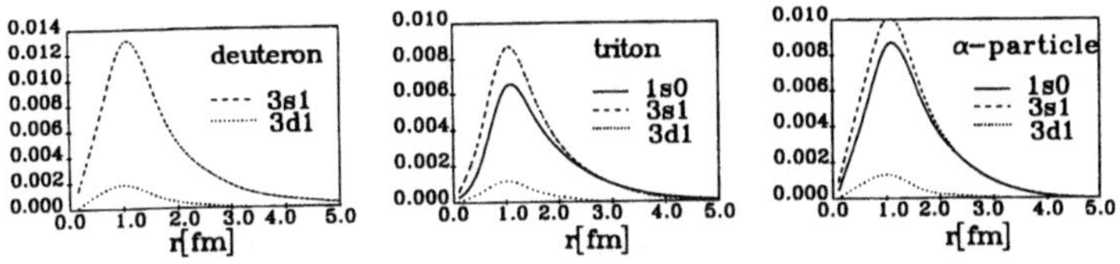

Figure 2. State dependent two-nucleon correlation functions.

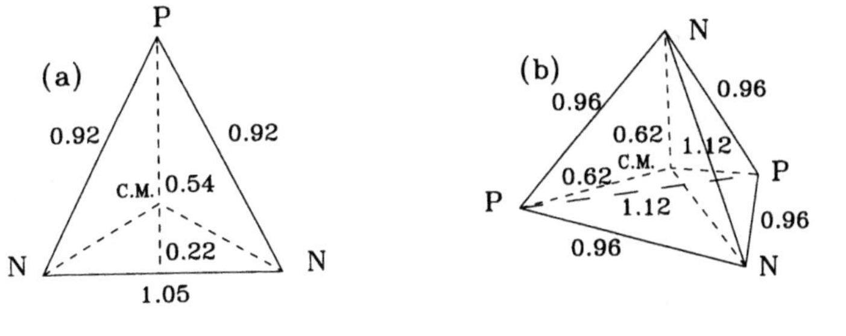

Figure 3. The most probable geometrical nucleon sites for (a) ^{3}H and (b) ^{4}He.

The largest ones are displayed in Fig. 2 for d, ^{3}H and ^{4}He based on AV14.

What are the most probable geometrical sites for the nucleons in ^{3}H and ^{4}He? The triangular and tetrahedral shapes are shown in Fig. 3; see also [9]. For the α-particle the pair distances are slightly larger than for ^{3}H. These numbers depend slightly on the chosen NN potential. If one puts the midpoints of spheres of radius 0.8 fm at the most probable sites, the overlap volumes of all pairs of nucleons to all triples of nucleons to the one of 4 nucleons are in the ratio 4.7 : 1 : 0.11. For ^{3}H one has 8.6 : 1. This very naive geometrical argument tells that in ^{4}He hardly any 4-body forces should be expected, but some 3-body force effects like in ^{3}H.

All that information indicates that nucleons in ^{3}H and ^{4}He are mostly well separated, interacting mainly by two-body forces, and one should not expect significant deviations from the free nucleon properties (like form factors). Of course with corresponding lower probabilties nucleons are closer together and one should see the deviations from the free nucleon properties using appropriate wave lengths of external probes.

2 Inclusive electron scattering on ^{3}He (^{3}H)

Inclusive electron scattering on ^3He (^3H) has been studied in [10] using the MT s-wave NN forces. FSI effects were emphazised. We used realistic NN forces and developped two approaches including FSI exactly. One relies on the separate determination of the *pd* and *ppn* breakup processes [11] and consecutive integration over all breakup configurations, the other on a type of optical theorem approach [12], which gives directly the two response functions. In both cases one has to solve an integral equation of the type

$$U = U^0 + t\; G_0\; P\; U \tag{7}$$

where the driving term, in the second approach for instance, has the form

$$U^0 = (1 + t\; G_0)\; \rho \mid \Psi > \tag{8}$$

with ρ being the density operator and $\mid \Psi >$ the target state. In the second approach [12] the longitudial response function is then

$$R_L(Q,\omega) = -\frac{1}{\pi}\; Im\; < \Psi \mid \rho^\dagger (1+P)\; G_0 \mid U > \tag{9}$$

The very good agreement of the numerical results in the two quite different approaches demonstrates the accuracy of our results. We determined in [13] the two responses R_L and R_T for $Q = 174, 250$ and 300 MeV/c and compared to Bates and Saclay data. The agreement for R_L is quite good, while for R_T theory underestimates the data slightly presumably caused by not yet taking MEC into account. For one example see Fig. 4.

The Coulomb sum rule S_L has been studied with the aim to extract the *pp* correlation function $C_{pp}(r)$ [14]. We confirmed [13] that independent of the choice of the NN forces $C_{pp}(r)$ for ^{3}He is nearly a factor 10 smaller than the correlation term extracted from the data. Since that term results as a small quantity out of cancellations between S_L, F_{ch} and $F_1^p(G_E^p)$ one has to expect great sensitivity to relativistic effects and MEC contributions. The study [15] indeed revealed that already the lowest order relativistic corrections to the non-relativistic density operator led to changes of nearly an order of magnitude in the right direction. Apparently this is not a small modification and a consistent relativistic framework is needed. Because of the cancellations and also the need of extrapolation in evaluating S_L improved data would be very desirable.

3 The Hypertriton

The number and quality of $\Lambda - N$ scattering data is still rather poor and therefore information from hypernuclei is needed to learn more about the hyperon-nucleon (YN) interaction. The hypertriton as the lightest hypernucleus is the

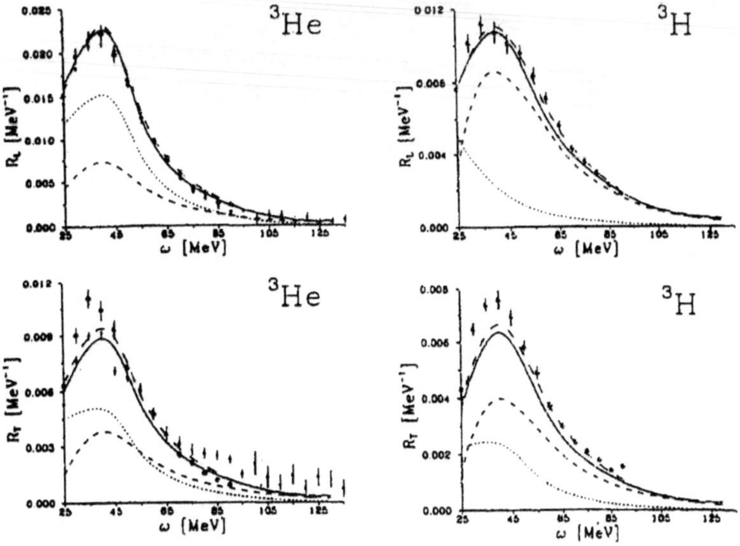

Figure 4. Responses at $Q = 250$ MeV/c for ^{3}He and ^{3}H, solid curve Bonn B, long dashed MT, dotted Nd and short dashed ppn contributions.

first nontrivial test case. We performed [16] fully converged calculations using realistic meson exchange baryon-baryon interactions including Λ-Σ conversion. The Faddeev equations take the form of coupled matrix equations

$$\begin{aligned} \underline{\psi}^{(12)} &= \frac{1}{E - \underline{H_0}} \underline{T_{12}} \, (1 - P_{12}) \underline{\psi}^{(13)} \\ \underline{\psi}^{(13)} &= \frac{1}{E - \underline{H_0}} \underline{T_{13}} \left(\underline{\psi}^{(12)} - P_{12} \underline{\psi}^{(13)} \right) \end{aligned} \tag{10}$$

where each of the two Faddeev amplitudes have two components

$$\underline{\psi^{(ij)}} = \begin{pmatrix} \psi_\Lambda^{(ij)} \\ \psi_\Sigma^{(ij)} \end{pmatrix} \tag{11}$$

and the free propagator, the NN and YN T-operators, $\underline{T_{12}}$ and $\underline{T_{13}}$ are 2×2 matrices. While $\underline{T_{12}}$ is diagonal $\underline{T_{13}}$ includes the $\Lambda N \leftrightarrow \Sigma N$ transitions. The two orthogonal parts of the total wavefunction are given as

$$\begin{aligned} \Psi_{NN\Lambda} &= \psi_\Lambda^{(12)} + (1 - P_{12}) \, \psi_\Lambda^{(13)} \\ \Psi_{NN\Sigma} &= \psi_\Sigma^{(12)} + (1 - P_{12}) \, \psi_\Sigma^{(13)} \end{aligned} \tag{12}$$

See [16] for the derivation and the way to handle that set precisely. The binding energy using the Nijmegen YN interaction [17] is -2.36 MeV (hardly dependent on the choice of the NN force), which equals the experimental value of

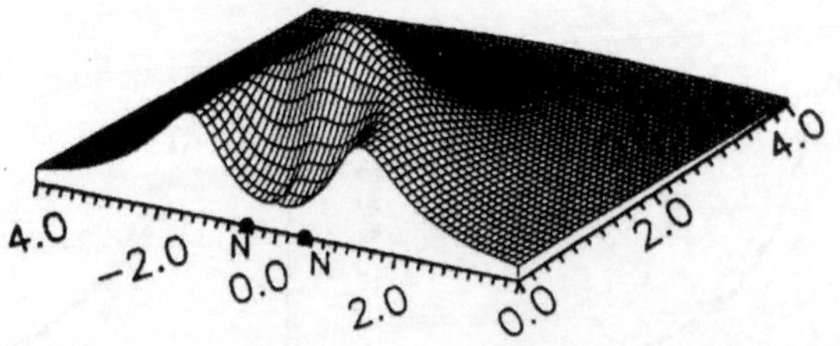

Figure 5. The probability for the site of Λ for given NN sites, as indicated.

-2.35 ± 0.5 MeV. The probability to find the hyperon as a Λ is 0.995 with little probability left for the Σ. The probability to find a deuteron inside the hypertriton is 0.987 in contrast to 0.448 in the triton. If one breaks the total potential and kinetic energy apart, one finds that the two nucleons provide -1.77 MeV binding energy and the rest comes from the Λ-Σ conversion, which thus plays a decisive role. The recent Jülich YN interaction [18] does not bind the hypertriton [16]. Other states for that $\Lambda(\Sigma)NN$ system with quantum numbers T = 0, J = 3/2 and T = 1, J = 1/2 or 3/2 turn out to be unbound [16].

The most probable spatial configurations are that Σ is very close to one or the other nucleon as a consequence of the large seperation energy including the mass difference between Σ, and Λ and the 2 nucleons form approximately an equilateral triangle. In contrast to a nucleon in ^{3}H the Λ is less confined to its most probable site as shown in Fig. 5.

4 Three-Nucleon Scattering

The last few years have seen the progress that $3N$ scattering can be treated precisely for any type of NN force and even including $3N$ forces [9]. Two recent benchmark calculations illustrate the maturity of handling that challenging problem. In one the Faddeev equations are solved in momentum space as integral equations and in configuration space as partial differential equations. The resulting breakup amplitudes were found [19] to agree perfectly well with each other as examplified in Fig. 6. In the other study [20] the S-matrix for elastic nd scattering as parametrised in form of eigenphases and mixing parameters has been compared below the deuteron breakup threshold using our way and the pair correlated hyperspherical harmonic basis method in configuration space. The perfect agreement is examplified in Table 2.

The results now shown are based on the solution of the Faddeev equation

$$T = t\,P + t\,P\,G_0 T \tag{13}$$

in momentum space. The operator T determines the transition operators for the breakup, $U_0 = (1 + P)T$, and for elastic nd scattering, $U = P\,G_0^{-1} + PT$. We use the most modern NN forces, Nijm 93, Nijm I, Nijm II [21], AV18 [22],

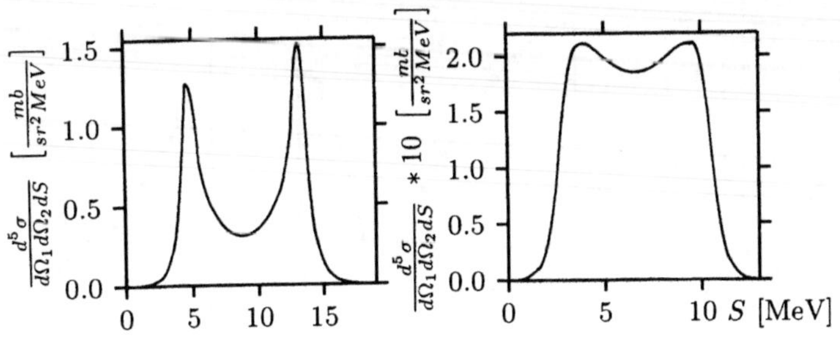

Figure 6. Breakup cross sections evaluated by two independent methods [19].

Table 2. Examples for S-matrix parameters evaluated by two methods [20].

eigenphase	Pisa	Bochum	mixing parameter	Pisa	Bochum
$\delta_{\frac{3}{2}0}$	−61.3	−61.3	ϵ	0.716	0.717
$\delta_{\frac{1}{2}2}$	1.54	1.55	ξ	1.01	1.01
$\delta_{\frac{3}{2}2}$	−2.77	−2.77	η	−0.246	−0.245

which are essentially phase equivalent and adjusted to the most recent Nijm PSA [23]. We shall also show predictions of the OBE Bonn B [24] potential. An updated version CD Bonn is coming up [24].

Let us first regard $3N$ observables based on the NN force picture only. The total nd cross section data are perfectly reproduced by theory as shown in Fig. 7a. Moreover there is no dependence on which of the NN forces is chosen.

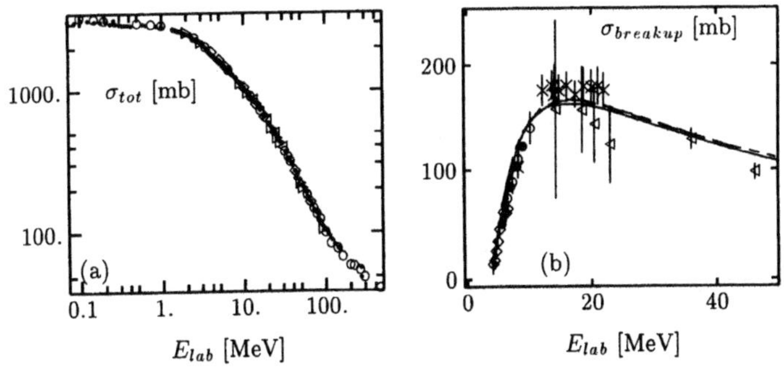

Figure 7. (a) Total nd cross section, data (•), theory (open symbols); (b) total nd breakup cross section, data in comparison to theory.

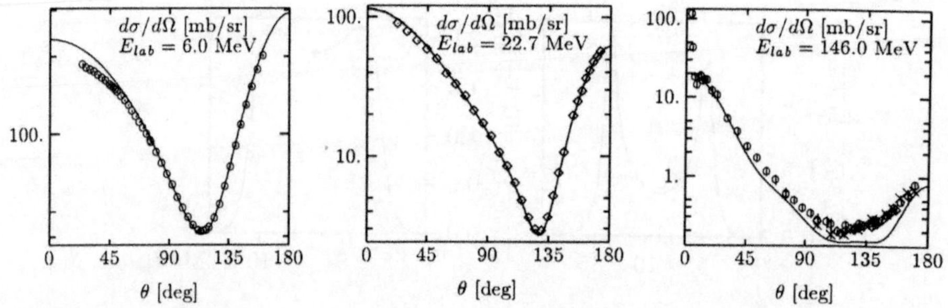

Figure 8. Examples for angular distributions in elastic Nd scattering.

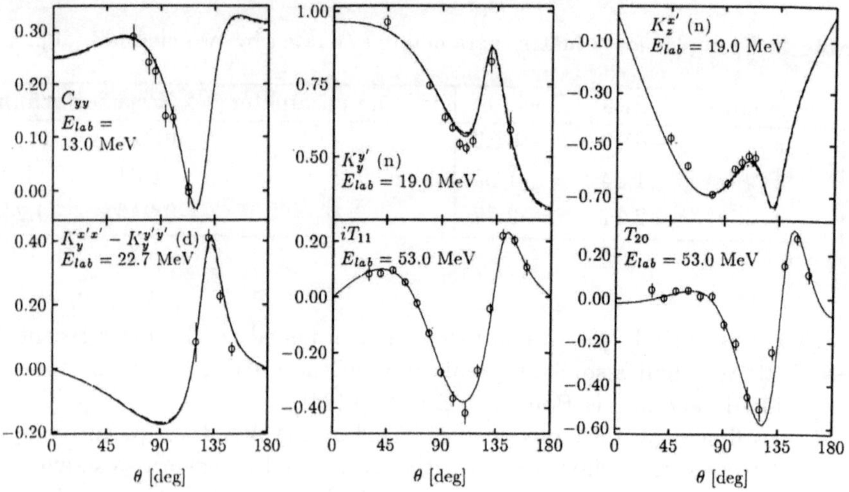

Figure 9. Examples for spin observables in elastic Nd scattering.

Also the angular distribution for elastic Nd scattering is very well described by theory. Examples are displayed in Fig. 8. Coulomb force effects are small and restricted mostly to forward angles, as the comparison of nd and pd data reveals. At the highest energy considered the discrepancy is possibly caused by neglected relativistic effects. This is under study. There are many spin observables in elastic scattering. In general we have very good agreement and we display a few examples in Fig 9. It would be very interesting to see experimental data for spin transfer coefficients and correlation coefficients also at higher energies. Up to now we found only one outstanding problem in elastic scattering, the analyzing power A_y, at low energies. An example is shown in Fig. 10. Since it is known that A_y depends sensitively on the ^3P_j NN force components, we allowed for changes in the ^3P_j NN phases of up to 3%, readjusted the Nijm I NN force to these modified phases and determined A_y again. The result is also shown in Fig. 10. Though A_y moves into the direction of the data the effect is much too small and the puzzle remains.

Let us turn to the breakup process. Fig. 7b shows the total breakup cross

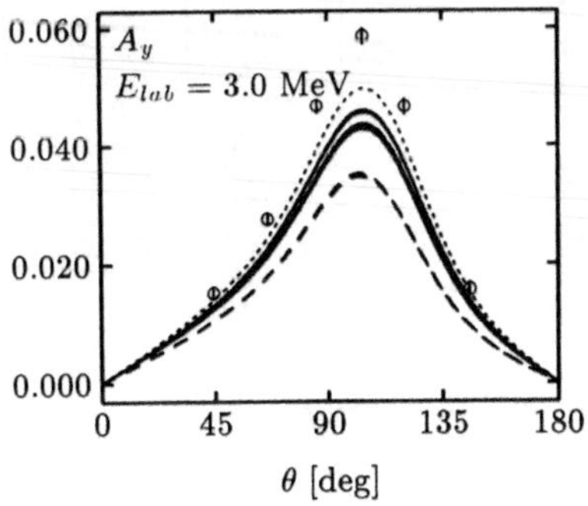

Figure 10. A_y in elastic nd scattering, Nijm I, II, 93, AV18, Bonn B (solid), Bonn B + TM 3NF's (dashed), Nijm I readjusted (dotted).

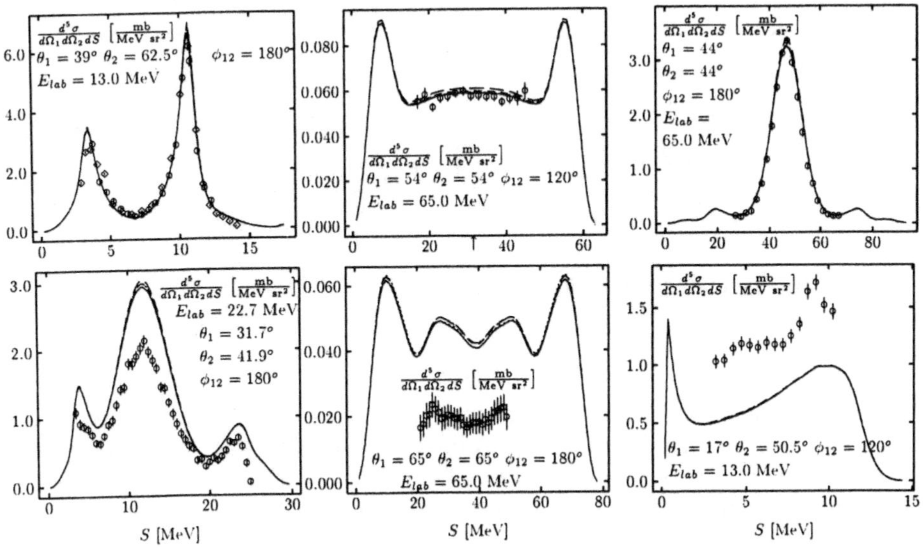

Figure 11. Examples for breakup cross sections with perfect agreement (upper row) and strong disagreement (lower row).

sections. Since we have excellent agreement to the elastic angular distribution and to the total cross section, the data for the total breakup should be updated and also generated at higher energies. Kinematically complete breakup cross sections exhibit a very controversial picture. There are cases with spectacular agreement and cases of strong disagreement. We display a few examples in Fig. 11. Unfortunately up to now no well established data basis exists and the existing data are mostly based on just one measurement.

We are comparing breakup cross sections for the modern NN forces covering all phase space. First results at 13 MeV are: all the NN force predictions agree within 10%. All of the 8-10% deviations occur in nn and np FSI peaks and can be traced back to different nn and np scattering lengths, especially for AV18,

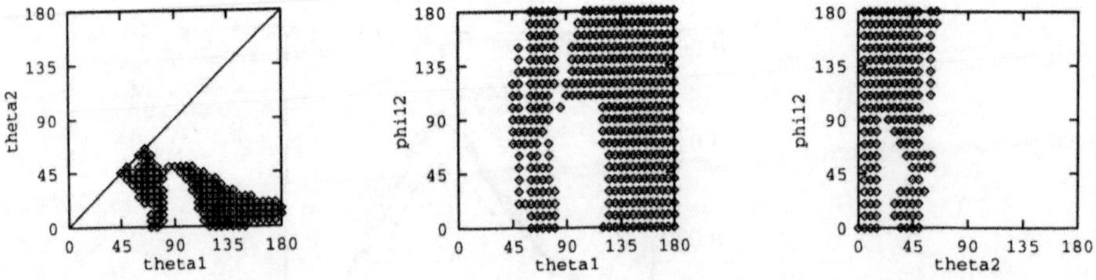

Figure 12. Insensitive regions in θ_1, θ_2 and ϕ_{12}.

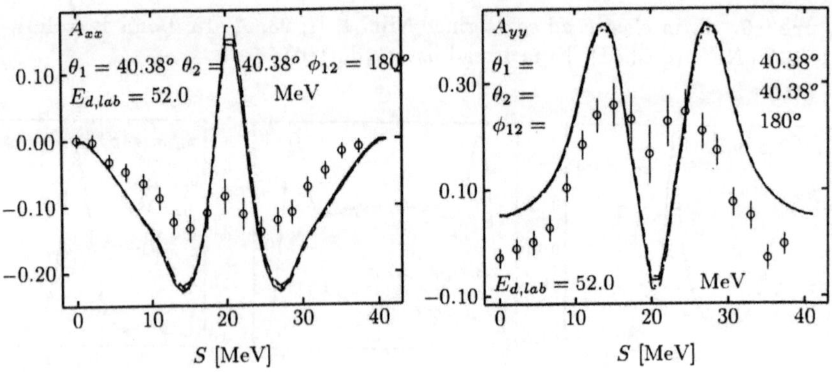

Figure 13. A_{xx} and A_{yy} for pd breakup [26].

which we use without electromagnetic terms. Fig. 12 displays in the space of the angular positions of two detectors, θ_1, θ_2 and $\phi_{12} = \phi_1 - \phi_2$, the region, where all the potential predictions coincide within 3% and are $\geq 1\ \frac{\text{mb}}{\text{MeV sr}^2}$. Precise measurements there would be of special interest, because deviations to experiment would challenge all present day forces. Ideally, however, would be the comparison of all breakup cross sections in theory and experiment (a "4 π" study, as done in the past [25], but based on oversimplified forces). This would be the very best test of the potential energy between three nucleons.

There are few spin observables in the breakup process. Analysing powers agree in general very well with theory and are not displayed. Recently measured tensor analysing powers A_{xx} and A_{yy} [26], however, are in striking disagreement with theory as shown in Fig. 13. More data are definitely needed.

Will the results described up to now be modified by the inclusion of a 3NF? Up to now we used the π-π, π-ρ and ρ-ρ exchange TM 3NF [8] and studied [27] its effect only at 3 and 14 MeV. The π-ρ 3NF acts opposite to the π-π force and cancels its effect partially. The ρ-ρ force is negligible. Since the 3NF's were used in combination with Bonn B, which leads to overbinding in the triton, we possibly even overestimate their effects in the $3N$ continuum. On top they are found to be small in general. They are negligible for σ_{tot}, $\frac{d\sigma}{d\Omega}$ and T_{20}, T_{21}, T_{22} in elastic scattering. Some effects are predicted for C_{xx}, C_{yy}, $K_z^{x'}$, $K_y^{y'}$, $K_x^{z'}$ and the n→d polarisation transfer coefficients. For A_y the effect worsens the description, see Fig. 10. For the breakup only the specific angular dependence of the FSI peak heights stick out. Contrary to the often expressed expectations, the space star configuration hardly shows a 3NF effect. For more details see [27]. In addition one should mention, that if one adjusts the 3NF to the ^{3}H binding energy one finds the phenomena of scaling with the ^{3}H binding energy[9] for some observables, not all. This feature is however restricted to very low energies well below 10 MeV.

The next "round" in theory and experiment will be more demanding. The pp Coulomb force should be included (together with modern NN forces) and theory should be pushed to the pion threshold. The Pisa group [28] made an important step forward by treating the pp Coulomb force exactly together with modern forces, though still only below the deuteron breakup threshold. More high quality experiments should be performed for cross sections and spin observables including "4π detection" for the breakup process. The $3N$ continuum will be an ideal test ground also for newly coming up dynamical models.

5 Outlook

Scattering of four nucleons upon each other in the Faddeev-Yakubovsky scheme or the equivalent AGS scheme has already some history [29]. More recently and using finite rank approximations to realistic NN forces interesting progress has been achieved [30]. Also in our group first steps are undertaken to solve the $4N$ Yakubovsky equations directly for realistic forces not using any seperabel approximation. The Pisa group with their configuration space method performed especially strong studies [31]. Four nucleon scattering will be of special interest in view of the resonances present in the $4N$ compound systems.

In view of the results and quantitative agreements with data achieved up to now one is temped to conjecture, that at least for the light nuclear systems and not too high energies (nonrelativistic regime) the nuclear Hamiltonian based on realistic NN forces and a small admixture of 3NF's (contributing a few percent to the total potential energy) describes the observables quantitatively well. The next round of supercomputers and improved algorithms will allow the verification of that claim.

References

1. W. Glöckle: Nucl. Phys. **A 381**, 343 (1982); H. Kamada, W. Glöckle: Nucl. Phys. **A 548**, 205 (1992) and Few-Body Systems Suppl. **7**, 217 (1994)
2. W. Glöckle, H. Kamada: Phys. Rev. Lett. **71**, 971 (1993)
3. M. Viviani, A. Kievsky, S. Rosati: Few-Body Systems **18**, 25 (1995)
4. J. Carlson, R. Schiavilla: Few-Body Systems Suppl. **7**, 349 (1994)
5. H.V. von Geramb: Private Communication
6. Y. Song, R. Machleidt: Contribution to the 14th International Conference on Few-Body Problems in Physics, Williamsburg 1994, page 189

7. A. Stadler et al.: Phys. Rev. **C** in press
8. S.A. Coon, M.T. Peña: Phys. Rev. **C48**, 2559 (1993)
9. W. Glöckle et al.: In: *Few-Body Problems in Physics.* AIP Press: New York 1995
10. E. van Meijgaard, J.A. Tjion: Phys. Rev. **C45**, 1463 (1992)
11. S. Ishikawa et al.: Nuovo Cim. **A107**, 305 (1994); J. Golak et al.: Phys. Rev. **C** in press
12. S. Ishikawa et al.: Phys. Lett. **B339**, 293 (1994)
13. J. Golak et al.: Bochum preprint 1995
14. D.H. Beck: Phys. Rev. Lett **64** 268 (1990)
15. R. Schiavilla, R.B. Wiringa, J. Carlson: Phys. Rev. Lett. **70**, 3856 (1993)
16. K. Miyagawa, W. Glöckle: Phys. Rev. **C48**, 2576 (1993); K. Miyagawa, H. Kamada, W. Glöckle, V.G.J. Stoks: Phys. Rev. **C** in press
17. P.M.M. Maessen, Th.A. Pijken, J.J. de Swart: Phys. Rev. **C40**, 2226 (1989)
18. A.G. Reuber, K.Holinde, J. Speth: Czech. J. Phys. **42**, 1125 (1992)
19. J.L. Friar et al.: Phys. Rev. **C** in press
20. D. Hüber et al.: Phys. Rev. **C51**, 1100 (1995)
21. V.G.J. Stoks et al.: Phys. Rev.**C49**, 2950 (1994)
22. R.B. Wiringa, V.G.J. Stoks, R. Schiavilla: Phys. Rev. **C51**, 38 (1995)
23. V.G.J. Stoks et al.: Phys. Rev. **C48**, 792 (1993)
24. R. Machleidt: Adv. Nucl. Phys. **19**, 189 (1989)189 and Private Communication
25. G.J.F. Blommestijn et al.: Nucl. Phys. **A365**, 202 (1981)
26. L.M. Quin et al.: Nucl. Phys. **A** in press
27. H. Witała et al.: Submitted for publication
28. A. Kievsky et al.: to appear in Phys. Rev. **C**
29. A.C. Fonseca: Few-Body Systems Suppl. **7**, 177 (1994)
30. E. Uzu et al.: Contribution to the 14th International Conference on Few-Body Problems in Physics, Williamsburg 1994, page 205
31. M. Viviani et al.: Contribution to the 14th International Conference on Few-Body Problems in Physics, Williamsburg 1994, page 840

Few-Body Systems Suppl. 8, 21–31 (1995)

Few-Body Systems

Correlated Hyperspherical Harmonic Functions for Few–Nucleon Systems

S. Rosati[1,2], M. Viviani[2], A. Kievsky[2]

[1] Phys. Dept., Pisa University, I-56100 Pisa, Italy

[2] INFN, Sezione di Pisa, I-56100 Pisa, Italy

Abstract. The correlated Hyperspherical Harmonic variational technique to describe the ground state and elastic scattering states below the break–up threshold for three– and four–nucleon systems is presented. Accurate numerical results are given for ground state properties and scattering observables of the reaction p–d and p–^{3}He. The agreement with other available results and with experimental analyses is satisfactory. A discussion on the present status of the non–relativistic model hamiltonians is included.

1 Introduction

The ground state of the three and four nucleon systems has been analyzed by many authors using different methods in the last few years. Among these approaches of particular relevance are the Faddeev and Faddeev–Yakubovsky integral equation methods [1, 2, 3], the Green Function Monte Carlo (GFMC) technique [4] and the variational method [5, 6, 7, 8], which have been applied to solve the Schroedinger equation. All of them have yielded extremely close results for the ground state properties of nuclei with $A = 3, 4$. More recently also N–d scattering states with realistic NN interactions have been calculated by either the Faddeev method in momentum space [9, 10] or the Kohn variational principle, in conjunction with the Schroedinger equation [10, 11]. The investigation on four nucleon scattering states presents large numerical difficulties and only a few results have been obtained so far.

At present, different realistic local and non local NN potentials, which fit very accurately the bound and scattering properties of two nucleons, are available. The problem of their adequacy in reproducing also the properties of systems with $A > 2$ is a crucial one. To this aim, the three and four nucleon systems represent good tests for these various potentials as they allow for very precise calculations. The first step is the study of the bound states; such a study has required in the past a lot of efforts but, at present, the situation is very satisfactory. The conclusion is that local potentials do not reproduce correctly the light nuclei binding energies. For non local potentials, as for example the Bonn potential [12], the triton binding energy is rather close to the experimental value. It is possible to reproduce the experimental binding energies of triton and alpha particle by introducing three–nucleon interaction (TNI) terms in the hamiltonian. Different choices have been proposed, but again a phenomenological choice might be the most appropriate. The bound state properties are insufficient to determine, in a phenomenological way, these terms. However, there is a large set of observables associated to scattering states of few nucleons and they should be used to this purpose.

Having in mind the previous points, we discuss in this paper the method for studying few–nucleon states by expanding the wave function on correlated hyperspherical harmonic (HH) functions. In Sect. 2 the introduction of the HH coordinates and HH functions is briefly recalled and the utility of multiplying these basis functions by an appropriate correlation factor is discussed. The bound or scattering state wave function is constructed as sum of channel amplitudes with increasing values of the orbital angular momenta of the particles. Each amplitude is then expanded in terms of the correlated HH functions. Some of the results obtained within such a method for the ground state of three– and four–nucleon systems are presented in Sect. 2, where a comparison is also done with the corresponding estimates from other techniques. Section 3 is devoted to the calculation of the nucleon–deuteron scattering below the breakup deuteron threshold. It has to be noticed that the coulomb repulsion is taken into account without any approximation, so that its effect in the scattering can be accurately estimated by comparing the n–d and p–d processes. The final section contains a brief discussion of the merits and the possible extensions of the presented method.

2 Calculation of the Ground States of Nuclei with $A = 3$, 4

For simplicity we will refer here explicitly to a three–nucleon system with total angular momentum J, J_z and total isospin T, T_z. The generalization to $A = 4$ is rather straightforward and is given in ref. [8]. Let us first introduce the Jacobi coordinates given by

$$\boldsymbol{x}_i = \boldsymbol{r}_j - \boldsymbol{r}_k \; , \quad \boldsymbol{y}_i = \sqrt{\frac{1}{3}}(\boldsymbol{r}_j + \boldsymbol{r}_k - 2\boldsymbol{r}_i) \; , \tag{1}$$

with i, j, k cyclic and $\boldsymbol{r}_p$ denoting the position of particle p. The w. f. of the system is written as a sum of three amplitudes

$$\Psi = \psi(\boldsymbol{x}_i, \boldsymbol{y}_i) + \psi(\boldsymbol{x}_j, \boldsymbol{y}_j) + \psi(\boldsymbol{x}_k, \boldsymbol{y}_k) \; , \tag{2}$$

where the spin–isospin dependence is implicitly understood. Each amplitude is antisymmetric with respect to the exchange of the two *reference particles* associated with the coordinate $\boldsymbol{x}$. The calculations performed within the Faddeev approach have shown that it is convenient to consider the contributions to the wave function from states with increasing values of the angular momenta of the particles. For a bound state the generic amplitude of Eq. (2) is therefore written as

$$\psi(\boldsymbol{x}_i, \boldsymbol{y}_i) = \sum_{\alpha=1}^{N_c} x_i^{\ell_\alpha} y_i^{L_\alpha} \Phi_\alpha(x_i, x_j, x_k) \times \left\{ \left[Y_{\ell_\alpha}(\hat{x}_i) Y_{L_\alpha}(\hat{y}_i) \right]_{\Lambda_\alpha} \left[s_\alpha^{jk} s_\alpha^i \right]_{S_\alpha} \right\}_{JJ_z} \left[t_\alpha^{jk} t_\alpha^i \right]_{TT_z} , \quad (3)$$

where x, y are the moduli of the Jacobi coordinate. Each channel α is specified by the quantum numbers ℓ_α, L_α, L_α, s_α^{jk}, t_α^{jk}, S_α where Λ_α is the total orbital angular momentum, s_α^{jk} (t_α^{jk}) is the spin (isospin) of the jk pair and S_α is the total spin. Y_α is the angular–spin–isospin function. The problem is to determine the number of channels to include in Eq. (3) in order to get an accurate description of the state. Moreover, the radial function $\Phi_\alpha(x_i, x_j, x_k)$ must be calculated for each channel. The method here discussed is based on the expansion of the radial functions on the so–called Hyperspherical Harmonic basis. To this end, the moduli of the Jacobi variables x_i, y_i, are replaced with the hyperspherical coordinates, given [13] by the hyperradius

$$\rho = \sqrt{x_i^2 + y_i^2} , \quad (4)$$

and the *hyperangular* variable

$$\cos\phi_i = x_i/\rho . \quad (5)$$

The radial function $\Phi_\alpha(x_i, x_j, x_k)$ is then expanded in terms of the HH basis as

$$\Phi_\alpha(x_i, x_j, x_k) = \sum_{n=0}^{N_\alpha} F_\alpha \frac{u_n^\alpha(\rho)}{\rho^{5/2}} Y_n^\alpha(\phi_i) , \quad (6)$$

where

$$Y_n^\alpha(\phi_i) = N_n^{\ell_\alpha, L_\alpha} P_n^{L_\alpha+1/2, \ell_\alpha+1/2}(\cos 2\phi_i) . \quad (7)$$

In the last expression $P_n^{a,b}$ are Jacobi polynomials, the integer n ranges from zero to infinity and N are normalization factors. If the factor F_α in Eq. (6) is a constant, the usual HH expansion is recovered. However, as discussed below, a different choice of F_α can be helpful for the construction of an accurate variational wave function.

The Rayleigh–Ritz variational principle is used to determine the functions $u_n^\alpha(\rho)$, namely

$$\langle \delta_u \Psi | H - E | \Psi \rangle = 0 , \quad (8)$$

where $\delta_u \Psi$ is the change in the wave function due to an infinitesimal variation of any of the functions $u_n^\alpha(\rho)$. From the last equation it is rather easy to obtain a set of coupled second–order differential equations of the form

$$\sum_{q'=1}^{N_u} \left(A_{qq'}(\rho) \frac{d^2}{d\rho^2} + B_{qq'}(\rho) \frac{d}{d\rho} + C_{qq'}(\rho) + \frac{m}{\hbar^2} E \, N_{qq'}(\rho) \right) U_{q'}(\rho) = 0 . \quad (9)$$

N_u	Volkov	S3	MTV
4	8.375	5.208	5.422
8	8.457	6.501	7.638
12	8.464	6.668	8.086
16	8.465	6.693	8.192
20	8.465	6.698	8.227
24		6.699	8.240
28		6.700	8.246
32		6.700	8.249
36			8.250
40			8.251
44			8.251
48			8.252

Table 1. Triton binding energies in MeV as obtained with the HH expansion for the Volkov [14], S3 [15] and MTV [16] central potentials. N_u is the number of the HH functions taken into account in the (uncorrelated) expansion.

E is the total energy of the system and the indices q and q' run over the channels α and the quantum number n. N_u is the number of coupled equations and it is given by

$$N_u = \sum_{\alpha=1}^{N_c} (N_\alpha + 1) , \tag{10}$$

and it corresponds to the number of terms included in the expansion of the wave function. The solution of Eq. (9) can be carried out by standard numerical methods [6].

In order to give a quantitative idea of the convergence of the HH expansion, corresponding to the choice $F_\alpha = 1$ in Eq. (6) for all the channels, the triton binding energy calculated with three central spin–independent potential models are listed in Table 1. The Volkov potential [14] is a rather soft one, the S3 potential of Afnan and Tang [15] is a sum of gaussian functions, while the Malfliet and Tjon potential MTV [16] is a sum of yukawian functions. Most of the results listed in Table 1 have been just obtained by other authors [17]. In this paper the number of HH taken into account has been increased in order to reach a better convergence for the MTV potential. By inspection of the table it can be concluded that the presence of a strong repulsion requires about 30–40 HH functions in order to reproduce all the important details of the radial function. In particular, the presence of a r^{-1} behavior in the potential at small distances causes a very slow rate of convergence. With these potential models, a larger number of HH functions is required for the ground state of the alpha particle.

Potential	ref.	B	$\langle T \rangle$	P_D	B_F	ref.
OBEPR	[12]	8.28	49.02	7.05	8.33	[23]
Nijmegen 78	[21]	7.63	40.24	7.89	7.63	[24]
Paris	[22]	7.46	43.41	8.55	7.46	[24]
AV18	[19]	7.61	46.66	8.50		
AV18 + UR	[19, 20]	8.49	51.34	9.31		

Table 2. Triton binding energy B in MeV, mean kinetic energy $\langle T \rangle$ in MeV and D–state probability P_D as obtained with the PHH expansion and five different NN potentials. The reference numbers of the potentials are given in the second column. For the sake of comparison also the binding energy values B_F calculated with the Faddeev approach (in the ref. listed in the last column) are reported.

For realistic state–dependent potentials, a fairly large number, about 14 (30) for $A = 3$ (4), of channels must be considered and therefore the number of equations is at least of some hundreds. Such a large value is due to the strong variability of the NN potential with the distances. In fact, the wave function of the system changes rapidly with their distance when two of the particles are rather close. To take care of this, the factor F_α in Eq. (3) can be chosen equal to a pair correlation function $f_\alpha(x_i)$ which is solution of a two–body Schroedinger–like equation [6, 18]. Such a choice of F_α characterizes the PHH expansion (P = pair). The CHH expansion [6] corresponds to F_α being a product of pair functions, as a Jastrow correlation factor. The PHH results obtained for the ground state of triton in correspondence to a few realistic NN potentials are given in Table 2. The numbers of channels and HH considered are $N_c = 12$ and $N_u = 60$, respectively. The TNI term used in connection with the AV18 potential [19] is the so–called Urbana model IX–b (UR) [20].

For the study of the ground state of the alpha particle, the procedure just outlined must be properly modified. Firstly, the wave function is written as a sum of 12 amplitudes corresponding to the even permutations of the particle. Moreover, for $A = 4$ there are two different sets of Jacobi coordinates. Only one of these can be used when the wave function is expanded in a complete basis, but, in order to reduce the number of basis functions, it can be convenient to use both the sets of coordinates. In the last case, two sets of amplitudes must be introduced. Table 3 presents the results for the ground state of the alpha particle obtained with the CHH expansion and $N_c = 23$ and $N_u = 164$. The potentials considered allow for a comparison with the corresponding estimates from other approaches. The binding energy values from the GFMC and Faddeev–Yakubovsky methods are also reported in the table. The CHH result, with TNI terms, has not yet reached convergence and a larger number of basis functions is required. A part from this, there is an overall agreement between the CHH results and the Faddeev–Yakubovsky ones, whereas some minor differences are present with GFMC. The reasons of these discrepancies

Potential	ref.	B	$\langle T\rangle$	P_D	B(ref. [24])	B(ref. [4])
AV8	[25]	25.34	98.22	14.28	25.31	25.8 ± 0.1
AV14 + C	[26]	23.93	94.75	14.20	23.92	24.2 ± 0.2
AV14 + TNI + C	[26, 27]	27.48	107.20	16.24		28.2 ± 0.2
SSC − A	[28]	25.66	80.61	10.56		

Table 3. Alpha particle binding energy B in MeV, mean kinetic energy $\langle T\rangle$ in MeV and D-state probability P_D as obtained with the CHH expansion. The term +C in the potential column means that also the coulomb repulsion has been taken into account.

must be object of further examination.

3 Scattering Processes Involving Three or Four Nucleons

Accurate experimental results have been obtained in the last years for reactions involving three and four nucleons [29, 30, 31, 32]. At the same time, new techniques have been developed to accurately solving the Schroedinger equation both for the bound and continuum states of few nucleon systems. The comparison between experiments and theory can therefore point out to the possible limits of the models and approximations adopted. In the standard approach to the problem, the nucleons interact through phenomenological two–body potentials in a non relativistic approximation. These potentials, in their more recent versions, contain charge dependent terms too. The role and the magnitude of the relativistic corrections are today extensively studied, but unfortunately there is not yet agreement among the different approaches to the problem.

The two–body interactions encounter some difficulties in reproducing the experimental data. This is particularly evident for local interactions, since they do not even reproduce the correct binding energy of the triton. Non local interactions, in particular the Bonn potential, can give a reasonable value of the triton energy. However, as an example, some phase shift parameters in the N–d scattering come out to be not correct [33]. In conclusion, in a non relativistic approach, a two–body NN interaction by itself does not reproduce accurately all the experimental data for the bound and scattering states with $A = 3$, 4.

As stated in the introduction, a possibility to overcome these difficulties is to include in the nuclear potential also three–nucleon interactions (TNI). They can be in some measure justified from a field theory approach. At the same time, the data on the continuum states of $A = 3$, 4 nucleons should be used to carefully test these terms and suggest the appropriate modifications. That program has not yet initiated but it will surely receive much attention in the near future.

Recently, using the Kohn variational principle, the correlated HH technique has been extended to calculate elastic scattering processes [7, 11]. It is worthwhile to notice that the Coulomb repulsion can be taken into account quite

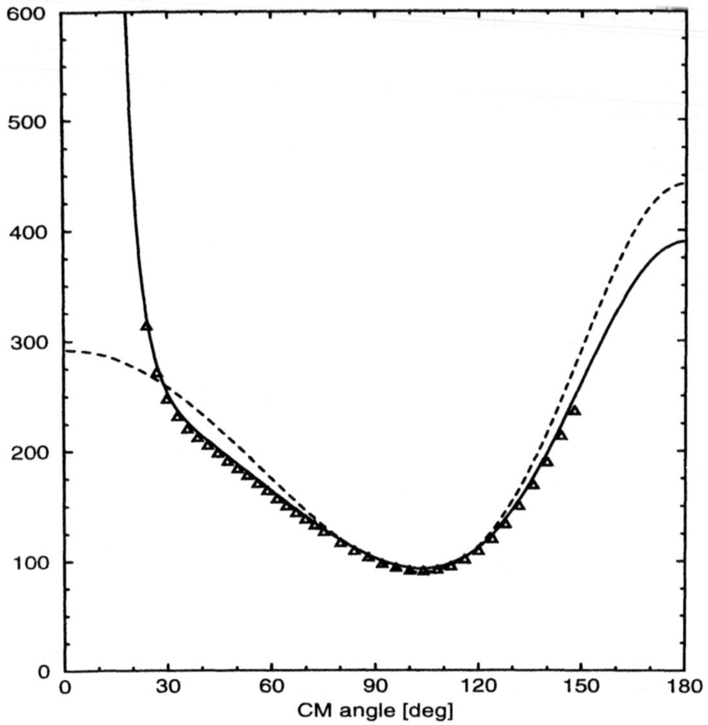

Figure 1. *p–d* (solid line) and *n–d* (dashed line) differential cross section (mb/sr), at lab. nucleon incident energy of 3 MeV, calculated with the AV18 [19] + UR [20] potential model. The experimental points (open triangles) are from ref. [29].

easily since the approach is a variational one and the potentials are not expanded in channels. In ref. [10] the *n–d* scattering below the deuteron break–up threshold has been studied by means of the Faddeev method (worked out by the Bochum group) and the PHH expansion basis in configuration space (used by the Pisa group). The phase–shift and mixing parameters up to J=9/2 have been calculated and the results given by the two methods are extremely close to each other.

Also the *p–d* elastic scattering has been evaluated in order to compare with the available accurate experimental results [29, 30]. The quantities to be compared are the scattering lengths, the differential cross sections at various energies, and the polarization observables. Moreover, the accuracy reached in the experiments allows for a phase–shift analysis with small relative errors [30], and thus, detailed checks can be done for each partial wave.

The differential cross section and the proton analysing power A_y, for a lab. proton incident energy of 3 MeV, are reported in Figs. 1 and 2, respectively. The potential models considered include different TNI interactions, chosen to

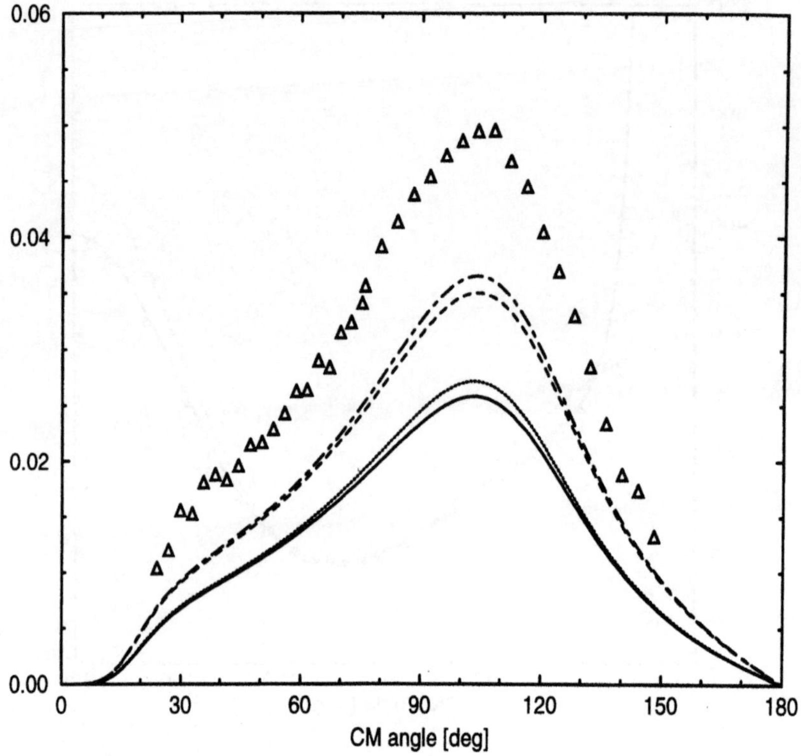

Figure 2. The proton analysing power A_y, for p–d scattering at lab. proton incident energy of 3 MeV, calculated with the following four potentials models: AV14 [26] (solid line), AV14+Brasil TNI [34] (dotted line), AV18 [19] (dashed line) and AV18+UR [20] (dashed–dotted line). The experimental points (open triangles) are from ref. [31].

reproduce the three–nucleon binding energy. This is an important point because of the scaling behaviour of the low energy $^2S_{1/2}$ (doublet) phase–shifts with the binding energy of the composite system. The comparison with the data shows a good agreement for the cross section but appreciable differences are evident in vector analysing powers. More in details, the calculated phase–shift and mixing parameters reveal slight differences from those extracted from the data. The small differences in the parameters corresponding to P–waves noticeably influence the analysing power. On the other hand, these parameters are mainly determined by the P–waves of the potential, therefore the sensitivity of A_y to them can be used for a fine tuning of the NN potential.

The extension of the method to the p–^{3}He elastic scattering, below the disintegration threshold of ^{3}He, is rather straightforward. In fact, there are not rearrangement channels for the process under examination, and thus, the procedure applied to the N–d scattering can be directly exploited. In this case however, as no many benchmark calculations can be found, two different

Potential	MT – [I/III]		AV14 + UR	
N_u	Kohn	Hulthen	Kohn	Hulthen
1	10.18	10.42	11.86	12.66
3	10.09	10.16	11.79	12.55
6	10.02	10.07	11.72	12.36
10	10.01	10.04	11.62	12.05
15	10.01	10.02	11.57	11.89
20	10.01	10.01	11.49	11.68
28			11.44	11.56
48			11.28	11.36
55			11.27	11.32

Table 4. Singlet $p-{}^3$He scattering length (in fm) for the MT–[I/III] [16] and AV14 [26] + Urbana VIII TNI [27] potentials as function of N_u.

variational principles have been employed to calculate the R–matrix elements and the wave function: the Kohn and the Hulthen variational principle. The two principles lead to essentially different set of equations. Therefore, if the solutions of the two set turn out to be close, this is a strong indication of their validity. The differences between the R–matrix element estimates obtained by the two variational principles have been used as a test of the accuracy of the calculation.

The ${}^1S_0$ scattering lengths obtained for the MT–[I/III] [16] and AV14 [26] + Urbana VIII TNI [27] potentials are reported in Table 4 as a function of N_u. As it can be seen by inspection of the table, a good convergence is obtained for both cases. When only a limited number of terms is included in the expansion, the estimates given by the two variational principles are quite different. However, as N_u is increased, they converge to the same value, ensuring the good quality of the calculation. The results obtained are also in reasonable agreement with the extrapolated experimental value of 10.2 ± 1.4 fm [35]. Calculations for different energies are in progress.

4 Conclusion

The method of expanding the wave function of few–nucleon systems on a correlated hyperspherical harmonic basis is quite successful for the calculation of the properties of the bound and elastic scattering states. The reason of the success is that the relative motion at short distances of a pair of particles in a nucleus with $A > 2$ is very similar to that for $A = 2$. This is clearly shown by the fact that the pair distribution function for a given state of two particles in the $A = 3, 4$ nuclei g.s. is almost equal, at small distances, to the one computed for the corresponding $A = 2$ case [36].

The results obtained for the ground state of nuclei with $A = 3$ coincide with those given by other accurate techniques. For the ground state of the alpha particle the estimate of the CHH approach is particularly relevant since only a very few calculations are available for realistic NN interactions. An interesting application for scattering states has been presented for the N–d system at energies below the breakup threshold and employing the Kohn variational principle. The use of the variational principle guarantees to have the best choice of the expansion coefficients of the w.f on a given basis functions.

Another important aspect of the method lies in the fact that it does not require any expansion in channels of the potential. As a consequence, the effect of the coulomb repulsion in scattering processes can be stressed, allowing for an accurate comparisons between theory and experimental p–d data. The application of the approach discussed in this paper to the study of breakup reactions and of the structures of p–shell nuclei is, at present, under investigation.

References

1. C.R. Chen, G.L. Payne, J.L. Friar, B.F. Gibson: Phys. Rev. **C31**, 266 (1985); J.L. Friar, B.F. Gibson, G.L. Payne: Phys. Rev. **C36**, 1138 (1987)
2. S. Ishikawa, T. Sasakawa, T. Sawada, T. Ueda: Phys. Rev. Lett. **53**, 1877 (1984); T. Sasakawa, S. Ishikawa: Few–Body Systems **1**, 3 (1986)
3. H. Kamada, W. Glöckle: Phys. Lett. **B292**, 1 (1992); W. Glöckle, H. Kamada: Nucl. Phys. **A560**, 541 (1993)
4. J. Carlson: Phys. Rev. **C36**, 2026 (1987); **C38**, 1879 (1988)
5. H. Kameyama, M. Kamimura, Y. Kukushima: Phys. Rev. **C40**, 974 (1989); H. Kameyama, M. Kamimura, Y. Kukushima: Nucl. Phys. **A508**, 17c (1990)
6. A. Kievsky, S. Rosati, M. Viviani: Nucl. Phys. **A551**, 241 (1993)
7. A. Kievsky, S. Rosati, M. Viviani: Nucl. Phys. **A577**, 511 (1994)
8. M. Viviani, A. Kievsky, S. Rosati: Few–Body Systems **18**, 25 (1995)
9. H. Witała, D. Hüber, W. Glöckle: Phys. Rev. **C49**, R14 (1994)
10. D. Hüber, W. Glöckle, J. Golak, H. Witała, H. Kamada, A. Kievsky, S. Rosati, M. Viviani: Phys. Rev **C51** (1995) 1100
11. A. Kievsky, M. Viviani, S. Rosati: Phys. Rev. **C**, in press
12. R. Machleidt, K. Holinde, C. Elster: Phys. Rep. **149**, 1 (1987)
13. M. Fabre de la Ripelle: Ann. Phys. (NY) **147**, 281 (1983)
14. A.B. Volkov: Nucl. Phys. **74**, 33 (1965)

15. I.R. Afnan, Y.C. Tang: Phys. Rev. **A175**, 1337 (1968)

16. R.A. Malfliet, J.A. Tjon: Nucl. Phys. **A217**, 161 (1969)

17. G. Erens, J.L. Visschers, R. van Wageningen: Ann. Phys. (NY) **147**, 281 (1971)

18. S. Rosati, M. Viviani, A. Kievsky: Few–Body Systems **9**, 1 (1990)

19. R.B. Wiringa, V.G.J. Stoks, R. Schiavilla: Phys. Rev. **51**, 38 (1995)

20. R.B. Wiringa: Private Communication

21. M.M. Nagels, T.A. Rijken, J.J. de Swart: Phys. Rev. **D17** (1978) 768

22. M. Lacombe, M. et al.: Phys. Rev. **C21** (1980) 189

23. T. Sasakawa: Few–Body Systems, Suppl. **1**, 104 (1986)

24. H. Kamada, W. Glöckle: Few–Body Systems Suppl. **7**, 217 (1994)

25. R. Schiavilla: Private Communication

26. R.B. Wiringa, R.A. Smith, T.L. Ainsworth: Phys. Rev. **C29** (1984) 1207

27. R.B. Wiringa: Phys. Rev. **C43**, 1585 (1991)

28. R. de Tourreil, D.W.L. Sprung: Nucl. Phys. **A201**, 193 (1973)

29. K. Sagara et al.: Phys. Rev. **C50**, 576 (1994)

30. L.D. Knutson et al.: Phys. Rev. Lett. **71**, 3762 (1993)

31. J.E. McAninch et al.: Phys. Rev. **C50**, 589 (1994)

32. M.T. Alley, L.D. Knutson: Phys. Rev. **C48**, 1890 (1993)

33. D. Hüber, J. Golak, H. Witała, W. Glöckle, H. Kamada: Ruhr Universitat Bochum preprint 1995, Few–Body Systems, in press

34. H.T. Coelho, T.K. Das, M.R. Robilotta: Phys. Rev. **C28**, 1812 (1983)

35. P.E. Tegner, C. Bargholtz: Astrophysical J. **272**, 311 (1983)

36. M. Viviani, A. Kievsky, S. Rosati: Proceedings of the Conference '*Perspectives on Nuclear Physics at Intermediate Energies*, Trieste (1995), to be published.

Few-Body Systems Suppl. 8, 32–43 (1995)

Few-Body Systems

Recent Progress in Path Integral Calculations of Nuclei

J. Carlson *

Theoretical Division, Los Alamos National Laboratory, Los Alamos, NM 87545, USA

Abstract. Path Integral methods have proven valuable in studying structure and dynamics in light nuclei. We illustrate the strengths and weaknesses of this approach, and review recent progress. In particular, we discuss calculations of the electromagnetic response of 3 and 4 body nuclei as well as the first calculations of the A=6 ground state and low-lying excitations. Physics issues which can be addressed in studies of light P-shell nuclei are also discussed.

1 Introduction

Path Integral methods have proven to be quite valuable in many areas of physics, enabling one to study non-perturbative properties of quantum systems in cases that would otherwise be difficult or impossible. The applications can be divided into two areas; lattice and continuum models. The former includes calculations of, for example, Hubbard[1] models of electrons and Monte Carlo Shell Model calculations.[2] In the continuum, path integral methods have proven most valuable in studies of bulk Helium,[3] Helium droplets,[4] and atoms and molecules.[5]

Path Integral (or Green's function Monte Carlo) methods are complimentary to the Faddeev[6] and Correlated Hyperspherical Harmonic[7] methods presented elsewhere in this volume. In one sense they are more limited, being primarily useful for calculating 'quasi-thermal' properties such as ground state observables, low energy scattering and excitations, and imaginary-time correlation functions (Euclidean Response). It is considerably more difficult to calculate, for example, scattering states above breakup threshold with Monte Carlo methods. On the other hand, Monte Carlo methods allow one to sample the 3A dimensional configuration space rather than explicitly integrating, hence one can treat larger systems than would otherwise be possible. Recently,

**E-mail address:* carlson@qmc.lanl.gov

the first microscopic calculations of the A=6 system with realistic two- and three-nucleon interactions have been performed.[8]

For calculations of the ground state, one computes:

$$\Psi_0 = \lim_{\tau\to\infty} \exp\left[-H\tau\right]\Psi_T, \tag{1}$$

where Ψ_T is a trial (variational) wave function, and Ψ_0 the exact ground state. In many applications, where the interaction depends only upon the spatial positions of the particles, the wave function can be represented as a real number for any given position of particles. Recently, some attention has been paid to electrons in external magnetic fields,[10] which requires a complex wave function. In these cases it is possible to treat problems with a very large number of particles, from 50 to several hundred, and even more in some cases. For approximate (fixed-node) calculations of many-fermion systems, the computational time scales as A to a small integer power.

For the nuclear case, though, the wave function is a many-component vector, each component representing the amplitude of a particular spin- and isospin-component of the A-particle wave function. The state-dependence of the interaction is crucial, particularly the one-pion-exchange potential (OPEP) at large distances. For a Hamiltonian of the form $H = T + V_{ij} + V_{ijk}$, the calculation proceeds by iterating equations of the form:

$$\Psi_{\tau+\delta\tau}(\mathbf{R}) = \int d\mathbf{R}' G(\mathbf{R},\mathbf{R}';\delta\tau)\Psi_\tau(\mathbf{R}'), \tag{2}$$

where the A-particle imaginary-time propagator G is given by:

$$\begin{aligned} G(\mathbf{R},\mathbf{R}';\delta\tau) &= \left[1-\delta\tau V_{ijk}(\mathbf{R})/2\right] G^0(\mathbf{R},\mathbf{R}';\delta\tau) \\ &\quad \mathcal{S}\prod_{i<j}\left(\frac{g^{ij}(\mathbf{r}_{ij},\mathbf{r}'_{ij};\delta\tau)}{g_0^{ij}(\mathbf{r}_{ij},\mathbf{r}'_{ij};\delta\tau)}\right)\left[1-\delta\tau V_{ijk}(\mathbf{R}')/2\right] + \mathcal{O}(\delta\tau^3). \quad (3)\end{aligned}$$

In this equation G^0 is the free A-body propagator (a gaussian) $\langle\mathbf{R}|\exp[-T\delta\tau]|\mathbf{R}'\rangle$, g_0^{ij} is the free two-body propagator $\langle\mathbf{r}_{ij}|\exp[-T_{ij}\delta\tau]|\mathbf{r}'_{ij}\rangle$, and g^{ij} is the interacting two-body propagator. The g^{ij} can either be approximated in a simple scheme,

$$g^{ij}(\mathbf{r}_{ij},\mathbf{r}'_{ij}) \approx \exp[-v_{ij}(\mathbf{r}_{ij})\delta\tau/2] g_0^{ij} \exp[-v_{ij}(\mathbf{r}'_{ij})\delta\tau/2], \tag{4}$$

or solved for explicitly in a partial wave expansion. The latter technique is particularly valuable in calculations of bulk helium and droplets, as it allows one to take a time step roughly 40-50 times larger than the naive approximation above. Recent experiments in nuclear physics suggest, however, that the gain is much smaller, roughly a factor of five. The difference is that the propagators are state-dependent at a long range, and hence three-body corrections to the propagator are much more important in nuclei than in interacting Helium atoms.

The problems in treating large nuclei are immediately apparent. If one keeps the explicit sums over spin and isospin, the computer time required for one iteration of the propagator grows exponentially in particle number. In principle it is possible to sample the spin and isospin degrees of freedom, but in practice a successful scheme has yet to be found. The primary difficulty is that different 'paths' in spin-isospin variables contribute with arbitrary signs and complex phases, and the cancellations in the sums are large. Hence, practical simulations are currently limited to $A \leq 7$.

In all of the calculations presented here, the full nuclear interaction is given by a sum of two- and three-nucleon strong interactions, the EM corrections, and possibly charge symmetry breaking (CSB) and charge independence breaking (CIB) interactions. The two-nucleon interaction (here either AV14[11] or AV18[12]) is approximated by a local V8 operator (including L·S interactions, but not L^2 and $L \cdot S^2$), and the difference between the full interaction and the local approximation is treated perturbatively. The three-nucleon and EM interactions can be treated exactly. For convenience, though, at present we have treated the EM corrections perturbatively for A=6 to calculate all members of the isospin multiplet simultaneously.

2 Three- and Four-Body Nuclei

2.1 Ground States

We have computed binding energies and various ground-state expectation values for the three- and four-nucleon systems. GFMC calculations of the three-body states agree quite well with Faddeev calculations. Using two-nucleon interactions alone, forces which underbind the triton by approximately 1 MeV underbind the alpha particle by approximately 5 MeV. We then introduce a three nucleon interaction consisting of two pieces, a two-pion exchange three-nucleon interaction (TNI) of the Fujita-Miyazawa form and a shorter (two-pion exchange) range phenomenological repulsive term. The three-nucleon interaction is expected to be significant because of the Delta, a relatively low-lying excitation of the nucleon. The shorter-range repulsive term is more phenomenological and is meant to simulate dispersive effects. This three-nucleon interaction, when adjusted to match the A=3 binding energy, can also describe the alpha particle binding energy quite successfully.[13]

Clearly this is not the complete story. Relativistic effects will play a role. Although the kinematic effects (boosts and Melosh spin rotations) appear to be quite modest, at least as far as the binding energies are concerned,[14] we have embarked on a more complete study which includes the lowest-order corrections to OPEP. Also, the TNI that we have included is very simple at short distances and may need improvement. We should emphasize, though, that these changes will likely be quite small on the scale of the two-nucleon interaction, say a few per cent.

Once the binding energies are reproduced successfully, other properties of the nucleus must be considered. In particular, the EM form factors are of primary importance and are extremely well-known experimentally. In order to calculate them, we introduce appropriate charge and current operators. Current conservation plays a valuable role here, providing an important constraint on the current operators. We have chosen the prescription advocated by Riska for the current operators; this prescription preserves current conservation at lowest order in p^2 and results in relatively hard form factors. The exchange-current corrections in the longitudinal channel are generally small, however in the transverse channel they can be quite significant even at moderate momentum transfers. This is to be expected as the former are relativistic corrections while the latter are necessary even in non-relativistic order.

The results for the three- and four-body form factors are presented in references [15] and [13]. The calculations are generally in good agreement with the data, although the description is not perfect. Errors are quite small at low-momentum transfers, typically a few per cent, but can be more significant in the region of the diffraction minima and beyond. Clearly studies of the deuteron (particularly T_{20}), triton, ^{3}He and ^{4}He form factors are an important tool for studying the structure of light nuclei.

2.2 Longitudinal and Transverse Response

Quasi-elastic electron scattering is a particularly valuable tool for studying nuclear dynamics. Traditionally this regime is described in terms of the propagation of single nucleons, and indeed that is a very important part of the physics. It is not the complete picture, however, as it is not possible to obtain a reasonable description of the longitudinal response in a single-nucleon picture.

Path integral techniques offer the possibility of studying the Euclidean response of light nuclei, defined as

$$E_L(k,\tau) = \int d\omega R_L(k,\omega) \exp[-(\omega - \omega_0)\tau], \tag{5}$$

where R_L is the standard 'real-time' response, and ω_0 is the ground state energy of the nucleus. The Euclidean response is essentially a finite energy sum-rule, with an inverse-energy parameter (τ) at your disposal. At $\tau = 0$, the Euclidean response is just a simple ground-state expectation value. As τ increases the Euclidean response measures lower energy states, finally just measuring the relevant form factors as $\tau \to \infty$.

The advantage of this formulation is that it is possible to directly evaluate this response in a path-integral calculation and to make direct comparisons with experiment. Also, as a consequence of the simple structure of the calculation, it is sometimes possible to understand the results quite simply in the imaginary-time regime. Of course, there is also a disadvantage; this type of sum rule is quite sensitive to shifts in strength, but not to strong oscillations in the response or to moderate changes in the width.

In order to present the non-trivial physics in the problem, we plot the response multiplied by the free-particle response: $\exp[k^2\tau/2m]E_L(k,\tau)$. This product is just a constant for an isolated nucleon. Since the real response has a width, the product actually grows exponentially with τ. In addition, we compute the elastic response (form factors) separately and subtract it to obtain the inelastic response.

Results for the longitudinal response of ^{3}He and ^{4}He at 400 MeV/c are shown in Fig. 1. For the case of ^{3}He, we show only the results of the experiment (indicated by a shaded band), and the 'full' calculation (with relativistic and two-body charge operator corrections). Results with a proton charge operator only are nearly identical. For the case of ^{4}He, we also show a PWIA response calculated from the momentum distribution and an 'extrapolated' response obtained by adding a high-energy tail to the experimental response consistent with the sum rules. The difference between the points with error bars and the 'extrapolated' curve give some idea of the uncertainty in the Euclidean response arising from unmeasured response at high energy. Of course, these uncertainties are confined to the small τ region.

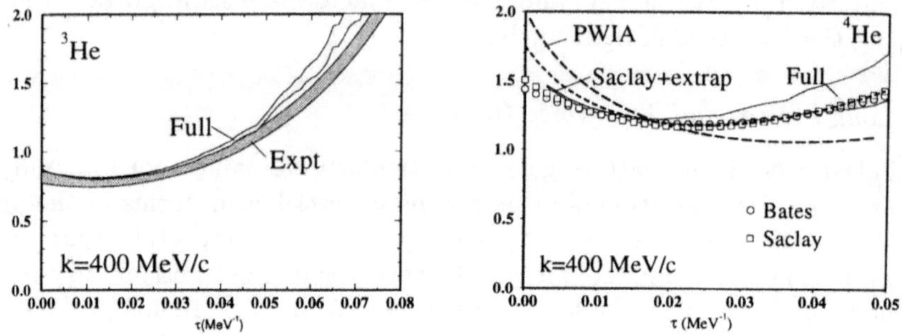

Figure 1. Longitudinal EM Response for ^{3}He and ^{4}He at 400 MeV/c

The results convincingly demonstrate the importance of final state interactions in obtaining the measured response. The PWIA approximation significantly overestimates the overall strength and the energy dependence is very different from what is observed experimentally. The difference is even more pronounced at 300 MeV/c. In this regime one is primarily sensitive to the charge-exchange component of the nuclear interaction; in imaginary time the charge propagates much faster than the nucleons. For the moderate momentum transfers considered here, the response is considerably suppressed in the region of the peak.

In the transverse channel, both the final-state interactions (FSI) and the two-body meson exchange currents (MEC) are vital. Results for ^{3}He and ^{4}He at 400 MeV/c are shown in Fig. 2. The agreement between theory and experiment is not perfect, but the energy dependence is well reproduced. In A=3, the curves labeled 'full' include both FSI and MEC, and are approximately 10% higher than the experimental results. This 10% discrepancy in cross section is not too different from that in the elastic channel at this momentum transfer.

The curves labeled 'impulse' contain the full FSI, but only the standard single-nucleon current operator. In ^{4}He, the impulse calculation obtains only about 2/3 of the total response, while the full calculation is slightly larger than the experimental result. In order to obtain such a large effect from exchange currents, the tensor correlations arising from OPEP must be included in both the initial and the final states.

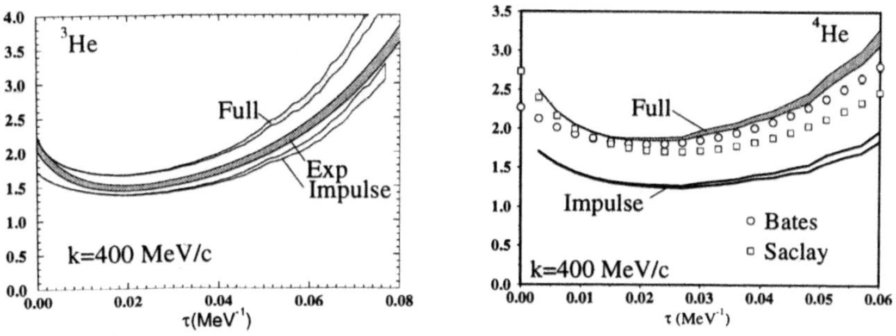

Figure 2. Transverse EM Response for ^{3}He and ^{4}He at 400 MeV/c

It is important to realize that there is a large change when going from A=2 to A=4. For the deuteron, FSI and MEC effects are generally quite small, principally because the rms separation between the nucleons is roughly 4 fm. In the alpha particle, though, the struck nucleon can interact with three other nucleons, each with an average rms separation of $\approx$ 2.4 fm. Both MEC and FSI are vital to understanding the experimental results; without both we cannot reconcile the enhanced T/L ratio in A$\geq$4. The calculations presented here are limited in that they are non-relativistic and do not include exchange currents associated with the three-nucleon interaction. More complete calculations are vital if we are to explain experimental results at higher momentum transfers.

2.3 Polarization Observables

Currently there is a great deal of experimental and theoretical interest in the polarization observables R'_T and R'_{LT} measured on three-body systems. In a simple PWIA model, one could use these experiments to extract the neutron charge (from R'_{LT}) and magnetic (from R'_T) form factors. Clearly it is important to determine at what momentum transfer the PWIA is accurate enough to allow a meaningful extraction of neutron form factors.

We have calculated the Euclidean polarization responses R'_{LT} and R'_{LT} for the three body systems at a momentum transfer of 400 MeV/c. The results are shown in Fig. 3. This is a lower momentum transfer than recent experiments, but will perhaps serve as a useful guide to understanding the importance of the FSI and MEC effects. At larger momentum transfers, the relativistic effects will become more important.

It is immediately apparent that the calculations with both FSI and MEC,

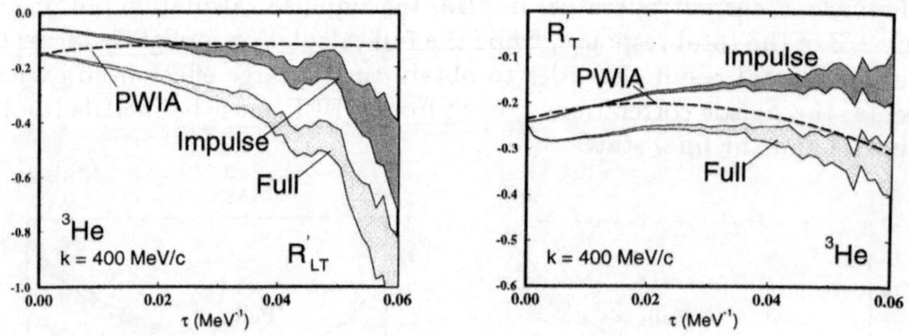

Figure 3. Euclidean Polarization Responses for ^{3}He at 400 MeV/c

Table 1. Ratios of integrated responses with full FSI to PWIA calculation for region shown in figure 3.

Current Operator	R_L	R_T	R'_{LT}	R'_T
Impulse	0.85(1)	0.95(1)	0.38(10)	1.14(2)
Full	0.85(1)	1.12(1)	0.93(10)	1.38(2)

(the curves labeled 'full' in the figure) are very different from the simple PWIA calculation. In the R'_{LT} figure, the sum rule of the PWIA happens to be similar to that of the full calculation. However, the energy dependence is drastically different. Based upon this figure, one can expect a large enhancement in R'_{LT} at low energies due to final state interactions. This effect is apparent even without MEC, as is shown in the curve labeled 'impulse'. The exchange currents increase this low-omega tail even more.

The R'_T curve is somewhat more difficult to interpret at first glance. Again, the total sum rules are not too different. However, the 'full' curve shows a dramatic energy dependence for very small τ, corresponding to energies near the delta resonance. Since these calculations only include static deltas through the exchange currents, they should not be taken too seriously. However, one would expect quite interesting physics in this regime.

In order to better understand the implications of these calculations, we have computed integrals of the response by taking sums of the imaginary time response at different values of τ. In order to estimate the response over the region of the peak, for example, we have computed the sum of R'_{LT} and R'_T with weights given by the $F(\tau)$ shown in Fig. 4. This corresponds to an integral of the response with a weight function $F(\omega)$ also shown in the figure. The ratios of the integrals of the unpolarized and polarized responses to the PWIA response are given in the table.

The longitudinal response is somewhat quenched in this regime, and is

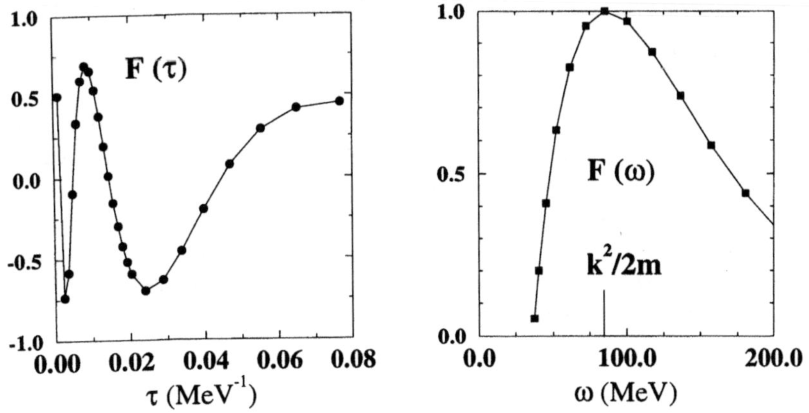

Figure 4. Weight functions in imaginary- and real-time corresponding to the sums given in Table 1 (see text).

nearly unaffected by the two-body corrections to the charge operator. The transverse response is slightly quenched for impulse current operators, and slightly enhanced when exchange currents are included. The calculated transverse response is roughly 10% larger than the experimental results at 400 MeV/c.

The R'_{LT} response is dramatically affected by final state interactions, in fact the integral over this region is only about 40% of the PWIA calculation when only impulse currents are included. R'_{LT} measures an interference between the longitudinal and transverse responses, and since the dominant one-pion-exchange interaction couples to the spin of the nucleon, there is an important probability for flipping the spin of the struck nucleon. The exchange currents, however, bring the R'_{LT} response part of the way back to the PWIA result. In the case of the R'_T response, the impulse response is somewhat enhanced compared to PWIA, and even more so when exchange currents are included. This enhancement is reduced when considering asymmetries rather than responses, since the transverse response is itself enhanced. A more complete study of the results in a variety of kinematic regimes will be published elsewhere.

Clearly, more detailed calculations are required; ideally relativistic calculations with a full treatment of MEC and FSI. The FSI and MEC are intimately related, and it may be possible in a complete treatment to understand these effects and extend the regime over which neutron form factors can be extracted. However, from these calculations it is clear that at these moderate momentum transfers, polarized electron scattering on A=3 is more likely to teach us about nuclear dynamics than about neutron form factors.

3 Beyond A=4

The lightest p-shell nuclei provide a very important check on our understanding of the nuclear interaction. Here, for the first time, one can directly study the $L \cdot S$ splitting in nuclear physics and the emergence of shell structure for larger systems. In order to test our understanding nuclear physics at a microscopic level, one must go beyond the simplest studies of three- and four-body ground states. One testing ground is scattering, the other is larger systems; both are necessary for a real test of the nuclear interaction.

3.1 A=5

Low-energy n-α scattering offers an interesting test of the $L \cdot S$ splitting in the nuclear hamiltonian. We have previously[9] reported results for the AV14 + TNI interaction, in table 2 we report results for the AV18 + TNI interaction. The TNI here has been readjusted to reproduce the 3- and 4-body binding energies with the AV18 interaction,[8] but its form is unchanged. The results shown here differ little from those obtained previously.

We have calculated the lowest-lying states with a specific boundary condition of a node in the scattering wave function at 12.5 fm α-n separation. This wave function can be smoothly matched to a real scattering wave function at that point, and hence we can directly compare our results to experiment. The importance of the boundary condition is that it serves to eliminate the low-lying excitations that would otherwise make Monte Carlo calculations difficult.

The J = 1/2 state is reproduced extremely well in these calculations, but the lower-lying $3/2^-$ state is approximately 0.5 MeV too high in energy. Thus we obtain approximately 2/3 of the experimental $L \cdot S$ splitting. One thing that remains to be checked is the effect of treating the higher-order NN interaction terms perturbatively. Although the effect is quite small in A=3 and 4, it could be somewhat larger here. Other possibly important effects are relativistic corrections and more realistic three-nucleon interactions. The latter can be most effectively checked by also studying a variety of A=6 states.

A variety of interesting physics remains to be studied in the five body system. We intend to calculate the low-lying s-wave states as well, and determine if a low-lying resonance exists in this model. From the calculations of the s- and p-wave states, it should be possible to study parity violation in the NN interaction solely within the context of few-body systems. Experiments currently under way will provide a stringent test of the extraction of parity-violating NN coupling constants which have been performed in larger nuclei.

Table 2. Expectation Values in A=4, 5, and 6

Nucleus (J)	$^4He(0)$	$^5He(3/2)$	$^5He(1/2)$	$^6He(0)$	$^6Li(1)$	$^6Li(3)$
E(Expt.)	-28.3	-27.2	-25.8	-29.3	-32.0	-29.8
E(Calc.)	-28.3(1)	-26.5(2)	-25.7(2)	-28.2(8)	-32.4(9)	-28.9(6)
$\bar{E}$	-28.3(1)	-26.5(2)	-25.7(2)	-27.3(4)	-31.1(4)	-28.2(3)
$\langle T \rangle$	118.(1)	122.(2)	117.(2)	146.(4)	143.(3)	138.(3)
$\langle v_{14} \rangle$	-142.(1)	-145.(2)	-140.(2)	-172.(4)	-173.(3)	-165.(3)
$\langle V_{ijk} \rangle$	-6.5(3)	-7.0(4)	-6.4(4)	-7.0(7)	-6.2(6)	-6.9(5)
$\langle v_{LS} \rangle$	-0.4(1)	-1.2(1)	-0.4(1)	-2.7(3)	-1.5(5)	-3.0(4)
$\langle v^{pp}_{coul} \rangle$	0.761(2)	0.745(3)	0.751(3)	0.724(8)	1.55(2)	1.57(2)

3.2 A=6

We have recently completed an initial set of calculations of the low-lying states in A=6. In a microscopic study, these states are extremely difficult to calculate. Not only are the number of spin-isospin states much larger, but the energy scales of the nuclear excitations are quite different. The lowest excited state of the alpha particle is at about 20 MeV, but in A=6 it takes only a tenth of that energy to break ^{6}Li into a constituent alpha and deuteron.

The results of our calculations are presented in figure 5 and table 2. The figures demonstrate the convergence of the calculation with increasing imaginary time τ, the error bands show the one standard deviation results of the Monte Carlo calculations. In the figure, the 0^+ state has been raised by 2.5 MeV for clarity.

The alpha particle converges extremely rapidly, precisely because there are no low-energy excitations for A=4. The six-body results, however, do not appear to be converged by $\tau = 0.06$ MeV^{-1}. This is not unexpected since there are important low-lying excitations. In order to estimate their effect, we have performed a fit to the Monte Carlo results, assuming that the calculated spectra at $\tau \geq 0.01$ MeV^{-1} is dominated by (1) the ground state, (2) the lowest-lying resonance in that channel, and (3) higher-energy strength at an average of 30 MeV. These fits are shown as solid lines in the figure. The results are not too sensitive to the exact energy assumed for the higher-lying strength. The extrapolated ground-state energy is given as E(Calc.) in the table, while $\bar{E}$ is the average over tau from 0.03 to 0.06 MeV^{-1}. $\bar{E}$ is a variational upper bound to the true ground state energy.

The table includes ground state energies and a variety of expectation values. The statistical errors are moderately large in these calculations, but overall the results are surprisingly good. The J=1 ground state energy is reproduced quite accurately, and the J=0 and 3 states are within an MeV of their experimental values. The kinetic and two-body potential energies are very similar in A=4 and 5, while in A=6 they are nearer to the sum of ^{4}He and the deuteron. The TNI expectation values are not dramatically different in A=4,5, and 6.

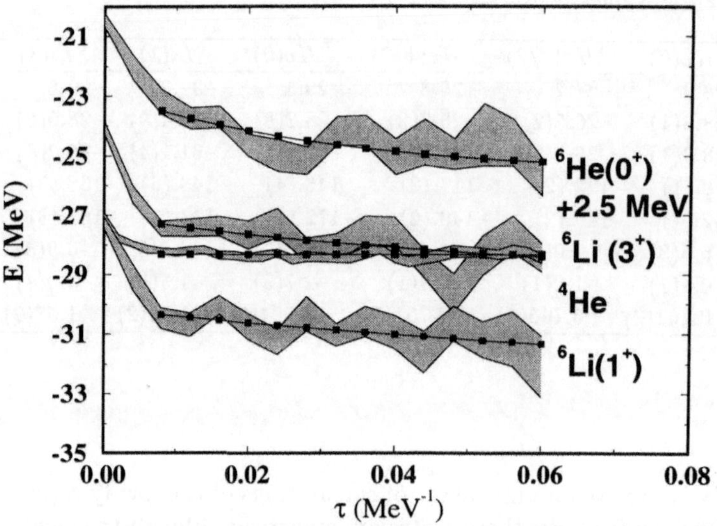

Figure 5. Energy as a function of inverse temperature (τ) for the alpha particle and the lowest-lying A=6 states

The Coulomb interaction between proton pairs is also shown in the table. The change in going from A=4 to 6 is a direct measure of the distortion of the alpha particle core as the number of nucleons increase. We see that there is roughly a 2% change in going from A=4 to the J=3/2$^-$ state of ^{5}He, and roughly a 5% change when comparing the alpha particle and ^{6}He. These distortions are significant, and indicate the need for going beyond a 3 body (αNN) model of ^{6}Li. In the future, we plan to compute overlaps of the 6-body wave functions in various decompositions. We also hope to use the five- and six-body systems to study the spin-isospin dependence of the shorter-range TNI, and to study relativistic Hamiltonians.

4 Conclusion

Calculations of light nuclei as systems of nucleons with simple models of the nuclear interaction and exchange currents are amazingly successful in describing the properties of few-body nuclear systems. They describe three-nucleon scattering and electromagnetic response quite well in A=3 and 4, and successfully describe the low-energy states of 5 and 6-body nuclei.

Nevertheless, many avenues of investigation remain open, including studies of relativistic effects, determination of the shorter-range part of the three-nucleon-interaction, and extensions of these methods to heavier nuclei. All of these are important in order to convince our experimental and theoretical colleagucs of the validity of this view of nuclear physics, and to understand its consequences for nuclear structure and dynamics.

Acknowledgement. All of the work presented here was performed in close collaboration with R. Schiavilla, R. B. Wiringa, B. S. Pudliner, and V. R. Pandharipande. Without their contributions none of this work would have been possible. We are also grateful to Argonne National Laboratory, the NERSC computer center at Livermore, and NCSA in Urbana for providing the computer resources necessary to carry out the calculations.

References

1. N. Furukawa, M. Imada: J. Phys. Soc. Jpn. **61**, 3331 (1992); Shiwei Zhang, J. Carlson, J.E. Gubernatis: Phys. Rev. Lett. **74**, 3652 (1995)
2. D.J. Dean et al.: Phys. Rev. Lett. **72**, 4066 (1994); C.W. Johnson et al.: Phys. Rev. Lett. **69**, 3157 (1992)
3. D.M. Ceperley: Rev. Mod. Phys. **67**, 279 (1995)
4. S.A. Chin: J. Low Temp. Phys. **93**, 921 (1993)
5. J.C. Grossman, L. Mitas: Phys. Rev. Lett. **74**, 1323 (1995)
6. W. Glöckle, H. Kamada: Phys. Rev. Lett. **71**, 971 (1993); W. Glöckle: Few-Body Systems Suppl. **8**, 9 (1995)
7. A. Kievsky, M. Viviani, S. Rosati: Nucl. Phys. **A551**, 241 (1993) and **A557**, 511 (1994); S. Rosati, M. Viviani, A. Kievsky: Few-Body Systems Suppl. **8**, 22 (1995)
8. B.S. Pudliner, V.R. Pandharipande, J. Carlson, R.B. Wiringa: Phys. Rev. Lett. **74**, 4396 (1995)
9. J. Carlson, R. Schiavilla: Few-Body Systems Suppl. **7**, 349 (1994)
10. G. Ortiz, D.M. Ceperley, R.M. Martin: Phys. Rev. Lett. **71**, 2777 (1993)
11. R.B. Wiringa, R.A. Smith, T.L. Ainsworth: Phys. Rev. **C29**, 1207 (1984)
12. R.B. Wiringa, V.G.J. Stoks, R. Schiavilla: Phys. Rev. **C51**, 38 (1995)
13. J. Carlson: Nucl. Phys. **A522**, 185c (1991)
14. J. Carlson, R. Schiavilla, V.R. Pandharipande: Phys. Rev. **C47**, 484 (1993); J.L. Forest, V.R. Pandharipande, J. Carlson, R. Schiavilla: Phys. Rev. **C**, in press
15. R.B. Wiringa: Phys. Rev. **C41**, 1585 (1991)

Few-Body Systems Suppl. 8, 44–48 (1995)

Few-Body Systems

Convergence Properties of the Adiabatic Expansion for Few–Nucleon Systems

M. Viviani, A. Kievsky

INFN, Sezione di Pisa, Piazza Torricelli, 2, I-56100 Pisa, Italy

Abstract. The convergence of the Adiabatic Hyperspherical Harmonic (AHH) functions as expansion basis for the few–nucleon wave function is studied. The case considered is the calculation of the ground state of nuclei with $A = 3$, 4, interacting via realistic potentials. The AHH basis is found to be very effective providing very fast convergence and it seems to be a promising tool for treating nuclei with $A > 4$.

1 Introduction

Recently, the correlated hyperspherical harmonic expansion technique has been successfully applied for calculating the ground and elastic scattering states of the three- and four-nucleon system with realistic NN potentials [1]. The accuracy of the method is ensured by *(i)* the flexibility of the Hyperspherical Harmonic (HH) expansion and *(ii)* the presence of appropriate correlation factors. However, the direct application of this technique to systems with $A > 4$ becomes more and more complex since the number of expansion terms greatly increases.

For this reason, we have considered the so–called adiabatic approximation [2, 3] as it represents an alternative and, perhaps, more effective expansion basis of the wave function. In the adiabatic approach, the Schroedinger equation is solved for fixed values of the hyperradius ρ to obtain a set of eigenfunctions $\Phi_m(\rho, \Omega)$, where Ω represents the hyperangular variables. These eigenfunctions are then used to expand the complete wave function. The advantage of this new basis is that the number of terms to be included for obtaining a good accuracy results to be rather small. The functions $\Phi_m(\rho, \Omega)$ can be calculated by expanding on the HH basis or, alternatively, on the correlated basis. Both methods have been explored and will be here discussed. The problem considered is the calculation of the ground–state properties of nuclei with $A = 3$, 4, interacting via semirealistic and realistic potentials.

The paper is organized in the following way: the AHH expansion basis is briefly outlined in Sect. 2, and the numerical results are reported in Sect. 3. The merits of the method and possible future applications are discussed in the final section.

2 The Adiabatic Hyperspherical Harmonic basis

The internal motion of an A-body system is described in terms of $N = A - 1$ Jacobi coordinates $(\boldsymbol{\xi}_1, \ldots, \boldsymbol{\xi}_N)$. In the framework of the HH approach to the problem [4], let us introduce as new variables the hyperradius ρ, $(\rho^2 = \sum_{q=1}^{N} \xi_q^2)$, the $2N$ angular variables $\hat{\xi}_q$ of the Jacobi coordinates and the $N-1$ hyperangles $(\phi_2, \ldots, \phi_N)$ related to the moduli of the Jacobi coordinates by

$$\xi_N = \rho \cos\phi_N \ , \tag{1}$$

$$\xi_q = \rho \sin\phi_N \cdots \sin\phi_{q+1} \cos\phi_q \ , \quad q = N-1, \ldots, 2 \ , \quad (\phi_1 = 0) \ . \tag{2}$$

The Laplace operator is then written as a sum of a purely hypercentral term and a hyperangular part, namely

$$\nabla^2 = \sum_{q=1}^{N} \nabla_q^2 = \frac{\partial^2}{\partial\rho^2} + \frac{D-1}{\rho}\frac{\partial}{\partial\rho} + \frac{1}{\rho^2} L^2(\Omega) \ , \tag{3}$$

with $D = 3N$ and $\Omega \equiv (\hat{\xi}_1, \ldots, \hat{\xi}_N, \phi_2, \ldots, \phi_N)$. The Adiabatic Hyperspherical Harmonic (AHH) functions are the eigenfunctions of the following operator,

$$\left\{ -\frac{\hbar^2}{M_N} \left[\frac{L^2(\Omega)}{\rho^2} - \frac{(D-1)(D-3)}{\rho^2} \right] + V(\rho, \Omega) \right\} \Phi_m(\rho, \Omega) = U_m(\rho) \Phi_m(\rho, \Omega) \ , \tag{4}$$

where $V(\rho, \Omega)$ is the total potential and the index m numbers the various eigenfunctions and the corresponding *eigenpotentials* $U_m(\rho)$. If the potential term is dropped out, the eigenfunctions are the usual HH functions $\mathcal{H}(\Omega)$; in this case, $U_m^{V=0}(\rho) \sim \rho^{-2}$. An explicit expression of the HH functions can be found in ref. [5].

The wave function of the system can be expanded in terms of the AHH functions

$$\Psi = \frac{1}{\rho^{(D-1)/2}} \sum_{m=1}^{M} u_m(\rho) \Phi_m(\rho, \Omega) \ , \tag{5}$$

and then, from the Schroedinger equation, the following set of coupled equations is obtained

$$-\frac{\hbar^2}{M_N} u_m''(\rho) + \sum_{n=1}^{M} [B_{mn}(\rho) u_n'(\rho) + C_{mn}(\rho) u_n(\rho)] + [U_m(\rho) - E] u_m(\rho) = 0 \ , \tag{6}$$

where

$$B_{mn}(\rho) = \langle\, \Phi_m(\rho, \Omega) \,|\, 2\frac{\partial}{\partial\rho} \,|\, \Phi_n(\rho, \Omega) \,\rangle_\Omega \ , \tag{7}$$

$$C_{mn}(\rho) = \langle\, \Phi_m(\rho, \Omega) \,|\, \frac{\partial^2}{\partial\rho^2} \,|\, \Phi_n(\rho, \Omega) \,\rangle_\Omega \ ; \tag{8}$$

the symbol $\langle\rangle_\Omega$ implies the integration over all the angular and hyperangular variables and the Φ_m are orthonormalized.

3 Results

In order to solve Eq. (4), the AHH functions $\Phi_m(\rho,\Omega)$ are expanded in the following way,

$$\Phi_m(\rho,\Omega) = \sum_{m'=1}^{K} a_{mm'}(\rho)\,\mathcal{Y}_{m'}(\rho,\Omega)\;, \tag{9}$$

where K is the number of terms included in the expansion and $\mathcal{Y}_m(\rho,\Omega)$ are antisymmetrical combination of HH functions, which can also include correlation factors F_m. Once the expansion given in Eq. (9) is substituted in Eq. (4), a $K\times K$ algebraic eigenvalue problem is obtained for the coefficients $a_{mm'}(\rho)$, which can be solved by using a standard numerical method.

The correlation factors F_m can be chosen either equal to 1, in which case one has the usual HH expansion, or equal to $f_m(|\boldsymbol{r}_i - \boldsymbol{r}_j|)$, where f_m is a pair–correlation function, obtaining in this case the PHH expansion (P≡pair) [1]. Another possibility is to choose F_m as a product of pair–correlation functions (Jastrow factor).

The eigenpotentials $U_m(\rho)$ have some important properties [6]. In all the cases here considered, the "lowest" potential (corresponding to $m=1$) is the only one having an attractive part at medium distances, whilst the other potentials contain larger repulsions. This fact guarantees a very rapid convergence of the expansion in the AHH function basis.

Once the functions $\Phi_m(\rho,\Omega)$ have been calculated, the M coupled differential equations given by Eq. (6) can be numerically solved. Let us refer with E_{MK} the energy calculated with this procedure. Alternatively, one can use the direct expansion of the wave function, namely,

$$\Psi_K = \frac{1}{\rho^{(D-1)/2}} \sum_{k=1}^{K} w_k(\rho)\mathcal{Y}_k(\rho,\Omega)\;. \tag{10}$$

In this case, K differential equations must be solved and let E_K indicate the corresponding energy. It can be proved [3] that $E_{MK} \geq E_K \geq E_{\rm exact}$ and that $E_{MK} \to E_K$ as $M \to K$. One of the main goal of this paper is to study the convergence of E_{MK} as M is increased.

The results obtained for the systems with $A=3$ and 4 with the AV14 potential [7], and using the PHH expansion in eq. (9), are presented in Table 1. As it can be seen by inspection of the table, just the first term of the expansion given in Eq. (5) is sufficient by alone to obtain a good estimate of the upper bound energy E_K. In all the examined cases, it has been found $E_{M=1,K} - E_K \approx 0.2$ MeV (when $K \gg 1$). Moreover, the limiting value E_K is almost reached by including only a small number of terms. To be noticed that the E_K values are obtained by solving $K=48$ ($K=81$) equations for the $A=3$ ($A=4$) system. However, it is possible to reach the same energies by solving only

M	$E_{MK}(A=3)$	$E_{MK}(A=4)$
1	−7.44	−20.91
5	−7.61	−21.05
9	−7.65	−21.08
13	−7.66	−21.09
E_K	−7.66	−21.10

Table 1. Convergence of the ground state energy E_{MK} (in MeV) for the $A = 3$, 4 systems in the case of the AV14 potential [7]. M is the number of functions $\Phi_m(\rho, \Omega)$ used in the expansion of the wave function. K is the number of the PHH functions used for expanding each function $\Phi_m(\rho, \Omega)$. Here, $K = 48$ for $A = 3$ and $K = 81$ for $A = 4$.

$M = 13$ coupled equations. The wave functions calculated by means of the two procedures are very nearly coincident, as it has been verified by evaluating various average quantities, as the mean kinetic energy, the mass radius, etc.

We have also explored the uncorrelated HH expansion (namely, we have used $F_m = 1$). Some of the results obtained are reported in Table 2. For the $A = 3$ system the calculation has been done with the realistic AV14 potential [7]. Again, the upper bound value E_K it is reached by solving only a small number M of differential equations, confirming the results obtained for central potentials in ref. [3]. For the four–nucleon system we have used the (central and spin–independent) S3 potential [8]. Also in this case, a quite satisfactory convergence is found. The calculation for $A = 4$ when using realistic potentials is in progress.

4 Conclusions

In this paper, we have studied the convergence properties of the AHH basis function for the triton and alpha–particle, extending the analysis given in ref. [3] for simple model potentials. The main conclusion we have arrived to is that the AHH functions represent a very effective basis for treating the problem of bound states. We have investigated two different methods for obtaining the AHH functions: *(i)* the expansion on a correlated basis and *(ii)* the expansion on the usual HH basis. In both cases, we have found a sizeable reduction of the number of coupled differential equations that must be solved.

The correlated expansion allows the reduction of the number K of terms needed to produce the AHH functions. However, the presence of correlations also implies the necessity to use complicate and time–consuming numerical procedure to evaluate the various integrals over Ω. This fact could be a major problem for the extension of the technique to larger nuclei.

On the other hand, the uncorrelated expansion requires the use of quite large K values, but it should allow an easy extension of the technique for

Potential	AV14	S3
M	$E_{MK}(A=3)$	$E_{MK}(A=4)$
1	−7.434	−27.10
5	−7.598	−27.30
10	−7.628	−27.35
20	−7.648	−27.36
30	−7.651	−27.37
40	−7.652	−27.37
E_K	−7.660	−27.37

Table 2. Convergence of the ground state energy E_{MK} (in MeV) for the $A = 3$, 4 systems in the case of the AV14 potential [7] ($A = 3$) and the S3 [8] potential ($A = 4$). M is the number of functions $\Phi_m(\rho, \Omega)$ used in the expansion of the wave function. K is the number of the HH functions used for expanding each function $\Phi_m(\rho, \Omega)$. Here, $K = 280$ in the case $A = 3$ and $K = 62$ for $A = 4$.

nuclei with $A > 4$. In the future, both techniques will be explored.

Acknowledgement. The authors would like to thank Profs. S. Rosati and M. Fabre de la Ripelle for useful discussions.

References

1. A. Kievsky, S. Rosati, M. Viviani: Nucl. Phys. **A577**, 511 (1994); M. Viviani, A. Kievsky, S. Rosati: Few Body Syst. **18**, 25 (1995)
2. J.M. Macek: J. Phys. **B1**, 831 (1968); M. Fabre de la Ripelle: C. R. Acad. Sci. Paris **274**, 104 (1972)
3. J.L. Ballot, M. Fabre de la Ripelle, J.S. Levinger: Phys. Rev. **C26**, 2301 (1982)
4. L.M. Delves: Nucl. Phys. **9**, 391 (1959); F.T. Smith: Phys. Rev. **120**, 1058 (1960); W. Zickendrath: Ann. of Phys. (NY) **53**, 18 (1965); Yu.A. Simonov: Sov. J. Nucl. Phys. **3**, 461 (1966)
5. M. Fabre de la Ripelle: Ann. Phys. **147**, 281 (1983)
6. M. Fabre de la Ripelle: Few Body Systems **14**, 1 (1993)
7. R.B. Wiringa, R.A. Smith, T.L. Ainsworth: Phys. Rev. **C29**, 1207 (1984)
8. L.R. Afnan, Y.C. Tang: Phys.Rev. **175**, 1337 (1968)

Few-Body Systems Suppl. 8, 49–53 (1995)

Few-Body Systems

Cross Sections and Vector Analyzing Powers in the Proton Induced Deuteron Breakup Reaction at 65 MeV: Star Configurations

M. Allet[1], K. Bodek[2], W. Glöckle[3], J. Golak[2], L. Jarczyk[2], St. Kistryn[1,2], J. Lang[1], R. Müller[1], S. Navert[1], O. Naviliat-Councic[1], J. Smyrski[2], J. Sromicki[1], E. Stephan[1*], A. Strzałkowski[2], H. Witała[2], J. Żejma[2]

[1] Institut für Teilchenphysik, Eidgenössische Technische Hochschule, Hönggerberg HPK, CH-8093 Zürich, Switzerland

[2] Institute of Physics, Jagellonian University, ul. Reymonta 4, PL-30059 Kraków, Poland

[3] Institut für Theoretische Physik II, Ruhr Universität, Universitätstraße 150, D-44780 Bochum, Germany

Abstract. New experimental cross-section and vector-analyzing-power data for three star configurations of the kinematically complete breakup reaction $^2\mathrm{H}(p,pp)n$ are compared to the rigorous Faddeev calculations involving four realistic NN potentials – Argonne AV14, Bonn B, Nijmegen and Paris. A satisfactory agreement between theory and experiment has been found. The effects of the Tucson-Melbourne three-body force included in the calculations are also discussed.

Understanding of the nucleon-nucleon (NN) interaction is one of the most fundamental goals of nuclear physics. Several meson-exchange models provide realistic two-nucleon ($2N$) potentials which describe satisfactorily the bulk of the $2N$ data. However, the $2N$ system cannot supply full information on the NN interaction and it is therefore necessary to study the three- or many-nucleon systems or the NN bremstrahlung. The biggest attention is drawn to the three-nucleon ($3N$) system. The nucleon-deuteron (Nd) breakup process is especially interesting – there the momenta of the outgoing nucleons are not integrated over the deuteron wave function what makes this process very rich in physical information. The possibility of choosing specific kinematical configurations for the outgoing nucleons allows one to concentrate on specific properties

* *On leave from:* Institute of Physics, University of Silesia, ul. Uniwersytecka 4, PL-40007 Katowice, Poland

of the NN interaction. Moreover, for the $3N$ system the observables can be calculated in an exact way by using the Faddeev formalism with both $2N$ and $3N$ forces [1], [2]. Thus a comparison of the theoretical predictions with the experimental data provides a direct test of the physical assumptions not obscured by computational approximations. An attempt is made to answer the following questions: Do calculations with the two-body realistic potentials describe $3N$ observables? Is it necessary to include genuine $3N$ forces in addition to $2N$ ones? Is the longe-range Coulomb force important?

The aim of the reported experiment was to provide accurate $3N$ observables in selected kinematical regions of the breakup reaction continuum where, according to model calculations, the $3N$-force effects are enhanced and, simultaneously, the sensitivity to details of the $2N$ potential is small [3]. This work is a continuation of the experiments where collinearity [4] and quasi-free scattering (QFS) [5] configurations have been measured. The kinematically complete measurement of the differential cross section $d^3\sigma/(d\Omega_1 d\Omega_2 dS)$ and the analyzing power A_y of the $^2H(p,pp)n$ reaction at 65 MeV incident proton energy was devoted to three configurations: symmetrical space star SSS ($\vartheta_1 = 54.0°, \vartheta_2 = 54.0°, \varphi_{12} = 120.0°$), symmetrical forward plane star FPS ($35.2°, 35.2°, 180.0°$), and symmetrical backward plane star BPS ($75.4°, 75.4°, 180.0°$).

The experiment was carried out at the Paul Scherrer Institute – Villigen, Switzerland. The Philips Cyclotron provided a transversally polarized proton beam with an energy of 65 MeV and an average intensity of about 250 nA. The beam polarization $|P_y|$ measured continuously with a carbon transmission polarimeter was about 0.80 and its sign was reversed every second. The beam was focused on a deuterium gas target cooled down to 77 K with windows formed specially to allow the detection of protons emitted out of the horizontal plane. Eight $\Delta E - E$ telescopes built of NE102A plastic scintillator were used for the detection of charged reaction products. Standard electronics and an on-line data aquisition system collected the coincidences between each pair of telescopes. Single events with rate reduced by a factor of 1000 were recorded simultaneously. The analyzing powers were calculated from the asymmetry of the spectra of coincidences corresponding to different beam spin states. The cross sections were determined from the same spectra by summing up over the spin states. The absolute normalization was achieved by a reference to the $p+^2H$ elastic scattering cross section. Accidental coincidences were measured simultaneously and subtracted in the off-line analysis.

Our experimental results are compared to theoretical predictions based on the rigorous solutions of the $3N$-Faddeev equations performed with four realistic potentials: Bonn B, Argonne AV14, Paris and Nijmegen. The calculations have been performed with $2N$-subsystem total angular momentum $j \leq 3$ for which full convergence is achieved. The Coulomb force was totally neglected. Both experimental results and theoretical calculations are presented as a function of the arc-length S of the kinematical curve in the E_2-E_1 energy plane. Before the comparison, the theoretical calculations were corrected to account for the finite experimental geometry and resolution as well as for the relativistic effects on the level of pure kinematics (see [4]).

One can notice a satisfactory agreement between measured and calculated observables (Fig. 1). This is particularly true for the case of the analyzing power which is very well described by theory. The cross section in the FPS configuration is dominated by a prominent bump which is a consequence of the QFS process affecting the measured configuration. The amplitude of this peak is very well reproduced by theory, however, the predicted width of this bump is larger than the experimental one by about 8%. The neglect of the Coulomb force is supposed to be the reason of this effect [5]. A very good agreement between theory and experiment is achieved for the cross section in case of the BPS configuration. The same statement for A_y is not yet justified because of experimental errors that are still quite large. For the SSS configuration the cross section is overestimated by about 8% in case of the Bonn B potential, which performed well in other configurations.

In addition, the rigorous calculations with included $3N$ force [2] (taken as the 2π-exchange Tucson-Melbourne (TM) model) were compared to the experimental data (Fig. 2). The Bonn B $2N$- and the TM $3N$-forces were allowed to act in all partial waves with two-body angular momenta $j \leq 2$. The πNN cutoff parameter Λ_π was set to 5.8μ ($\mu = 139.6$ MeV). The Bonn B potential with such $3N$ force overbinds the triton by about 2 MeV, therefore the $3N$-force effects are overestimated. In case of the BPS and FPS configurations no significant $3N$-force effects have been found. At the SSS point the addition of the $3N$ force shifts the theoretical cross section by about 15% away from the data (Fig. 2b). It has been shown that the $3N$ force adjusted to the experimental triton binding energy ($\Lambda_\pi = 4.55\mu$) gives smaller effect on the breakup observables [2]. But anyhow these effects tend to increase the disagreement between theory and experiment in the SSS point. On the other hand, the role of the Coulomb force is unknown. As indicated by model calculations where the nucleon interaction was truncated to the S-waves only, while the full Coulomb potential was included [6], the Coulomb force may decrease the cross section in the SSS configuration and thus compensate the $3N$ force enhancement. Drawing conclusions concerning $3N$-force effects in the analyzing powers is a more delicate task. Surprisingly, they generally agree better with the theory based on the $2N$-interaction only than do the cross sections. Thus, there is less space for $3N$-force effects. On the other hand, higher partial waves ($j \leq 3$) are more important in A_y than in the cross section so the present calculations performed in the truncated angular momentum space ($j \leq 2$) should be extended to at least $j_{max} = 3$.

Acknowledgement. This work was supported by the Polish Comittee for Scientific Research under Grant No. 2 P302 104 06 and by the Swiss National Foundation.

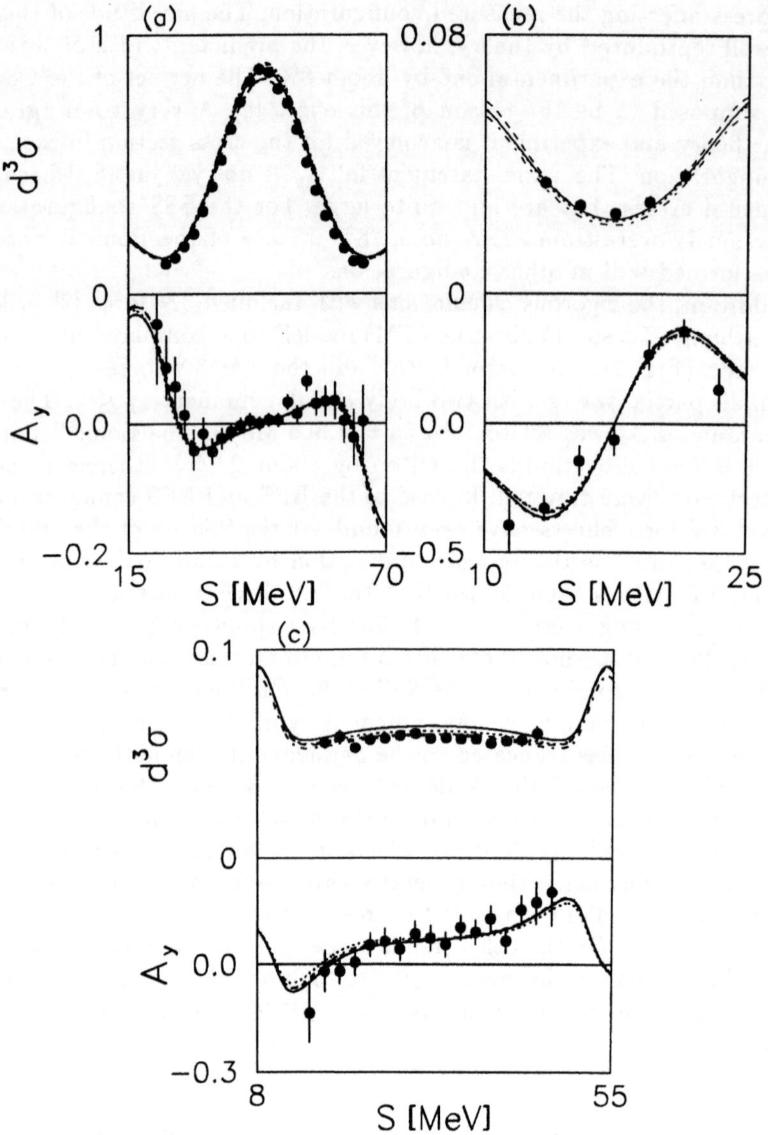

Figure 1. Experimental cross sections and analyzing powers for the (a) FPS, (b) BPS and (c) SSS configurations. Error bars represent the total (systematic+statistical) experimental errors. Curves represent calculations obtained using four $2N$ potentials: Bonn B (solid), Paris (dashed), Argonne AV14 (dotted) and Nijmegen (dash-dotted).

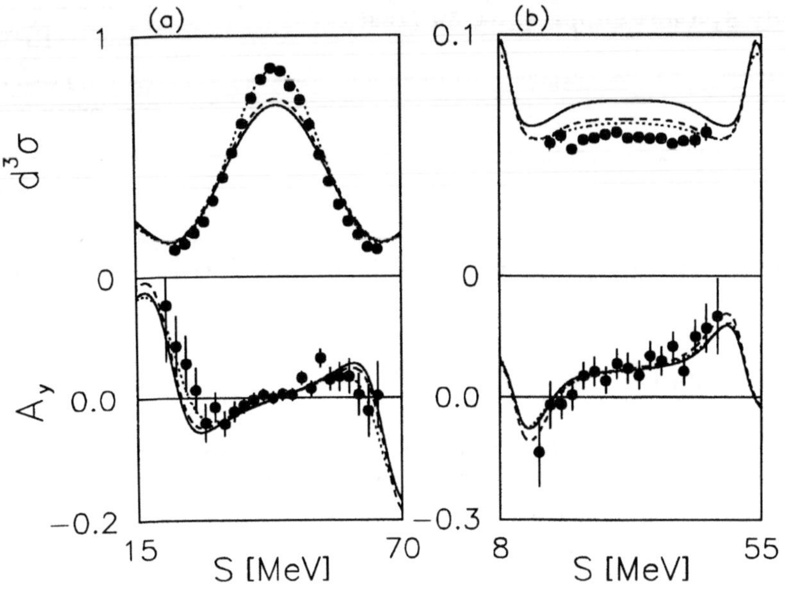

Figure 2. Cross sections and analyzing powers for the (a) FPS and (b) SSS configurations compared to the Faddeev calculations including $2N$ potential only with $j \leq 3$ (dotted line), $2N$ potential with $j \leq 2$ (dashed line), and $2N$+$3N$ potentials with $j \leq 2$ (solid line). Bonn B and TM potentials were chosen for $2N$ and $3N$ interactions, respectively.

References

1. H. Witała, T. Cornelius, W. Glöckle: Few-Body Systems **3**, 123 (1988)
2. D. Hüber, H. Witała, W. Glöckle: Few-Body Systems **14**, 171 (1993).
3. W. Meier, W. Glöckle: Phys. Lett. **138B**, 329 (1984)
4. M. Allet, K. Bodek, W. Hajdas, J. Lang, R. Müller, O. Naviliat-Councic, J. Sromicki, J. Zejma, L. Jarczyk, St. Kistryn, J. Smyrski, A. Strzałkowski, W. Glöckle, J. Golak, H. Witała, B. Dechant, J. Krug, P.A. Schmelzbach: Phys. Rev. **C50**, 602 (1994)
5. M. Allet, K. Bodek, W. Hajdas, J. Lang, R. Müller, S. Navert, O. Naviliat-Councic, J. Sromicki, J. Zejma, L. Jarczyk, St. Kistryn, J. Smyrski, A. Strzałkowski, W. Glöckle, J. Golak, D. Hüber, H. Kamada, H. Witała: Few-Body Systems (in print)
6. E.O. Alt, M. Rauh: Few-Body Systems **17**, 121 (1994)

Few-Body Systems Suppl. 8, 54–59 (1995)

Few-Body Systems

Single Particle Spectral Function for Finite Nuclei

A. Polls [1], H. Müther [2], W.H. Dickhoff [3]

[1] Departament d'Estructura i Constituents de la Matèria, Universitat de Barcelona, Diagonal 647, E-08028 Barcelona, Spain

[2] Institut für Theoretische Physik, Universität Tübingen, Auf der Morgenstelle 14, D-72076 Tübingen, Germany

[3] Department of Physics, Washington University, St. Louis, MO 63130, USA

Abstract. The effects of short-range correlations on the momentum and energy distribution of nucleons in finite nuclei are investigated using realistic nucleon-nucleon interactions. The calculations are performed in the framework of the Green function theory. The spectral functions are calculated from the momentum and energy dependent nucleon self-energy by solving a Dyson equation. The nucleon-nucleon correlations induced by the short-range and tensor components of the interaction yield an enhancement of the momentum distribution at high momenta. These high momentum components should be observed mainly in nucleon knockout reactions like $(e, e'p)$ leaving the final nucleus in a state of high excitation energy.

1 Introduction

Microscopic description of nuclear systems requires a careful treatment of nucleon-nucleon (NN) correlations. In particular, the correlations induced by the short range repulsion as well as the tensor component present in any realistic NN interaction are the main responsible for the fact that simple independent particle model yield an unsatisfactory description of nuclear systems. It is therefore convenient to find nuclear observables sensitive to NN correlations. In this sense, the single particle spectral function and the momentum distribution, in spite of being one-body quantities, are nuclear properties experimentally accessible that provide information on the effects of NN correlations in a clear manner. In adition, the knowledge of the spectral function $S(\mathbf{k}, E)$ is an essential ingredient for the description of different nuclear processes [1]. In particular, the analysis of the recent high resolution $(e, e'p)$ experiments needs a realistic $S(\mathbf{k}, E)$ and has motivated a lot of theoretical effort towards its microscopic determination [2, 3]. The physical meaning of the spectral function is rather simple. The hole spectral function $S(\mathbf{k}, E)$ gives the probability of removing a particle with momentum **k** from the target system of A particles leaving the resulting $(A-1)$ system with an energy $E^{A-1} = E_0 - E$, where E_0 is the ground state energy of the target. $S(\mathbf{k}, E)$ gives access to all single particle observables of nuclei and through the energy sum-rule it is also possible to evaluate the binding energy if the Hamiltonian contains only two-body forces.

The present available techniques to calculate the spectral function are based either on correlated basis function theory or on the perturbation expansion of the one body propagator. Due to the difficulties implied by the treatment of finite nucleus, at both the formal and computational level, the microscopic calculations have been mainly performed for nuclear matter [4, 5] and for very light nuclei [6, 7, 8]. In this contribution we present a method to calculate the spectral function and the corresponding momentum distribution directly for finite nuclei in the framework of the Green function approach [9, 10]. Calculations are performed for the ^{16}O assuming a realistic meson exchange potential for the NN interaction [11]. Special atention is paid to the effect of correlations on the spectral function at different energies. We find clear signatures of the short range NN correlations by studying the spectral function at very negative energies, which in knockout experiments correspond to the excitation energies of around 100MeV and more in the remaining nucleus.

2 Spectral Function in Finite Nuclei

For finite systems it is convenient to introduce a partial wave decomposition of the spectal function

$$S_{lj\tau}(k,E) = \sum_n \mid \langle \Psi_n^{A-1} \mid a_{klj\tau} \mid \Psi_0^A \rangle \mid^2 \delta(E - (E_0^A - E_n^{A-1})) \tag{1}$$

where $a_{klj\tau}$ denotes the annihilation operator for a nucleon with orbital angular momentum l, total angular momentum j, isospin τ and momentum k. The momentum distribution

$$n_{lj\tau}(k) = \langle \Psi_0 \mid a^\dagger_{klj\tau} a_{klj\tau} \mid \Psi_0^A \rangle \tag{2}$$

is obtained by integrating the spectral function $S_{lj\tau}(k,E)$ over the excitation energies of the $A-1$ system. Then taking into account the degeneracy factors of each orbital one obtains the total momentum distribution

$$n(k) = \sum_{l,j,\tau} (2j+1) n_{lj\tau}(k). \tag{3}$$

The spectral function for the various partial waves, $S_{lj\tau}(k,E)$ can be obtained from the imaginary part of the corresponding single-particle Green function $g_{lj}(k,E)$. Note that in order to simplify the notation here and in the following we have dropped the isospin quantum number τ.

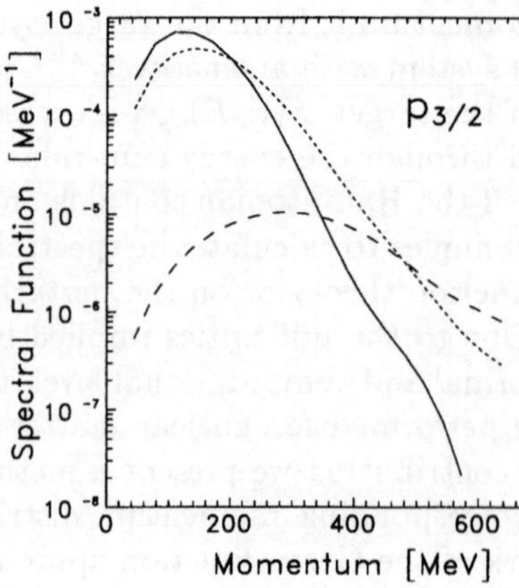

Figure 1. The $p_{3/2}$ spectral function as a function of momentum at fixed energies corresponding to -50 (solid line), -100 (short-dashed), and -250 (long-dashed) MeV.

The single particle propagator can be obtained by solving the Dyson equation

$$g_{lj}(k_1, k_2; E) =$$

$$= g_{lj}^{(0)}(k_1, k_2; E) + \int dk_3 \int dk_4 g_{lj}^{(0)}(k_1, k_3; E) \Delta\Sigma_{lj}(k_3, k_4; E) g_{lj}(k_4, k_2; E), \quad (4)$$

where $g^{(0)}$ refers to a Hartree-Fock propagator and Σ_{lj} represents contributions to the real and imaginary part of the irreducible self-energy which go beyond the Hartree-Fock approximation used to derive $g^{(0)}$. The key point of this method is the calculation of the self-energy. In the present approach, the Hartree-Fock contribution to the self-energy has been evaluated using a G-matrix interaction calculated for nuclear matter at a given fixed energy and density. The additional terms to the self-energy, $\Delta\Sigma_{lj}(k, k'; E)$, incorporate the relevant second order diagrams in this G-matrix which contain the two-particle-one-hole and two-hole-one-particle terms. The imaginary part of the $2h1p$ term is crucial for the calculation of the hole part of the spectral function. As a single particle basis one includes the Hartree-Fock bound states and for the states at positive energy plane waves are employed with corresponding single particle energies [12].

The Dyson equation can also yield discrete solutions. These solutions are obtained by solving an eigenvalue problem. The eigenvector corresponding to these discrete states yields the quasi-hole wave function in momentum space, which still needs to be normalized by the spectroscopic factor

$$Z_{\alpha_{qh}} = \left(1 - \frac{\partial \Sigma(\alpha_{qh}, \alpha_{qh}; E)}{\partial E} \mid \epsilon_{qh}\right)^{-1}, \quad (5)$$

where α_{qh} corresponds to the quasihole single-particle quantum numbers. In the present approach, the $2h1p$ imaginary part of the self-energy is zero at the

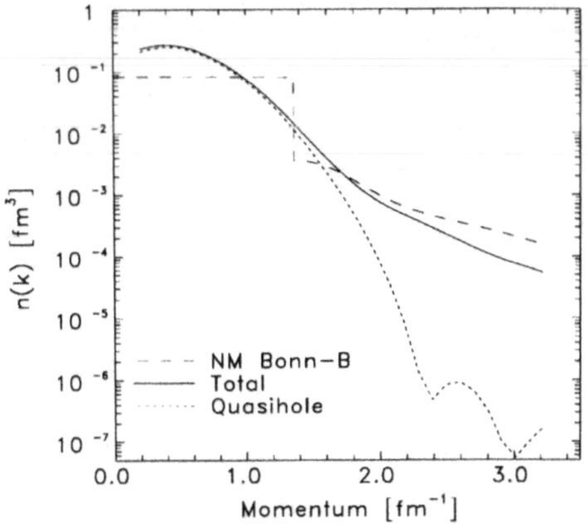

Figure 2. The total momentum distribution (solid line) for ^{16}O with the OBE-B potential compared with the n(k) for nuclear matter (long-dashed line). In this figure, n(k) is normalized such that $\int d^3k n(k)$ yields 1.

quasihole energies and therefore, $S_{lj}(k, E)$ shows δ-functions of strength Z_{lj} at those energies. In our discussion of the hole spectral function we can therefore distinguish the contributions originating from the quasihole states from the ones of the continuum of $2h1p$ configurations [10].

The momentum dependence of $S_{p3/2}(k, E)$ in the continuum is shown in Fig. 1 at three different energies. From these results it is clear that an important change in the momentum content of the single-particle strength occurs with increasing excitation energy. At higher excitation energy one finds more high momentum components. The total momentum distribution (continuous line) is shown in Fig. 2. The short dashed line stands for the quasi-hole contribution, which for ^{16}O arises from the partial waves $s_{1/2}, p_{3/2}, p_{1/2}$. The momentum distribution at small momenta is dominated by the quasi-hole contribution whereas the high momentum components are given by the high energy continuum part of the spectral function. For illustration, Fig. 2 also shows $n(k)$ of infinite nuclear matter at the saturation density calculated with the same potential.

Fig. 3 shows the local density distribution obtained in the HF (long dashes) compared to the density distribution obtained from the spectral function (solid line). Also displayed are the contributions of the quasihole states to this total density (short-dashed line) and the density originating from the occupation of states with l larger than 1, which account for 4% of the total number of particles. One finds that the depletion of states with $l = 0$ yields a reduction of the density in the centre of the nucleus. The difference between the density distribution from the quasi-hole approximation (dotted line) and the total results is the contribution from the continuum part of the spectral function. The quasi-hole contribution of the $s_{1/2}, p_{3/2}$ and $p_{1/2}$ partial waves accounts for a 89% of the total number of particles and taking into account their contribution in the

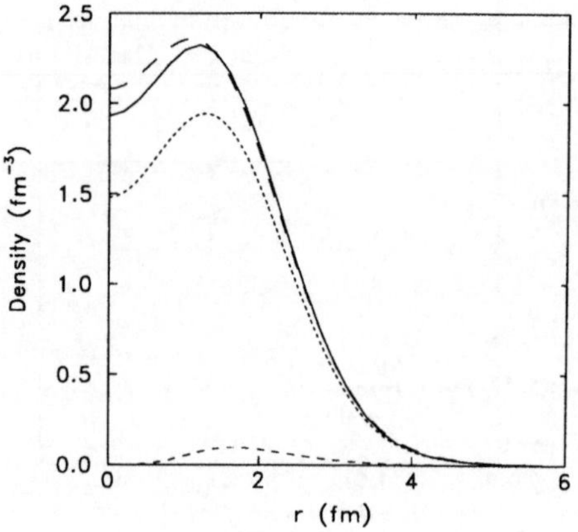

Figure 3. Density distribution of nucleons in ^{16}O as a function of the distance from the center of the nucleus. In this figre, $\rho(r)$ is normalized such that $\int drr^2\rho(r)$ yields 16.

continuum one gets the 96%.

3 Summary and Conclusions

We have presented a suitable method for microscopic studies of the energy and momentum dependence of the single particle spectral function in finite nuclei in the framework of the Green function theory. The explicit calculations have been performed for ^{16}O using a realistic meson exchange potential for the NN interaction. The approximations made in the method are specially devised for the treatment of the short range correlations. These correlations yield strong enhancement of the momentum distribution at high momenta. This enhancement should be observed in nucleon knockout reactions with large energy transfer leaving the final nucleus at high excitation energy. Non negligible contributions to the momentum distribution are found in partial waves which are unoccupied in the simple shell model. Presently we are extending the calculations to lighter (^4He) and to heavier (^{40}Ca) nuclei for which experiments are currently in progress.

Acknowledgement. This research project has partially been supported by DG-ICYT PB92/0761 (Spain), the contract SFB382 of the *Deutsche Forschungsgemeinschaft*, the EEC-contract CHRX-CT93-0323 and the U.S. NSF under Grant No. PHY-9307484.

References

1. A.E.L. Dieperink, P.K.A. de Witt Huberts: Ann. Rev. Nucl. Part. Sci. **40**, 239 (1990)

2. W.H. Dickhoff, H. Müther: Rep. Prog. Phys. **55**, 1947 (1992)

3. C. Mahaux, R. Sartor: Adv. Nucl. Phys. **20**, 1 (1991)

4. O. Benhar, A. Fabrocini, S. Fantoni: Nucl. Phys. **A505**, 267 (1989)

5. A. Ramos, A. Polls, W.H. Dickhoff: Nucl. Phys. **A503**, 1 (1989)

6. H. Morita, T. Suzuki: Prog. Theor. Phys. **86**, 671 (1991)

7. H. Meier-Hajduk et al.: Nucl. Phys. **A395**, 332 (1983)

8. C. Ciofi degli Atti, E. Pace, G. Salmè: Phys. Rev. **C21**, 805 (1980)

9. H. Müther, W.H. Dickhoff: Phys. Rev. **C49**, R17 (1994)

10. H. Müther, A. Polls, W.H. Dickhoff: Phys. Rev. in press

11. R. Machleidt: Adv. Nucl. Phys. **19**, 1 (1989)

12. M. Borromeo et al.: Nucl. Phys. **A539**, 189 (1992)

Few-Body Systems Suppl. 8, 60–66 (1995)

Few-Body Systems

Few-Body Problems in Atomic Physics

Ingvar Lindgren, Hans Persson, Sten Salomonson, Per Sunnergren

Department of Physics, Chalmers University of Technology and Göteborg University, Göteborg, Sweden

Abstract. By means of new experimental techniques heavy ions with a few electrons can now be studied with improved accuracy. In combination with new computational schemes such data can be used for testing QED at strong fields. In this paper we discuss in particular recent QED calculations on lithium-like uranium and on some helium-like ions and compare them with corresponding experimental results. The agreement between theory and experiments is generally very good. The helium-like ions are of particular interest in this respect. Here, it is possible to determine the two-electron QED contribution independently of the much larger one-electron part. Only a moderate improvement in the experimental accuracy is needed in order to make it possible, for the first time, to test QED effects at strong fields beyond the lowest-order Lamb shift.

1 Introduction

The experimental situation concerning heavy-ion spectroscopy has improved drastically in recent years. EBIT and Super-EBIT facilities together with storage rings with improved cooling facilities have made it possible to study highly charged ions with increasing precision. Also on the theoretical side a corresponding progress has taken place. New techniques have been developed for relativistic many-body calculations with high accuracy. For heavy, highly charged ions also quantum- electrodynamical (QED) effects are of increasing importance. Here, the field from the nucleus is so strong that standard techniques for QED calculations cannot be used. During the last years, significant progress have been made in such calculations, and the corresponding effects can now for the first time be tested. For a recent review of the field, see [1]

Measurements on hydrogen-like ions are of particular interest, since the theoretical non-QED value can be determined essentially exactly by solving the Dirac equation, and such ions are now being studied up to uranium [2]. A disadvantage is here that the effect of the nuclear effect is so large (about 200eV for the $1s$ state in hydrogen-like uranium) that it largely masks the details of the QED effect.

The next simplest systems are the alkali-like ions with one electron outside a closed rare-gas core. Such systems behave mainly as single-electron system, although the valence electron moves in a non-coulombic field. In addition, of course, there are many-body effects, which have to be considered, when high accuracy is desired. One of the most accurate measurements on a highly charged alkali-like ion is that on the $2s$-$2p$ transitions in Li-like uranium, recently reported from Berkeley [3]. The experimental transition energy is here 280eV with an accuracy of 0.1eV. The corresponding theoretical transition energy obtained from relativistic many-body theory is 322eV. The discrepancy, about 42eV, is due to the ***Lamb shift***, which thus can obtained with high relative accuracy in this case. The comparison between the experimental and theoretical results can here serve as a good test of QED calculations in strong fields, and a few calculations of this kind have recently been performed [4, 5]. This will be discussed further below.

Also helium-like systems with two electrons are of great principal interest, since nowadays very accurate calculations can be performed on such systems. Here, the many-body part, i. e. the electron correlation, is more important than for alkali-like systems. Recently, there has also been a significant progress in the experimental situation concerning such systems, also for high nuclear charges. Using the Super- EBIT facility at Lawrence Livermore National Laboratory, Marrs et al. [6] have experimentally determined the *two-electron* contribution to the ground-state energy of some of these ions by comparing the binding energy of helium-like and hydrogen-like ions of the same elements. Such data are particularly interesting for testing calculations of the small but significant two-electron QED contribution, since the dominating one-body parts are completely eliminated. Another interesting feature of such measurements is that the nuclear size has an almost negligible effect on the results. We have for the first time performed complete second-order calculations of the two-electron contribution to the Lamb shift of helium-like ions and compared them with the new experimental data, as will be further discussed below.

For light elements, for which $Z\alpha$ is small, the procedure for QED calculations was developed long ago. (Z is here the nuclear charge and $\alpha \approx 1/137$ the fine-structure constant). The starting point for such calculations is free-electron QED, and the potential from the nucleus is treated as a perturbation. This leads to a $Z\alpha$ expansion, which is rapidly converging, and the calculations are mainly analytical [7]. For heavy, highly charged ions, on the other hand, where $Z\alpha$ approaches unity, the potential felt by the electron is so strong that a $Z\alpha$ expansion is not meaningful. In this case it is more efficient to start from single-electron states generated in the external field. This technique has been developed for Lamb-shift calculations of hydrogen-like ions in a sequence of papers by Mohr [8]. The accuracy is here quite high. The uncertainty lies mainly in the nuclear effect and in the omission of higher-order effects. For heavy, highly charged ions with more than one electron, it is necessary to consider QED effects *as well as* many-body effects. Several groups have recently made important progress is this field [4, 5, 1].

In the following we shall decsribe the present status of QED calculations on

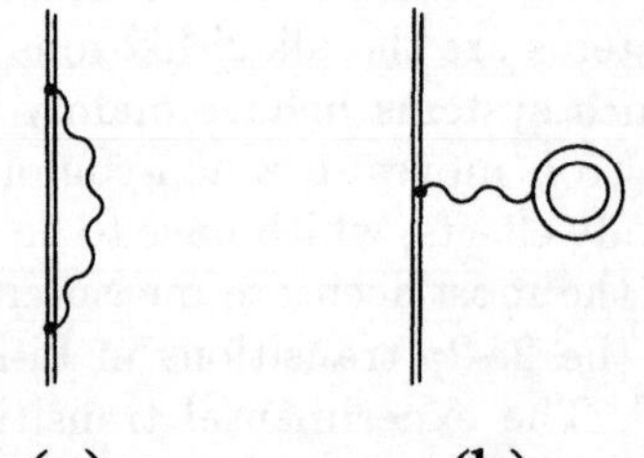

Figure 1. The first-order Lamb shift for a single bound electron - the electron self energy (a) and the vacuum polarization (b).

highly charged ions with few electrons, and, in particular, discuss some recent calculations on lithium-like uranium and on some helium-like ions and compare them with corresponding experimental results.

2 Lamb Shift of Lithium-like Uranium. Single-electron Lamb Shift

As mentioned in the introduction, it is for QED calculations on heavy, strongly charged ions necessary to start from single-electron states generated in the potential of the nucleus (and possibly also the remaining electrons). The single-electron states are then generated by solving the Dirac equation with the external potential $u(r)$,

$$h^{\rm ext}|i\rangle = [\boldsymbol{\alpha}\cdot\mathbf{p} + \beta m + u(r)]|i\rangle = \epsilon_i|i\rangle$$

The equations are solved by the method of discretization [9]. Finite nuclei are used with the experimentally determined radius and a uniform charge distribution. The effect of the charge distribution on the final result is found to be negligible.

The first-order Lamb-shift diagrams, the *self-energy* and *vacuum- polarization* diagrams, are shown in Fig. 1. Using a numerical set of single-electron states, these diagrams can be evaluated using the so-called Furry interaction picture. Both diagrams are infinite and have to be *renormalized.*

The renormalized vacuum polarization can be fairly well respresented by the *Uehling potential* [10], but there exist nowadays more accurate methods for calculating this effect at strong fields [11, 12]. For the self-energy part, the renormalization implies that the free-electron self energy, which is already included in the physical electron mass, has to be removed. Due to the spherical symmetry of the atomic problem, the external-field self energy is conveniently evaluated by means of an expansion in spherical waves. The contributions for each wave is finite, but the expansion diverges logarithmically. By evaluating the free-electron part in the same fashion, renormalization can be performed separately for each partial wave. Then the renormalized contributions form a *convergent* series. This is the so-called *partial-wave renormalization*, which

Table 1. A comparison with the experimental value on the $2p_{1/2}$-$2s_{1/2}$ transition in lithium-like uranium. (eV)

	This work	Blundell [14]
Vacuum polarization (VP)	12.586	12.56
Self-energy (SE)	-54.32	-54.24
Combined SE and VP	-0.19	
Reducible Breit-Coulomb correction	0.04	
Nuclear polarization and recoil	0.10	0.10
Relativistic MBPT	322.33	322.41
Total	280.55	280.83
Experiment, ref. [3]	280.59(9)	

was recently introduced independently by us and the Oxford group [4, 13]. The advantage of this method is that it is numerically very stable, since no infinities are involved.

In Table 1 we present our results [4, 15], and compare them with those of Blundell [5], obtained with a different technique. The agreement with the experimental result is extremely good - probably too good, considering the fact that higher-order (two-photon) single-electron contributions to the Lamb shift has been calculated only partially. The agreement between the two theoretical calculations is also very good, although the computational techniques involved are quite different. This indicates that calculations of this kind can now be performed with high accuracy and high reliability.

3 Helium-like Ions. Two-electron Lamb Shift

We shall here present calculations on helium-like ions and compare them with recent experimental results. Calculations to order $(Z\alpha)^3$ of the QED contributions to the energy of helium-like ions were made long ago by Sucher and Araki [16]. For heavy, highly charged ions it is necessary to take into account the nuclear potential interaction to all orders by generating the electronic states in the external potential, as described above. We have recently applied this technique also to heavy ions with two electrons.

The second-order two-electron diagrams are shown in Fig. 2. Diagrams (a), the "*ladder*" diagram, and (b), the "*crossed-photon*" diagram, represent the non-radiative part of the interaction. This is largely taken care of in the many-body calculation, but not exactly. The crossed-photon diagram is not included in such a calculation, and also some partcd of the ladder diagram is left out. These parts, called the *Araki-Sucher effect*, have recently been calculated for arbitrary nuclear charges by us [17] and by Blundell et al. [18].

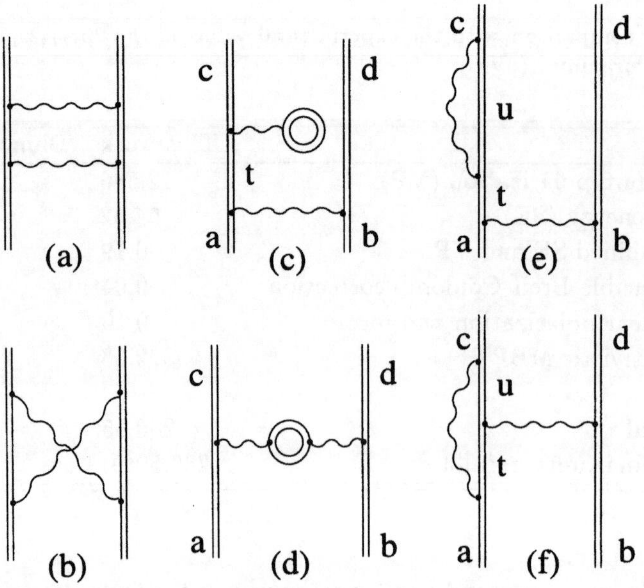

Figure 2. Feynman diagrams representing the two-photon contribution to the electron-electron interaction for helium-like ions. The first row, the "ladder" (a) and the "crossed-photon" (b) diagram, represent the non-radiative part and the remaining diagrams the two-electron Lamb shift, i.e., vacuum polarization (c,d) and self energy (e,f), respectively.

The remaining diagrams in Fig. 2 (c-f) represent the radiative effects. The second line (c and d) represents the two-electron vacuum polarization and the last line (e and f) the two-electron self energy. The latter part, which dominates, can to first approximation be regarded as a *screening* of the single-electron Lamb shift and estimated by means of a modification of the nuclear potential [19, 5, 4]. In this way, however, it is not possible to treat the exchange effect and the effect due to transverse photons in an exact way. Furthermore, in order to get high accuracy it is important to include also the smaller vacuum-polarization contribution. We have for the first time evaluated both these effects in a complete manner [20].

In Table 2 we present the various contributions to the two-electron part of the energy for some helium-like ions with nuclear charges ranging from 32 to 83 and compare them with recent experimental results of Marrs et al. [6]. The over-all uncertainty of our theoretical results can be estimated to be 0.2eV, including numerical uncertainties and left-out higher-order terms.

The agreement between the experimental and theoretical results supports the accuracy of the experimental and the many-body results, but the accuracy of the experimental results is not yet high enough to seriously test the QED part of the calculation. On the other hand, an improvement of only half an order of magnitude is required here in order to achieve –for the first time– an

Table 2. Various components of the two-electron contribution to the ground-state energy of some helium-like ions. Comparison between theory and experiment (in eV).

Nucl. charge	1st order MBPT	2nd MBPT	$\geq$ 3rd MBPT	Araki-Sucher	Lamb shift	Total theory	Experiment Ref. [6]
32	567.61	-5.22	0.02	0.03	-0.42	562.0	562.5 ± 1.6
54	1036.56	-7.04	0.03	0.16	-1.51	1028.2	1027.2 ± 3.5
66	1347.45	-8.64	0.03	0.36	-2.57	1336.6	1341.6 ± 4.3
74	1586.93	-9.97	0.04	0.55	-3.54	1574.0	1568 ± 15
83	1897.57	-11.86	0.04	0.86	-4.96	1881.7	1876 ± 14

experimental test of QED contributions for heavy elements beyond the lowest-order Lamb shift.

References

1. I. Lindgren, I. Martinsson, R. Schuch (eds): *Heavy-Ion Spectroscopy and QED Effects in Atomic Systems*, Proceedings of the Nobel Symposium 85, Saltsjöbaden, Sweden 29 June - 3 July, 1992, Physica Scripta **T46**, (1993)
2. H.F. Beier et al.: Phys. Lett. **A184**, 435 (1994)
3. J. Schweppe et al.: Phys. Rev. Lett. **66**, 1434 (1991)
4. I. Lindgren, H. Persson, S. Salomonson, A. Ynnerman: Phys. Rev. A **47**, R4555 (1993)
5. S.A. Blundell: Phys. Rev. A**46**, 3762 (1992)
6. R.E. Marrs, S.R. Schneider, Th. Stöhlker: Phys. Rev. A (submitted)
7. G. Drake: Can. J. Phys. **66**, 586 (1988)
8. P J Mohr: Ann. Phys. (NY) **88**, 2652 (1974)
   P.J. Mohr: Phys. Rev. A **26**, 2338 (1982)
   P.J. Mohr: Phys. Rev. A **46**, 4421 (1992)
9. S. Salomonson, P. Öster: Phys. Rev. A **40**, 5548 (1989)
10. E.A. Uehling: Phys. Rev. **48**, 55 (1935)
11. G. Soff, P. Mohr: Phys. Rev. A **38**, 5066 (1988)
12. H. Persson, I. Lindgren, S. Salomonson, P. Sunnergren: Phys. Rev. A **48**, 2772 (1993)
13. I.P. Grant, H. Quiney: J. Phys. **B27**, L299 (1994)

14. S.A. Blundell: Phys. Rev. A**47**, 1790 (1993)

15. A. Ynnerman, J. James, I. Lindgren, H. Persson, S. Salomonson: Phys. Rev. A **50**, 4671 (1994)

16. H. Araki: Prog. Theor. Phys. **17**, 619 (1957)
J. Sucher: Phys. Rev. **109**, 1010 (1958) and Ph. D. Thesis, Columbia University (1957), unpublished

17. I. Lindgren, H. Persson, S. Salomonson, L. Labzowsky: Phys. Rev. A **51**, 1167, (1995)

18. S. Blundell, P.J. Mohr, W.R. Johnson, J. Sapirstein: Phys. Rev. A **48**, 2615 (1993)

19. K.T. Cheng, W.R. Johnson, J. Sapirstein: Phys. Rev. Lett. **66**, 2960 (1991)

20. H. Persson, S. Salomonson, P. Sunnergren, I. Lindgren: to be published

Few-Body Systems Suppl. 8, 67–78 (1995)

Few-Body Systems

Molecular Potentials and Relativistic Effects

U. Kaldor* †

School of Chemistry, Tel Aviv University, 69978 Tel Aviv, Israel

Abstract.
The Fock-space coupled-cluster method, a powerful and efficient scheme for the incorporation of electron correlation in atomic and molecular systems, is described, and representative applications are reviewed. The molecular potentials of alkali-metal dimers in their ground and excited states are calculated. A relativistic coupled-cluster method, starting from the four-component Dirac-Coulomb-Breit Hamiltonian, is applied to calculate transition energies of heavy atoms (Au is given as an example) and to determine the ground-state configurations (not known experimentally) of the superheavy elements 104 and 111. These are found to be different from the ground states of elements above them in the periodic table, due to relativistic effects. Finally, using a Douglas-Kroll transformation of the relativitic wave function, the potential function of AuH is investigated. Very good agreement with known experimental data is obtained in all cases. Large relativistic effects on the structure and spectra of heavy atoms and molecules are observed. Nonadditivity of relativistic and correlation energies is demonstrated.

1 Introduction

Accurate prediction of atomic levels and molecular potentials requires high-order inclusion of correlation effects. The coupled-cluster (CC) method [1, 2, 3] provides one of the best routes to the calculation of these effects. The Fock-space variant of the method is particularly suitable for calculating transition energies of atoms and molecules, including ionization potentials, excitation energies, and electron affinities, as well as molecular potentials for open-shell ground and excited molecular states. A large number of such calculations has been performed in recent years, and the method has proven accurate for a variety of atomic and molecular systems [4].

*Supported by the U.S.-Israel Binational Science Foundation and by the German-Israeli Foundation for Scientific Research and Development

† *E-mail address:* kaldor@chemib2.tau.ac.il

Relativistic effects are very important in heavy atoms and in molecules built from them. Relativistic and correlation effects are strongly interconnected, and must therefore be calculated simultaneously. A relativistic coupled-cluster (RCC) method has therefore been developed, starting from the Dirac-Coulomb-Breit Hamiltonian, which is correct to second order in the fine-structure constant α. The one-electron Dirac-Fock-Breit orbitals are calculated, and correlation is then included by Fock-space coupled cluster. This method has been applied to a variety of atomic systems, including Ne [5], Xe [6], Au [7], the alkali atoms Li–Fr [8], few-electron ions [9], Pr^{3+} and U^{4+} [10], Yb and Lu [11], and the superheavy elements Lr [11], 104 [12], and 111 [13]. Calculated properties (ionization potentials, excitation energies, electron affinities) give very good agreement with available experimental results. For the superheavy elements, the main issue was the determination of the ground-state configuration.

To facilitate relativistic calculations for molecules we replace the four-component Dirac-Fock-Breit function by the one-component Douglas-Kroll (DK) orbitals [14, 15, 16, 17], again including correlation by the CC method. The DK-CC method has been applied to AuH and its ions [18].

Section 2 gives a brief description of the methods used in these calculations. Representative applications are reviewed in Sect. 3. They include the alkali-metal dimers as an example of nonrelativistic molecular potentials, few-electron ions to demonstrate the nonadditivity of relativistic and correlation corrections, investigation of heavy atoms (Au, elements 104 and 111) to show what information the RCC method can provide, and the AuH molecule as an example of relativistic effects on molecular structure and potentials.

2 Methodology

2.1 The Fock-Space Coupled Cluster Method

The Fock-space coupled cluster method has been described before [4], and only a brief outline is given here. The wave operator Ω, which transforms the Hartree-Fock function Φ_0 to the exact Ψ, is given in the coupled-cluster scheme by [19]

$$\Omega = \{\exp(S)\} = 1 + S + \frac{1}{2}\{S^2\} + \cdots , \tag{1}$$

where the curly braces indicate normal ordering with respect to a reference determinant. S is the excitation operator describing *connected* single, double, ... , excitations

$$S = S_1 + S_2 + \cdots = \sum_{ij}\{a_i^\dagger a_j\}s_j^i + \frac{1}{2}\sum_{ijkl}\{a_i^\dagger a_j^\dagger a_l a_k\}s_{kl}^{ij} + \cdots \tag{2}$$

where $a_i^\dagger$ and a_j are creation and annihilation operators, and s_j^i, s_{kl}^{ij}, ... , are excitation amplitudes. The summation is carried out over connected terms only. The equations determining the excitation amplitudes in a complete model space may be derived from the generalized Bloch equation [19]

$$[S, H_0] = (QV\Omega - \Omega V_{\text{eff}})_{\text{conn}} \tag{3}$$

$$V_{\text{eff}} = PV\Omega P \tag{4}$$

where P and Q are the usual projection operators, and H_0 and V result from the partitioning of the Hamiltonian,

$$H = H_0 + V\,. \tag{5}$$

The excitation operator S is defined in the Fock-space coupled cluster approach with respect to a closed-shell reference determinant. In addition to the traditional decomposition into terms with different total (l) number of excited electrons, S is partitioned according to the number of valence holes (m) and valence particles (n) to be excited with respect to the reference determinant,

$$S = \sum_{m\geq 0}\sum_{n\geq 0}\Big(\sum_{l\geq m+n} S_l^{(m,n)}\Big). \tag{6}$$

The upper indices in the excitation amplitudes reflect the partitioning of the Fock space into sectors, which correspond to the different numbers of electrons in the physical system. This partitioning allows for partial decoupling of the open-shell CC equations, since the equations in each sector do not involve excitation amplitudes from higher sectors. The eigenvalues of the effective Hamiltonian

$$H_{\text{eff}} = H_0 + V_{\text{eff}} \tag{7}$$

in a sector give directly the correlated energies in that sector with respect to the correlated (0,0) reference state. These transition energies may be ionization potentials, electron affinities, or excitation energies, according to the presence of valence holes and/or valence particles.

The lower index l in (6) is truncated at l=2. The resulting coupled cluster with single and double excitations (CCSD) scheme involves the fully self-consistent, iterative calculation of all one- and two-body virtual excitation amplitudes, and sums all diagrams with these excitations to infinite order.

2.2 *The Relativistic Hamiltonian*

We start from the projected Dirac-Coulomb (DC) or Dirac-Coulomb-Breit (DCB) Hamiltonian advocated by Sucher [20],

$$H_+ = H_0 + V, \tag{8}$$

where (in atomic units)

$$H_0 = \sum_i \Lambda_i^+ h_D(i) \Lambda_i^+, \tag{9}$$

$$h_D(i) = c\boldsymbol{\alpha}_i \cdot \mathbf{p}_i + c^2(\beta_i - 1) + V_{\text{nuc}}(i) + U(i), \tag{10}$$

$$V = \sum_{i<j} \Lambda_i^+ \Lambda_j^+ (V_{\text{eff}})_{ij} \Lambda_j^+ \Lambda_i^+ - \sum_i \Lambda_i^+ U(i) \Lambda_i^+ . \tag{11}$$

Here h_D is the one-electron Dirac Hamiltonian. An arbitrary potential U is included in the unperturbed Hamiltonian H_0 and subtracted from the perturbation V. This potential is chosen to approximate the effect of the electron-electron interaction; in particular, it may be the Dirac-Fock self-consistent-field potential. The nuclear potential V_{nuc} includes the effect of finite nuclear size. Λ_i^+ are projection operators onto the positive energy states of the Dirac Hamiltonian h_D. Because of their presence, the Hamiltonian H_+ has normalizable, bound-state solutions. This approximation is known as the no-(virtual)-pair approximation, since virtual electron-positron pairs are not allowed in intermediate states. The form of the effective potential V_{eff} depends on the gauge used. In Coulomb gauge it becomes (in atomic units, correct to second order in the fine-structure constant α) [21]

$$V_{\text{eff}} = \frac{1}{r_{12}} + B_{12} + O(\alpha^3), \tag{12}$$

where the frequency-independent Breit interaction is

$$B_{12} = -\frac{1}{2r_{12}} \left[\boldsymbol{\alpha}_1 \cdot \boldsymbol{\alpha}_2 + (\boldsymbol{\alpha}_1 \cdot \mathbf{r}_{12})(\boldsymbol{\alpha}_2 \cdot \mathbf{r}_{12})/r_{12}^2\right]. \tag{13}$$

In q-number theory the Dirac-Coulomb-Breit Hamiltonian H_+ is rewritten in terms of normal-ordered products of the spinor operators, $\{r^+s\}$ and $\{r^+s^+ut\}$ [20, 22]

$$\begin{aligned} H &= H_+ - \langle 0|H_+|0\rangle \\ &= \sum_{rs} f_{rs}\{r^+s\} + \frac{1}{4}\sum_{rstu} \langle rs||tu\rangle \{r^+s^+ut\}, \end{aligned} \tag{14}$$

where f_{rs} and $\langle rs||tu\rangle$ are, respectively, elements of one-electron Dirac-Fock and antisymmetrized two-electron Coulomb-Breit interaction matrices over Dirac four-component spinors. The effect of the projection operators Λ^+ is now taken over by the normal ordering, denoted by the curly braces in the equation above, which requires annihilation operators to be moved to the right of creation operators as if all anticommutation relations vanish. The Fermi level is set at the top of the highest occupied positive energy state, and the negative energy states are ignored.

2.3 The Douglas-Kroll Transformation

The Douglas-Kroll (DK) transformation [14, 15, 16, 17] seeks to decouple the upper and the lower components of the no-pair equation (8)–(11). This leads to a transformed Dirac operator, correct to second order in the external potential. New one- and two-electron terms are added to the nonrelativistic Hamiltonian.

The two-electron term may be ignored in a first approximation, since its influence on the valence properties of molecules with heavy atoms has been found to be small in the cases considered so far [17]. The implementation of the simplified spin-free no-pair Hamiltonian into nonrelativistic *ab initio* program packages requires only a modification of the one-electron integrals. The latter is done approximately by a matrix transformation method [15]. The original papers may be consulted for more detail.

3 Applications

3.1 *Potential Functions of Alkali-Metal Dimers*

The first example of the (nonrelativistic) coupled-cluster method involves the potential functions of the alkali-metal dimers Li_2 [23], Na_2 [24], and K_2 [25] in their ground and excited states. For Li_2 we start from the Li_2^{2+} ion, which stays closed-shell at all internuclear separations. The nine lowest states of the molecule are calculated by adding two electrons in the $2\sigma_g$, $3\sigma_g$, $2\sigma_u$, $1\pi_u$, and $1\pi_g$ orbitals. It should be noted that including all determinants obtained this way in the P space would lead to a very broad energy span, with Q-space states such as $2\sigma_g 4\sigma_g$ having energies within this span. These states will become "intruders" and spoil the convergence of the method. To avoid the difficulty, we resort to the so-called incomplete model space method [26], which allows individual selection of P determinants.

Table 1. Molecular constants of Li_2

State		R_e(Å)	T_e(eV)	D_e(eV)	ω_e(cm^{-1})
$X^1\Sigma_g^+$	exp	2.673	—	1.056	351.4
	CCSD	2.67	—	1.061	351
$a^3\Sigma_u^+$	CCSD	4.05	1.023	0.038	75
$b^3\Pi_u$	exp	2.591	1.394	1.510	345.7
	CCSD	2.58	1.375	1.533	349
$A^1\Sigma_u^+$	exp	3.108	1.744	1.160	255.5
	CCSD	3.10	1.750	1.158	257
$^3\Sigma_g^+$	CCSD	3.06	2.032	0.876	252
$^1\Sigma_g^+$	CCSD	3.56	2.530	0.378	137
$B^1\Pi_u$	exp	2.936	2.534	0.370	270.7
	CCSD	2.94	2.557	0.350	269
$^1\Pi_g$	CCSD	4.02	2.726	0.181	97

The calculated spectroscopic constants of eight states of Li_2 (the ninth calculated state, $^3\Pi_g$, is unbound) are shown and compared with experiment in Table 1. These constants include the equilibrium internuclear separation R_e, the binding energy D_e, the adiabatic excitation energy T_e, and the harmonic frequency ω_e. All calculated values are very close to known experimental quantities, the errors being less than 0.01 Å for the internuclear separation, 0.02 eV for the binding and excitation energies, and 3 cm^{-1} for the frequencies. Similar resulats were obtained for Na_2 and K_2.

3.2 Few-Electron Ions

The relativistic coupled-cluster method based on the Dirac-Coulomb-Breit Hamiltonian was applied to ions with 2–5 electrons and with nuclear charges of 2, 4, 10, 18, and 50 [9]. Moderate-size bases were used (typically $24s16p12d$ Gaussian-type orbitals). The aim was not highly accurate solutions of the DCB equations, which would have required higher-l basis functions. Rather, we wanted to assess the importance and additivity of the relativistic and correlation effects on total and transition energies. The results for Sn ($Z = 50$) are shown in Table 2.

Very strong relativistic effects are observed, as expected. Correlation effects, while smaller, are still substantial, particularly for the total energies. The point to notice is the strong nonadditivity of relativistic and correlation effects. Additivity of these effects would yield correlation energies which do not depend on the starting one-electron orbitals. The results show, however, very strong dependence. The Breit term increases the correlation energy for the two-electron ion by about 70%, while the four-electron correlation is decreased by one-third when relativity is included.

3.3 Gold and Eka Gold (Element 111)

The gold atom represents a local maximum in the periodic table in the size of relativistic effects [27]. Differences between relativistic and nonrelativistic ionization potentials and excitation energies in the atom are several eV, and the order of the first two excited states is reversed by relativity. Correlation effects also play an important role in the level structure of the atom, which provides therefore a stringent test for a method aspiring to provide accurate prediction of energy levels in heavy atoms.

The four-component RCC method has been applied to the atom, starting from the closed-shell species Au^+ and Au^- [7]. The basis included $21s17p11d7f$ Gaussian-type orbitals. The $4spdf5spd$ electrons were correlated, so that core-valence polarization effects are accounted for. The ionization potential, excitation energies, and electron affinity of the atom were calculated in the nonrelativistic (NR), Dirac-Coulomb (DC), and Dirac-Coulomb-Breit (DCB) schemes. The results are presented and compared with experiment [28] in Table 3.

The results demonstrate the very large relativistic effects (3–4 eV in some cases) on the transition energies. All the calculated relativistic values are in very

Table 2. Non-relativistic, Dirac-Coulomb, and Dirac-Coulomb-Breit energies of Sn (hartree a.u.). The last two columns show second-order (E_2) and coupled-cluster (E_{CCSD}) corrections to the total energies, or the corrected ionization potentials and fine-structure splitting.

		E_{SCF}	E_2	E_{CCSD}
Total energy, Sn^{48+}	NR	−2468.800	−0.0443	−0.0446
	DC	−2556.314	−0.0452	−0.0455
	DCB	−2554.613	−0.0770	−0.0773
Total energy, Sn^{46+}	NR	−3047.190	−0.3714	−0.6519
	DC	−3159.833	−0.2763	−0.4137
	DCB	−3157.780	−0.3201	−0.4542
Ionization potential, Sn^{47+}	NR	292.850		292.857
	DC	305.604		305.611
	DCB	305.442		305.452
Ionization potential, Sn^{45+}	NR	271.210		275.083
	DC	286.567		286.433
	DCB	286.235		286.108
^2P fine structure, Sn^{45+}	NR	0		0
	DC	9.649		9.515
	DCB	9.484		9.356

good agreement with experiment (0.1 eV or better). The Breit effect is rather small, except for the fine-structure splittings. These agree with experiment within a few meV when the Breit term is included.

The success of the gold calculation led us to consider eka gold, element 111 [13]. The main issue here is the nature of the ground state. Relativity stabilizes s and p orbitals relative to d. Even in gold, the $d^9s^2\ {}^2D$ state is lowered relativistically by over 3 eV, putting it much closer to the $d^{10}s$ ground state than in the lighter coinage metals Cu and Ag. The question arose whether the energy lowering in the next atom in the group, element 111, would be sufficient to make d^9s^2 its ground state. RCC calculations show that this is indeed the case (Table 4). The $6d^97s^2\ {}^2D_{5/2}$ state, which nonrelativistic CC puts at 5.4 eV above the $6d^{10}7s$ configuration, is reduced relativistically to 3.0 eV *below* this configuration, thus becoming the ground state of the atom. It is also interesting to notice the very large relativistic contraction of the orbitals (Fig. 1).

Table 3. Transition energies in Au (eV), calculated in the Dirac-Coulomb, Dirac-Coulomb-Breit, and nonrelativistic schemes. IP is the ionization potential, EA denotes electron affinity, and EE — excitation energy relative to the ground state. FS denotes fine-structure splittings.

			DC	DCB	NR	Expt.
IP	$5d^{10}6s$	$^2S_{1/2}$	9.101	9.086	6.981	9.22
EE	$5d^9 6s^2$	$^2D_{3/2}$	2.661	2.669	5.301	2.658
	$5d^9 6s^2$	$^2D_{5/2}$	1.115	1.150	5.301	1.136
	FS	^2D	−1.546	−1.519	0	−1.522
	$5d^{10}6p_{1/2}$	$^2P_{1/2}$	4.723	4.720	3.313	4.632
	$5d^{10}6p_{3/2}$	$^2P_{3/2}$	5.193	5.184	3.313	5.105
	FS	^2P	0.470	0.466	0	0.473
EA	$5d^{10}6s^2$	$^1S_0$	2.278	2.269	1.267	2.31

Table 4. CCSD excitation energies (EE), electron affinity (EA), and ionization potentials (IP) of element 111 (eV).

	Transition	DC	DCB	NR
EE	$6d^9 7s^2\ ^2D_{5/2} \rightarrow 6d^9 7s^2\ ^2D_{3/2}$	2.719	2.687	0
	$6d^9 7s^2\ ^2D_{5/2} \rightarrow 6d^{10} 7s\ ^2S_{1/2}$	3.006	2.953	−5.430
EA	$6d^9 7s^2\ ^2D_{5/2} \rightarrow 6d^{10} 7s^2\ ^1S_0$	1.542	1.565	6.484
IP	$6d^9 7s^2\ ^2D_{5/2} \rightarrow 6d^8 7s^2\ ^3D_4$	10.57	10.60	22.98
	$6d^9 7s^2\ ^2D_{5/2} \rightarrow 6d^9 7s\ ^3D_3$	12.36	12.33	0.92
	$6d^9 7s^2\ ^2D_{5/2} \rightarrow 6d^{10}\ ^1S_0$	15.30	15.23	−0.44

3.4 The Ground State of Element 104

The ground state of element 104 has been a subject of long debate. Lighter group 4 elements (Ti, Zr, Hf) have d^2s^2 configurations. It has been suggested [29] that the relativistic stabilization of the $7p_{1/2}$ orbital would yield a $7s^2 7p_{1/2}^2$ ground state for 104. Recent multiconfigurations Dirac-Fock (MCDF) calculations [30] found this state is rather high in energy; the MCDF ground state is $6d7s^2 7p$, found to lie 0.24–0.5 eV below the lowest state of the $6d^2 7s^2$. The MCDF function includes several hundred configurations, corresponding to all distributions of the four external electrons in the $6d$, $7s$, and $7p$ orbitals, accounting for all near-degeneracy effects (non-dynamic correlation). However, short-range or dynamic correlation is largely left out. The latter effect is known to favor d over s and p orbitals. In fact, previous MCDF calculations for atomic Y, La and Lu showed errors of about 0.5 eV, with the $(n-1)d$ configuration

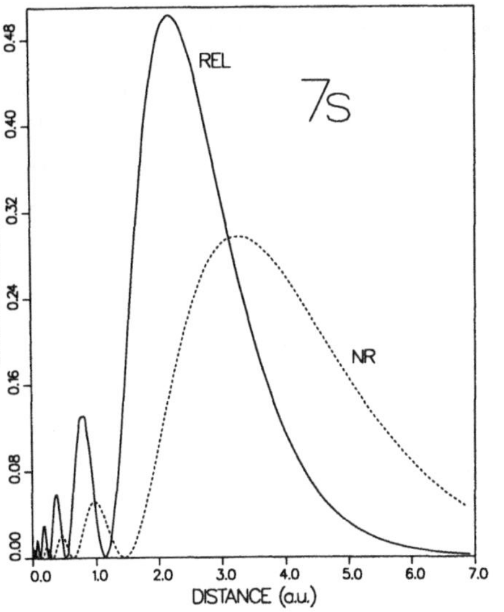

Figure 1. Relativistic and nonrelativistic densities of element 111 7*s* orbital

coming in too high relative to *np*. The MCDF verdict on the nature of the 104 ground state could not therefore be taken as definitive.

Table 5 shows the ionization potentials and excitation energies of some states of 104 and its cation. All the calculated transition energies show monotonic dependence on the amount of correlation included. Of particular interest is the difference between the $6d^27s^2\ ^3F_2$ state and the $6d7s^27p$ state, the contenders for the position of ground state. The former goes from 0.6 eV above the latter at low dynamic correlation to 0.27 eV below it in our best calculation. The Breit interaction adds ~0.02 eV to this value [12], and a rough estimate of basis-set truncation effects puts the $6d^27s^2$ state about 0.32 eV below the $6d7s^27p$. The MCDF values for all calculated energies fall between the CCSD results with l going up to 2 and 3, which is indeed where they should be, considering what correlation effects they include.

3.5 The AuH *Molecule*

The structure and chemistry of molecules with heavy atoms are strongly influenced by both relativistic and correlation effects. The gold hydride is taken

Table 5. Calculated ionization potentials and excitation energies of element 104 and its cation (eV). *5df6spd7sp* electrons correlated, unless otherwise noted.

State			MCDF	CCSD, with virtual orbitals up to $l=$					
				2[a]	3[a]	3	4	5	6
						104+			
IP	$7s^26d_{3/2}$	$^2D_{3/2}$	13.47	13.37	13.95	14.05	14.20	14.33	14.36
EE	$7s^26d_{5/2}$	$^2D_{5/2}$		0.79	0.81	0.87	0.90	0.92	0.92
	$7s^27p_{1/2}$	$^2P_{1/2}$		1.56	2.05	2.15	2.27	2.38	2.40
	$7s^27p_{3/2}$	$^2P_{3/2}$		3.48	4.01	4.12	4.25	4.38	4.40
						104			
IP	$7s^26d^2$	$J=2$	5.30	5.15	5.65	5.76	5.90	5.99	6.01
EE	$7s^27p6d$	$J=2$	−0.24	−0.60	−0.11	0.03	0.17	0.25	0.27
	$7s^26d^2$	$J=3$		0.52	0.53	0.58	0.60	0.61	0.61
		$J=0$		0.38	0.50	0.57	0.64	0.65	0.65
		$J=2$		0.69	0.78	0.86	0.97	0.94	0.95
		$J=4$		0.93	0.94	1.01	0.99	1.05	1.05

[a] $5f$ electrons not correlated.

as an example. Both nonrelativistic and relativistic calculations were done, the latter using the Douglas-Kroll approximation (Sect. 2.3). Correlation was included by the coupled-cluster method. The potential curves (Fig. 2) show very large relativistic effects, with RCC results close to experiment. Thus, the equilibrium distance is reduced by relativity from 1.75 to 1.525 Å, which compares well with the experimental 1.524 Å. The harmonic frequency ω_e goes from 1565 to 2288 cm^{-1}, again close to experiment (2305 cm^{-1}). A substantial increase occurs in the binding energy too.

4 Conclusion

The Fock-space coupled-cluster method has been applied to a variety of problems in atomic and molecular structure, in both the nonrelativistic and relativistic domains. Good agreement with experiment has been observed whenever experimental data are available. In particular, atomic transition energies and molecular spectroscopic constants determined by the method proved highly reliable. We therefore consider the method to have predictive capabilities, and applied it to superheavy atoms where data are not available.

Acknowledgement. Various parts of the work described here were carried out with the collaboration of Dr. E. Eliav (previously Ilyabaev) of Tel Aviv University, Professor Y. Ishikawa of the University of Puerto Rico, and Professor B. A. Heß of the University of Bonn. The author is grateful to all of them.

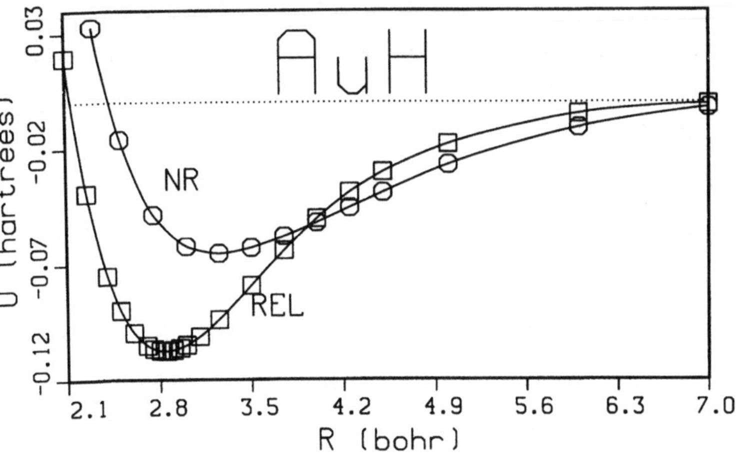

Figure 2. Relativistic and nonrelativistic potentials of AuH

References

1. J. Hubbard: Proc. Roy. Soc. (London) **A240**, 539 (1957); *ibid.* **A243**, 336 (1958); F. Coester: Nucl. Phys. **7**, 421 (1958); F. Coester, H. Kümmel: Nucl. Phys. **17**, 477 (1960); H. Kümmel, K.H. Lührmann, J.G. Zabolitzky: Phys. Rept. **36**, 1 (1978).

2. J. Čížek: J. Chem. Phys. **45**, 4256 (1969); Adv. Chem. Phys. **14**, 35 (1969); J. Paldus, J. Čížek, I. Shavitt: Phys. Rev. A **5**, 50 (1972); J. Paldus: J. Chem. Phys. **67**, 303 (1977).

3. R.J. Bartlett: J. Phys. Chem. **93**, 1697 (1989).

4. D. Mukherjee, S. Pal: Adv. Quantum Chem. **20**, 291 (1989); U. Kaldor: Theor. Chim. Acta **80**, 427 (1991); C.M.L. Rittby, R.J. Bartlett: *ibid.* **80**, 469 (1991); and references cited therein.

5. E. Ilyabaev, U. Kaldor: J. Chem. Phys. **97**, 8455 (1992).

6. E. Eliav, U. Kaldor, Y. Ishikawa: Int. J. Quantum Chem. Symp. **28**, 205 (1994).

7. E. Eliav, U. Kaldor, Y. Ishikawa: Phys. Rev. A **49**, 1724 (1994).

8. E. Eliav, U. Kaldor, Y. Ishikawa: Phys. Rev. A **50**, 1121 (1994).

9. E. Eliav, U. Kaldor, Y. Ishikawa: Chem. Phys. Lett. **222**, 82 (1994).

10. E. Eliav, U. Kaldor, Y. Ishikawa: Phys. Rev. A **51**, 225 (1995).

11. E. Eliav, U. Kaldor, Y. Ishikawa: Phys. Rev. A in press.

12. E. Eliav, U. Kaldor, Y. Ishikawa: Phys. Rev. Lett. **74**, 1079 (1995).

13. E. Eliav, U. Kaldor, P. Schwerdtfeger, B.A. Heß, Y. Ishikawa: Phys. Rev. Lett. **73**, 3203 (1994).

14. M. Douglas, N.M. Kroll: Ann. Phys. (NY) **82**, 89 (1974).

15. B.A. Heß: Phys. Rev. A **32**, 756 (1985); *ibid.* **33**, 3742 (1986).

16. G. Jansen, B.A. Heß: Phys. Rev. A **39**, 6016 (1989).

17. R. Samzow, B.A. Heß, G. Jansen: J. Chem. Phys. **96**, 1227 (1992).

18. U. Kaldor, B.A. Heß: Chem. Phys. Lett. **230**, 1 (1994).

19. I. Lindgren, J. Morrison: *Atomic Many-Body Theory*, 2nd ed. Berlin: Springer-Verlag 1986.

20. J. Sucher: Phys. Rev. A **22**, 348 (1980); Phys. Scr. **36**, 271 (1987).

21. I. Lindgren: In *Many-Body Methods in Quantum Chemistry*, (Lecture Notes in Chemistry vol. 52), ed. U. Kaldor, p. 293. Heidelberg: Springer-Verlag 1989; Nucl. Instrum. Methods Phys. Res. Sec. B **31**, 102 (1988).

22. Y. Ishikawa, H.M. Quiney: Phys. Rev. A **47**, 1732 (1993); Y. Ishikawa: *ibid.* **42**, 1142 (1990); Y. Ishikawa, R. Barrety, R.C. Binning: Chem. Phys. Lett. **121**, 130 (1985).

23. U. Kaldor: Chem. Phys. **140**, 1 (1990).

24. U. Kaldor: Israel J. Chem. **31**, 345 (1991).

25. E. Ilyabaev, U. Kaldor: J. Chem. Phys. **98**, 7126 (1993).

26. G. Hose, U. Kaldor: J. Phys. B **12**, 3827 (1979); B. Jeziorski, H.J. Monkhorst: Phys. Rev. A **24**, 1668 (1981).

27. P. Pyykkö: In *The Effects of Relativity in Atoms, Molecules, and the Solid State*, ed. S. Wilson, I.P. Grant, B.L. Gyorffy, p. 1. New York: Plenum Press 1991.

28. C.E. Moore: *Atomic energy levels*, Natl. Bur. Stand. (U.S.) Circ. No. 467. Washington: U.S. GPO 1948.

29. O.L. Keller: Radiochim. Acta **37**, 169 (1984).

30. V.A. Glebov, L. Kasztura, V.S. Nefedov, B.L. Zhuikov: Radiochim. Acta **46**, 117 (1989); E. Johnson, B. Fricke, O.L. Keller, C.W. Nestor Jr., T.C. Tucker: J. Chem. Phys. **93**, 8041 (1990).

Few-Body Systems Suppl. 8, 79–83 (1995)

Few-Body Systems

Strong Correlation in Three-Particle Continuum $A^{++} + e + e$: Theory and Visualization

A.K. Kazansky, V.N. Ostrovsky*

Institute of Physics, The University of StPetersburg, 198904 StPetersburg, Russia

Abstract. The correlations in the three-particle continuum $A^{++} + e + e$ are very strong near the threshold of three-body breakup due to the long-range nature of the Coulomb interaction. The angular distributions are calculated within a new method (Extended Wannier Ridge Model) which effectively reduces the amount of dynamic variables. The formation of the correlation pattern (as the particles recede) is visualized and discussed in physically appealing terms. The method is applied to the double photoionization process $(\gamma, 2e)$. The correlation patterns depend drastically on the orbital momentum of the residual core A^{++}. The results compare favourably with the recent experimental data.

1 Introduction

We develop a new method for description of the angular correlations in the system of three Coulomb particles above the three-body breakup threshold. The method is oriented on the application to atomic physics where the two typical features are well known to exist.

1. The particles can be classified according to their masses as heavy (atomic nuclei or cores) and light (electrons). We consider the case of two electrons in the field of the residual ion A^{++}. The latter is treated as infinitely heavy center of force. It is important that the ion generally bears some angular momentum L_i and spin S_i.

2. The long-range nature of Coulomb interaction is of fundamental importance leading to the particle correlation in the final state. Near the threshold of three-body breakup the electrons are slow and they have enough time to feel the long-range forces. The correlations are particularly strong in this domain which presents the principal challenge for the theoretical

*E-mail address: ostr@niif.spb.su

treatment. The typical processes are the electron impact ionization of atom $(e, 2e)$ and the double photoionization $(\gamma, 2e)$. The latter case is particularly convenient for the theory since only few partial waves in the *two-electron continuum* are populated due to the selection rules. It became accessible for the detailed experimental study only in the recent years because of low efficiency and difficulties with the source of radiation (synchrotron).

The physical mechanism of the two-electron escape processes was revealed in 1953 by Wannier [1] who stressed importance of what is now named the *Wannier ridge*: the region $r_1 = r_2$ in the configuration space ($\boldsymbol{r}_1$, $\boldsymbol{r}_2$ are the electron vectors relative the atomic nucleus). If the system slides from the ridge in the near-threshold regime, then one of the electrons is captured into the highly-excited Rydberg state $(A^{++} + e)$ and the double ionization does not occur. Additionally, because of the Coulomb repulsion, the electron emission in the opposite directions is favoured ($\theta_{12} = \pi$, where θ_{12} is the angle between $\boldsymbol{r}_1$, $\boldsymbol{r}_2$). Later it was recognized that when the electron pair in continuum has non-zero orbital momentum, then the electrons are repelled from the configuration $\theta_{12} = \pi$ due to the centrifugal-type forces. However, only semiempirical theory was available which proved to be in disagreement with the experimental data for the angular correlations of the ejected electrons.

We put forward the *extended Wannier ridge model* [2,3] (EWRM) for the dynamical calculations of the electron correlation in various partial waves of the two-electron continuum. An important feature of the present approach is that, along with the numerical output, it allows one to *visualize formation of the correlation pattern* while the electrons recede from the core.

2 Extended Wannier Ridge Model

It is convenient to use the collective coordinates: the hyperradius $R = \sqrt{r_1^2 + r_2^2}$, the hyperangle $\alpha = \tan^{-1}(r_1/r_2)$, the interelectron angle θ_{12} and the set of Euler angles $\Omega = \{\alpha_E, \beta, \gamma\}$ which specify orientation in space of the triangle formed by the three particles. The novel computational strategy allows us to effectively reduce the number of variables in the six-dimensional two-electron configuration space.

The Ω-dependence is factorised in the form of symmetrised [4] Wigner functions $D^{(L\pm)}_{MK}(\Omega)$:

$$\Psi_{LM}(R, \alpha, \theta_{12}, \Omega) = \qquad (1)$$
$$R^{-5/2}(\sin 2\alpha)^{-1}\left[\phi^{+}_{LK}(R, \alpha, \theta_{12}) D^{L+}_{-M\,K}(\Omega) + \phi^{-}_{LK}(R, \alpha, \theta_{12}) D^{L-}_{-M\,K}(\Omega)\right] .$$

Here the total angular momentum L and its projection M specify the state of the active electron pair in the *final-state two-electron correlated continuum*. The functions $\phi^{\pm}_{LK}$ obey [4] the system of the second order differential equations in R, α, θ_{12} coordinates. A judicious reduction of the amount of variables is desirable to make the numerical calculations feasible and the physical mechanisms visualizable.

In the near-threshold domain the dynamical variables α and θ_{12} are approximately separated [5]. The Wannier ridge (which is expected to give the major contribution in this regime) corresponds to $\alpha = \frac{1}{4}\pi$. Further from the threshold this region of configuration space retains its importance provided the excess energy E (above the double-ionization threshold) is *equally shared* between the receding electrons. This constraint is imposed by the detector setup which is operative in the major part of the current experiments. Therefore as a rule[1] in EWRM we confine the calculations to the ridge $\alpha = \frac{1}{4}\pi$.

The next simplifying assumption is the semiclassical treatment of the motion over R. Mathematically this implies transition from the second order to the first order derivatives in R. After that the *effective time* $\tau(R)$ is introduced by the relations[2]

$$d\tau = \frac{dR}{p_R R}, \qquad \frac{1}{2}p_R^2 - \frac{Z_0}{R} = E, \qquad Z_0 = 2^{3/2}\left(Z - \frac{1}{4}\right). \tag{2}$$

The change of variable $R \Rightarrow \tau$ casts the problem into terms of the *non-stationary* Schrödinger equation with a single coordinate θ_{12} $(0 < \theta_{12} < \pi)$. The *effective potential* depends on both the 'time' and coordinate. Although generally the *system of coupled equations* appears for the functions $\phi^{\pm}_{LK}$, the equations proves to be decoupled for low L values considered below. The effective potential contains the electron-electron repulsion term calculated at the Wannier ridge and also the effective centrifugal repulsion. The form of the latter term depends on L (see explicit expressions in refs. [2,3,8]).

The photoionization process populating the two-electron continuum states is treated within the first order of the perturbation theory over the radiation field. This is equivalent to introducing the *non-homogeneous ('source') terms* in the Schrödinger equation for $\phi^{\pm}_{LK}$ proportional to $\boldsymbol{F}(\boldsymbol{r}_1 + \boldsymbol{r}_2)\,\psi_0$, where $\boldsymbol{F}$ is the electric field vector in the light wave, ψ_0 is the *initial* state of the active electron pair (the dipole-length form is used for the interaction).

To summarize, the three-body problem is effectively reduced to the propagation of the one-dimensional wave packet (in the coordinate θ_{12}) in the τ-dependent potential. The wave packet is fed by the source term in the core region where ψ_0 differs from zero appreciably. The electron-electron repulsion tends to shift the packet to the $\theta_{12} = \pi$ edge, whereas the effectve centrifugal potential acts in the opposite direction. The role of the latter effect depends on which particular partial L-wave is under consideration.

It seems that the problem cannot be simplified further in a reliable way. Therefore the *grid* (Cranck-Nicholson) *numerical method* is employed to simulate the wave packet propagation.

---

[1] Note however a modification of EWRM [6] with approximate dynamical treatment of all essential variables R, α and θ_{12} and the simplified scheme [7] for the analysis of the excess energy partition between the receding electrons.

[2] The R-motion proceeds with the momentum p_R in the effective Coulomb potential $-Z_0/R$, see Eq. (2). The semiclassical approximation is known to be well applicable to this field even for low energies.

3 Results and Discussion

The angular correlation of electrons is directly observed in the triple differential cross sections (TDCS). Following current experiments we consider double photoionization of the inert gas atoms. We have applied EWRM to almost all the situations where the experimental data are available. The good agreement has been achieved in all cases considered, i.e. in a broad range of excess energies E above the threshold, for various inert gas atoms and different symmetries of the residual doubly-charged ions A^{++}. The formation of the correlation patterns was visualized [2,3,8] using the three-dimensional plots which show the evolution of the electron density distribution $\left|\phi^{\pm}_{LK}(\theta_{12})\right|^2$ as the hyperradius R increases, i.e. as the electrons fly apart. Unfortunately the paper length limitations do not allow us to exemplify hereafter the graphical output.

3.1 Double Photoionization of Helium Atom

Here the electrons are initially in $s^2\ ^1S^e$ state and the final state in the two-electron continuum possesses $^1P^o$ symmetry. A single component ϕ^{+}_{11} is populated. The wave packet has one maximum which is shifted towards the edge $\theta_{12} = \pi$ and narrowed as R increases. The peak in TDCS becomes more narrow when the excess energy E decreases. The electron density turns zero at the edges $\theta_{12} = 0$ and $\theta_{12} = \pi$ because of the electron-electron repulsion and the centrifugal type barrier respectively. The detailed comparison [2,3] (with full account for the polarization properties of ionizing light and the geometry of the experiments) shows good agreement with the available experimental data even for quite high excess energies (at least up to ~20 eV).

3.2 Double Photoionization of Xenon and Neon Atoms

The current experiments for these species detect the electrons leaving the residual ions in the $^1S^e$ state. The initial state of the active electron pair is $p^2\ ^1S^e$, and the final state has again $^1P^o$ symmetry. An elementary reasoning shows that the initial state wave function ψ_0 calculated at the Wannier ridge possesses *a node* at $\theta_{12} = \frac{1}{2}\pi$. This effect can be considered as *kinematic* initial state correlation[3] which appears exclusively due to the coupling of two one-electron p-orbitals into S state (such a node does not appear for s^2 configuration in He). The node in the source term generates the *two-maxima* structure for the wave packet in the core region. Although the packet is modified during the propagation, it retains the two-maximum pattern in the resulting TDCS [3] in agreement with the experiment.

[3]The *dynamic* initial state correlations also exist being taken into account in some calculations within EWRM.

3.3 Double Photoionization of Krypton Atom

The experiment was carried out with the residual ion in $^3P^e$ state. Initially the electron pair is also in $^3P^e$ state; in the final state the selection rules allow (coherent) population of two partial waves in the two-electron continuum: $^3P^o$ and $^3D^o$. In the former wave the centrifugal-type barrier is absent whereas in the latter case the barrier is even more repulsive than for $^1P^o$ symmetry considered above. Therefore the correlation patterns in the related partial waves are drastically different [8]. In the observable total TDCS the $^3P^o$ and $^3D^o$ contributions interfere.

4 General Outlook

The implications of the theoretical description of the physical object varies with the progress of the computing facilities. In the early era our understanding of the reality was confined to the simple models allowing explicit analytical solution. The progress of the computers has allowed us to tackle much more complicated problems which cannot be reduced to simple models. The new basis for qualitative understanding of underlying physical mechanisms seems to be *visualization* of the numerical calculations 'hidden' in the computer. This goal can be considered as having a value comparable to that of the numerical output related to with the experimental data. These idea is one of the principal motivations for developing the models presented here.

Acknowledgement. This work was supported by the grant No 93-02-14888 from the Russian Foundation for Fundamental Research and by the grant NX 2300 from the International Science Foundation.

References

1. G.H. Wannier: Phys. Rev. **90**, 817 (1953)
2. A.K. Kazansky, V.N. Ostrovsky: J. Phys. B **27**, 447 (1994); J. de Physique IV **3**, 159 (1993)
3. A.K. Kazansky, V.N. Ostrovsky: J. Phys. B **28**, 1453 (1995); Phys. Rev. A **51**, 3712 (1995)
4. A.K. Bhatia, A. Temkin: Rev. Mod. Phys. **36**, 1050 (1964); S.I. Nikitin, V.N. Ostrovsky: J. Phys. B **18**, 4349 (1985)
5. A.K. Kazansky, V.N. Ostrovsky: J. Phys. B **26**, 2231 (1993); Phys. Rev. A **48**, R871 (1993)
6. A.K. Kazansky, V.N. Ostrovsky: Phys. Rev. A **51**, 3698 (1995)
7. A.K. Kazansky, V.N.Ostrovsky, L.Yu. Sergeeva: J. Phys. B **27**, 5197 (1994)
8. A.K. Kazansky, V.N. Ostrovsky: J.Phys.B **28**, L333 (1995)

Few-Body Systems Suppl. 8, 84–88 (1995)

Few-Body Systems

Faddeev Approach to Scattering in Atomic and Mesic Atomic Systems

Andrei A. Kvitsinsky[1,2], Chi-Yu Hu[2], James S. Cohen[3]

[1] Department of Mathematical & Computational Physics, Institute for Physics, University of St. Petersburg, 198904 St. Petersburg, Russia

[2] Physics Department, California State University, Long Beach, CA 90840, USA

[3] Theoretical Division, Los Alamos National Laboratory, Los Alamos, NM 87545, USA

Abstract. Applications of the modified Faddeev equations to three-body Coulomb scattering problems are described. Recent results include the $e^- + \mathrm{He}^+$ elastic scattering as well as scattering and fusion in flight in muonic atomic systems $p + p\mu, d + d\mu, t + t\mu$ and $d + t\mu$. An oscillating energy dependence of the rate of fusion-in-flight for the symmetric systems is found, and a profound peak is established in the reaction $d + t\mu \rightarrow {}^4\mathrm{He} + n + \mu$ at $E = 76.3$ eV. Effects due to the hyperfine interaction between nuclei and muon are also studied.

1 Introduction

The Faddeev approach [1], while being a key method of nuclear physics for already three decades, was not of much use in atomic physics. This state of affairs has began to change recently, when first applications [2] – [5] of the approach to benchmark scattering problems of atomic physics $e^- + \mathrm{Ps}$ and $e^\pm + \mathrm{H}$ demonstrated that it is a very suitable and powerful method to study the three–body Coulomb scattering problems. With maturing of the method, the range of problems within its reach has been growing rapidly. This work deals with some recent advances.

The Faddeev equations for the components of the wave function $\Psi = \psi_1 + \psi_2 + \psi_3$, modified for long-range Coulomb forces by Merkuriev [6, 1], are

$$\left(H_0 + V_\alpha + \sum_{\beta \neq \alpha} V_\beta^{(l)} - E \right) \psi_\alpha = -V_\alpha^{(s)} \sum_{\beta \neq \alpha} \psi_\beta, \qquad (1)$$

where the Coulomb potentials V_α are decomposed into the long– and short–

range parts $V_\alpha^{(l)}$ and $V_\alpha^{(s)}$ by making use of a cutoff function which vanishes in the three–body sector of configuration space, where all particles are far apart. Basic feature of the equations is that different channels of a three–body system are distributed over different Faddeev components ψ_α and asymptotically are decoupled due to the cutoff. This greatly simplifies setting asymptotic conditions for scattering problems and underlies computational efficiency of the Faddeev approach. The numerical solution of the Faddeev equations makes use of the total-angular-momentum representation [7], which lowers the dimensionality of the equations, and a quintic spline expansion on each variable [3]. This enables us to solve scattering problem directly, without any intermediate approximations, with an easy–to–control accuracy. Details of the numerical procedure of solving Eqs. (1) can be found in refs. [3, 8] . We now describe some recent results obtained by this approach.

2 Elastic $e^- + \mathrm{He}^+$ Scattering

Electron collisions with ionized helium is a classical problem of atomic physics which is a little bit more complicated than e^-+H scattering because of Coulomb interaction between electron and the ion. The problem has been previously studied by means of the exchange approximation (E) and methods of polarized orbitals (PO) and exchange polarization (EP), see ref. [9]. Our results in Table 1 are in between the PO and EP results at lower energies but get much closer to the PO ones at higher energies.

Table 1. s–wave singlet phase shifts of $e^- + \mathrm{He}^+$ scattering.

k	present	PO	EP	E
0.2	0.4335	0.4404	0.4164	0.3828
0.4	0.4238	0.4301	0.4057	0.3721
0.6	0.4166	0.4151	0.3907	0.3568
0.8	0.3970	0.3986	0.3736	0.3396

3 Muonic Atom Collisions

The Faddeev approach also provides an efficient alternative to the standard Born-Oppenheimer adiabatic expansion method in the theory of mesic atomic systems. Of much interest are elastic and spin-flip transitions in the symmetric mesic atomic systems

$$a + a\mu(\uparrow\downarrow) \rightarrow \left\{ \begin{array}{l} a + a\mu(\uparrow\downarrow) \\ a + a\mu(\uparrow\uparrow) \end{array} \right. , \quad a = p, d \text{ or } t; \tag{2}$$

and the non-symmetric collisions

$$d + t\mu \rightarrow \begin{cases} d + t\mu \\ t + d\mu \end{cases} \tag{3}$$

In particular, the Faddeev approach allows explicit inclusion of the hyperfine splitting of the target mesic atom in (2). Our s-wave cross sections of the processes (2), (3) tend to fall in between earlier calculations done using a large standard adiabatic expansion [10] and those done using a two-state improved adiabatic expansion [11], but generally are closer to the former.

In Fig. 1 the spin-flip cross sections with and without hyperfine splitting (hfs) are compared for $p + p\mu$ scattering. They are in fair agreement at energies several times the hyperfine splitting ($\Delta E = 0.128$ eV for $p\mu$) but differ greatly at lower energies, where the hyperfine effects are amplified by a low–energy virtual state of the $pp\mu$ system. In the elastic transitions interplay of the hyperfine

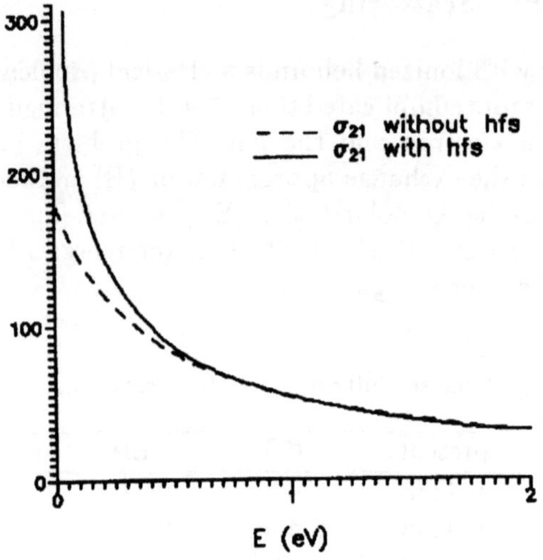

Figure 1. s-wave cross section (in units of 10^{-20} cm^2) for $p\mu(\uparrow\uparrow) + p \rightarrow p\mu(\uparrow\downarrow) + p$.

and virtual state effects is stronger. As a result, neglect of the hyperfine splitting becomes a reasonable approximation for elastic cross sections at higher collision energies (> 2 eV) than for the inelastic cross section.

A unique feature of the processes (2), (3) is the fusion-in-flight reactions, for instance

$$p + p\mu \rightarrow d + e^+ + \bar{\nu}_e + \mu\,,\; d + d\mu \rightarrow \begin{cases} t + p + \mu \\ {}^3\mathrm{He} + n + \mu \end{cases},\; t + t\mu \rightarrow {}^4\mathrm{He} + 2n + \mu\,. \tag{4}$$

The rate for fusion in flight is given by $\lambda_f = \rho_f\, A\, N_0$ where A is the nuclear constant, N_0 is the density of liquid hydrogen, and ρ_f is the probability for the nuclei to coincide given by

$$\rho_f = \int |\Psi(\mathbf{r}_1, \mathbf{r}_2, \mathbf{r}_3)|^2 \, \delta(\mathbf{r}_1 - \mathbf{r}_2) \, d\mathbf{r}_3 \ , \tag{5}$$

where Ψ is the wave function of the scattering channel. The quantities ρ_f are shown in Fig. 2 for the symmetric collisions (4). The resonance amplification of

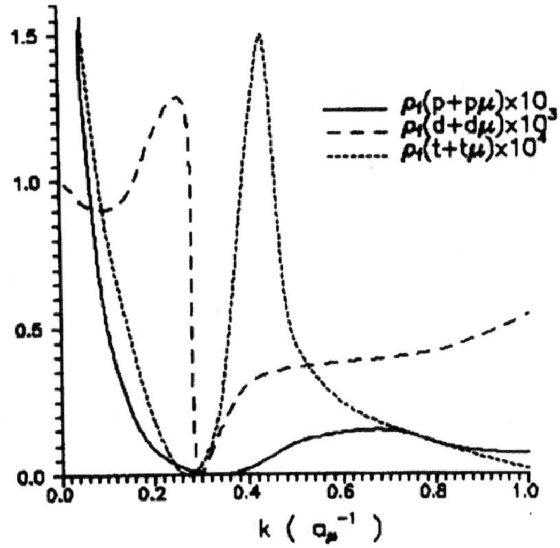

Figure 2. Factors ρ_f for the symmetric fusion-in-flight reactions.

the fusion-in-flight reactions in $p + p\mu$ and $t + t\mu$ collisions near the threshold is due to low-energy virtual states of these systems. For all systems in Fig. 2 there is an oscillating pattern, with exact zeros of ρ_f being accompanied by nearby maxima. A more involved analysis shows that the zeros and maxima occur when one of the eigenphase shifts of S-matrix goes through $\pi/2$. A similar effect also is found in the reaction $d + t\mu \to {}^4\mathrm{He} + n + \mu$ which has a profound resonance at 76.3 eV, in excellent agreement with earlier analysis of ref. [12]. The rate $\lambda_f = 3.40 \times 10^6 \ \mathrm{s}^{-1}$ of fusion in flight at this energy is about 7.5 times faster than the rate of muon decay $0.455 \times 10^6 \, \mathrm{s}^{-1}$.

4 Conclusions

Comparison with well-established methods shows that the method of modified Faddeev equations is a new powerful tool of atomic and mesic atomic physics. It enables one to deal in a simple, uniform and accurate way with a wide variety of systems and processes which otherwise are usually treated by very different methods (variational, adiabatic expansions, close coupling etc.). It would be of much interest to bring the method up to the most challenging problems of the three-body Coulomb scattering: collisions with several open thresholds, resonances in mesic atomic systems with charged targets (like $d+\alpha\mu$), break-up reactions etc. Work in this direction is now in progress.

References

1. L.D. Faddeev, S.P. Merkuriev: *Quantum Scattering Theory for Several Particle Systems.* Dordrecht: Kluwer (1993)
2. A.A. Kvitsinsky, J. Carbonell, C. Gignoux: Phys. Rev. A **46**, 1310 (1992)
3. A.A. Kvitsinsky, C.-Y. Hu: Phys. Rev. A **47**, 994 (1993).
4. A.A. Kvitsinsky, J. Carbonell, C. Gignoux: Phys. Rev. **A51**, 2997 (1995)
5. A.A. Kvitsinsky, A. Wu, C.-Y. Hu: J. Phys. B **28**, 275 (1995)
6. S.P. Merkuriev: Ann. Phys. (NY) **130**, 395 (1980)
7. A.A. Kvitsinsky, V.V. Kostrykin, S.P. Merkuriev: Fiz. Elem. Chastits At. Yadra **21**, 1301 (1990) [Sov. J. Part. Nucl. **21**, 553 (1990)]
8. C.-Y. Hu, A.A. Kvitsinsky: Hyperfine Interact. **82**, 59 (1993)
9. I.H. Sloan: Proc. Royal Soc. **281**, 151 (1966)
10. L. Bracci et al.: Muon Catal. Fusion **4**, 247 (1989)
11. J.S. Cohen: Phys. Rev. A **43**, 4668 (1991)
12. V.S. Melezhik: Nucl. Phys. **A550**, 223 (1992)

Few-Body Systems Suppl. 8, 89–92 (1995)

Few-Body Systems

Precise Nonvariational Calculation of Resonant States of Helium with the Correlation Function Hyperspherical Harmonic Method

S. Berkovic, R. Krivec*, V.B. Mandelzweig

Racah Institute of Physics, Hebrew University, Jerusalem 91904, Israel

Abstract. Direct solution of the Schrödinger equation for resonant S states of the Helium atom is obtained with the help of the complex rotation method (which reduces the resonant problem to that of bound states) and the correlation function hyperspherical harmonic method (CFHHM). In the CFHH method the bound state solution is a product of a correlation function and of a smooth factor expanded into hyperspherical harmonic functions. Given a proper correlation function, chosen from physical considerations, the method generates resonant wave functions, accurate in the whole range of interparticle distances. Calculated energies are compared to those obtained by variational and other precise computations.

The doubly excited states of the Helium atom provide a good testing ground for various methods because of their strongly correlated nature. This also necessitates a full three-body approach, which in turn is even more important to obtain precise wave functions than to obtain energies alone.

This paper presents the first CFHHM calculation of resonance states employing the Complex Coordinate Rotation (CCR) method (see the review [1] and references therein). The advantage of CCR is that with the complex rotation

$$r_{ij} \rightarrow r_{ij} e^{i\theta} \tag{1}$$

it transforms the continuum resonance wave function into a square integrable one. Therefore, the existing CFHHM bound state code can be used, after a relatively straightforward transformation to the complex domain, to obtain the resonance wave function. We start from the usual CFHH method [2] which separates the wave function into two factors

$$\Psi = \chi \phi \tag{2}$$

**Permanent address*: J. Stefan Institute, Jamova 39, POB 100, 61111 Ljubljana, Slovenia

Table 1. Energy of the ground state of the He atom (a.u.), represented as $E_r + i\Gamma/2$, calculated with the dilation parameter $\theta = 0.001$ radians, for the infinite-mass nucleus. The cusp parameters $\gamma = 2$, $\delta = -0.5$ are used. The second entry for each K_m is the standard CFHHM calculation.

K_m	N	$-E_r$	$\Gamma/2$
0	1	2.976759502	-9.53×10^{-4}
		2.976760455	
4	4	2.910266623	-2.45×10^{-4}
		2.910266865	
8	9	2.903217750	-1.07×10^{-4}
		2.903217858	
12	16	2.903907607	-5.50×10^{-5}
		2.903907661	
16	25	2.903704370	-3.25×10^{-5}
		2.903704404	
20	36	2.903741487	-2.01×10^{-5}
		2.903741506	
28	64	2.903727429	-9.36×10^{-6}
		2.90372743	
32	81	2.903724358	-6.80×10^{-6}
		2.903724364	
36	100	2.903725180	-5.05×10^{-6}
		2.903725184	

where the factor χ contains the singular (cusps) and cluster features of the wave function, and ϕ is a smooth function expandable in hyperspherical harmonics (HH).

The equation for ϕ after expansion in HH results in an infinite coupled system of ordinary differential equations in one variable (ODE). The ODE are truncated to a finite number of equations, N, which is determined by the maximum global angular momentum used, K_m. K_m is increased until convergence is achieved. As the effective Hamiltonian is nonhermitian, the energy eigenvalue obtained by solving the ODE does not have the variational property and its error is proportional to the error of the wave function. However, the expectation value obtained with the calculated wave function will have the variational property and will therefore be more accurate.

In this work we obtained preliminary results on the energy eigenvalue in order to estimate the potential precision of our combined CFHH-CCR method. The calculation of the expectation values of different operators, requiring substantial modifications in the expectation value program to accomodate complex

Table 2. Energy of the $^1S^e$ state of the He atom (a.u.), for the infinite-mass nucleus. Uncertainties in the last figure depend on the θ point density and do not reflect the accuracy of the CFHHM. The last lines of the table represent theoretical values by different methods and two experimental values.

K_m	N	E_r	$\Gamma/2$	θ_{opt}
20	36	-0.764320(2)	0.002538(1)	-0.056(1)
28	64	-0.77076(2)	0.00238(5)	-0.060(5)
36	100	-0.7737008(2)	0.002373(2)	-0.069(1)
44	144	-0.77520(5)	0.0024(1)	-0.14(8)
48	169	-0.77571(2)	0.00243(3)	-0.19(2)
	ref. [7]	-0.7778	0.00229	
	ref. [8]	-0.778035	0.002295	
	ref. [9]	-0.77841	0.00227	
	ref. [10]	-0.77787	0.00227	
	ref. [11]	-0.777858	0.002285	
	ref. [4]	-0.777872	0.00227	
	ref. [12]	-0.77735	0.002205	
	ref. [13]	-0.778164±0.00147	0.002536±0.000276	
	ref. [14]	-0.779634±0.00110	0.002536±0.000276	

wave functions, is in progress.

We first calculated the ground state of the Helium atom in order to compare with the standard CFHH method. The results are displayed in Table 1 showing that the discrepancy between the energies for $\theta = 0.001$ radians and $\theta = 0$ radians (see the previous CFHHM calculations of the Helium atom, ref. [3] and references therein) decreases smoothly towards 0 as N is increased. Moreover, the intrinsic precision in both cases is the same.

The results for the lowest resonance, the $^1S^e$ state, are displayed in Table 2. Because the electrons in this state are located on average far from the nucleus, the asymptotic parametrization

$$\chi = \exp(f), \ f = -\gamma(r_{13} + r_{23}) - \delta r_{12}; \ \gamma = \sqrt{\frac{M|E|}{M+1}}, \ \delta = 0 \tag{3}$$

was used [4].

θ_{opt} was determined from the condition of maximum curvature of the E_r *vs* E_i plots, where E_r and E_i are functions of θ (see Eq. (1)). This prescription is approximately equivalent to the prescriptions $\partial E_r/\partial\theta = 0$, $\partial E_i/\partial\theta = 0$ (see the discussion in ref. [5]). Different prescriptions and their relation to the virial theorem even for incomplete basis sets are discussed in [1] and [6]. An example of the E_r *vs* E_i plot is shown in Fig. 1.

Based on the CFHHM calculations of the Helium atom [3], we are expecting one to two digit improvement in accuracy of energy as given by $\langle H \rangle$.

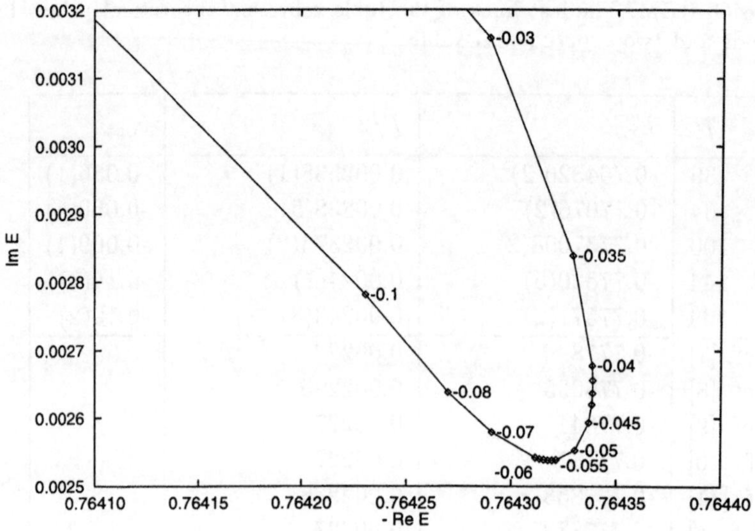

Figure 1. The dependence of the complex energy of the $^1S^e$ state on θ for $K_m = 20$. The calculated points on the curve are labelled by the θ values (radians).

References

1. Y.K. Ho: Phys. Rep. **99**, 1 (1983)
2. M.I. Haftel, V.B. Mandelzweig: Ann. Phys. **189**, 29 (1989)
3. R. Krivec, M.I. Haftel, V.B. Mandelzweig: Phys. Rev. **A44**, 7158 (1991)
4. P. Froelich, S.A. Alexander: Phys. Rev. **A42**, 2550 (1990)
5. N. Moiseyev, P.R. Certain, F. Weinhold: Mol. Physics **36**, 1613 (1978)
6. R. Yaris, P. Winkler: J. Phys. **B11**, 1475 (1978)
7. D. H. Oza: Phys. Rev. **A33**, 824 (1986)
8. A. K. Bhatia, A. Temkin: Phys. Rev. **A11**, 2018 (1975)
9. A. Macias, A. Riera: Phys. Lett. **A119**, 28 (1986)
10. Y.K. Ho: Phys. Rev. **A23**, 2137 (1981); J. Phys. **B12**, 387 (1979)
11. K.T. Chung, B.F. Davie: Phys. Rev. **A26**, 3278 (1982)
12. J. Tang, S. Watanabe, M. Matsuzawa: Phys. Rev. **A46**, 2437 (1992)
13. P. J. Hicks, J. Comer: J. Phys. **B8**, 1866 (1975)
14. F. Gelbart, R.J. Tweed, J. Peresse: J. Phys. **B9**, 1739 (1976)

Few-Body Systems Suppl. 8, 93–96 (1995)

Few-Body Systems

A Coupled–Channel Formalism for Breakup Reactions

C. Chandler*

Department of Physics and Astronomy, University of New Mexico, Albuquerque, NM 87131, USA

Abstract. Coupled-channel methods have been employed to calculate reaction probabilities for a number of breakup reactions, despite the fact that all the methods assume forms of wave functions which do not have the asymptotic behavior necessary to describe such reactions directly. This contribution provides a general conceptual framework for these methods, shows how to extend them to include asymptotic behavior appropriate to breakup reactions, calls attention to a possible reason for their practical success, and points out unresolved problems.

Coupled-channel, or close-coupling, methods have been employed to calculate cross sections for deuteron induced nuclear reactions [1, 2] and for electron [3] and positron [4] induced atomic reactions, including breakup. The nuclear and the atomic formalisms are different in important ways but are both based on three-body models and on partial discretization of channel Hamiltonians. As these calculations are remarkably successful in fitting experimental data, the methods deserve to be better understood. This contribution shows how all these methods fit into general multichannel scattering theory, indicates how to extend them to include asymptotic behavior appropriate to breakup reactions, calls attention to a possible reason for their practical success, and points out unresolved theoretical issues.

Consider a three-body system with Hamiltonian $H = K + V + U$. K is the kinetic energy, V the sum of short-range interactions between the particles, and U the sum of Coulomb interactions. The amplitude for a breakup reaction with initial two-cluster channel α has the form

$$T_{0\alpha}(\mathbf{p}_0,\mathbf{q}_0;\mathbf{q}_\alpha) = \langle \mathbf{p}_0,\mathbf{q}_0 | K_0^{(-)\dagger} W_\alpha^{(+)} + W_0^{(-)\dagger}\,(E^+ - H)^{-1}\, W_\alpha^{(+)} | \mathbf{q}_\alpha, \psi_\alpha \rangle\,, \quad (1)$$

where $|\mathbf{p}_0,\mathbf{q}_0\rangle$ is a plane-wave state for Jacobi momenta $(\mathbf{p}_0,\mathbf{q}_0)$ and $|\mathbf{q}_\alpha,\psi_\alpha\rangle$ is a state with two particles bound (wave function ψ_α) and the third moving with momentum $\mathbf{q}_\alpha$ relative to the bound pair. The operators $W_0^{(-)}$ and $W_\alpha^{(+)}$

*E-mail address: chandler@unm.edu

are defined by $W_0^{(-)} = HK_0^{(-)} - K_0^{(-)}H_0$ and $W_\alpha^{(+)} = HK_\alpha^{(+)} - K_\alpha^{(+)}H_\alpha$, where H_0 and H_α are the usual channel Hamiltonians, and play a role analogous to that of ordinary potentials.

In Eq. (1) it is understood that channel α is a fixed representative of the class of (physically indistinguishable) two-cluster channels that differ only by interchange of identical particles. The symmetrization operators required by the Pauli principle are contained in the operators $K_0^{(-)}$ and $K_\alpha^{(+)}$ (as are the distorted waves required by the Coulomb interactions).

The question of how best to choose $K_0^{(-)}$ is unresolved for atomic systems [5, 6], but for the deuteron-nucleus systems the best choice seems to be given by $K_0^{(-)} = N_0^{1/2}\, Y\, \bar{K}_0^{(-)}$, with

$$\langle \mathbf{r}_p, \mathbf{r}_n | \bar{K}_0^{(-)} | \mathbf{p}_p, \mathbf{q}_n \rangle = \langle \mathbf{r}_p | \chi_p^{(-)}(\mathbf{p}_p) \rangle \, \langle \mathbf{r}_n | \chi_n^{(-)}(\mathbf{p}_n) \rangle \, . \tag{2}$$

Here Y is the total (anti)symmetrization operator dictated by the Pauli principle, and N_0 is an associated normalization factor. The target nucleus is assumed to remain fixed, $(\mathbf{r}_p,\ \mathbf{r}_n)$ are the position vectors of the proton and neutron, and $(\mathbf{p}_p,\ \mathbf{p}_n)$ are corresponding asymptotic momentum vectors. The state $|\chi_p^{(-)}(\mathbf{p}_p)\rangle$ is a distorted wave state (with incoming scattering boundary conditions) for the proton interacting with the nucleus, and $|\chi_n^{(-)}(\mathbf{p}_n)\rangle$ is a distorted wave state (with incoming scattering boundary conditions) for the neutron interacting with the nucleus.

Similarly, $K_\alpha^{(+)} = N_\alpha^{1/2}\, Y\, \bar{K}_\alpha^{(+)}$ for two-cluster channels α, with

$$\langle \mathbf{x}_\alpha, \mathbf{y}_\alpha | \bar{K}_\alpha^{(+)} | \mathbf{p}_\alpha, \mathbf{q}_\alpha \rangle = \langle \mathbf{x}_\alpha | \mathbf{p}_\alpha \rangle \, \langle \mathbf{y}_\alpha | \chi_\alpha^{(+)}(\mathbf{q}_\alpha) \rangle \, . \tag{3}$$

Here $|\chi_\alpha^{(+)}(\mathbf{q}_\alpha)\rangle$ is a distorted wave state (with outgoing scattering boundary conditions) for the bound pair interacting as a point particle with the remaining particle. The operators $K_\alpha^{(-)}$ are similarly defined with incoming scattering boundary conditions.

While they are not usually so formulated, all the various coupled-channel methods are equivalent to approximating $(E^+ - H)^{-1}$ in Eq. (1) by $(E^+ - PHP)^{-1}P$, where P is a projection operator defined as follows. For each two-cluster channel β let $I_\beta = \sum_{\ell=1}^{\nu_\beta} \int d\mathbf{q}_\beta \ |\mathbf{q}_\beta, \phi_{\beta\ell}\rangle\langle \mathbf{q}_\beta, \phi_{\beta\ell}|$, where the functions $\langle \mathbf{p}_\beta | \phi_{\beta\ell}\rangle$ are square integrable, orthonormal, and include any relevant two-body bound states ψ_β for channel β [7]. The states $|\mathbf{q}_\beta, \phi_\beta\rangle$ that are not physical asymptotic states of the system are usually called pseudo states. For the breakup channel let $I_0 = \sum_{\ell=1}^{\nu_0} \int d\lambda \ |\lambda, X_{0\ell}\rangle \, \langle \lambda, X_{0\ell}|$, with $|\lambda, X_{0\ell}\rangle = \int d\mathbf{p}_0 \, d\mathbf{q}_0 \, \delta(\lambda - T(\mathbf{p}_0, \mathbf{q}_0)) \, X_{0\ell}(\mathbf{p}_0, \mathbf{q}_0) \, |\mathbf{p}_0, \mathbf{q}_0\rangle$, $T(\mathbf{q}_0, \mathbf{p}_0)$ being the total kinetic energy (in the center of mass frame). The normalization condition $\langle \lambda', X_{0\ell'} | \lambda, X_{0\ell}\rangle = \delta(\lambda' - \lambda)\, \delta_{\ell'\ell}$ is to be satisfied. Then P is defined as the projection of the entire three-body Hilbert space onto the subspace formed by linear combinations of functions in the ranges of the operators $J_\beta^{(-)} = K_\beta^{(-)} I_\beta$ and $J_0^{(-)} = K_0^{(-)} I_0$.

The coupled-channel equations can now be derived using the methods of [8], with the result

$$T_{0\alpha}(\mathbf{p}_0, \mathbf{q}_0; \mathbf{q}_\alpha) \approx$$
$$\langle \mathbf{p}_0, \mathbf{q}_0 | K_0^{(-)\dagger} W_\alpha^{(+)} + \sum_\beta W_0^{(-)\dagger} J_\beta^{(-)} G_\beta(E^+) M_{\beta\alpha}(E^+) | \mathbf{q}_\alpha, \psi_\alpha \rangle, \quad (4)$$

$$M_{\beta\alpha}(z) = J_\beta^{(-)\dagger} W_\alpha^{(+)}$$
$$+ \sum_\gamma \left[J_\beta^{(-)\dagger} W_\gamma^{(-)} I_\gamma G_\gamma(z) - (J_\beta^{(-)\dagger} J_\gamma^{(-)} - \delta_{\beta\gamma} I_\gamma) \right] M_{\gamma\alpha}(z), \quad (5)$$

where $G_\beta(z) = (z - I_\beta H_\beta I_\beta)^{-1} I_\beta$. Note that the Hamiltonian $I_\beta H_\beta I_\beta$ is partially discretized. The sums in Eqs. (4) and (5) range over the physically distinct channels that have been used in defining the operator P. Equations (5), or their differential counterparts, are the coupled-channel equations that are to be solved numerically.

It is important to note that Eqs. (5) differ from the usual coupled-channel equations in that functions in the range of $J_0^{(-)}$, which have the correct asymptotic behavior for the breakup channel, have been included in defining the projection operator P. If these functions are omited the usual equations [7] result.

Inclusion of the functions in the range of $J_0^{(-)}$ has interesting consequences. In the absence of Coulomb interactions it has been proved [8] that Eqs. (5) are well behaved as $z \to E^+$ and their solution provides a uniquely determined transition amplitude when inserted into Eqs. (4). In fact, the on-energy-shell matrix elements of the operators $M_{\beta\alpha}$ are themselves precisely the transition amplitudes for the Hamiltonian PHP. Consequently, if the range of the operator I_0 is rich enough, solution of Eqs. (5) (including the new terms represented by $J_0^{(-)}$) should yield accurate breakup amplitudes, further substitution into Eqs.(4) then being unnecessary. Extension of the proofs [8] to include Coulomb interactions has not been done but is not expected to be difficult.

Much more troublesome is the tacit assumption that the approximate breakup amplitudes determined by Eqs. (4) and (5) converge to the exact amplitudes when the number of basis functions included in the operators I_β increases and, consequently, $P \to Y$. While the results of numerical calculations [1, 2, 3, 4] are consistent with such convergence, there is presently no proof that it must occur. Some theorems in [8] suggest, however, that, again in the absence of Coulomb interactions, the desired convergence can be proved if

$$\mathrm{s} - \lim_{\nu_\beta \to \infty} (z - H_\beta) I_\beta (z - H_\beta)^{-1} = 1\,, \quad (6)$$

a condition that appears to be satisfied by the method of Austern and coworkers [1]. The lack of a general convergence proof is certainly the most fundamental theoretical issue that remains to be resolved about the coupled-channel approach to breakup calculations.

Why have the coupled-channel calculations yielded such good numerical results for breakup amplitudes, given that the they have corresponded to projection operators P that do not include functions with the correct asymptotic behavior for breakup amplitudes. It is evident from Eqs. (4) that the coupled-channel approach can be accurate only if $\langle \mathbf{q}_0, \mathbf{p}_0 | W_0^{(-)\dagger}$ can be accurately approximated by a sum of terms of the form $\langle \mathbf{q}_0, \mathbf{p}_0 | W_0^{(-)\dagger} J_\beta^{(-)}$. In other words, the coupled-channel approach requires the condition $\langle \mathbf{q}_0, \mathbf{p}_0 | W_0^{(-)\dagger} \approx \langle \mathbf{q}_0, \mathbf{p}_0 | W_0^{(-)\dagger} P$ for success. In the example of deuteron-nucleus scattering in which nuclear recoil effects are neglected $W_0^{(-)}$ is proportional to the interaction between the neutron and the proton that make up the deuteron. It is sufficient, therefore, to approximate $\langle \mathbf{q}_0, \mathbf{p}_0 | W_0^{(-)\dagger}$ accurately only over the coordinate space region in which the distance r_{np} between the neutron and proton is small. The fact that functions in the range of $J_\alpha^{(-)}$ have the wrong behavior as $r_{np} \to \infty$ is unimportant, as Austern and coworkers have pointed out [1, 9], and this accounts for the success of the method.

Finally, it is to be noted that the considerations presented here are not intrinsically restricted to three-body models. All of the discussion extends immediately to N-body systems.

Acknowledgement. It is a pleasure to thank the Electronic Structure of Materials Centre and the Department of Physics at Flinders University for their hospitality during this work, as well as the Australian-American Educational Foundation (Fulbright Senior Scholar Program) and the US National Science Foundation (Grant PHY-9204252) for their support.

References

1. N. Austern, Y. Iseri, M. Kamimura, M. Kawai, G. Rawitscher, M. Yahiro: Phys. Rep. **154**, 125 (1987)
2. Y. Fujiwara, Y.C. Tang: Prog. Theo. Phys. **91**, 631 (1994)
3. I. Bray, A.T. Stelbovics: Comput. Phys. Commun. **85**, 1 (1995)
4. J. Mitroy, A.T. Stelbovics: J. Phys. B **27**, 3257 (1994)
5. E.O. Alt, M. Mukhamedzhanov: Phys. Rev. A **47**, 2004 (1993)
6. H. Klar, D.A. Konovalov, I.E. McCarthy: J. Phys. B **26**, L711 (1993)
7. Z.C. Kuruoglu: Phys. Rev. A **44**, 7307 (1991)
8. C. Chandler, A.G. Gibson: J. Funct. Anal. **52**, 80 (1983); J. Math. Phys. **25**, 1841 (1984); *ibid.*, **33**, 3477 (1992); *ibid.*, **35**, 1487 (1994)
9. N. Austern, M. Yahiro, M. Kawai: Phys. Rev. Lett. **63**, 2649 (1989)

Few-Body Systems Suppl. 8, 97–101 (1995)

Few-Body Systems

$^{2}\vec{\mathrm{H}}(\vec{d},p)^{3}\mathrm{H}$ Reaction Calculations at 20–120 keV by the Four–Body Yakubovsky Equations with the Paris Potential

Eizo Uzu[1], Shinsho Oryu[1], Makoto Tanifuji[2]

[1] Department of Physics, Faculty of Science and Technology, Science University of Tokyo, Noda, Chiba 278 Japan

[2] Department of Physics and Research Center of Ion Beam Technology, Hosei University, Chiyoda-ku, Tokyo 102 Japan

Abstract. We analyze $^{2}\vec{\mathrm{H}}(\vec{d},p)^{3}\mathrm{H}$ reaction measurements by using the four-body Yakubovsky integral equations and simple Coulomb correct ion methods. The adopted potential is the Paris-potential (PEST-1) for the $^{1}S_0$ and $^{3}S_1$-$^{3}D_1$ states. The [3+1] and the [2+2] sub-amplitudes are given by the energy dependent pole expansion of rank-3, and also the Hilbert-Schmidt one. We included $1/2^+$ and $1/2^-$ states for ^{3}He and ^{3}H nuclei which may occur in the intermediate subnuclei as part of the four-nucleon system. The angular momentum between two deuterons in the initial state is retained up to 2, and also the maximum value between ^{3}H and p in the final state is 4. Therefore, the total spins and parities of the four-nucleon system are 0^+, 0^-, 1^+, 1^-, 2^+, 2^-, 3^+, 3^-, and 4^+. The calculated total cross section, differential cross section, and tensor analyzing powers at energies from 20 to 120 keV are in very good agreement with the experimental data, but the vector analyzing power is still very small. Ratios of the double polarized deuteron scattering cross sections to the unpolarized one (σ_{pol}/σ_0) were calculated. It was found that the cross section is suppressed by a factor one half at higher energies.

1 Introduction

It has been suggested that the double polarized $\vec{d}+\vec{d} \rightarrow {}^{3}\mathrm{He}+n$ and ${}^{3}\mathrm{H}+p$ reactions could be suppressed in the low energy region, because of total spin and parity conservation. Then, the mixed plasma of ^{2}H and ^{3}He could create a neutron and triton free "clean" fusion reactor, when they are polarized [1]. Unfortunately, we have no double polarized deuteron scattering experiments $\vec{D}(\vec{d},p)^{3}\mathrm{H}$ in the energy range from 20 to 120 keV. However, there are several polarized deuteron beam experiments with unpolarized targets. In order to investigate the above mentioned speculations, we would like to calculate the four-nucleon system by four-body Yakubovsky equation with the Paris potential rigorously.

It is well-known that very low energy ^{3}He+$d \rightarrow {}^4$He+p and ^{3}H+$d \rightarrow {}^4$He+n reactions mainly occur through the $3/2^+$ resonance state at several hundreds keV. Therefore if these low energy reactions with polarized ^{3}He, ^{3}H, and ^{2}H beams occurred selectively through $3/2^+$ resonance, then the nuclear fusion cross section could be 50 % bigger than the case of unpolarized fuels, because the $1/2^+$ state of five nucleon system may be abandoned [1].

It is very hard to examine this speculation with experiments in such a very low energy regime, even more in the polarized plasma. In order to examine it theoretically, several numerical investigations have been attempted on $^2\vec{\mathrm{H}}(\vec{d},p)^3$H reactions so far [2],[3]. Recently we presented a result of a microscopic four-body calculation which fit the total cross section and the tensor analyzing powers T_{20}, T_{20}, T_{20} of the ^{2}H$(\vec{d},p)^3$H reaction recently obtained by Tsukuba [4] and Köln [5] groups. However, the differential cross section and the vector analyzing power were not very well fit to the experimental results. Especially, our calculation for the vector analyzing power is too small.

On the other hand, a success of the invariant-amplitude analysis [6] seems to indicate that only a small number of partial waves contribute to the reaction at such extreme low energies. It suggests that the P-wave NN-forces must be included in the A_y or iT_{11} calculations even in the four-body system.

2 The Four-Body Calculation

The numerical calculation was carried out using the four-body Faddeev-Yakubovsky integral equations [7] with the Ernst-Shakin-Thaler separable expansion of the Paris potential or the PEST1 (rank-1) potential for the $^1S_0$ and $^3S_1$-$^3D_1$ states [8]. The equations are given by the 2×2 coupled integral equations after symmetrization of the 18-coupled Faddeev-Yakubovsky integral equations

$$\begin{pmatrix} M_{\alpha\alpha} & M_{\alpha\beta} \\ M_{\beta\alpha} & M_{\beta\beta} \end{pmatrix} = \begin{pmatrix} \pm E & F \\ 2F^\dagger & 0 \end{pmatrix} + \begin{pmatrix} \pm E & F \\ 2F^\dagger & 0 \end{pmatrix} \begin{pmatrix} H & 0 \\ 0 & G \end{pmatrix} \begin{pmatrix} M_{\alpha\alpha} & M_{\alpha\beta} \\ M_{\beta\alpha} & M_{\beta\beta} \end{pmatrix} \quad (1)$$

where $M_{\alpha\alpha}$, $M_{\alpha\beta}$, $M_{\beta\alpha}$ and $M_{\beta\beta}$ are the scattering amplitudes, and α and β indicate [3+1] and [2+2] particle channels, respectively. E is a four-body potential for $\alpha \rightarrow \alpha$ with the plus sign for bosons and minus for fermions, F is for $\beta \rightarrow \alpha$, $F^\dagger$ is $\alpha \rightarrow \beta$ potentials, and H and G are [3+1] and [2+2] propagators, respectively. Let us define the relative momentum $\boldsymbol{q}_\beta$ for the initial state, and $\boldsymbol{q}_\alpha$ for the final state. We take the z-axis to be in the $\boldsymbol{q}_\beta$ direction, and the y-axis the $\boldsymbol{q}_\beta \times \boldsymbol{q}_\alpha$ direction. The partial wave amplitude is given by

$$\langle \tilde{J}\tilde{J}_z, s\nu; \boldsymbol{q}_\alpha | M_{\alpha\beta} | jj_z, \bar{j}\bar{j}_z; \boldsymbol{q}_\beta \rangle$$
$$= \frac{1}{\sqrt{2}} \sum_{L_\beta L_\alpha} (-1)^{M_\alpha} \sqrt{2L_\beta+1}\sqrt{2L_\alpha+1} \sqrt{\frac{(L_\alpha - |M_\alpha|)!}{(L_\alpha + |M_\alpha|)!}} P_{L_\alpha}^{M_\alpha}(\cos\theta)$$

$$\times \sum_{JS_\beta S_\alpha} \langle \tilde{J}\tilde{J}_z s\nu | S_\alpha S_{\alpha z}\rangle \langle jj_z \bar{j}\bar{j}_z | S_\beta S_{\beta z}\rangle \langle L_\alpha M_\alpha S_\alpha S_{\alpha z} | JJ_z\rangle \langle L_\beta 0 S_\beta S_{\beta z} | JJ_z\rangle$$
$$\times \quad \langle JJ_z 00 | JJ_z\rangle \langle N_\alpha J; N_\alpha^I I; q_\alpha | M_{\alpha\beta} | N_\beta J; N_\beta^I I; q_\beta\rangle \qquad (2)$$

where j and j_z are the incident deuteron spin and its z-component, and the target spin and z-component are $\bar{j}$, $\bar{j}_z$, respectively. $|N_\beta J\rangle$ and $|N_\alpha J\rangle$ are the state-vectors of the initial and the final spin-angular momentum space, and $|N_\beta^I I\rangle$, $|N_\alpha^I I\rangle$ are the corresponding isospin state vectors, respectively. Because the two particles in the initial state are identical, the scattering amplitude must be symmetrized which leads to

$$\langle \tilde{J}\tilde{J}_z, s\nu; \boldsymbol{q}_\alpha | M_{\alpha\beta} | jj_z, \bar{j}\bar{j}_z; \boldsymbol{q}_\beta\rangle_s$$
$$= \langle \tilde{J}\tilde{J}_z, s\nu; \boldsymbol{q}_\alpha | M_{\alpha\beta} | jj_z, \bar{j}\bar{j}_z; \boldsymbol{q}_\beta\rangle + \langle \tilde{J}\tilde{J}_z, s\nu; \boldsymbol{q}_\alpha | M_{\alpha\beta} | jj_z, \bar{j}\bar{j}_z; -\boldsymbol{q}_\beta\rangle. \qquad (3)$$

Therefore, the differential cross section is given by

$$\sigma_0(\theta) = \frac{1}{(2j+1)(2\bar{j}+1)} \frac{q_\alpha}{q_\beta} \sum_{\tilde{J}_z \nu j_z \bar{j}_z} |\langle \tilde{J}\tilde{J}_z, s\nu; \boldsymbol{q}_\alpha | M_{\alpha\beta} | jj_z, \bar{j}\bar{j}_z; \boldsymbol{q}_\beta\rangle_s|^2. \qquad (4)$$

The analyzing powers are defined as follows,

$$T_{K\kappa}(\theta) = \frac{1}{(2j+1)(2\bar{j}+1)\sigma_0(\theta)} \frac{q_\alpha}{q_\beta} \sum_{\tilde{J}_z \nu j_z} \sum_{\bar{j}_z \bar{j}'_z} \langle \tilde{J}\tilde{J}_z, s\nu; \boldsymbol{q}_\alpha | M_{\alpha\beta} | jj_z, \bar{j}\bar{j}_z; \boldsymbol{q}_\beta\rangle_s$$
$$\times \quad \langle \bar{j}\bar{j}_z | \tau_{K\kappa} | \bar{j}'\bar{j}'_z\rangle \langle \tilde{J}\tilde{J}_z, s\nu; \boldsymbol{q}_\alpha | M_{\alpha\beta} | jj_z, \bar{j}'\bar{j}'_z; \boldsymbol{q}_\beta\rangle_s^*. \qquad (5)$$

In this work we included the $1/2^+$ and $1/2^-$ states for the ^{3}He and ^{3}H nuclei and omitted $3/2^+$ $3/2^-$, $5/2^+$ ... excited states which might occur in the intermediate subnuclei as part of the four nucleon system. The equivalence between the particle and the target is fully taken into account. The angular momentum between two deuterons in the initial state is adopted up to $L_\beta = 2$, and also the maximum value of the final state angular momentum is $L_\alpha = 4$. Therefore, the total spins and parities of the four-nucleon system are 0^+, 0^-, 1^+, 1^-, 2^+, 2^-, 3^+, 3^-, and 4^+. The convergence of the energy dependent pole expansion (EDPE) for the [3+1] subamplitudes and Hilbert-Schmidt separable expansion (HSE) for the [2+2] ones are almost exact in the rank-3 in our case [9]. Therefore our calculations were carried out by the rank-3 approximation for those sub-amplitudes. The technique used to solve the four-body equation uses to the generalized separable expansion method given by one of the authors [10] which is exact in the on-shell and the half-on-shell cases. These calculations include the Coulomb penetration effect when both the charged particles are close together, and the modification of the phase shift by the Coulomb repulsive force.

3 Results

The results for the total cross section (Fig. 1) are in very good agreement with the experimental data gathered by the Los Alamos group [11]. The differential cross sections (Fig. 2) at 30 keV nearly match the Los Alamos data. The tensor analysing powers give good fits to the experimental data at 30 keV (Fig. 3). This disagreement may be due to the effect of higher spin states of [3+1], more channels of [2+2] subsystems, and also the insufficient Coulomb corrections. The vector analyzing power iT_{11} is smaller than the experimental value and wrong sign (Fig. 4). The NN-interactions of the $^3P_0$, $^1P_1$, $^3P_1$, and $^3P_2$ states should also be taken into account. The result for the ratio of the $^2\vec{\mathrm{H}}(\vec{d},p)^3\mathrm{H}$

Figure.1

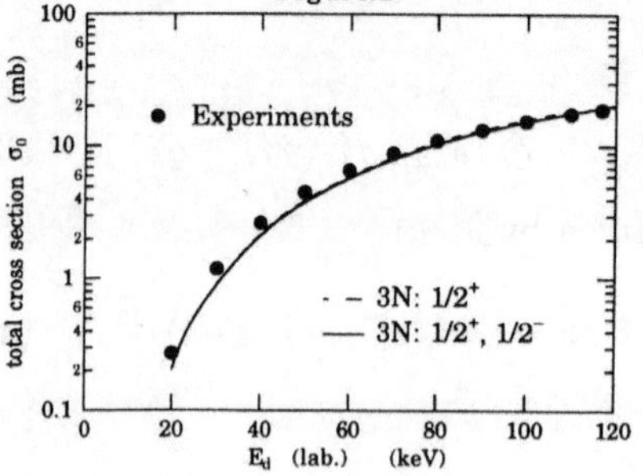

Figure.2

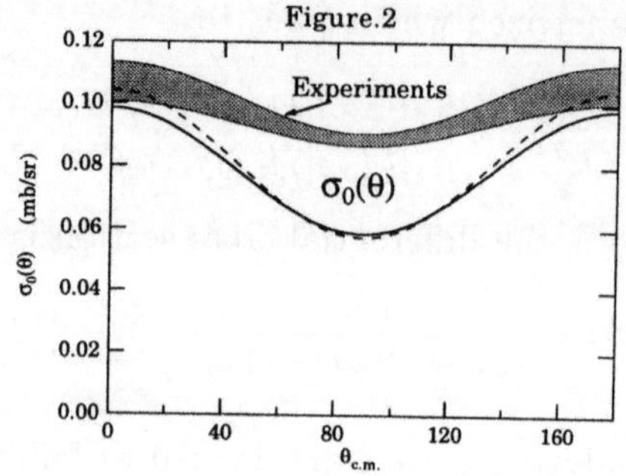

Figure.3

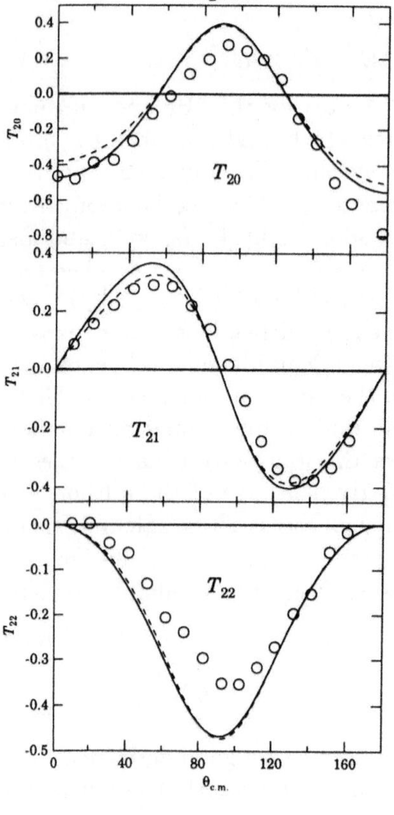

Figure.4

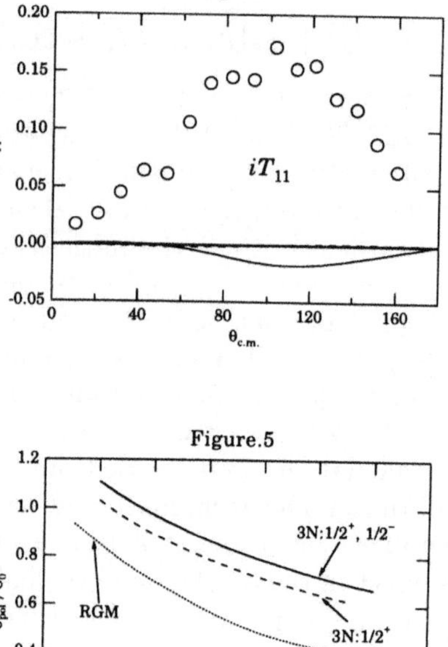

Figure.5

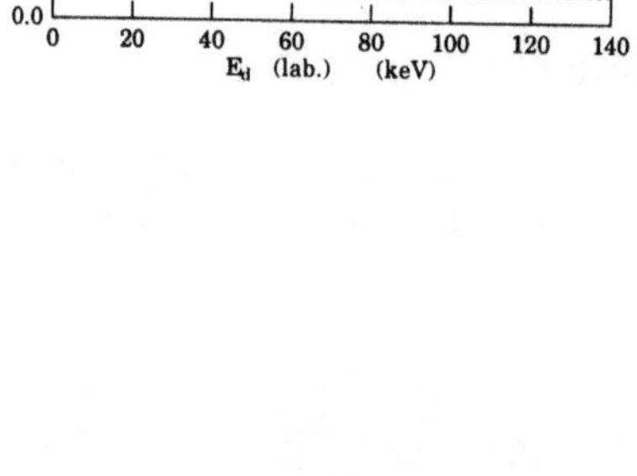

reaction cross section to the $^2H(d,p)^3H$ cross section is shown in the figure by σ_{pol}/σ_0 (Fig. 5). The half of the cross section is suppressed at higher energies. Our present result is similar to the RGM calculation [2].

Acknowledgement. This work was performed using FACOM M780/10, /10S of INS Computer Center in Tokyo University, and also FACOM M1800/20 of RCNP Computer Center in Osaka University.

References

1. R.M. Kulsrud, H.P. Furth, E.J. Valeo, M. Goldhaber: Phys. Rev. Lett. **49**, 1248 (1982)
2. H.M. Hofmann, and D. Fick: Phys. Rev. Lett. **52**, 2038 (1984); D. Fick, and H.M. Hofmann: Phys. Rev. Lett. **55**, 1650 (1985)
3. K.F. Liu, J.S. Zhang, and G.W. Shuy: Phys. Rev. Lett. **55**, 1649 (1985); J.S. Zhang, K.F. Liu, and G.W. Shuy: Phys. Rev. Lett. **57**, 1410 (1986)
4. Y. Tagishi, N. Nakamoto, K. Katoh, J. Togawa, T. Hisamune, T. Yoshida, Y. Aoki: Phys. Rev. **C46**, R1155 (1992)
5. B. Becker, R. Randermann, B. Polke, S. Lemaître, R. Reckenfelderbäumer, P. Niessen, G. Rauprich, L. Sydow, H. Paetz gen. Schieck: Few-Body Sys. **13** (1992), 19; *ibid.*: Phys. Lett. B276 (1992), 290
6. Makoto Tanifuji: Phys. Lett. **B289**, 233 (1992)
7. O.A. Yakubovsky: Sov. J. Nucl. Phys. **5**, 937 (1967)
8. J. Haidenbauer and W. Plessas: Phys. Rev. **C30**, 1822 (1984); *ibid.*: **C32** 1424 (1985)
9. Eizo Uzu, Shinsho Oryu, and Makoto Tanifuji: Prog. Theor. Phys. **90**, 937 (1993)
10. S. Oryu: Prog. Theor. Phys. Suppl. **61**, 180 (1977); Phys. Rev. **C27**, 2500 (1983)
11. Ronald E. Brown, Nelson Jarmie: Phys. Rev. **C41**, 1391 (1990)

Few-Body Systems Suppl. 8, 102–106 (1995)

Few-Body Systems

Integral Equation Calculations for the Photodisintegration of ^{4}He

G. Ellerkmann [1], W. Sandhas [1], S.A. Sofianos [2]

[1] Physikalisches Institut, Universität Bonn, Endenicher Allee 11-13, D-53115 Bonn, Germany

[2] Physics Department, University of South Africa, P.O. Box 392, Pretoria 0001, South Africa

Abstract. Results obtained by solving AGS-type integral equations for the photodisintegration of ^{4}He, employing the Malfliet-Tjon potential, are compared with the latest experimental data. At low energies good agreement between theory and experiment is found. At higher energies electric quadrupole contributions have to be taken into account, but are not sufficient to fully remove the discrepancies between theory and experiment. In order to get some feeling for the sensitivity to the underlying potential, we in addition show calculations based on the Yamaguchi potential. They differ from the MT results in a way which resembles the trends known from triton photodisintegration.

1 Introduction

The photodisintegration of ^{4}He into $n+^3$He or $p+^3$H has been for a long time a rather controversial topic in few-body physis. Early data appeared consistent with the picture of a low-energy giant dipole resonance, a picture also supported by shell model and resonating group calculations. Generalizing the four-nucleon Alt-Grassberger-Sandhas (AGS) formalism [1], exact integral equations have been derived by Casel and Sandhas for the ^{4}He photodisintegration amplitude which led to a completely different prediction [2,3]. The corresponding calculations, in fact, showed a flat non-resonant behaviour, a prediction confirmed by practically all recent measurements [4,5].

The remaining discrepancies suggested to repeat the separable-potential calculations of refs. [2,3] employing, however, a more realistic nuclear interaction as, e.g., the Malfliet-Tjon (MT I+III) potential [6]. For the treatment of the two- and three-body subsystem amplitudes in the kernel of the AGS four-body equations, the W-matrix approximation [7] and the energy-dependent pole expansion (EDPE) [8] were chosen. In the pure nuclear case these two approximations have been shown to lead to rather accurate results.

Our calculations demonstrate the necessity of incorporating meson exchange currents via Siegert's theorem [9]. At low energies there is quite a difference between the plane wave (Born) approximation and the full solution of the integral equation, by which the final state interaction is taken into account. Only with both these contributions good agreement between theory and experiment is achieved. At higher energies the plane wave approximation appears justified, but E2 contributions turn out to be important, and the incorporation of meson exchange currents remains essential.

2 Formalism

The properly antisymmetrized amplitude for the photodisintegration of the ^{4}He bound state $|\psi_{IV}\rangle$ into a three-body bound state $|\psi_{III}\rangle$ and a nucleon of relative momentum $|\boldsymbol{q}\rangle$ is given by

$$M(\boldsymbol{q}) = 2\,{}^{(-)}\langle \boldsymbol{q}; \psi_{III}|H_{\text{em}}|\psi_{IV}\rangle. \tag{1}$$

Replacing the scattering state ${}^{(-)}\langle \boldsymbol{q}; \psi_{III}|$ by the channel state $\langle \boldsymbol{q}|\langle\psi_{III}|$, one gets the corresponding plane wave (Born) approximation $B(\boldsymbol{q})$. At low energies the process under consideration takes place primarily via electric dipole transition. Assuming point-like charges, the E1 operator is given by

$$H'^{(1)}_{\text{em}} = -\frac{e}{m}\sum_{j=1}^{4} \frac{1+\tau_z^j}{2}\,\hat{\epsilon}_\lambda \cdot \boldsymbol{p}_j\,, \tag{2}$$

where $\hat{\epsilon}_\lambda$ denotes one of the two polarization directions of the incident photon. Applying Siegert's theorem [9], Eq. (2) is replaced by

$$H^{(1)}_{\text{em}} = -i\,e\,(E_f - E_i)\sum_{j=1}^{4} \frac{1+\tau_z^j}{2}\,\hat{\epsilon}_\lambda \cdot \boldsymbol{x}_j\,, \tag{3}$$

with $(E_f - E_i)$ being the difference of the final state energy and the binding energy of $|\psi_{IV}\rangle$ in the initial state. After going over to Jacobi coordinates, Eqs. (2) and (3) each become a sum of three terms, one of them depending only on the relative momentum or coordinate of the two outgoing fragments. Since we take into account S-wave subsystems only, the two other (internal) terms vanish.

Taking over the AGS reduction technique developed for the four-nucleon collision problem [1], one arrives at the set of integral equations [2]

$$\mathcal{M}(\boldsymbol{q}) = \mathcal{B}(\boldsymbol{q}) + \int d^3q'\; \mathcal{V}(\boldsymbol{q},\boldsymbol{q}')\,\mathcal{G}_0(\boldsymbol{q}')\,\mathcal{M}(\boldsymbol{q}'), \tag{4}$$

i.e., at an effective two-body matrix equation for an off-shell extension $\mathcal{M}(\boldsymbol{q})$ of the amplitude (1). This relation is almost identical to the set of integral equations for four-nucleon rearrangement processes [1]. Its kernel is built up by the effective potential $\mathcal{V}$ and the Green's function $\mathcal{G}_0$ of this case, found in detail in [1,14]. The only difference is that now the inhomogeneous term is given by an off-shell extension $\mathcal{B}(\boldsymbol{q})$ of the plane wave (Born) amplitude $B(\boldsymbol{q})$ rather than the potential $\mathcal{V}$.

In order to determine the 3- and 4-nucleon bound states entering the inhomogeneity $\mathcal{B}(\boldsymbol{q})$ of Eq. (4), one has first to solve the homogenous AGS equations [1]. The final step then consists in solving the corresponding inhomogenous equation (4) in the continuous spectrum.

At higher energies we include besides the dipole operator (3) the electric quadrupole operator in its Siegert form

$$H_{\text{em}}^{(2)} = \frac{e}{2}(E_f - E_i)\sum_{j=1}^{4}\frac{1+\tau_z^j}{2}\,\hat{\epsilon}_\lambda\cdot\boldsymbol{x}_j\,\boldsymbol{k}_\gamma\cdot\boldsymbol{x}_j\,, \tag{5}$$

with $\boldsymbol{k}_\gamma$ being the photon momentum.

3 Results

For the MT I+III potential, the $3N$ binding energy is 8.595 MeV, the 4N binding energy obtained in EDPA is 30.1 MeV or, with higher expansion terms, 30.2 MeV. The corresponding energies for the Yamaguchi potential are 9.97 MeV and 39.1 MeV.

Figure 1 shows our ^{4}He(γ,n)^{3}He total cross section results. Due to the complicated cut structure of the integral equation kernel, the full calculations, which take into account the final state interaction, could only be performed below the three-fragment break-up threshold. In order to achieve good agreement with the data [4,11,12] the full solution of Eq. (4) including meson exchange currents is needed (solid curve).

In Fig. 2 the corresponding results obtained for the Yamaguchi potential are presented. They lie below the MT curves, in fact too low as compared to the data. A similar trend was observed in ^{3}H photodisintegration calculations [10].

Figures 3 and 4 show our total cross section results for the process ^{4}He(γ,p)^{3}H at higher photon energies, taking into account also the E2 operator (for ^{4}He(γ,n)^{3}He the E2 contributions are negligible). The agreement with the newer experimental data [13] is quite remarkable. The potential dependence is seen to be fairly diminished at these energies, an observation also made in ^{3}H photodisintegration calculatons [10].

Figures 5 and 6 show our results for the differential cross sections in comparison with the data of ref. [13]. Evidently, the E2 contributions are essential for achieving better agreement with experiment. We expect the remaining disagreement to be removed by including higher order electric or magnetic multipoles, higher partial waves in the effective $3N$ potential, or the final state interaction also at these energies.

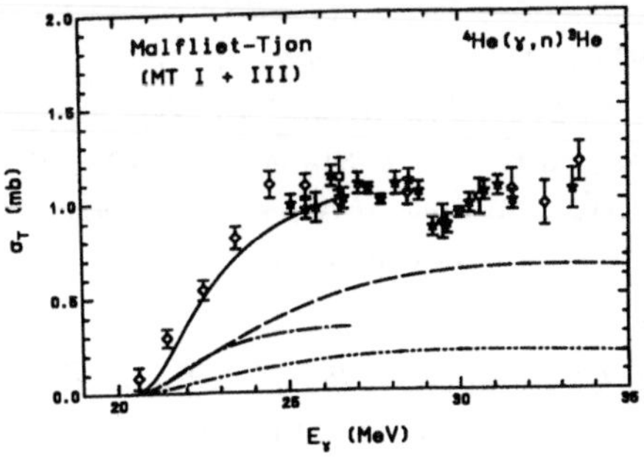

Figure 1. Total cross section for ^{4}He$(\gamma,n)^3$He at low energies: Full solution with exchange currents (——) and without (– · –); plane wave approximation with exchange currents (– –) and without (– · · –); the data are from [4] ◇, [11] ⋆, and [12] □.

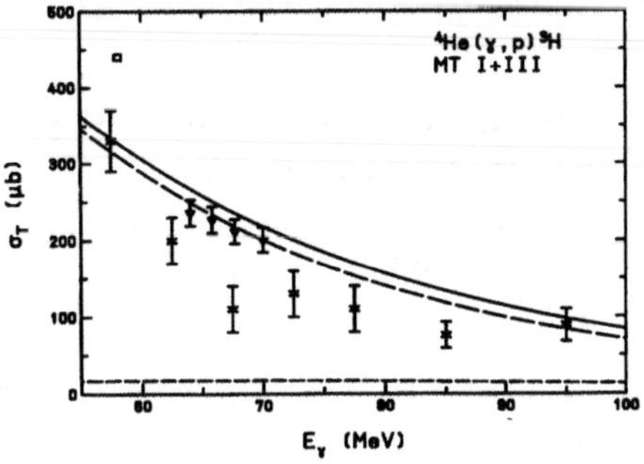

Figure 3. Total cross section for ^{4}He$(\gamma,p)^3$H at higher energies: Plane wave approximation including E1 and E2 (——). Contributions of E1 (— —) and E2 (– – –); the data are from [13] ▽, [15] ×, and [16] □.

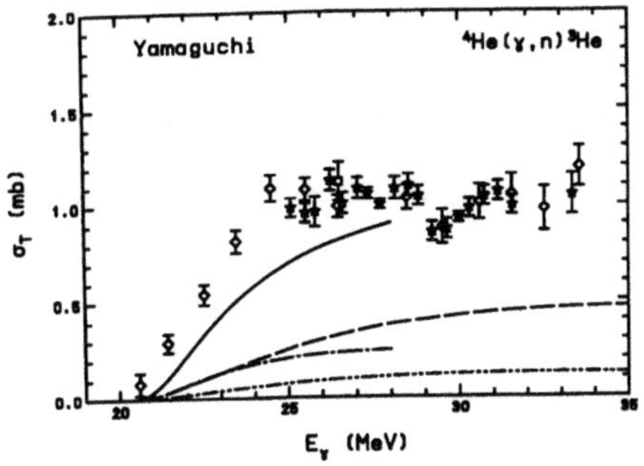

Figure 2. Same as Fig. 1, but with the Yamaguchi potential.

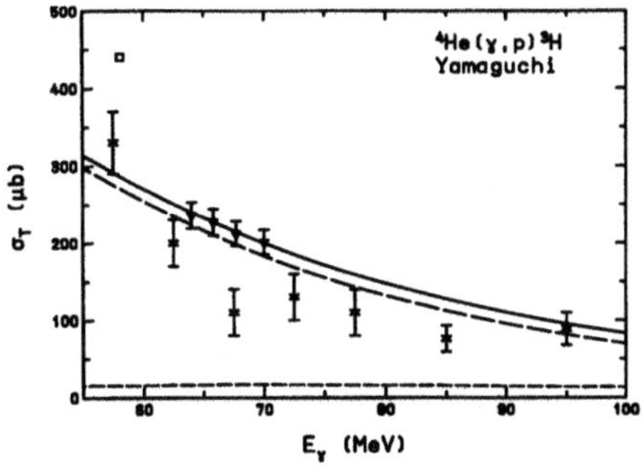

Figure 4. Same as Fig. 3, but with the Yamaguchi potential.

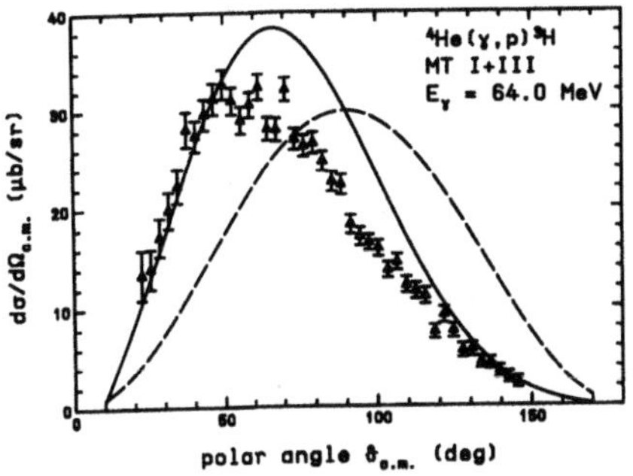

Figure 5. Differential cross section: Plane wave approximation including E1 and E2 (——). Contribution of E1 only (– –); the data are from [13] △.

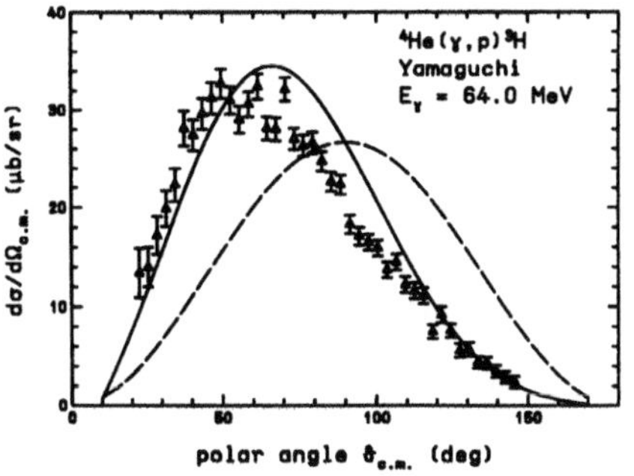

Figure 6. Same as Fig. 5, but with the Yamaguchi potential.

References

1. P. Grassberger, W. Sandhas: Nucl. Phys. **B2**, 181 (1967); E.O. Alt, P. Grassberger, W. Sandhas: Phys. Rev. C **1**, 85 (1970) and in *Few Particle Problems in the Nuclear Interaction*, edited by I. Slaus et al. (North-Holland, Amsterdam, 1972), p. 299; W. Sandhas: *Progress in Particle Physics*, edited by P. Urban [Acta Physica Austriaca, Suppl. XIII, 679 (1974)]; Czech. J. of Phys. **B25**, 251 (1975)

2. A. Casel, W. Sandhas: *Proc. Ninth Int. Conference on the Few-Body Problem*, Vol. I: Contributed Papers, ed. M.J. Moravcsik (Eugene, Oregon, 1980), p. 33; see also W. Sandhas: Few-Body Systems Suppl. **1**, 64 (1986); W. Böttger, A. Casel, W. Sandhas: Phys. Lett. **92B**, 11 (1980)

3. A. Casel, W. Sandhas: Czech. J. of Phys. **B36**, 300 (1986)

4. B.L. Berman, D.D. Faul, P. Meyer, D.L. Olson: Phys. Rev. C **22**, 2273 (1980)

5. J.R. Calarco, B.L. Berman, T.W. Donnelly: Phys. Rev. C **27**, 1866 (1983)

6. R.A. Malfliet, J.A. Tjon: Nucl. Phys. **A127**, 161 (1969)

7. E.A. Bartnik, H. Haberzettl, W. Sandhas: Phys. Rev. C **34**, 1520 (1986)

8. S.A. Sofianos, N.J. Mc Gurk, H. Fiedeldey: Nucl. Phys. **A318**, 295 (1979);

9. A.J.F. Siegert: Phys. Rev. **52**, 787 (1937)

10. W. Schadow, J. Richter W. Sandhas: *Proceedings of the 14th International IUPAP Conference on Few-Body Problems*, 695 (1994)

11. L. Ward, D.R. Tilley, D.M. Skopik, N.R. Roberson, H.R. Weller: Phys. Rev. C **24**, 317 (1981)

12. J. Asai, G. Feldman, R.E.J. Florizone, E.L. Hallin, D.M. Skopik, J.M. Vogt, R.C. Haight, S.M. Sterbenz: Few-Body Systems Suppl. **7**, 136 (1994)

13. R.T. Jones, D.A. Jenkins, P.T. Debevec, P.D. Harty, J.E Knott: Phys. Rev. C **43**, 2052 (1991)

14. S.A. Sofianos, H. Fiedeldey, W. Sandhas: Phys. Rev. C **32**, 400 (1985)

15. A.N. Gorbunov: Proc. of the P.N. Lebedev Physics Institute, **71**, 1 (1976)

16. R. Bernabei et al.: Phys. Rev. C **38**, 1990 (1988)

Few-Body Systems Suppl. 8, 107–120 (1995)

Few-Body Systems

Spin Observables in High Energy Electron Scattering

S. Kowalski

Bates Linear Accelerator Center, Laboratory for Nuclear Science and Department of Physics, Massachusetts Institute of Technology, Cambridge, MA 01239, U.S.A.

1 Introduction

The electromagnetic probe has long been an important tool for studying nucleon and nuclear structure and dynamics. High energy electron scattering can probe distances of less than 1 fm where quark degrees of freedom are expected to play an observable role. A new generation of medium energy accelerators, operating CW with energies of 0.5 – 6 GeV, has recently come into operation. These enhanced capabilities, particularly the 100% duty factor and kinematic reach, will dramatically improve our ability to carry out coincidence studies. Storage ring facilities at MIT-Bates and NIKHEF make possible unique studies with thin polarized and unpolarized internal nuclear targets at relatively high luminosity.

In this review, we summarize the status and results of some recent experiments which exploit the measurement of spin observables to study nucleon and nuclear structure. The use of polarized beams, polarized targets, recoil polarimetry and coincidence measurements of reaction products out-of-the scattering plane allows us to take full advantage of the capabilities of electron scattering.

2 Neutron Form Factors

The electromagnetic structure of the nucleon is of fundamental importance to our understanding of models of the nucleon and of nuclei. The neutron electric form factor, G_E^n, is the least well known of the four nucleon electromagnetic form factors.

The elastic scattering of unpolarized electrons from unpolarized nucleons ($I = \frac{1}{2}$), at a momentum transfer Q, involves a measurement of the cross section

$$\frac{d\sigma}{d\Omega} = \sigma_M\, f_{rec}^{-1}\, [(G_E^N)^2 \; + \; \tau(G_M^N)^2\, \{1 \; + \; 2(1+\tau)\tan^2\frac{\theta}{2}\}]$$

where $\tau = -Q^2/4M_N^2$ and f_{rec}^{-1} is a kinematic recoil factor. A Rosenbluth separation of the electric and magnetic form factors, $G_E^N(Q^2)$ and $G_M^N(Q^2)$, allows for reasonable accuracy only when the two amplitudes are comparable. In the nucleon case the magnetic form factor dominates over the electric one at high momentum transfers. As a result, only the magnetic form factor is relatively well known over an extended range in momentum transfer. For the proton, reasonable knowledge of G_E^p exists only up to ~ 4 GeV^2. For the neutron, which is charge neutral, G_E^n is very small and as a result is very poorly known for all momentum transfers.

At present, much of our information on G_E^n comes from measurements of $A(Q^2)$ from electron-deuteron elastic and quasi-elastic scattering. The neutron electric form factor has recently been extracted from some precise new measurements [1] on $A(Q^2)$. The results for G_E^n are dependent upon the potential model which is used to describe the deuteron. The resulting systematic dependences are as large as ~50%.

Spin observables will allow us to make more accurate measurements of G_E^n. For the nucleon case, experiments with both polarized beams and targets (or alternatively recoil polarimetry) are required. The differential cross section for inclusive scattering of longitudinally polarized electrons from a spin$\frac{1}{2}$ target may be written [2] as

$$\frac{d\sigma}{d\Omega d\omega} = \Sigma \pm \Delta(\theta^*, \phi^*)$$

where ω is the energy transfer and the angles θ^* and ϕ^* define the target spin direction relative to Q and the scattering plane as shown in Fig. 1. The plus (minus) sign corresponds to positive (negative) helicity of the incident electrons. The spin-independent cross section Σ has the usual Rosenbluth form

$$\Sigma = 4\pi\sigma_M[v_L R_L(q,\omega) \;+\; v_T R_T(q,\omega)],$$

where q is the momentum transfer, v_L and v_T are kinematic factors, R_L and R_T are the longitudinal and transverse response functions. These response functions are dependent on bilinear combinations of the nucleon electromagnetic form factors. The spin-dependent cross section Δ involves two other response functions:

$$\Delta = -4\pi\sigma_M \left[v_{T'} R_{T'}(q,\omega) \cos\theta^* + 2 v_{TL'} R_{TL'}(q,\omega) \sin\theta^* \cos\phi^*\right],$$

where $v_{T'}$ and $v_{TL'}$ are kinematic factors independent of nuclear structure.

Experiments measure the asymmetry in the cross section under reversal of electron helicity: $A = \Delta/\Sigma$. The experimental asymmetry A_{exp} is related to A by $A_{exp} = P_e P_T A$, where P_e and P_T are the electron beam and target polarizations, respectively. The electron polarization $P_e = 0.40$ typically.

In the case of a polarized neutron target, the asymmetry may be expressed in terms of the electric and magnetic form factors as

$$A = \frac{-2\tau v_{T'}(G_M^n)^2 \cos\theta^* \;-2\sqrt{2\tau(1+\tau)}\; v_{TL'} G_M^n G_E^n \sin\theta^* \cos\phi^*}{(1+\tau)v_L(G_E^n)^2 + 2\tau V_T (G_M^n)^2}$$

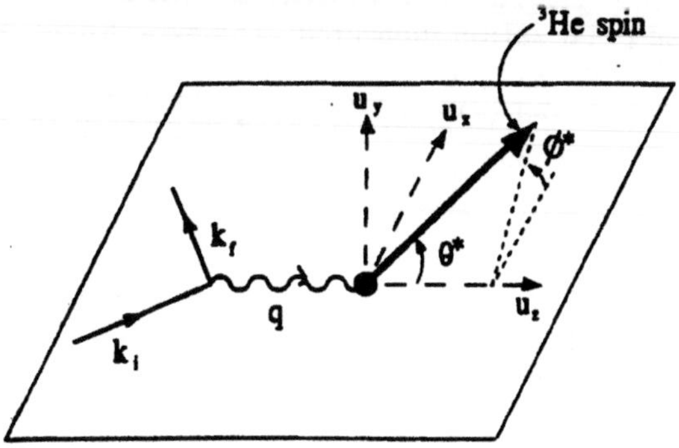

Figure 1. Kinematic variables for electron scattering from polarized targets $\vec{q}$. The vector u_z is along the direction of momentum transfer $\vec{q}$. The vector u_y is normal to the electron scattering plane and $u_x = u_y \times u_z$ lies in the scattering plane. The target polarization direction is defined by the angles (θ^*, ϕ^*) as shown.

A measurement of the asymmetry for $\theta^* = \pi/2$ involves the interference term $G_M^n\, G_E^n$ which is directly sensitive to the small electric form factor and the relative sign.

Several experiments have been designed to exploit the measurement of this interference term and provide precise new information on G_E^n. These include quasi-elastic scattering from the deuteron and from polarized ^{3}He.

2.1 $^2\mathrm{H}(\vec{e}, e'\vec{n})p$

In this experiment the electric form factor of the neutron, G_E^n, is determined by scattering longitudinally polarized electrons from deuterium quasi- elastically and measuring the transverse polarization component, P_x, of the recoiling neutron. In the impulse approximation, it has been shown [3] that the neutron polarization components are given by:

$$P_x = -2\sqrt{\tau(1+\tau)}\; G_M^n\; G_E^n/I_o,$$
$$P_z = \frac{E+E'}{M}\; \sqrt{\tau(1+\tau)}\; (G_M^n)^2\; tan^2\frac{\theta}{2}/I_o,$$

where $\tau = Q^2/4M^2$ and the unpolarized cross section

$$I_o = (G_E^n)^2 + \tau(G_M^n)^2\; [1 + 2(1+\tau)\; \tan^2\theta/2].$$

The neutron polarization component, P_x, in the scattering plane normal to the neutron momentum is seen to be directly proportional to G_E^n.

In the Bates experiment [4], a longitudinally polarized electron beam, with an energy of 868 MeV, was incident on a liquid deuterium target (0.845 g/cm^3). The scattered electrons were momentum-analyzed in a magnetic spectrometer.

The recoil neutrons were detected in coincidence with the scattered electrons, and their transverse polarization measured. The experiment operated at a luminosity of 3×10^{36} cm^{-2} s^{-1}, using a pulsed beam with a duty factor of 1%. Fig. 2 displays the measured value of G_E^n. It is in agreement with the Galster parameterization.

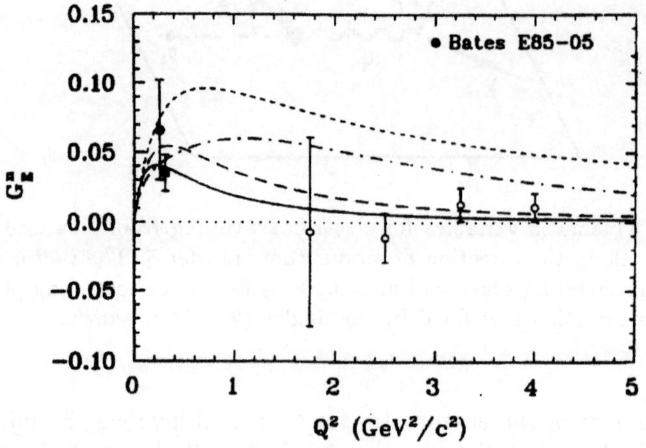

Figure 2. Recently reported values of G_E^n: quasielastic $^2H(\vec{e}, e'\vec{n})$ reaction (shaded circle), exclusive $^3\vec{He}(\vec{e}, e'n)$ by Meyerhoff et al. [27] (shaded square), and the inclusive quasielastic $^2H(e, e')$ results of Lung et al. [28] (open circles). The parameterizations are from the work of Galster et al. [29] $G_E^n = -\tau\mu_n(1 + 5.6\tau)^{-1}$ (long dashes), Platchkov et al. [1] where the Paris potential fit is shown (solid), the Gari-Krümpelmann VMD-PQCD model 3 (dash-dot) and the parameterization $G_E^n = -\tau G_M^n$ (short dashes) where $F_{1n} = 0$.

The expressions for P_x and P_z can be used to calculate the ratio P_x/P_z:

$$\frac{P_x}{P_z} = -\left[\frac{2M}{E + E'}\right] \frac{G_E^n}{G_M^n} \tan^{-1}\theta/2$$

A measurement of this ratio would be largely insensitive to both the electron polarization and polarimeter analyzing power. The longitudinal polarization can be measured with the same polarimeter after precessing the neutron spin by $\pi/2$ with a dipole magnet.

Further experiments are planned at Bates at higher Q^2 using extracted CW beams from the South Hall Ring. Similar experiments have been carried out at MAINZ and are also planned for CEBAF at higher Q^2.

2.2 $^3\vec{He}(\vec{e}, e')$

^{3}He is a very interesting nucleus for electromagnetic studies. It can be polarized and one expects that spin-dependent scattering from ^{3}He would be sensitive to neutron electromagnetic form factors. To a good approximation, the two protons in the ground-state are in a singlet configuration.

For the case of ^{3}He, ($I = \frac{1}{2}$), in analogy with the nucleon case, the asymmetry is given by

$$A = -\frac{v_{T'} R_{T'} \cos\theta^* + 2v_{TL'} R_{TL'} \sin\theta^* \cos\phi^*}{v_L R_L + v_T R_T}$$

By choosing the target spin to be along $\boldsymbol{q}(\theta^* = 0°)$ or normal to $\boldsymbol{q}(\theta^* = 90°)$, one can select the asymmetry to be proportional to $R_{T'}$ or $R_{TL'}$, respectively.

All calculations show that the spin-dependent asymmetry for $\boldsymbol{q}$ parallel to the target spin is proportional to $(G_M^n)^2$. In comparison to a target of polarized neutrons, ^{3}He dilutes the asymmetry since proton scattering contributes to the unpolarized cross section. In addition, small amplitudes in the ^{3}He wave function which are due to polarized protons also contribute to the asymmetry.

Similarly, the spin-dependent asymmetry for $\boldsymbol{q}$ perpendicular to the target spin is sensitive to the neutron electric form factor. The calculations of Blankleider and Woloshyn [5] showed that the neutron contribution dominated allowing a direct measurement of G_M^n and G_E^n. The more recent calculations [6,7,8] for the response function $R_{TL'}$ show a substantial proton contribution which significantly reduces the sensitivity to G_E^n.

The first measurements of spin-dependent inclusive electron scattering from a polarized ^{3}He target were carried out at Bates in 1990 [9,10]. A high-precision measurement [11] was performed at Bates in 1993 using the Caltech target. The beam energy was 370 MeV, and 25 μA beams at 37% polarization were typical. The target polarization was ~38%. Fig. 3 shows the measured asymmetry, $A_{T'}$, over the quasi-elastic peak. The neutron magnetic form factor, G_M^n, has been extracted from this asymmetry based on recent PWIA calculations using spin-dependent spectral functions. The result at $Q^2 = 0.19$ GeV^2 is in agreement with the dipole parameterization, $\mu_n G_D$, and the ratio is $(G_M^n/\mu_n G_D)^2 = 0.998 \pm 0.117 \pm 0.059 \pm 0.030$, where the errors are statistical, systematic, and model dependence, respectively. The result for G_M^n is shown in Fig. 4, together with other experimental data. This experiment is the first measurement for the neutron magnetic form factor using spin- dependent inclusive electron scattering.

The same measurement [26] also yielded results for the $A_{TL'}$ asymmetry. The experimental result is $A_{TL'}^{exp} = 1.52 \pm 0.55 \pm 0.15\%$ and the theoretical prediction is , $A_{TL'}^{thy} = 2.1 - 2.9\%$. The theoretical variation is due to uncertainties in the wavefunction of ^{3}He, nucleon form factors and off-shell prescriptions. Given the sizeable proton contribution, it is unlikely that this can be used to provide accurate information on G_E^n. There is some expectation that measurements at higher Q^2 could provide useful data for G_E^n.

Further inclusive measurements of quasi-elastic asymmetries at high-Q^2 are planned at Bates and CEBAF. The MAINZ program on polarized ^{3}He emphasizes exclusive measurements.

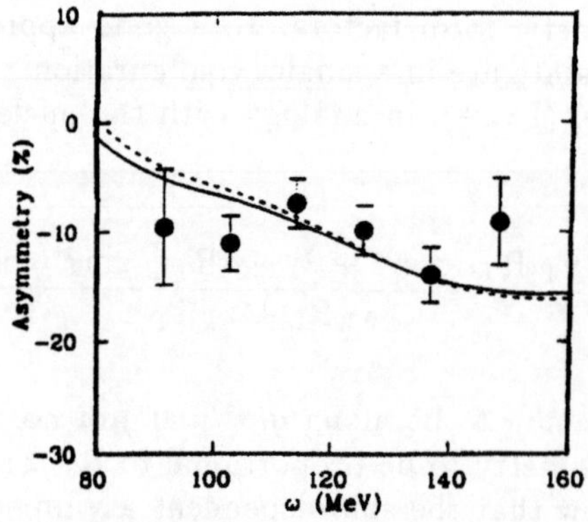

Figure 3. The measured transverse asymmetry $A_{T'}$ as a function of energy loss ω. The dashed line is the calculation by Salmè et al. [7], and the solid line is the calculation by Schulze et al. [8].

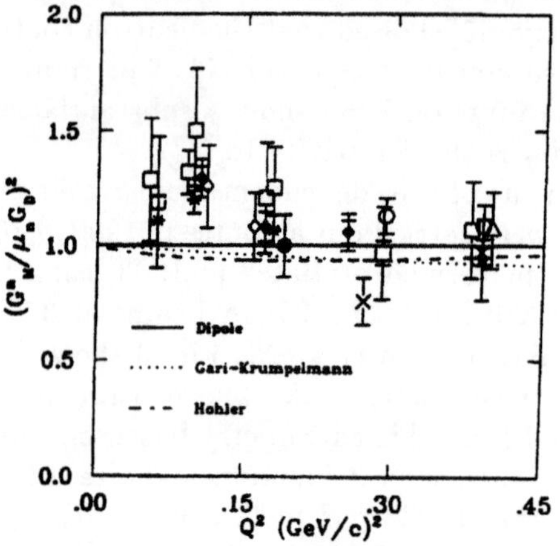

Figure 4. The square of the neutron magnetic from factor $(G^n_M)^2$, in units of the standard dipole parameterization, $(\mu_n G_D)^2$, in the low $-$ Q^2 region. The solid circle is from the data of Gao et al. [11]. The other data points are defined in ref. [11].

3 Out-of-Plane Spectrometry

Exclusive measurements involving electron scattering have been used for many years to study excitations in the nuclear continuum and to provide information on momentum distributions and medium effects in nuclei. An out-of-plane electron scattering capability is essential to fully exploit the potential of coincidence experiments. It has been shown that such a capability would give access to new observables and thus permit the isolation of important and otherwise inaccessible amplitudes.

The $A(e, e'x)B$ cross section corresponding to the reaction depicted in Fig. 5 can be written in the one-photon exchange approximation as [12];

$$\frac{d\sigma}{d\Omega} = \Sigma(\theta^*, \phi^*) + h\ \Delta(\theta^*, \phi^*),$$

where

$$\Sigma = \sigma_M[v_L R_L + v_T R_T + v_{TL} R_{TL} \cos\phi^* + v_{TT} R_{TT} \cos 2\phi^*]$$

and

$$\Delta = \sigma_M\ v_{TL'}\ R_{TL'} \sin\phi^*.$$

The polar angles θ^* and ϕ^* define the direction of the decay product (e.g. a proton) relative to the direction of the momentum transfer and the scattering plane. The cross section is separated into helicity dependent and independent parts ($h = \pm 1$ is the electron helicity). In the usual case, where unpolarized targets are used and the polarization of the decay proton is not measured, the cross section depends on five nuclear response functions.

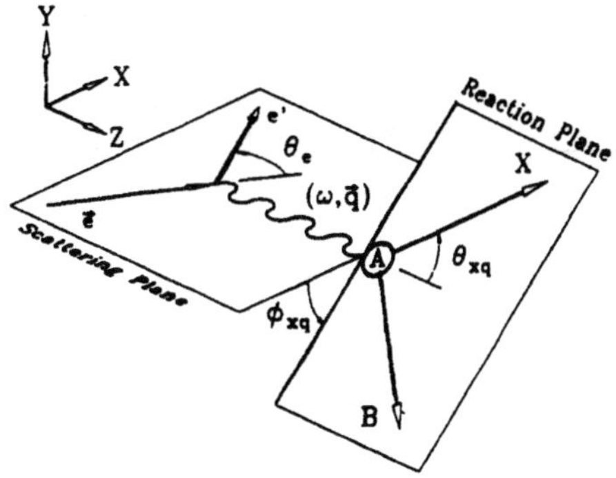

Figure 5. Kinematic definitions for the $A(\vec{e}, e'x)B$ reaction.

These five response functions are bilinear combinations of the transverse and longitudinal components of the nuclear current with respect to the direction of the momentum transfer $\vec{q}$. The goal of any experiment is to effect an accurate isolation of each response. R_L and R_T are readily determined using a Rosenbluth separation. R_{TL} and R_{TT} can be separated by measurements at values of ϕ^* on a cone centered on $\vec{q}$. The "fifth" structure function, $R_{TL'}$ - a new observable, breaks the symmetry for decay particles scattering above and below the plane and can be measured if the incident electrons are longitudinally polarized and decay particles are observed out-of-plane.

Interference terms typically contribute at a level of (1-20)% to the total cross section. As a result, there are major experimental challenges in performing measurements with high statistical and systematic precision. The most straightforward approach involves sequential measurements with a spectrometer positioned at different θ^* and ϕ^*.

At MIT-Bates, we are implementing a unique system of four small magnetic spectrometers (OOPS) which will allow for simultaneous measurements [13]. The basic idea is to arrange the spectrometers in a cone symmetrically about the momentum transfer $\boldsymbol{q}$. It is believed that the simultaneous measurement of several asymmetries will allow for the isolation of the interference structure functions with very high accuracy. The required control of kinematic variables and other experimental parameters is expected to be greatly reduced in comparison with the usual technique of sequential measurements. Large out-of-plane angles are readily accommodated. This is essentially impractical with typical large magnetic spectrometers. The Bates system will be fully operational in 1996.

3.1 $N \rightarrow \Delta$ transition

The electroexcitation of the $\Delta(1232)$ resonance is predominantly M1. In a naive quark model, the nucleon and delta are each made up of three 1-S quarks and the transition is a pure spin-isospin flip excitation. If the quark wavefunction has a non-vanishing D-state component, C2 and E2 transition amplitudes would also be allowed. Such admixtures can arise in bag models through bag deformation or through color magnetic effects from one-gluon exchange in a more fundamental QCD picture. The isolation of quadrupole amplitudes in the $N \rightarrow \Delta$ transition is of fundamental importance to our understanding of the nucleon.

The first nucleon resonance has been studied with photoproduction and with inclusive and coincidence electron scattering. Rosenbluth separations have been used to extract longitudinal and transverse cross-sections. Experimental values of E2/M1 (EMR) and C2/M1 (CMR) ratios can have contributions from "background" amplitudes such as Born terms and the tails of nearby resonances. There have been recent results from the LEGS collaboration for the $^1\mathrm{H}(\vec{\gamma}, p)\pi^0$ reaction [14] and from Bonn [15]. A consistent model description of the LEGS data at $Q^2 = 0$ with pion photoproduction data has not yet been achieved for any EMR value. In the Bonn experiment, a surprisingly large C2/M1$= 13 \pm 2\%$ at the four-momentum transfer $Q^2 = 0.127\mathrm{GeV}^2$ was reported. Both experiments suggest the importance of background terms and reinforce the need to confirm these results.

Model calculations suggest that the R_{LT} response function is sensitive to CMR, while other observables appear to be more sensitive to background terms. An experiment has just been completed at Bates which measured the coincidence yields for the $^1\mathrm{H}(e, e'p)\pi^0$ reaction at $Q^2 = 0.127\mathrm{GeV}^2$ at residual energies $W = 1170$, 1232 and 1280 MeV. A liquid H_2 target was used together with the MEPS and OHIPS spectrometers for electron and proton detection, respectively. OHIPS was also instrumented with a focal plane proton polarimeter which was used for these studies. Very careful attention was given to the optical charactersitics of the spectrometers, luminosities, and detector efficiencies.

The data will allow an extraction of absolute cross-sections and the response functions R_T (all W values) and R_{LT} (W = 1170 and 1232), as well as the proton polarization P_n at W = 1232 MeV. The preliminary cross sections are within the range spanned by current models. P_n , which is believed to be mainly sensitive to background terms, shows a large preliminary value $P_n \sim -0.3$.

3.2 Fifth Structure Functions: $^2H, ^{12}C$

We have recently completed experiments involving the measurement of the "fifth" structure function in nuclei. They involved the out-of-plane coincident detection of protons, using longitudinally polarized electrons. The reactions studied were $^2H(\vec{e}, e'p)$ and $^{12}C(\vec{e}, e'p)$. The objective was to test model descriptions of these nuclei and, in particular, to isolate the contributions of background terms from the resonant ones.

The measurements were carried out with a single OOPS module. Data were acquired at quasi-free kinematics at $Q^2 = 3.3$ fm^{-2}. Results [16] for ^{12}C, dominated by counting statistics, are compared with theoretical calculations [17] in Fig. 6. They are in good agreement with the model calculations. When the OOPS system is fully operational, we expect to continue these measurements with much improved accuracy.

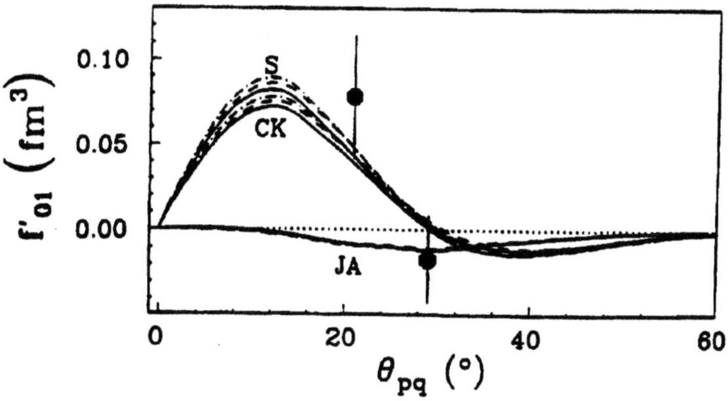

Figure 6. The fifth structure function is compared to theoretical predictions in the impulse approximation for three optical potentials.

4 Proton Recoil Polarimetry

Recently, the first series of coincidence $(e, e'\vec{p})$ reactions with hydrogen and deuterium targets were carried out at Bates. The proton focal plane polarimeter (FPP) was constructed by a UVa, W&M and MIT collaboration [18] and is installed in the OHIPS spectrometer. Physics issues which will be addressed include the quadrupole amplitudes in the $N \rightarrow \Delta$ transition via the $^1\mathrm{H}(\vec{e}, e'\vec{p})\pi^0$ reaction, the electromagnetic structure of the deuteron and sensitive tests of nucleon-nucleus dynamics in complex nuclei.

The initial measurements were of the spin transfer coefficients in elastic scattering of polarized electrons from hydrogen and in quasielastic scattering from the deuteron at $q^2 = 0.38$ and $0.5\ \mathrm{GeV}^2$. Preliminary analysis indicates that, in quasifree kinematics, the spin transfer coefficients on deuterium are consistent with those on hydrogen at the level of a few percent. All three components of the proton polarization were measured in these experiments and appear to be consistent with modern two-body theories. Subsequent measurements will explore the deuteron away from quasifree kinematics where the influence of other amplitudes may be enhanced.

The polarimeter was also used in some initial $N \rightarrow \Delta$ studies. A measurement was made of $\mathrm{R}^{\mathrm{n}}_{\mathrm{LT}}$ with unpolarized beams at $Q^2 = 0.127\mathrm{GeV}^2$ to complement the measurements of $\mathrm{R_T}$ and $\mathrm{R_{LT}}$ at the same Q^2. These data are now being analyzed.

5 Electroweak Physics

Experiments designed to probe the structure of the neutral currents continue to be of great interest to atomic, nuclear and particle physicists. All of the data are in impressive agreement with the Standard Model [19] of Glashow, Weinberg, and Salam. The best value for the weak mixing angle, a free parameter in the theory, is $\sin^2\theta_{\mathrm{w}} = 0.233 \pm 0.002$ [20] as determined from measurements of the Z-boson mass.

The neutral currents are known to be parity violating and this aspect of their structure may be studied using polarized electrons. The parity violating asymmetry may be defined as

$$\mathrm{A} = (\sigma_{\mathrm{R}} - \sigma_{\mathrm{L}})/(\sigma_{\mathrm{R}} + \sigma_{\mathrm{L}}),$$

where $\sigma_{\mathrm{R}}(\sigma_{\mathrm{L}})$ is the differential cross section for the scattering of electrons with right (left) helicity. Parity violating asymmetries were first observed in the scattering of polarized electrons from deuterons at SLAC [21] and in the spectra of heavy atoms [22] at low energies.

The first Bates parity experiment involved elastic scattering from $^{12}\mathrm{C}$. In the case of a spinless and isoscalar nucleus the electromagnetic scattering is described by a single form factor and the asymmetry, A, may be written at the tree-level [23] as

$$\mathrm{A} = \bar{\gamma}\frac{3}{2}\mathrm{G_F}\mathrm{Q}^2(\sqrt{2}\ \pi\ \alpha)^{-1}$$

where $\mathrm{G_F}$ is the Fermi coupling constant, α is the fine structure constant, Q is the momentum transfer, and $\bar{\gamma}$ is the parity violating constant for an axial vector coupling to the hadronic matter. The asymmetry is completely independent of the electromagnetic form factor. It is a direct consequence of our assumption that the weak and electromagnetic currents couple to exactly the same operators. The cancellation is Q^2 independent and a precision test could be sensitive to possible extensions of the SM.

The ^{12}C experiment was carried out at $Q = 150$ MeV/c. A value of $\bar{\gamma} = 0.136 \pm 0.032 \pm 0.009$ was measured [24] which is consistent with the prediction of the SM where, $\bar{\gamma}_{SM} = 0.155 \pm 0.002$.

Recently, there has been a lot of experimental and theoretical interest in strange quark contributions to hadronic matrix elements. New experiments using parity violating electron scattering as a neutral current probe of hadronic structure are underway at MIT-Bates and are also planned for MAINZ and CEBAF. They are designed to measure the neutral weak form factors G_E^z, G_M^z and G_A^z of the proton to provide an unambiguous signature for the presence of strange quarks in the nucleon.

The parity violating asymmetry for elastic scattering from a proton target can be written as a sum of three terms which reflect the interference between the electromagnetic and the neutral weak interactions:

$$A = \frac{G_F Q^2}{\pi\alpha\sqrt{2}} \frac{[\epsilon G_E^p G_E^z + \tau G_M^p G_E^z - \frac{1}{2}(1 - 4\sin^2\theta_w)\, \epsilon' G_M^p G_A^z]}{\epsilon (G_E^p)^2 + \tau (G_M^p)^2}$$

where $\tau = Q^2/4m_p^2$, $\epsilon = [1 + 2(1+\tau)\tan^2\theta/2]^{-1}$ and $\epsilon' = \sqrt{(1-\epsilon^2)\tau(1+\tau)}$.

The first two terms dominate at forward angles and the latter two terms contribute at backward angles. The term including the axial vector form factor, G_A^z is suppressed by the factor $(1 - 4\sin^2\theta_w)$.

The form factors $G_{E,M}^z$ contain contributions from strange quarks, $G_{E,M}^s$. The SAMPLE [25] experiment at MIT-Bates is designed to measure G_M^z and extract the contribution of strange quarks to the static anomalous magnetic moment of the proton.

SAMPLE will measure backward angle scattering ($\bar{\theta} \sim 150°$) at E=200 MeV and $Q^2 = 0.1$ GeV^2. A large solid angle (2 sr) air Čerenkov detector (10 mirrors) will be used to detect the scattered electrons from a 40 cm long liquid hydrogen target and 40μA of polarized beam ($P_e = 40\%$).

The predicted asymmetry is $\sim 8 \times 10^{-6}$. The contribution of the axial vector term to the asymmetry is $\sim 20\%$ and the weak radiative corrections to this term are about $20 - 30\%$ with large uncertainty. The goal is to measure the asymmetry to a statistical accuracy of 5%. We expect to achieve an overall error in G_M^s of $\Delta G_M^s = 0.22$. Theoretical predictions fall in the range $-1 < G_M^s(0) < 0$.

It would also be useful to have a direct measurement of the axial vector term. In quasi-elastic electron- deuteron scattering, at the same kinematics, the strange quark contributions to the proton and neutron add incoherently and approximately cancel in the asymmetry. However, the axial vector term and the uncertainty in the corresponding hadronic radiative correction contribute the same fraction to the asymmetry. Combining the two measurements puts constraints on the radiative corrections and improves the determination of the strange magnetic form factor. It is planned to do the deuterium measurements following those on the proton.

6 Internal Target Physics

The Bates South Hall Ring was designed and constructed to provide high duty factor extracted beams and an internal target capability for nuclear research. In the internal target mode, beam energies of (0.3 – 1.0) GeV and circulating currents of 80 mA are available with very long storage times. Windowless storage cell targets of polarized ^{1}H, ^{2}H and ^{3}He can be used with densities of $10^{14} - 10^{16}$ atoms/cm^2. Luminosities up to 10^{33}cm^{-2}s^{-1} will be possible.

The proposed physics program will emphasize spin- dependent electron scattering. A major component will be neutron electromagnetic form factor measurements on ^{2}H and ^{3}He in both inclusive and exclusive channels over the range $Q^2 = 0.1 - 0.8$ GeV^2. Other measurements are designed to measure small amplitudes in the structure of ^{3}He and to study the $N \rightarrow \Delta$ transition. These experiments would complement efforts with extracted beams.

The Bates Large Acceptance Spectrometer Toroid (BLAST) project is an initiative to develop and construct a large acceptance detector optimized for internal target physics with the SHR. BLAST is a non-focussing magnetic spectrometer with eight copper coils arranged in a toroidal configuration. Two opposing sectors would be instrumented with wire chambers, scintillation counters, and Čerenkov counters for tracking, time-of- flight, and particle identification.

At NIKHEF, an internal target experiment has recently been completed involving electron scattering from a tensor polarized deuterium target. The target was a long storage cell fed by an atomic beam source. The beam energy was 565 MeV and the stored current was 60 mA for an effective luminosity of $\approx 10^{31}$cm^{-2}s^{-1}. Scattered electrons were detected in a Cerenkov counter and recoil nucleons in a large hadron array. An accurate measurement was made of t_{20} at $Q = 1.7$fm^{-1}. The tensor asymmetry for the $^2\vec{\mathrm{H}}(e, e'p)$ reaction was also measured for target polarizations perpendicular and parallel to $\vec{q}$. These initial results show that the internal target technique has a lot of potential for some very interesting physics.

7 Summary

The recent experiments at MIT-Bates and MAINZ, exploiting the use of polarized electrons and the measurement of spin observables, point to the importance of such studies in understanding nucleon and nuclear structure. In many instances we find both increased sensitivity and new observables which allow the accurate determination of small amplitudes. A major component of the proposed future program at MIT-Bates, NIKHEF, MAINZ and CEBAF, will involve the use of polarized beams, polarized targets and recoil polarimetry.

Modern electron accelerators, operating at 100% duty factor, will provide the ultimate capability for coincidence experiments. The stretcher/storage ring facilities at MIT- Bates and NIKHEF were designed to operate with internal targets. Internal target experiments with polarized beams and polarized targets at high luminosities represents an important new capability for nuclear research.

The next decade promises to be a very exciting one for electronuclear physics. For the first time the full power of the electron probe can be applied in the broadest sense to study nucleons and nuclei.

I would like to acknowledge the help of my colleagues at MIT-Bates and those at other laboratories who provided me with data and information about the research reported here. This work was supported in part by the U.S. Department of Energy under cooperative agreement NO. DE-FC02-94ER40818.

References

1. S. Platchkov et al.: Nucl. Phys. **A510**, 740 (1990)
2. T.W. Donnelly, A. Raskin: Ann. Phys. (NY) **169**, 247 (1986)
3. R. Arnold et al.: Phys. Rev. **C23**, 363 (1981)
4. T. Eden et al.: Phys. Rev. **C50**, R1 (1994)
5. B. Blankleider and R.M. Woloshyn: Phys. Rev. **C29**, 538 (1984)
6. J.L. Friar et al.: Phys. Rev. **C43**, 2310 (1990)
7. C. Ciofi degli Atti, E. Pace, G. Salmé: Phys. Rev. **C46**, R1591 (1992)
8. R. Schulze, P. Sauer: Phys. Rev. **C48**, 38 (1993)
9. C.E. Jones et al.: Phys. Rev. **C47**, 110 (1993)
10. A.K. Thompson et al.: Phys. Rev. Lett. **68**, 2910 (1992)
11. H. Gao et al.: Phys. Rev. **C50**, R546 (1994)
12. S. Boffi, C. Giusti, F.D. Pacati: Phys. Rep. **226**, 1 (1993) ; S. Boffi, C. Giusti, and F.D. Pacati: Nucl. Phys. **A435**, 697 (1985)
13. C.N. Papanicolas: Bates Proposal #87-09; S. Dolfini et al.: Bull. Am. Phys. Soc. **33**, 1578 (1988)
14. A. Sandorfi: BNL - 60923 (1994)
15. B. Schoch: Bull. Am. Phys. Soc. **39**, 1033 (1994)
16. J. Mandeville et al.: Phys. Rev. Lett. **72**, 3325 (1994)
17. J.R. Comfort, B.C. Karp, Phys. Rev. **C21**, 2162 (1980) ; D.F. Jackson, I. Abdul-Jalil, J. Phys.**G6**, 481 (1980); M.M. Giannini, G. Ricco, Ann. Phys. (NY) **102**, 458 (1976)
18. R.W. Lourie, V. Burkert: Bates Proposal #89- 03 (1989)

19. S.L. Glashow: Nucl. Phys. **22**, 579 (1961); S. Weinberg: Phys. Rev. Lett. **19**, 1264 (1967); Phys. Rev. Lett. **27**, 1688 (1971); A. Salam: Rev. Mod. Phys. **52**, 525 (1980)

20. S. Fanchiotti, A. Sirlin: Phys. Rev. **D35**, 785 (1987)

21. C.Y. Prescott et al.: Phys. Lett. **84B**, 524 (1979)

22. M.C. Noecker et al.: Phys. Rev. Lett. **61**, 319 (1988)

23. G. Feinberg: Phys. Rev. **D12**, 3575 (1975); J.D. Walecka: Nucl. Phys. **A285**, 345 (1977)

24. P.A. Souder et al.: Phys. Rev. Lett. **65**, 694 (1990)

25. R.D. McKeown, D. Beck et al.: Bates Proposal #89-06

26. J.-O. Hansen et al.: Preprint LNS-94-80, submitted for publication.

27. M. Meyerhoff et al.: Phys. Lett. **B327**, 201 (1994)

28. A. Lung et al.: Phys. Rev. Lett. **70**, 718 (1993)

29. S. Galster et al.: Nucl. Phys. **B32**, 221 (1971)

Few-Body Systems Suppl. 8, 121–132 (1995)

Few-Body Systems

Pion Nucleus Scattering

G.R. Smith * †

TRIUMF, 4004 Wesbrook Mall, Vancouver, B.C. V6T 2A3 Canada

Abstract.

Recent results from pion–nucleus scattering will be presented, with emphasis on light systems. In particular, recent studies of $\pi^{\pm}$p differential cross sections, partial total cross sections, and analyzing powers will be highlighted and compared to previous data where they exist. The new data make a strong case for a re–evaluation of many long–accepted results. For example, they point to a lower value of the πNN coupling constant (which to a great extent removes the Goldberger–Triemann discrepancy), a change in the mass and width of the P_{33} πN partial wave (the Δ^{++}), isospin breaking and charge splitting, different values for the πN scattering lengths, and a slightly higher value for the $\pi N \Sigma$ term, which is an explicit measure of chiral symmetry breaking and can be related to the strange sea quark content of the proton.

Other recent work in the πN system will be summarized briefly, including studies of $H(\pi^{\pm}, 2\pi)$ at TRIUMF. Besides providing more precise values of the $\pi\pi$ scattering lengths and chiral symmetry breaking parameter within the context of soft–pion theory, the new $(\pi, 2\pi)$ data are being analyzed within a Chew–Low framework in order to compare $\pi\pi$ scattering lengths, amplitudes, and phase shifts more directly to the predictions of chiral perturbation theory.

1 Introduction

Despite the fact that it is now some twenty-odd years since intense beams of pions have been available from the three meson factories (TRIUMF, LAMPF, and PSI), major studies of pion–proton interactions have taken place in the last few years. The renewed interest in this field stems in large part from the refinement and new predictive capabilities [1] of chiral perturbation theory (ChPT) [2]. ChPT provides an attractive and powerful link between quarks and intermediate energy nuclear physics experiments.

QCD describes systems of massless quarks which exhibit chiral symmetry: the interactions are the same for left- and right-handed quarks. The successes

* *E-mail address:* smith@triumf.ca

† *WWW page:* `http://www.triumf.ca/chaos/chaos.html`

of perturbative QCD rely on an expansion in terms of the strong coupling constant, which is small at high energies. For $q^2 << \Lambda_{QCD}$, however, the coupling constant is large and the series diverges. ChPT essentially overcomes this problem by rewriting the QCD Hamiltonian in terms of a piece involving massless quarks, and a perturbative piece involving the quark masses, which are small enough that convergence is achieved even at small q^2. In such a prescription, the QCD Lagrangian is replaced by a mathematically equivalent effective Lagrangian. Chiral symmetry provides tight constraints and accurate predictions for many soft processes [3].

As a result of the substantial refinements made in recent years to the original versions of ChPT, πN physics programs have become major thrusts at several laboratories. At TRIUMF, extensive measurements of exclusive H($\pi^{\pm}$, 2π) reactions have just been performed. The resulting many-fold differential cross sections will be compared to the models of Johnson, Oset [5], and Bolokhov [4]. In particular from Bolokhov's model, values for the $\pi\pi$ scattering lengths, scattering amplitudes, phase shifts, and cross sections may be obtained and compared to the predictions [3] of ChPT. In parallel, a Chew-Low analysis [6] will be performed to obtain these same observables in a model-independent treatment [7].

In the $\pi\pm$p elastic channel, several new experiments [8, 9, 10, 11, 12, 13, 14, 15] have been performed to measure the differential cross section ($d\sigma/d\Omega$). Unfortunately, discrepancies as large as 20% can still be found between some of the new experiments, most of which purport to have ~2% uncertainties. Agreement with older experiments [16, 17] which formed the initial database and hence defined the canonical partial wave analyses is also in general poor at energies below the resonance. As a result, attention has turned to measurements of the $\pi^{\pm}$p analyzing power (A_y), which can be used as a filter to weed out inconsistent $d\sigma/d\Omega$ results using single energy partial wave analysis (PWA). The A_y and surviving $d\sigma/d\Omega$ results will then form reliable input to a PWA from which πN scattering lengths and the πN coupling constant can be more accurately determined. The observables are parameters required by ChPT for the determination [18] of the $\pi N \Sigma$ term, which constitutes an explicit measure of chiral symmetry breaking. The Σ term, in turn, can be used [18] to measure the strange sea quark content of the proton.

Other important elements of this program include the measurement of $\pi^{\pm}$p scattering in the Coulomb-nuclear interference region, from which the real part of the isospin even pion-nucleon scattering amplitude and thus the Σ term can be determined directly. Such data, as well as measurements of the $d\sigma/d\Omega$ and A_y for π^-p single charge exchange can also be used to measure the degree of charge splitting and isospin breaking in πN scattering.

2 $\pi^{\pm}$p Elastic Cross Section Experiments

A simple but reliable experimental arrangement has been employed at TRIUMF for several precise $d\sigma/d\Omega$ experiments [9, 10, 15] in the elastic πN channel.

The setup consists of an array of scintillation counter telescopes for detection of scattered pions, and associated recoil proton scintillation counters. The time-of-flight (TOF) of coincident pions and protons is used to obtain a clean signature of πp scattering. The incident beam is counted directly by in-beam scintillation counters. Early results [9] spanned the energy interval from 67 to 140 MeV, and considerable discrepancies were observed between the measured cross sections and those obtained in much earlier, pre-meson factory experiments [16, 17], as well as with the PWAs based on those earlier data. The TRIUMF TOF results agree well, however, with the results of Frank et al. [8] obtained at LAMPF using a technique employing multi-wire proportional chambers (MWPC) and scintillation counter range telescopes for the detection of pions scattered from a LH_2 target.

The TRIUMF TOF experimental setup has been used most recently to acquire data from 140 to 267 MeV for both pion polarities [15]. Data were acquired with both solid (CH_2) and liquid hydrogen targets, using coincidence as well as singles triggers. A precision of typically 1.5% was achieved. At bombarding energies of 190 MeV and above, reasonable agreement with the older results of Bussey et al. [16] were observed. However, at lower energies the new results are systematically lower than the Bussey et al. results. For example the new data are about 10% lower than the Bussey et al. data at 140 MeV for both π^+ and π^-, well outside the uncertainties of each experiment. The new data are found to be consistent, however, with the earlier TRIUMF results [9] using essentially the same apparatus at 140 MeV, as well as with LAMPF data at 267 MeV [19].

A variation on this experimental technique has been applied at TRIUMF in order to extend the measurements down to 30 MeV [11] as well as to acquire forward angle data between 87 and 140 MeV [14]. Scattered pion scintillation counter telescopes were used as before, only the recoil proton was detected in an active scintillation counter target. The large pulse height deposited by protons stopping in the target provided the signature of πp scattering. The angle of the planar target is adjusted such that the proton trajectories lie within the target plane. The results from this technique agree well with those acquired with our earlier recoil technique [9, 15] described above, as well as with the LAMPF data of Frank, et al. [8]. In general the data are overpredicted by the canonical PWA solutions KH80 [20], KA84 and KA85 [21].

Beyond the measurements of Frank, et al. [8] (30 to 90 MeV) and Sadler, et al. [19] (267 to 500 MeV), no new elastic measurements have been performed at LAMPF. However, measurements of SCX $d\sigma/d\Omega$ have recently been carried out and extensive measurements of A_y in the SCX channel are planned for the summer of 1995.

At PSI a magnetic spectrometer has been employed [12] to measure $\pi^{\pm}$p scattering from a CH_2 target for bombarding energies near 33, 45, and 69 MeV. The results were normalized to μC scattering at forward angles (for angles $> 65°$, the spectrometer solid angle was assumed to be the same as that determined at forward angles). Although at 33 MeVand to a lesser extent 45 MeV the PSI results are in reasonable agreement with those from TRI-

UMF [11] and LAMPF [8], and are considerably overpredicted by, for example, the KA85 PWA solution (up to 30% discrepancies), their 69 MeV results disagree violently with the TRIUMF results in shape and magnitude. In fact, the 69 MeV PSI results are reasonably well described by the PWA solution KA85. The TRIUMF results are overpredicted by the same solution by more than 20% in the backward hemisphere. It is worth pointing out the the TRIUMF 67 MeV results consist of measurements employing three independent techniques [9, 10, 11] and were acquired on two separate pion channels. As a result the degree of confidence in the TRIUMF results is quite high. At present this disturbing experimental discrepancy between TRIUMF and PSI remains unexplained. However, the general conclusions drawn from the PSI data are consistent with those found by both the TRIUMF and LAMPF experiments, namely, that the canonical PWA solutions [20, 21] seriously overpredict the measured cross sections, especially at the lowest energies, where the problem is most acute for π^+p scattering. The PSI group performed [12] a PWA constrained to describe their data and found that the S_{31} and S_{11} phase shifts are about one degree smaller than predicted by KA85, which is a lot considering that at these energies these phase shifts have a magnitude near 5°.

The same Karlsruhe/Tuebingen/PSI collaboration has employed their magnetic spectrometer with a LH_2 target to make pioneering measurements of $\pi^\pm$p scattering into the Coulomb-nuclear interference region at 32 and 45 MeV [13]. Their results extend to c.m. scattering angles as small as 12°. An earlier experiment [22] by that collaboration used a range technique to obtain results at 55 MeV. They find the same overprediction of their small angle results by, e.g. KH80, which is characteristic at larger angles. In order to use these data to extract the $\pi N \Sigma$ term, an extrapolation must be made to t=0. More than three bombarding energies are required in order to perform such an extrapolation reliably, however, the results of this experiment do support a large (≥ 60 MeV) value for the Σ term, which in turn implies an $s\bar{s}$ content in the proton of ~10%.

Measurements of π^+p partial total cross sections have been performed recently at TRIUMF [23] and LAMPF [24] (π^-p has also been explored). These experiments measure the integral cross section for pion scattering into angles greater than a given cutoff angle, typically 20° or 30°, and thereby avoid model dependent extrapolations to 0° characteristic of true total cross section experiments [25, 26]. No total cross section experiments have been performed since 1978. Although results from the two laboratories initially seemed to be in disagreement, recent corrections [27] to the TRIUMF results (in particular the beam momentum calibration) have brought the two into good agreement. Both results are overpredicted by PWA solutions like KH80. Both results are consistent with the integral of the TRIUMF $d\sigma/d\Omega$ results [11].

3 $\pi^{\pm}$p Elastic Analyses

One of the most fundamental and important numbers to come out of analyses of πN scattering data is the πN coupling constant, $g^2/4\pi$. This number has been the subject of considerable controversy since the Nijmegen group published values in 1987 [28] and 1990 [29] from a nucleon-nucleon (NN) PWA considerably lower than the long-accepted value of 14.3 derived [20] in 1980 from a πN PWA. The VPI PWA group also obtained a low value in 1990 [30] from an analysis of a greatly expanded πN database relative to what was available in 1980. The VPI group has worked on improving their techniques for extracting $g^2/4\pi$ by including fixed-t dispersion relations and other dispersion constraints and has just come up with [31] their most reliable value, which is still some 5% lower than the 1980 value of 14.3. Even some of the loudest critics of the lower $g^2/4\pi$ values are now obtaining [32] similar results, and although it is fair to say that most workers are settling on a new value for $g^2/4\pi$ near 13.7, there is still some argument in the community over this issue.

It is interesting to note [33] that the lower values of $g^2/4\pi$ significantly reduce the Goldberger-Triemann discrepancy [34], $\Delta_{\pi N} = 1 - (g_A(q^2=0)m_\pi)/(2f_\pi f_{\pi NN})$, where g_A is the nucleon axial-vector form factor, which is well known [35] from neutron beta decay (1.261±0.004), f_π is the pion decay constant determined [36] from $\pi \longrightarrow \mu\nu$ decay (92.6±0.2 MeV), and $f_{\pi NN}$ the πNN coupling constant. In the chiral limit $\Delta_{\pi N} = 0$, chiral perturbation theory predicts a 1% discrepancy. With $g^2/4\pi$=14.3, $f^2_{\pi NN}/4\pi$=0.0790 and $\Delta_{\pi N} = 0.049$. With $g^2/4\pi$=13.49, $f^2_{\pi NN}/4\pi$=0.0745 and $\Delta_{\pi N} = 0.019$.

The canonical PWA solutions are highly constrained by only a few older (pre-meson factory) experiments, in particular Bussey et al. [16] ($d\sigma/d\Omega$) and Carter et al. [25] (σ_{tot}), which purport to have zero systematic uncertainties. Since these measurements are observed to be at variance with many of the newer experiments [8, 9, 11, 14, 15, 26, 24, 23], a new PWA was performed [31] recently at VPI which 'de-emphasized' the Bussey [16] and Carter [25] experiments. The solution which resulted (SP95) from this effort provides a much improved description of the meson factory results. Most partial waves are found [15] to be slightly smaller than in the past. Evidence for charge splitting was observed [15] (different P_{33} phases were used for π^+p and π^-p), and the Δ^{++} was found [15] to be narrower and higher in mass than previous analyses had determined. In addition, the πNN coupling constant was determined [31] to be 13.7±0.1.

Isospin breaking in low energy πN scattering has been studied [37] recently by using the triangle identity $f_{CEX} = 1/\sqrt{2}(f_+ - f_-)$, which relates the charge exchange amplitude to those for π^+p and π^-p elastic scattering. A coupled channels potential model was used to fit recent charge exchange and elastic differential cross sections at low energies. A 7% discrepancy was observed in the S-wave beyond the effects of hadronic mass differences and Coulomb corrections.

The impact of recently published low energy ($T_\pi < 100$ MeV) $\pi^{\pm}$p $d\sigma/d\Omega$ measurements has also been explored [38] recently by means of an extended threshold K-matrix expansion. A good description of the data published up to early 1995 was obtained. The S_{31} and P_{33} phase shifts obtained are about 1° smaller than those given by KH80, and the isospin 3/2 S-wave scattering length a_3 was found to be 30% smaller than KH80.

4 $\pi^{\pm}$p Analyzing Powers

Although at resonance energies there exist precise measurements [39] of the analyzing power for πN scattering, there is great interest in exploring this observable at low energies ($T_\pi < 100$ MeV). The analyzing power is a sensitive tool which can be used to pin down the smaller partial waves, especially the S-waves, at low energies as a consequence of the strong S-P interference which is most pronounced near bombarding energies of 50 MeV. PWA can be used to find those $d\sigma/d\Omega$ measurements which are consistent with the measured A_y, effectively providing a filter which can be used to choose from among the many conflicting $d\sigma/d\Omega$ measurements which are present in the πN database. As an example, the only low energy A_y data reported to date [40] (four angles measured at 67 MeV by a Karlsruhe/PSI collaboration) have been compared to single energy PWA solutions constrained to describe either the new PSI $d\sigma/d\Omega$ data [12] at that energy, or the TRIUMF $d\sigma/d\Omega$ data [11], or the Bertin et al. $d\sigma/d\Omega$ data [17]. The results of such a comparison clearly indicate a preference for the TRIUMF $d\sigma/d\Omega$ data [11].

Together with the surviving cross sections, low energy A_y data will make a more precise determination of low energy amplitudes possible, which will improve considerably the accuracy with which the $\pi N \Sigma$ term can be determined, and consequently the $s\bar{s}$ content of the proton. The sensitivity of the πN observables near 50 MeV to the πN scattering lengths and thus the Σ term has been explored by Locher and Sainio [41]. Although only a minor impact is predicted for $d\sigma/d\Omega$, a fantastic sensitivity is predicted for the analyzing power for backward angle π^-p scattering, where the S-P interference dominates.

In order to both filter out the incorrect low energy $d\sigma/d\Omega$ data and to provide the A_y measurements needed in the S-P interference region where the sensitivity to the $\pi N \Sigma$ term is greatest, a unique detector ensemble [42] has recently been commissioned at TRIUMF which addresses the considerable experimental challenges posed by such measurements. The detector, dubbed CHAOS, is essentially a 360° magnetic spectrometer capable of simultaneously measuring multiple charged particles emitted at all scattering angles within ±7° of the horizontal plane. The device achieves a momentum resolution of ~1% and covers about 10% of 4π sr. A sophisticated, multi-tiered trigger system permits beam intensities of up to 5 MHz to be employed. A polarized target has been built for this device and A_y measurements are currently in progress. The target is situated in the center of a large cylindrical dipole magnet and is surrounded by four concentric cylindrical wire chambers which track charged particles (entering and) emanating from the target. Particle identification and track multiplicity are provided by a ring of scintillation counters and lead-glass Cerenkov counters which surround the outermost wire chamber. Approximately

5000 channels of information are delivered by the detector. The main detector elements are depicted in Fig. 1. There are many advantages of this experimental approach. The fact that all angles are acquired simultaneously improves the reliability of the shape of the angular distribution, and makes such an experiment feasible since otherwise the time spent acquiring data and polarizing the target several times at each of many angles would be prohibitive. Also, since data are acquired simultaneously for $-180 < \theta < 0$ as well as for $0 < \theta < 180$, left-right asymmetries may be used to determine the analyzing power as well as the more usual technique of reversing the target polarization.

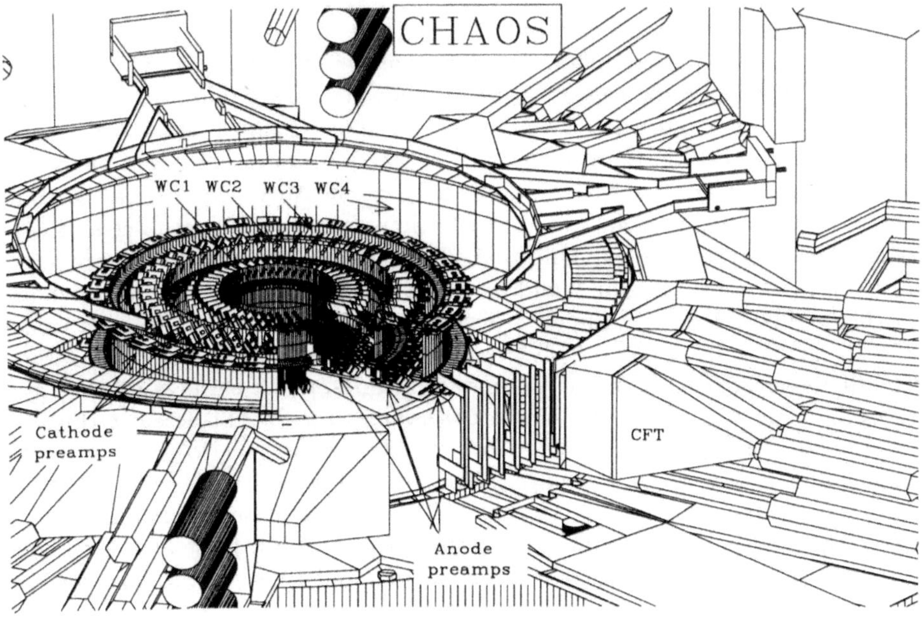

Figure 1. Details of the central region of the CHAOS detector are presented. The detector array is actually completely cylindrical– in this figure parts of each detector component are cut away for clarity. The individual components are discussed in the text.

5 $H(\pi^{\pm}, 2\pi)$

There have been several recent experiments devoted to studies of $H(\pi^{\pm}, 2\pi)$ which can be used to test predictions of ChPT. Sevior, et al. have reported [43] measurements of $H(\pi^{+}, \pi^{+}\pi^{+})n$ very near threshold which were utilized to obtain a value for the isospin 2 S-wave $\pi\pi$ scattering length $a_2^0 = -0.040 \pm 0.001 m_\pi^{-1}$ obtained within the framework of an effective Lagrangian model [44], compared to the ChPT prediction [3] of $a_2^0 = -0.042 \pm 0.002\, m_\pi^{-1}$.

At LAMPF, exclusive measurements of $H(\pi^+, \pi^0\pi^+)p$ have recently been acquired [45] from 190 MeV to 260 MeV. These workers obtain $a_2^0 = -0.041 \pm 0.003\, m_\pi^{-1}$, again within the framework of soft pion theory. Total cross sections and threshold matrix elements were also obtained. Differential cross sections have not yet been reported- the limited statistics of this experiment would make this difficult.

In order to acquire the statistics necessary for extracting many-fold differential cross sections and as a result subject $(\pi, 2\pi)$ data to a Chew-Low analysis, described below, the CHAOS detector at TRIUMF has been employed to make extensive measurements of exclusive $(\pi, 2\pi)$ reactions from 220 MeV to 300 MeV. For incident π^-, the $H(\pi^-, \pi^+\pi^-)n$ and the $H(\pi^-, \pi^-\pi^0)p$ channels were measured simultaneously by detecting the two charged particles in the final state in coincidence. Likewise for incident π^+, the $H(\pi^+, \pi^+\pi^+)n$ and the $H(\pi^+, \pi^+\pi^0)p$ channels were measured simultaneously. Approximately 10,000 $(\pi, 2\pi)$ events were recorded at each energy for the channels with two charged pions in the final state. By measuring both reaction channels for both incoming pion polarities from near threshold up to $T_\pi = 300$ MeV over a broad angular range, it is possible to isolate the contributions from the various diagrams contributing to this process [4]. One goal of these measurements is to improve the values for the S-wave isospin 0 and 2 $\pi\pi$ scattering lengths a_0^0 and a_2^0, which are predicted [3] within the framework of chiral perturbation theory to have values of 0.20 ± 0.01 and -0.042 ± 0.002, respectively. In addition, these data should permit the extraction of $\pi\pi$ scattering amplitudes, phase shifts, and cross sections as a function of energy. These observables can be extracted within the framework of the model developed by Bolokhov et al. [4]. However, with the large number of events now acquired with CHAOS, a model independent Chew-Low analysis [6] can also be performed. The $\pi\pi$ interaction observables can be determined [7] in this technique by extrapolating to $t = m_\pi^2$ the measured doubly differential cross section $\partial^2\sigma/\partial s \partial t$ where s and t denote the $\pi\pi$ invariant mass and four-momentum transfer to the nucleon, respectively.

A sample missing mass spectrum acquired with CHAOS for the 280 MeV $\pi^+ p \longrightarrow \pi^+\pi^- n$ reaction is shown in Fig. 2. The $\pi^+\pi^-$ events are almost free of background, and are centered at the neutron mass with a width of only 6 MeV. The analysis of these data is expected to be completed early in 1996.

To illustrate the capabilities of the CHAOS detector in the context of $(\pi, 2\pi)$ measurements, a typical event recorded by the spectrometer for the $\pi^+ n \longrightarrow \pi^+\pi^- p$ reaction is shown in Fig. 3. The incident beam was 400 MeV/c π^+ and a liquid deuterium target was in place at the center of the spectrometer, which was set to a field of 0.5T. The figure shows the incoming beam trajectory, which is reconstructed using the hits in WC1 and WC2 in conjunction with the known incident beam momentum. The three reaction products are also reconstructed and labeled in the drawing with a track number (1,2,3). The momentum reconstruction uses the hits in all four wire chambers for each track, as well as the result of particle identification provided by the CFT blocks. The missing energy reconstructed for this event is less than 5 MeV. The interaction vertex and scattering angle of each outgoing track are obtained by extrapolating

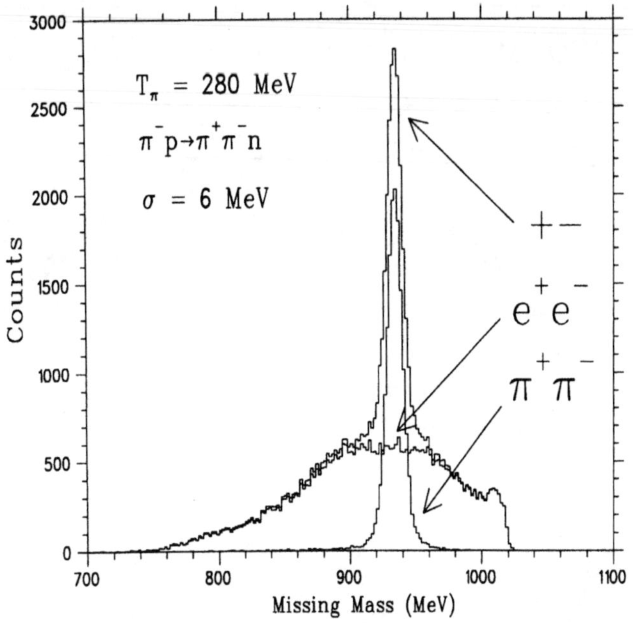

Figure 2. A missing mass spectrum of $(\pi, 2\pi)$ events acquired with CHAOS is shown for three particle identification conditions: opposite polarity tracks, tracks identified as e^+e^-, and tracks identified as the $\pi^+\pi^-$ events of interest.

each trajectory inside the WC1 radius and intersecting it with the incident beam trajectory.

6 Conclusions

Many new measurements of πN scattering observables have appeared in the last few years and several more crucial experiments are planned for the near future. The main motivation for these new efforts is to measure the parameters and test the predictions of chiral perturbation theory. The new results imply smaller πN elastic S & P waves, a smaller πN coupling constant, charge splitting (in the P_{33} wave), a narrower and higher mass Δ^{++}, isospin breaking, and a large value for the πN Σ term with a corresponding strange sea quark content in the proton of ~10%. In the pion production channel, the large amount of data now available should permit a model independent (Chew-Low) analysis which will provide $\pi\pi$ scattering lengths, as well as phase shifts, scattering amplitudes and cross sections as a function of energy.

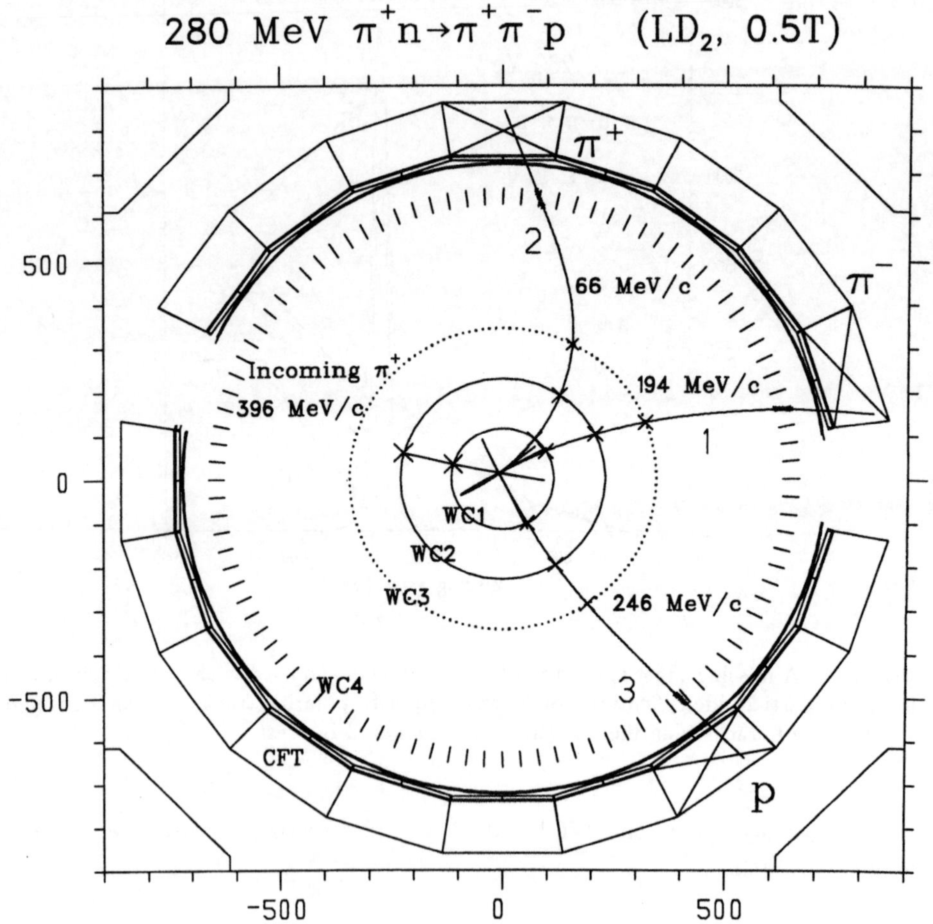

Figure 3. A typical event recorded by the spectrometer is shown. The axes are in mm. Struck anodes are indicated by a cross, as are the CFT blocks which contained signals for this event. The innermost two circles correspond to the 720 anodes per chamber of WC1 and WC2. The dotted circle represents the 144 anodes of WC3. The 100 radial line segments near an average radius of 640 mm correspond to the 100 cells of WC4, each of which has 8 drift anodes. Outside WC4 are the 18 CFT blocks, each 18° wide and comprised of a 3 mm scintillator (ΔE_1), followed by a 13 mm scintillator (ΔE_2), followed by a 120 mm lead glass Čerenkov detector. One CFT block is removed on the left of the figure to let the incident beam into the spectrometer, and another is removed on the right to permit the beam to exit. The drift chambers (WC3 and WC4) are deadened in these regions as well. Not shown in the figure are the 360 cathode strips per chamber in WC1 and WC2, 576 cathode strips in WC3, 200 resistive wires in WC4, and in-beam scintillation counters at the entrance and exit of the spectrometer.

References

1. J. Gasser, H. Leutwyler: Ann. Phys. **158**, 142 (1984); Nucl. Phys. **B250**, 465 (1985)
2. S. Weinberg: Phys. Rev. Lett. **17**, 616 (1966)
3. J. Gasser, H. Leutwyler: Phys. Lett. **125B**, 321 (1983); **125B**, 325 (1983)
4. A.A. Bolokhov, V.V. Vereshchagin, S.G. Sherman: Nucl. Phys. **A530**, 660 (1991)
5. E. Oset, M.J. Vicente-Vacas: Nucl. Phys. **A454**, 637 (1986)
6. G.F. Chew, F.E. Low: Phys. Rev. **113**, 1640 (1959)
7. J.P. Baton, G. Laurens, J. Reignier: Phys. Lett. **33B**, 525 (1970)
8. J.S. Frank et al.: Phys. Rev. **D28** 1569 (1983)
9. J.T. Brack et al.: Phys. Rev. **C34**, 1771 (1986)
10. J.T. Brack et al.: Phys. Rev. **C38**, 2427 (1988)
11. J.T. Brack et al.: Phys. Rev. **C41**, 2202 (1990)
12. Ch. Joram et al.: Phys. Rev. **C51**, 2159 (1995)
13. Ch. Joram et al.: Phys. Rev. **C51**, 2144 (1995)
14. J.T. Brack et al.: Phys. Rev. **C51**, 929 (1995)
15. M. Pavan: PhD Thesis. Univ. of British Columbia 1995 (unpublished)
16. P.J. Bussey et al.: Nucl. Phys. **B58**, 363 (1973)
17. P.Y. Bertin et al.: Nucl. Phys. **B106**, 341 (1976)
18. J. Gasser et al.: Phys. Lett. **213B**, 85 (1988)
19. M.E. Sadler et al.: Phys. Rev. **D35**, 2718 (1987)
20. R. Koch, E. Pietarinen: Nucl. Phys. **A336**, 331 (1980)
21. R. Koch: Z. Phys. **C29**, 597 (1985); Nucl. Phys. **A448**, 707 (1986); R. Koch, M. Hutt: Z. Phys. **C19**, 119 (1983)
22. U. Wiedner et al.: Phys. Rev. Lett. **58**, 648 (1987)
23. E. Friedman et al.: Phys. Lett. **B231**, 39 (1989); E. Friedman et al.: Nucl. Phys. **A514**, 601 (1990); E. Friedman et al.: Phys. Lett. **B254**, 40 (1991)
24. B.J. Kriss et al.: πN Newsletter No. 8, p. 17, Oct. 1993, Boulder, Co.

25. A.A. Carter et al.: Nucl. Phys. **B26**, 445 (1971)

26. E.Pedroni et al.: Nucl. Phys., **A300** 321 (1978)

27. E. Friedman: Private Communication (1995)

28. V.G.J. Stoks: Phys. Rev. Lett. **59**, 2255 (1987)

29. J.R. Bergervoet: Phys. Rev. **C41**, 1435 (1990)

30. R.A. Arndt: Phys. Rev. Lett. **65**, 157 (1990)

31. R.A. Arndt: Phys. Rev. **C49**, 2729 (1994)

32. F.G. Markopoulou-Kalamara, D.V. Bugg: Phys. Lett. **B318**, 565 (1993); D.V. Bugg, R. Machleidt, Los Alamos preprint nucl-phys:9404017

33. R. Timmermans: Private Communication (1993)

34. S.A. Coon, M.D. Scadron: Phys. Rev. **C42**, 2256 (1990)

35. P. Bopp et al.: Phys. Rev. Lett. **56**, 919 (1986)

36. B.R. Holstein: Phys. Lett. **B244**, 83 (1990)

37. W.R. Gibbs, L. Ai, W.B. Kaufmann: Phys. Rev. Lett. **74**, 3740 (1995)

38. N. Fettes, E. Matsinos: ETH Zurich preprint ETHZ-IPP PR-95-3 (1995)

39. M.E. Sevior et al.: Phys. Rev. **C40**, 2780 (1989)

40. C. Joram et al.: πN Newsletter No. 8, p. 30, Oct. 1993, Boulder, Co.

41. M. Locher, M. Sainio: Czech. J. Phys. **B39**, 943 (1989)

42. G.R. Smith et al.: Nucl. Inst. Meth. (in print) (1995)

43. M.E. Sevior et al.: Phys. Rev. **D48**, 3987 (1993)

44. M.G. Olsson, L. Turner: Phys. Rev. Lett. **20**, 1127 (1968); M.G. Olsson, E.T. Osypowski, L. Turner: Phys. Rev. Lett. **38**, 296 (1977)

45. E. Frlez: PhD Thesis, Univ. of Virginia, Los Alamos LA-12663-T (unpublished) (1993)

Few-Body Systems Suppl. 8, 133–144 (1995)

Few-Body Systems

Recent Developments in one and two Pion Production in Elementary Reactions and Few–Body Systems

E. Oset, F. Cano, J.A. Gómez Tejedor, E. Hernández, M.J. Vicente Vacas

Departamento de Física Teórica and IFIC, Centro Mixto Universidad de Valencia-CSIC, 46100 Burjassot (Valencia), Spain

Abstract. In this talk we cover several issues concerning pion production. The first one is the $pp \rightarrow pp\pi^0$ reaction where an alternative explanation based on the off shell extrapolation of the πN amplitude is offered. A recent model for the $\gamma N \rightarrow \pi\pi N$ reaction is presented and a new kind of exchange current is constructed from it which allows one to address problems like double Δ photoproduction from the deuteron. Finally the $(\gamma, \pi\pi)$ reaction in nuclei is studied and shown to be highly sensitive to the renormalization of the pions in nuclei.

1 The $pp \rightarrow pp\pi^0$ Reaction at Threshold.

This reaction has been addressed in different works [1, 2] and shown to be in strong disagreement with very precise data obtained at the Indiana Cyclotron [3]. The starting point is a model which contains the impulse approximation, diagram 1a) and the rescattering term, diagram 1b).

The first mechanism involves the usual Yukawa vertex and is only possible due to the initial and final state interaction. The second one involves the πN scattering amplitude in one of the vertices which is given in terms of an effective Lagrangian

$$\delta H = 4\pi \frac{\lambda_1}{m_\pi} \bar{\psi} \boldsymbol{\phi}\boldsymbol{\phi} \psi + 4\pi \frac{\lambda_2}{m_\pi^2} \bar{\psi} \boldsymbol{\tau} \boldsymbol{\phi} \times \dot{\boldsymbol{\phi}} \psi \tag{1}$$

where $\lambda_1 = 0.0075$ and $\lambda_2 = 0.053$ for on shell pions at threshold. The $\phi_3 \times \dot{\phi}_3$ combination is null and hence the second term in eq. (1) does not contribute for the $\pi^0 p \rightarrow \pi^0 p$ amplitude.

Hence only the λ_1 term contributes, which gives a small contribution given the smallness of λ_1. Actually λ_1 is strictly zero in the limit of exact chiral symmetry. A way out of the puzzle was offered in [4] in terms of a relativistic exchange current involving σ exchange between positive and negative energy

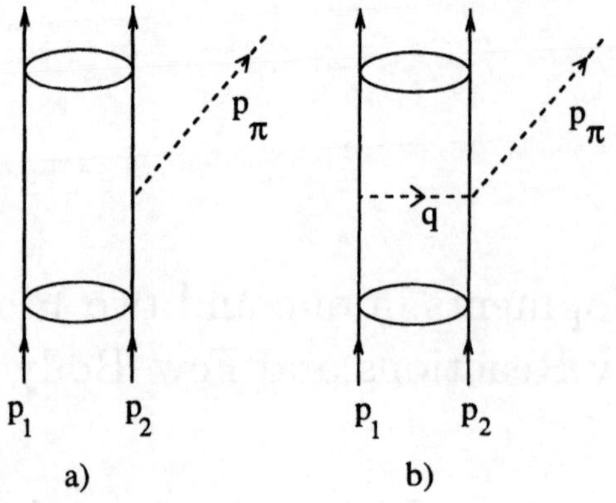

Figure 1. Feynman diagrams considered in the $pp \to pp\pi^0$ reaction near threshold. a) Born term; b) rescattering term.

components. This exchange is tied to the scalar potential, V_s, of the relativistic potential of Walecka type $V_S + \gamma^0 V_V$. This interpretation which can not be excluded a priori, does not have however a strong support, because the exchange current is generated from a relativistic form of the $NN \to NN$ amplitude which is unique for on shell nucleons. However, there are an infinite amount of possible forms of the amplitude which are equivalent for on shell nucleons but which provide quite different off shell extrapolations or extensions to the states of negative energy, as is the case in [4]. One typical example is the γ_5 matrix which is equivalent to $\not{q}\gamma_5/2m$ for on shell nucleons but which provides different extensions in the negative energy sector. The work of [4] is based on a particular choice for the NN amplitude but one can obtain quite different results for the $pp \to pp\pi^0$ reaction if one starts from other expressions which are equivalent for on shell nucleons. More concretely, recent models [5] which consider σ exchange as a correlated two pion exchange would provide a different answer in the negative energy sector than the ordinary σ exchange.

We have followed here a different idea suggested in [6]. The idea is that the πN amplitude appearing in the rescattering term, fig. 1b) is not the on shell πN amplitude but the half off shell amplitude and this one is much bigger than the on shell one in all existing models of the off shell extrapolation.

In a recent work [7] we have carried out detailed calculations of the $pp \to pp\pi^0$ cross section using two different off shell extrapolations for λ_1, one from Hamilton [8] which contains a σ exchange with the $\sigma\pi\pi$ and σNN vertices and a short range piece, which approximately cancel on shell. The other one, from Banerjee [9], is based on PCAC and has also been extensively used (we use a regularized version of it to have a well behaved Fourier Transform). The two curves can be seen in fig. 2, where we appreciate the large enhancement of the off shell amplitude for momenta around 400 MeV which are the relevant ones for the present problem (the upper curve corresponds to [9] and the lower one to [8]).

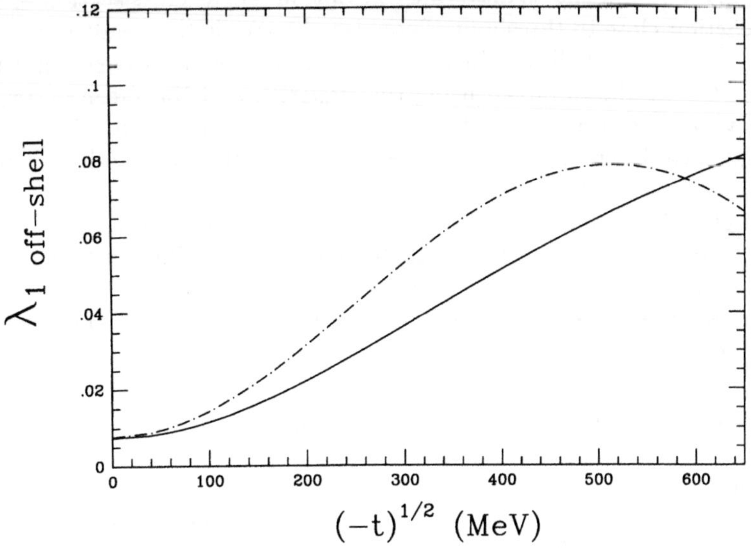

Figure 2. Off shell extrapolation of the πN isoscalar amplitudes in the Hamilton [8] (solid line) and current algebra model [9] (dashed-dotted line).

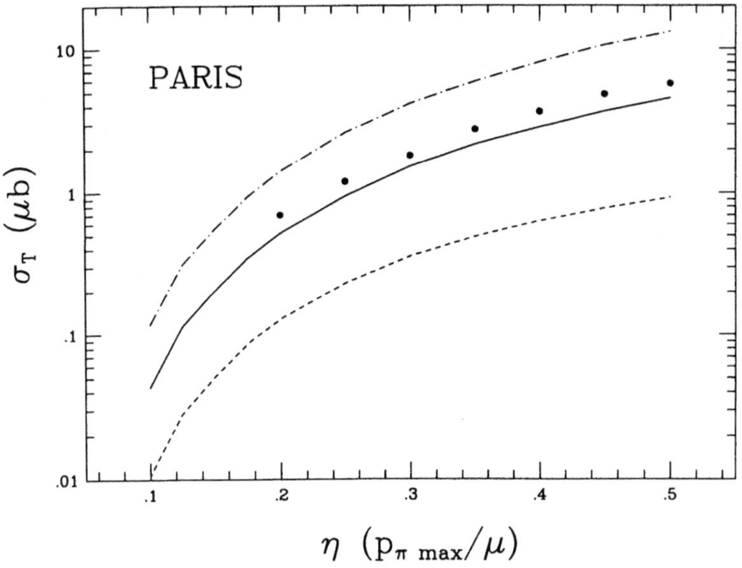

Figure 3. Cross section for $pp \to pp\pi^0$ near threshold using the Paris NN potential for the initial and final state interaction. Dashed line: with on shell πN isoscalar amplitude. Solid line: with off shell πN amplitude from Hamilton's model [8]. Dashed dotted line: same with the current algebra πN extrapolation [9].

In fig. 3 we show the results which we obtain for the cross section of the $pp \to pp\pi^0$ reaction close to threshold. The dashed curve is the result obtained using the $\pi N \to \pi N$ on shell amplitude, while the dash-dotted curve and the solid line give the results obtained using the Banerjee and Hamilton extrapolations respectively. The results are obtained using the Paris potential for the initial and final state interaction and the results are similar, although a little higher, if the Bonn potential is used.

It is clear from the results that one can not be very assertive, because there are still uncertainties on the off shell extrapolation, but the results plotted there clearly show that the off shell extrapolation by itself can easily explain the experimental data, offering a down to earth explanation of the discrepancies with experiment based on the use of the on shell πN amplitude.

2 Double Pion Photoproduction on the Nucleon.

A recent model for the $\gamma p \to \pi^+\pi^- p$ reaction has been presented in [10]. The model is schematically depicted in fig. 4, where the baryonic intermediate states stand for $N, \Delta, N^*(1440)$ and $N^*(1520)$. In total 67 Feynman diagrams appear when these states are considered, although some terms found to be negligible are not taken into account. At energies below $E_\gamma = 800$ MeV many of those terms are also negligible and a simplified model, fig. 5, which contains only 20 terms, is sufficiently good, and this is the one used for the study of the other isospin channels.

The results for the $\gamma p \to \pi^+\pi^- p$ reaction are shown in fig. 6. Four different lines are depicted there which we explain below.

The dominant term in the $\gamma p \to \pi^+\pi^- p$ amplitude is the one involving the $\gamma N \Delta \pi$ Kroll-Ruderman term (diagram (i) in fig. 5). It just happens that the amplitude depicted in fig. 5(p), with the excitation of the $N^*(1520)$ resonance which later on decays in $\Delta^{++}\pi^-$, has a piece in the amplitude with exactly the same spin and momentum structure as the dominant term of fig. 5(i) and gives rise to a strong interference even if individually it is a small term. This is one of the surprising and interesting accidents of this reaction which allows one to determine the signs of the s- and d-wave amplitudes for the $N^*(1520) \to \Delta\pi$ reaction, something which poses new challenges to quark models of the hadrons [11]. To envisage this we write here the basic ingredients of these amplitudes which are the couplings $N^*(1520) \to N\gamma$ and $N^*(1520) \to \Delta\pi$:

$$-i\delta H_{N^* N\gamma} = i g_\gamma \boldsymbol{S}^\dagger \cdot \boldsymbol{\varepsilon} + g_\sigma (\boldsymbol{\sigma} \times \boldsymbol{S}) \cdot \boldsymbol{\varepsilon}$$

$$-i\delta H_{N^* \Delta\pi} = -[f + \frac{g}{\mu^2}(\boldsymbol{S}_\Delta \cdot \boldsymbol{q})^2] T^{\dagger\lambda} \qquad (2)$$

where $\boldsymbol{S}, \boldsymbol{T}$ are spin, isospin transition operators from spin, isospin 1/2 to 3/2, $\boldsymbol{S}_\Delta$ the ordinary spin matrix for spin 3/2 and ϵ is the photon polarization vector. The amplitudes g_γ and g_σ are fitted to the experimental $N^*(1520)$ helicity amplitudes, and the strength of the s-wave and d-wave partial widths

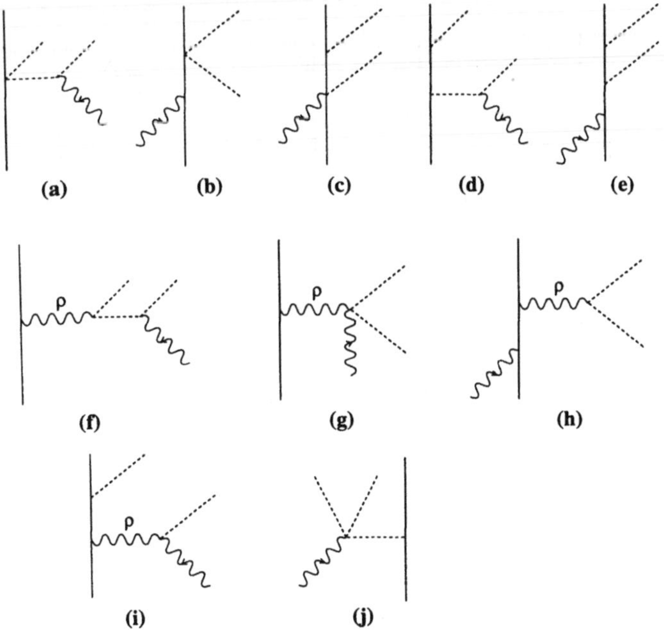

Figure 4. Classification of the Feynman diagrams into one point, two point and three point diagrams. Continuous straight lines: baryons. Dashed lines: pions. Wavy lines: photons and ρ-mesons (marked explicitly).

in the $N^*(1520) \to \Delta\pi$ decay are also taken from experiment. This gives, however, four possible solutions:

$$a)\ f=-0.19\ ,\ g=0.18\quad ;\quad b)\ f=1.03\ ,\ g=-0.18\ ;$$
$$c)\ f=-1.03\ ,\ g=0.18\quad ;\quad d)\ f=0.19\ ,\ g=-0.18 \tag{3}$$

The interference of the amplitudes is between the Kroll-Ruderman term of fig. 5(i), and the $N^*(1520)$ term of fig. 5(p) through the s-wave decay amplitude of the $N^*(1520) \to \Delta\pi$. Thus one finds a combined amplitude

$$i\frac{f^*}{\mu}\boldsymbol{S}\cdot\boldsymbol{q}_+ D_\Delta\left[e\frac{f^*}{\mu}-(g_\gamma-g_\sigma)\left(f+\frac{5}{4}g\frac{\boldsymbol{q}^2}{\mu^2}\right)D_{N^*}\right]\boldsymbol{S}^\dagger\cdot\boldsymbol{\varepsilon} \tag{4}$$

where $\boldsymbol{q}_+$ is the momentum of the π^+ and D_Δ, D_{N^*} are the propagators of the Δ and $N^*(1520)$ respectively.

One can see in fig. 6 that out of the four possible solutions of eq. (3) only one, solution (b), gives rise to good agreement with experiment [12] . The reason of the interference can be seen in eq. (4). Since $g_\gamma - g_\sigma = 0.157 > 0$ and $(f+\frac{5}{4}g\boldsymbol{q}^2/\mu^2) > 0$ one finds constructive interference below the $N^*(1520)$ energy and destructive interference above that energy and this interference is what gives rise to the peak in the cross section, which does not appear without the $N^*(1520)$ contribution.

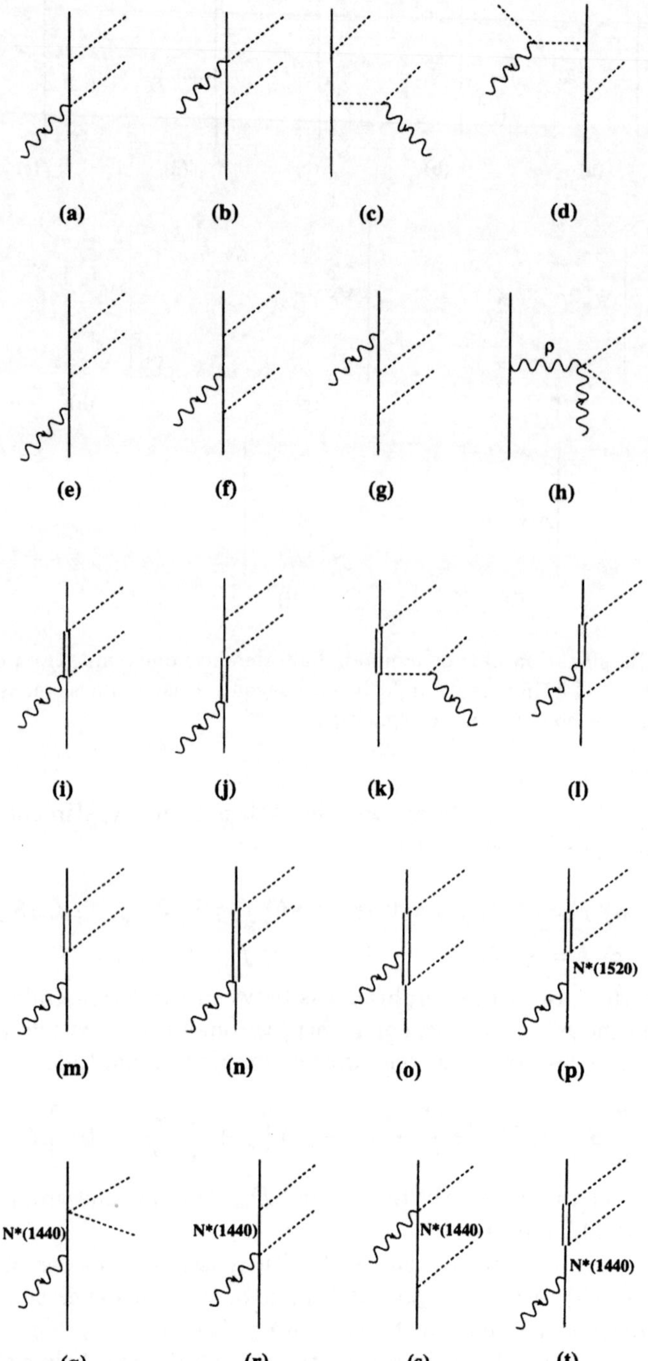

Figure 5. Feynman diagrams for the $\gamma N \to \pi\pi N$ reactions.

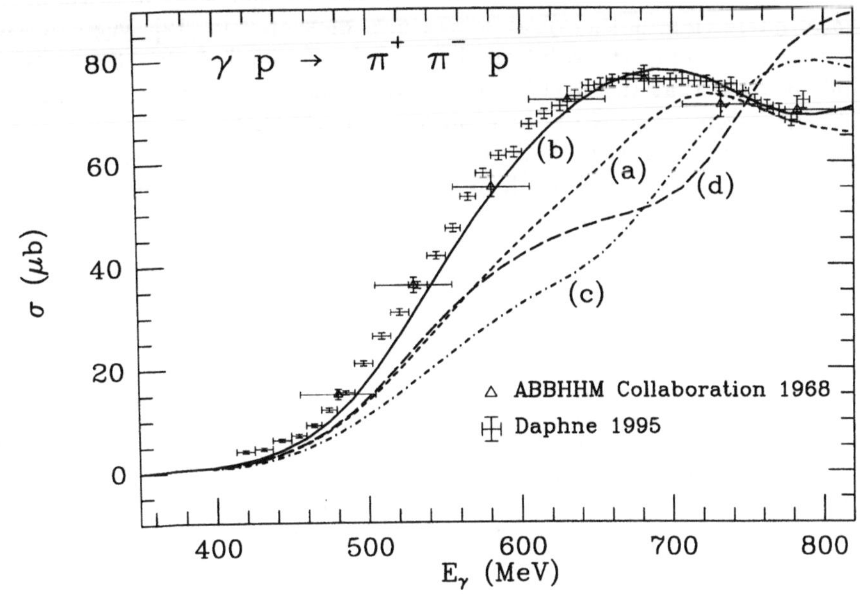

Figure 6. Total cross section for the $\gamma p \to \pi^+\pi^- p$ reaction. See text for explanation.

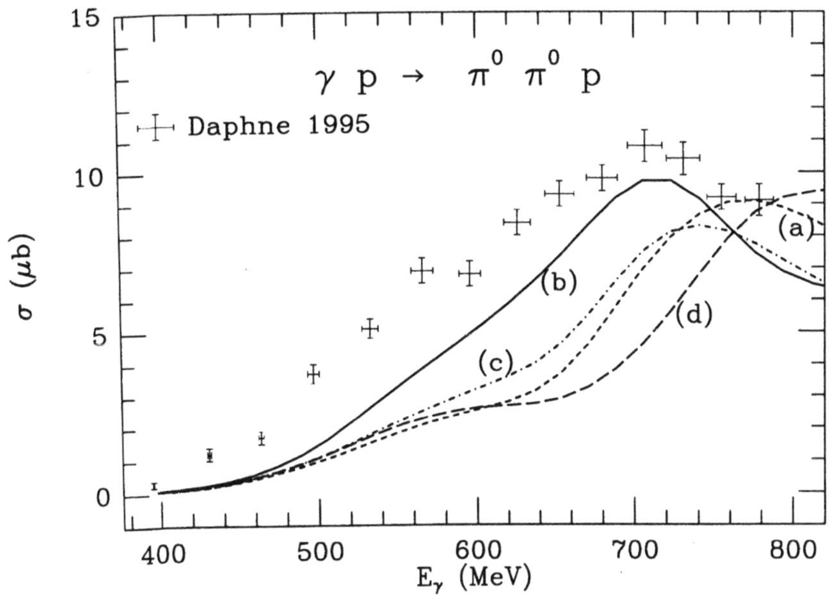

Figure 7. Total cross section for the $\gamma p \to \pi^0\pi^0 p$ reaction. See text for explanation.

We have calculated the cross section for other channels. One of them, particularly interesting, is the $\gamma p \to \pi^0 \pi^0 p$ also measured in [12]. We can see in fig. 7 that again only the solution (b) is the one that best fits the data. We should mention that the TAPS collaboration has also preliminary data [13] which agree even better with our results in this channel.

For reasons of space we shall not show here the results for other channels. Only we mention that we find disagreement with the experimental results in the $\gamma p \to \pi^+ \pi^0 n$ channel for which we do not yet envisage an explanation.

3 Double Δ Production in the $\gamma d \to p n \pi^+ \pi^-$ Reaction

In a recent paper [14] we have tackled the reaction $\gamma d \to \Delta^{++} \Delta^-$ which has been measured recently [15].

The excitation of two Δ necessarily requires the participation of two nucleons. We need some exchange currents mechanism which generates this transition. The previous model for the $\gamma p \to \pi^+ \pi^- p$ reaction offers us a clue on how to proceed. Let us take the diagrams of fig. 5(i), 5(p), which together with the pion pole term (diagram (k) of fig. 5), give already a fair description of the reaction. We can attach the π^- produced to a second nucleon and produce a second Δ. The diagrams which appear are shown in fig. 8, providing a fair model for this reaction up to energies of around $E_\gamma \simeq 800$ MeV, where the elementary model for the $\gamma p \to \pi^+ \pi^- p$ reaction starts having deficiencies.

The results for the $\gamma d \to \Delta^{++} \Delta^-$ reaction can be seen in fig. 9 compared with the experiment. The dashed line omits the contribution of the $N^*(1520)$ while the solid one takes it into account. We can see that there is a good agreement with experiment up to around $E_\gamma = 800$ MeV and the data hint at discrepancies from there on. The consideration of the $N^*(1520)$ term seems to improve the results a bit.

4 Double Pion Photoproduction in Nuclei.

We have studied the $(\gamma, \pi^+ \pi^-)$ inclusive reaction in nuclei in order to see renormalization effects due to the interaction of pions with the nucleus. Such effects were predicted in [16] and observed in [17] for the $(\pi, 2\pi)$ reaction in nuclei and they could only be magnified here since the photons penetrate deeper than the pions. We evaluate the cross section by assuming the nucleus to be made of bits of infinite nuclear matter at each point of the nucleus. This is the essence of the local density approximation, which given the fact that the photons explore the whole nuclear volume, is a very good approximation [18].

The cross section for this process in the "improved" impulse approximation would be

$$\sigma = \frac{\pi}{k} \int d^3 r \int \frac{d^3 p}{(2\pi)^3} \int \frac{d^3 q_1}{(2\pi)^3} \int \frac{d^3 q_2}{(2\pi)^3} \sum_\alpha \overline{\sum} \sum |T_\alpha|^2$$

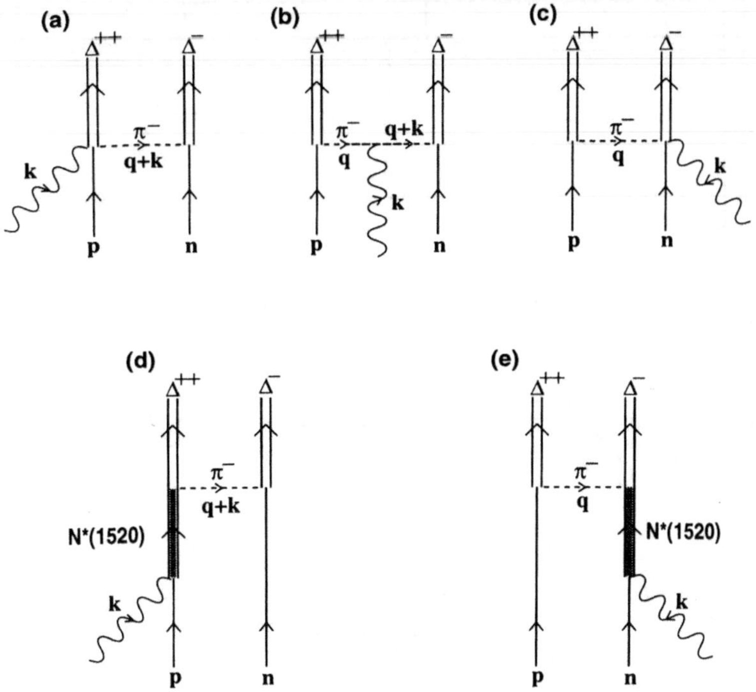

Figure 8. Terms considered in our model for the $\gamma d \to \Delta^{++}\Delta^{-}$ reaction.

$$n_\alpha(\boldsymbol{p})[1 - n_\alpha(\boldsymbol{k} + \boldsymbol{p} - \boldsymbol{q}_1 - \boldsymbol{q}_2)]\frac{1}{2\omega(q_1)}\frac{1}{2\omega(q_2)}$$

$$2\pi\delta(k^0 + E(p) - \omega(\boldsymbol{q}_1) - \omega(\boldsymbol{q}_2) - E(\boldsymbol{k} + \boldsymbol{p} - \boldsymbol{q}_1 - \boldsymbol{q}_2)) \tag{5}$$

where $\boldsymbol{p}$ is the momentum of the nucleons from the Fermi sea, $\boldsymbol{k}$ the momentum of the photon and $\boldsymbol{q}_1, \boldsymbol{q}_2, \omega(\boldsymbol{q}_1), \omega(\boldsymbol{q}_2)$ the momenta and energies of the two pions. The variable α indicates isospin channels, and $n_\alpha(\boldsymbol{p})$ is the nucleon occupation number in the Fermi sea. Since

$$2\int \frac{d^3p}{(2\pi)^3}\, n_\alpha(\boldsymbol{p}) = N \text{ or } Z \tag{6}$$

one is essentially multiplying by N or Z (and summing) the individual $\gamma N \to \pi^+\pi^- N$ cross sections, with two important modifications: i) the Fermi motion is explicitly taken into account ii) the Pauli blocking factor $[1-n]$ is explicitly considered there. The formulation of the problem is easily done in terms of the selfenergy of the photon, $\Pi(\boldsymbol{k})$ and the relationship

$$\sigma = -\frac{1}{k}\int Im\Pi\,(\boldsymbol{k}, \rho(\boldsymbol{r}))\, d^3r \tag{7}$$

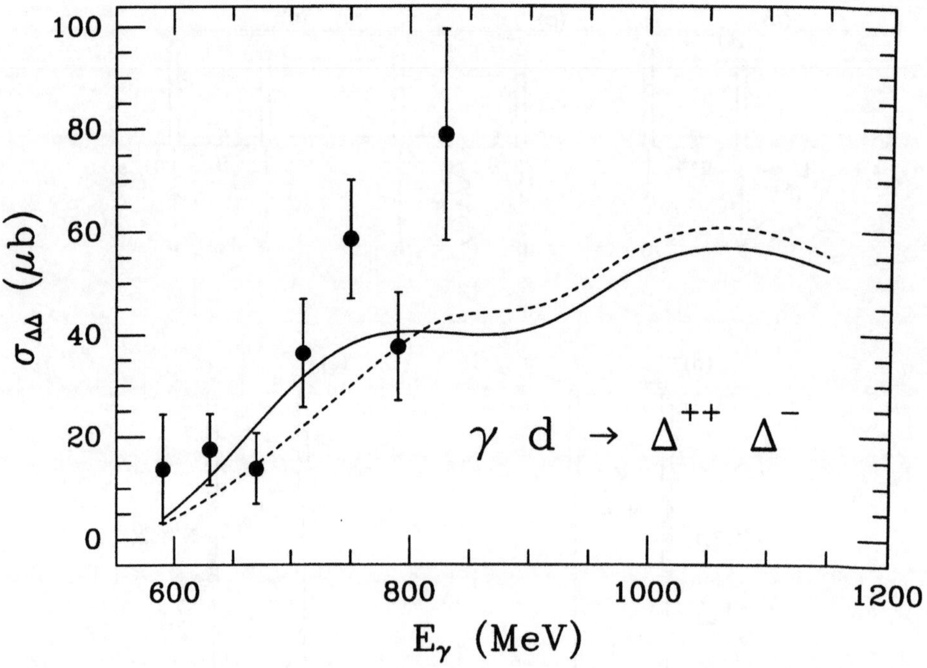

Figure 9. Results of the model compared to the data of ref. [15]. Dashed line, omitting the $N^*(1520)$ terms. Solid line, results including all terms of the model of Fig. 8.

which leads to eq. (5). The formulation in terms of a selfenergy diagram allows for systematic corrections. An important one arises from the consideration of the renormalization of the pion propagators, which essentially substitutes $\omega(\boldsymbol{q})$ in eq. (5) by $\tilde{\omega}(q)$, the pion dispersion relation in the medium. Simultaneously we also renormalize the Δ and pion propagators in the amplitudes for the model of $(\gamma, \pi^+\pi^-)$. In addition one must distort the pion waves by the imaginary part of the pion selfenergy due to absorption in order to take into account the absorption of the pions in their way out of the nucleus.

In fig. 10 we can see the results of the evaluation of [19]. The curve labelled 1 is the impulse approximation $N\sigma_n + Z\sigma_p$ multiplied by the pion absorption factor. The curve 2 is a pure scaling of the cross section in the deuteron $(A\sigma_d/2)$, while curve 3 is the result from our calculations. The differences between 1 and 3 indicate effects of renormalization, which increase the cross section by a factor from 4 to 5. It is also interesting to see that in spite of the absorption of the final pions, the cross section obtained is still larger than A/2 times the cross section in the deuteron.

It would be interesting to perform such experiments, easily implementable at Mainz as a complement to the $(\gamma, \pi\pi)$ program on the nucleon, and in general to the whole program of reaction mechanisms in photonuclear interactions.

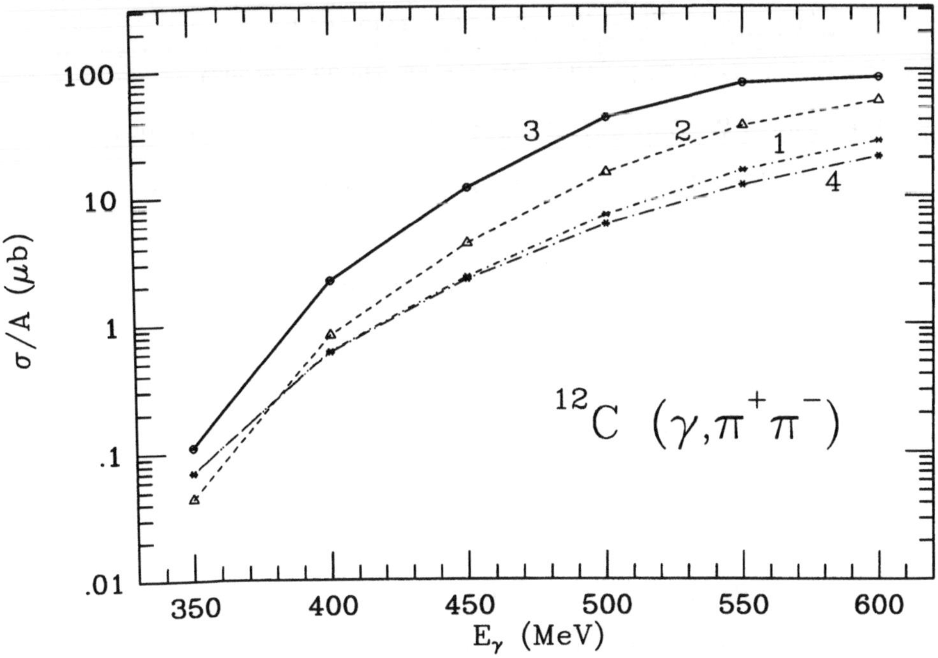

Figure 10. Cross section per nucleon for the $^{12}C(\gamma,\pi^+\pi^-)$ reaction. Dot-dashed line (labelled 1): Simple impulse approximation. Dashed line (labelled 2): Scaling of the deuteron cross section. Continuous line (labelled 3): Full model. Long dashed-dotted curve (labelled 4): cross section renormalizing the internal pion and delta lines in the amplitudes and ignoring the renormalization of the external pion lines.

5 Summary

We made a short review of reactions producing one or two pions in elementary processes, in deuteron and ^{12}C. The off shell dependence of the isoscalar πN amplitude was shown to be very important in the $pp \rightarrow pp\pi^0$ reaction. On the other hand a rather complete model for the $\gamma N \rightarrow \pi\pi N$ reaction is made, which is bound to play a similar role as the $\gamma N \rightarrow \pi N$ reaction when the energy of the experiments go up to the GeV regime as in Mainz. We offered an example of exchange currents produced from that model in the $\gamma d \rightarrow \Delta^{++}\Delta^-$ reaction. Finally we also showed important renormalization effects in the $(\gamma,\pi^+\pi^-)$ reaction in nuclei, which once more stresses the role played by pion physics in photonuclear reactions in general.

References

1. G.A. Miller, P.U. Sauer: Phys. Rev. **C44**, R1725 (1991)

2. J.A. Niskanen: Phys. Lett. **B289**, 227 (1992)

3. H.O. Meyer et al: Phys. Rev. Lett. **65**, 2846 (1990)

4. T.S.H. Lee, D.O. Riska: Phys. Rev. Lett. **70**, 2237 (1993)

5. H.C. Kim, J.W. Durso, K. Holinde: Phys. Rev. **C49**, 2355 (1994)

6. F. Hachenberg, H.J. Pirner: Ann. Phys. (NY) **112**, 401 (1978)

7. E. Hernández, E. Oset: Phys. Lett. **B350**, 158 (1995)

8. G. Hamilton: *High Energy Physics*, Ed. E.H.S. Burhop, **Vol. 1**: Academic Press, New York (1967) p. 194

9. M.K. Banerjee, J.B. Cammarata: Phys. Rev. **D18**, 4078 (1978)

10. J.A. Gómez Tejedor, E. Oset: Nucl. Phys. **A571**, 667 (1994)

11. J.A. Gómez Tejedor, F. Cano, E. Oset: in preparation

12. A. Braghieri, L. Murphy et al: to be published

13. H. Ströher: Private Communication

14. J.A. Gómez Tejedor, E. Oset, H. Toki: Phys. Lett. **B346**, 240 (1995)

15. M. Asai et al: Z. Phys. **A344**, 335 (1993)

16. E. Oset, M.J. Vicente Vacas: Nucl. Phys. **A454**, 637 (1986)

17. N. Grion et al: Nucl. Phys. **A492** (1989) 509

18. R.C. Carrasco, E. Oset: Nucl. Phys. **A536**, 445 (1992)

19. J.A. Gómez Tejedor, M.J. Vicente Vacas, E. Oset: Nucl. Phys. **A588**, 819 (1995)

Few-Body Systems Suppl. 8, 145–149 (1995)

Few-Body Systems

Plans for Nucleon–Nucleon Bremsstrahlung Measurements with AGOR

N. Kalantar-Nayestanaki, A. Bakker, H. Huisman, J.G. Messchendorp, S. Schadmand, O. Scholten, J. Zijlstra

Kernfysisch Versneller Instituut, Zernikelaan 25, 9747 AA Groningen, The Netherlands
(for the collaboration of SALAD and TAPS)

Abstract. The plans for nucleon-nucleon bremsstrahlung measurements below the particle-production threshold are outlined. The main goal of these measurements is to study the off-shell behavior of the nucleon-nucleon interaction. This is done by measuring cross sections and analyzing powers of the production of high energy photons in one of the most fundamental processes, i.e. proton-proton scattering, and also in more complicated reactions such as proton-nucleus collisions. A proton detection system designed to handle extremely high rates at small angles is discussed. An extension of this system can be used to measure more exclusive observables such as spin-transfer coefficients.

1 Introduction

Nucleon-nucleon bremsstrahlung is the most fundamental reaction used in studying the off-shell behavior of the nucleon-nucleon interaction as it involves the strong interaction between two nucleons, and the well-known electromagnetic interaction. In addition, the cross section for nucleon-nucleon bremsstrahlung is the basis of all calculations which deal with photon production in heavy-ion collisions. These calculations rely heavily on theoretical predictions of the nucleon-nucleon bremsstrahlung process, which in turn resort to approximations such as the soft-photon approximation that, to leading order, contains no off-shell information. Recently, the study of off-shell effects by nucleon-nucleon bremsstrahlung below the particle-production threshold has been receiving increasing attention both theoretically and experimentally. The reason for this is the success of modern potential-model calculations on the one hand and the advent of high-precision detection systems on the other [1]. Polarized-proton beams and high-flux neutron beams have also delivered a boost to the experimental efforts. Recent cross-section and analyzing-power data from TRIUMF [2] clearly show the need for the inclusion of off-shell effects in the calculations.

With even higher accuracies achievable nowadays, one can hope to study, in detail, the properties of the off-shell nucleon as it propagates in an intermediate state as well as the non-nucleonic degrees of freedom (such as Δ-excitation) which can admix even at energies below the pion-production threshold. Recent calculations have shown that, including the latter effect and depending on the choice of the $N\Delta\gamma$ coupling constants and kinematical conditions, the cross section may be enhanced by up to 30% for an incident proton energy of 280 MeV [3].

2 Experimental Plans at KVI

At KVI, a program is launched to study the nucleon-nucleon bremsstrahlung starting with the proton-proton system. The first goal of the experiment is to measure absolute cross sections and settle the question of normalization raised by the TRIUMF group [2]. Using polarized protons, high precision analyzing powers will also be measured. The measurements will be done with the new superconducting cyclotron, AGOR, which is capable of delivering (polarized) protons with energies of up to 200 MeV. To probe matrix elements involving intermediate states that are far off the mass shell, we need to measure high energy photons. This in turn restricts the detection of the protons to small laboratory angles. However, at small angles one is limited by the quality of the beam and the very high rate of proton-proton elastic scattering (for the forward detector considered here, this rate is about 1.4×10^7 counts/s). The high-quality polarized proton beam at AGOR overcomes the first limitation. The problem of high rates can be dealt with if a highly segmented detection system is utilized.

The Small-Angle Large-Acceptance Detector (SALAD) designed and built for precise measurements of the absolute cross sections and analyzing powers is shown in Fig. 1. It is a detection system with almost cylindrical symmetry which measures charged particles between 6° and 19° for the whole azimuthal angular range and proton energies of up to 130 MeV. The upper limit of the range of the polar angles increases to about 26° for a smaller region of the azimuthal angles. The large solid angle (about 500 msr) is mandated by the very small cross sections of the nucleon-nucleon bremsstrahlung process. To allow the beam free passage, we will use a detector designed with a central hole. To determine the coordinates of the proton, we will utilize two wire chambers, the first one with three planes (x, y, and 45°) to overcome ambiguities in multi-prong events. The wire spacing in these chambers is 2 mm. However, to reduce the number of read-out channels, every two adjacent wires go through one amplifier-discriminator channel. The anode-cathode spacing is 4 mm. The angular resolution obtained with these chambers is better than 0.5°. The first wire chamber with a hole in its center has now been made and tested in a proton beam at IUCF (see Fig. 2). A high detection efficiency was observed even at high rates of more than 100 kHz on each wire.

Plastic scintillators are used to measure the energy of the protons: a thick

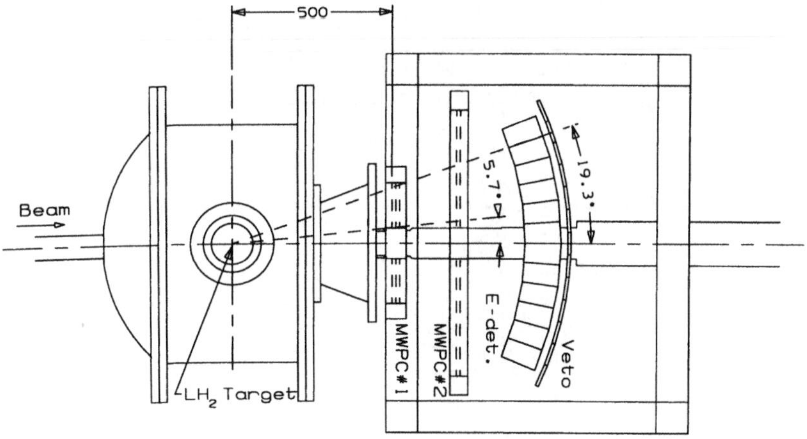

Figure 1. Top view of SALAD. The size of the first wire chamber is 38 × 38 cm^2. The total number of scintillators is 50 (24 for the energy detector and 26 for the veto detector; half of them are read out from the top and the other half from the bottom). TAPS will be placed at the backward angles (not drawn). The beam pipe goes through the center of all detectors. The special shape of the scattering chamber presents little wall thickness at forward and backward angles.

and segmented layer detecting the protons originating from bremsstrahlung events is backed by a thin (also segmented) layer vetoing protons of higher energies coming from elastic events. For the measurement of the position and the energy of the photons, we are planning to use the existing photon detection system, the Two-Arm Photon Spectrometer (TAPS), which will be placed at backward angles. TAPS is equipped with 384 BaF_2 crystals providing an angular resolution of better than 6° for our geometry. Depending on its configuration, TAPS can cover up to about 35% of the full 4π solid angle. If one keeps the cylindrically-symmetric geometry at the backward angles as well, one covers a solid angle of about 20% of 4π with it. With the setup depicted in Fig. 1 along with TAPS, coplanar and non-coplanar events will be measured simultaneously. The kinematics of the events is overdetermined as we measure energies and coordinates of all the particles involved in this reaction.

The target which will be used in these measurements is a liquid hydrogen target with very thin windows made out of organic materials to reduce the background from the windows. These foils are being tested at this moment for strength and porosity at lower temperatures. A 6 mm thick disk of LH_2 (about 43 mg/cm^2) will be placed in a 10 nA beam of (polarized) protons. A rough estimate for count rates of coplanar geometry for proton and photon bins of 5° yields cross-section rates of about 2000 counts/hr. This large rate is of

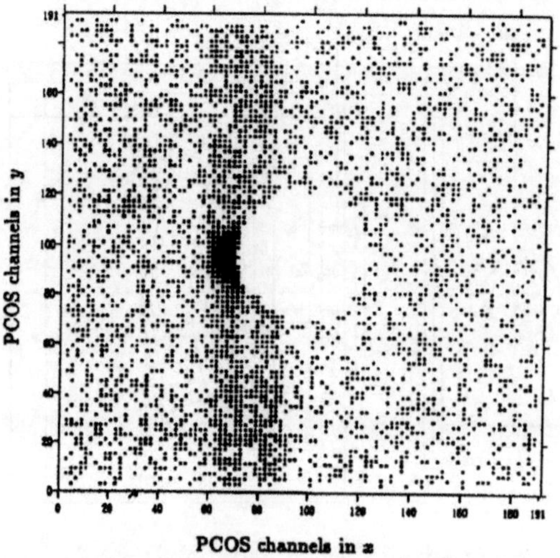

Figure 2. An $x - y$ spectrum measured with the small wire chamber. The geometry, the gates and the thresholds were so chosen to enhance the accidental rate in order to illuminate the whole chamber. The shadows of the circular collimator and the trigger scintillators in front and at the back of the chamber are clearly observed outside the hub. No cuts are made on the scintillators in generating this spectrum.

course the result of exploiting the full cylindrical symmetry of the system. A similar estimate gives better than 10% accurate analyzing-power measurements in about a week of beam time.

The proton-neutron bremsstrahlung can, in principle, be extracted from proton-deuteron bremsstrahlung measurements. Here, the meson-exchange currents (MEC) give a contribution of about 50% to the cross section, which allows for an investigation of MEC effects. The extension to proton-nucleus bremsstrahlung measurements will provide information about the reaction mechanism, an essential ingredient in heavy-ion reaction studies.

As TAPS is also capable of detecting charged particles such as electrons and positrons, another experiment is planned to measure e^+e^- pairs which originate from virtual bremsstrahlung. Here, the emphasis will be more on higher invariant masses thereby studying the longitudinal part of the electromagnetic operator. It should be noted that compared to real photon bremsstrahlung, and depending on the invariant mass, the cross section is lower by at least a factor α, the fine structure constant, making the experiment even more challenging.

Finally, to enhance smaller effects which are overwhelmed in cross-section and analyzing-power measurements, exclusive observables such as spin-transfer coefficients might need to be measured. In order to measure the polarization of the outgoing protons, which is necessary for the determination of the spin-transfer coefficients, the detection system should be supplemented with a polarimeter with which we will measure the left-right asymmetry in a secondary scattering by the proton. With the addition of an extra wire chamber and some scintillators, the present detection system can be extended to include this polarimeter. The low energy of the outgoing proton carrying the polarization information mandates a large size for the polarimeter and yields a low efficiency of about 2×10^{-4}.

3 Summary

The study of the nucleon-nucleon bremsstrahlung process has two important aspects. The first one is to confront the potential-model calculations in an effort to refine them and to investigate effects such as the modification of the nucleon form factor and virtual excitation of the Δ as the nucleon goes off the mass shell. The second aspect is to provide accurate cross-section data to be used in the calculations performed for nucleus-nucleus collisions. To avoid dealing with too many uncertainties, one should probe the nucleon with energies below particle-production threshold. After the commissioning of the superconducting cyclotron, AGOR, at KVI, we will use the 200 MeV polarized protons to study proton-proton bremsstrahlung in the course of 1996. A large solid angle detector suitable to detect protons at forward angles coupled with TAPS located in the backward hemisphere offers a unique configuration to span a large part of the phase space where off-shell effects are expected to be very large. Accurate cross sections and analyzing powers will be measured. The program will continue with measuring proton-deuteron bremsstrahlung to extract proton-neutron bremsstrahlung information. The next step would be to extend the measurement to proton-nucleus bremsstrahlung, and further in the future one can hope, with the extension of the present detection system, to measure other spin observables such as spin-transfer coefficients.

Acknowledgement.
The authors would like to thank the IUCF staff for their hospitality during the stay of N.K. and A.B. in Bloomington, and for their help in setting up for the test measurements on the wire chamber. In particular, we are grateful for the invaluable assistance of K. Solberg, T. Rinckel and A. Eads.

References

1. Summary of the Contributions of the "Workshop on NN Bremsstrahlung and Polarization Observables", Groningen, 16–18 November (1992)

2. K. Michaelian et al.: Phys. Rev. **D41**, 2689 (1990)

3. F. de Jong, K. Nakayama, V. Herrmann, O. Scholten: Phys. Lett. **B333**, 1 (1994), and references therein.

Few-Body Systems Suppl. 8, 150–155 (1995)

Few-Body Systems

First Results on Proton–Proton Scattering and Near Threshold Two Meson Production at COSY

R. Maschuw for the EDDA [1] and MOMO [2] collaborations

Institut für Strahlen- und Kernphysik, Universität Bonn, Nussallee 14-16, 53115 Bonn, Germany

1 Introduction

The new Cooler Synchrotron COSY at the Forschungszentrum Jülich, Germany, has taken up operation and provides a (cooled) proton beam with maximum energy of 2.5 GeV. Experimental areas are available for internal beam experiments using the circulating COSY beam even during acceleration as well as for external experiments at dedicated beamlines for the extracted beam. The two experiments reported here, EDDA and MOMO, represent one of each type of these experiments.

2 EDDA: *pp*-elastic scattering excitation functions

The EDDA experiment is dedicated to high precision measurements of proton-proton elastic scattering excitation functions in the energy range from 0.5 GeV to 2.5 GeV. This applies to the measurement of the spin averaged cross section, and four more spin observables i.e. the analyzing power and three spin correlation parameters. Apart from providing precision data for an advanced global phase shift analysis, particularily in the energy domain above 1.5 GeV where existing data are sparse, the experiment will be highly sensitive to the existence or nonexistence of dibaryonic resonances coupling to the elastic channel. Such a structure should show up in the different observables with characteristic significance. It requires high statistics data with small energy binning over a wide energy range but with minimum systematic uncertainties. This is provided by running EDDA as an internal experiment where the proton target - a CH_2-fiber or polarized atomic gas target - intercepts the recirculating COSY beam during acceleration. Thus a full excitation function is measured during each acceleration cycle whereas statistical accuracy is achieved by averaging over many cycles. Elastic *pp*-scattering events are identified by kinematic coincidence i.e. coplanarity with the beam $(\phi_1 - \phi_2) = 180^0$ and the polar angle correlation of the two outgoing protons $\tan\theta_1 \times \tan\theta_2 = 1/\gamma_{cm}^2$ (Lorentz γ).

The EDDA detector is a cylindrical, high granularity multilayer scintillator hodoscope enclosing the COSY beam pipe. It covers the angular range from 30^0 to 150^0 c.m. for elastic *pp*-scattering, i.e. $\sim 85\%$ of the full solid angle. The inner layer, mounted tightly onto the beam pipe consists of opposingly, helically wound scintillating fibers of 2.5 mm diameter, which are individually read out by 42 16-fold phototubes (Hamamatsu R4760) upstream the proton target. The outer layer is a scintillator barrel made from overlapping triangular shaped scintillator bars parallel to the beam pipe surrounded by scintillator semi-rings. The scintillator bars are measuring the azimuthal angle whereas the polar angles are determined by the rings having increasing widths downstream correspondingly. With the dimensions of bars and rings choosen a proton track from the target always penetrates two adjacent modules. Thus the angular resolution is significantly improved over the geometric granularity by fractional light output analysis of adjacent modules giving $\Delta\theta = 1^0$ for the polar angular resolution and $\Delta\phi = 3^0$ for the azimuth angle. Together with the spatial resolution of the scintillating fiber layer a vertex reconstruction of a few millimeter accuracy is provided. This is particularly essential when the more extended polarized atomic beam target is used. However, unpolarized data have already been taken using the outer layer detector only, using a $3\mu m \times 4\mu m$ CH_2-filament target.

With COSY having achieved its design energy, the excitation function of the spin averaged cross section for elastic *pp*-scattering from 0.5 GeV to 2.5 GeV kinetic energy has been measured during acceleration. Energy dependent luminosity was controlled by two monitor systems: one was a current measurement of secondary electrons knocked off from the aluminized filament target by use of a dedicated fast picoamperemeter, the second was the detection of δ-electrons from the target with PIN-diodes implemented in the COSY beam pipe. Both monitor signals were in complete accordance with the detector trigger rate allowing a precise relative normalisation of the data at different energies. Raw data from the EDDA detector are shown in Fig. 1: as the COSY beam momentum rises during acceleration from 1.6 GeV/c to 3.2 GeV/c a clear *pp*-scattering peak is observed at the corresponding values of the angular correlation parameter $\tan\theta_1 \times \tan\theta_2 = 1/\gamma_{cm}^2$ each containing the entire angular distribution at this particular energy. The small and smooth background from quasifree *pp*-scattering off protons in the ^{12}C-nucleus has been measured with a pure carbon fiber target and can easily be subtracted.

From these data angular distributions have been deduced in the off-line evaluation, a few of them are shown in Fig. 2. Due to still insufficient statistics in this first EDDA run a rather coarse energy binning of 20 MeV width was chosen. However the agreement with existing precision data at 793 MeV recently measured [3] is excellent, proving the validity of the multi-pass technique during acceleration. Even this first measurment already provides a significant

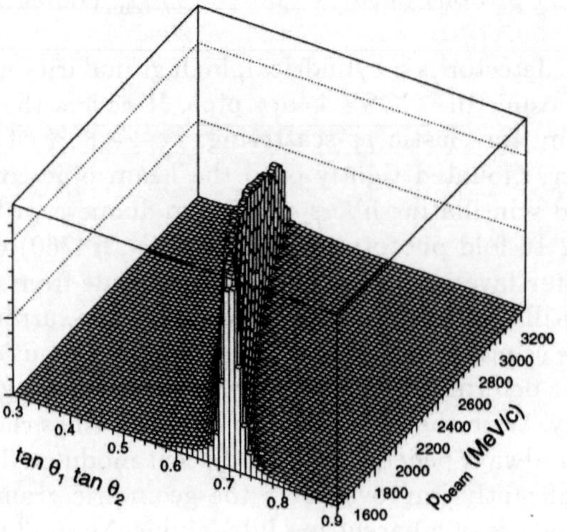

Figure 1. angular correlated *pp*-scattering events at varying proton momentum during COSY acceleration

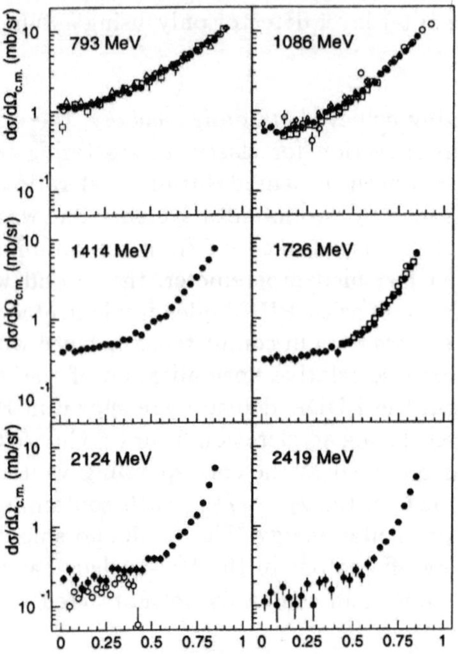

Figure 2. Some angular distributions of spin averaged *pp*-scattering cross section as measured by EDDA (solid dots, preliminary) and as available from literature at indicated energies (open symbols)

improvement over existing data filling up different angular and energy regions with lacking experimental data.

Precision measurements of the spin averaged cross section with high statistics will be performed during this summer. With a polarized COSY beam and a polarized proton target currently under development at Bonn, measurement of polarization observables of *pp* elastic scattering is expected to start in 1996.

3 MOMO: the reaction $pd \rightarrow {}^3\text{He}\ \pi^+\pi^-$ near threshold

The MOMO experiment focusses on near-threshold two-meson production in the reactions $pd \rightarrow {}^3\text{He}\ \pi^+\pi^-$ and $pd \rightarrow {}^3\text{He}\ K^+K^-$. It is dedicated to investigate meson-meson interactions at low relative energies (T $\leq$ 50 MeV) and to probe into questions of meson-nucleon resonances and KK - molecules. The experiment takes advantage of the high quality (cooled) external COSY beam and the existing magnetic spectrometer BIG KARL. The setup shown in Fig. 1 consists of a liquid deuterium target (LD2) housed in the MOMO scattering chamber, a high granularity scintillating fiber meson detector near the target and the large acceptance magnetic spectrometer instrumented with multiwire drift chamber (MWDC) focal plane detectors followed by a scintillator hodoscope.

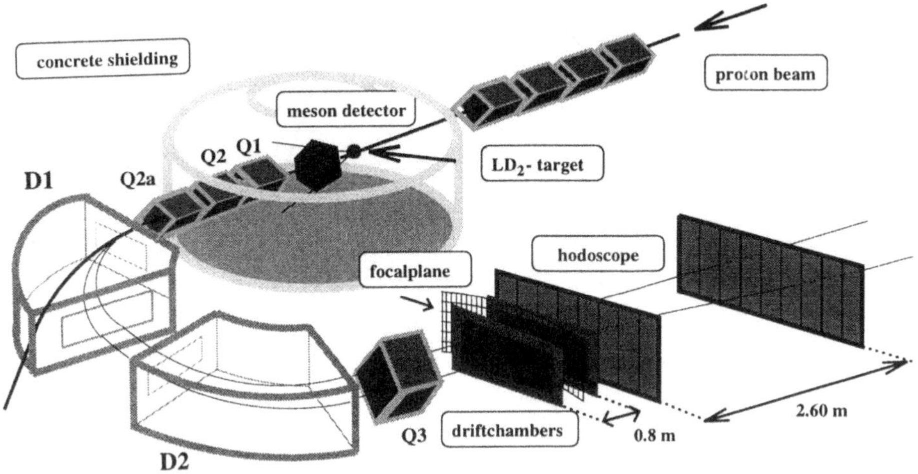

Figure 3. MOMO experimental setup

The forward going ^{3}He-particles are momentum analyzed in the focal plane of the spectrometer by a stack of two chambers each containing six MWDC-planes. With a ^{3}He-track resolution of better the 300 μm the corresponding momentum resolution is $\Delta p/p \approx 10^{-4}$. Particle identification is provided by energy loss against time of flight measurement ($\Delta E/TOF$) in the scintillator hodoscope downstream the focal plane. The MOMO meson detector in front of the magnetic spectrometer consists of 672 scintillating fibers of 2.5 mm diameter, arranged in three planes tilted by 60^0 against each other. Its purpose is to determine the momentum direction of the two outgoing mesons $\pi^+\pi^-$ or K^+K^- which are required to be coplanar with respect to the ^{3}He-missing momentum. The fibers are read out by 16-fold phototubes (Hamamatsu R4760) followed by an adapted PCOS III data acquisition system. The opening angle of this vertex detector is $\pm 45^0$, providing a full solid angle acceptance for the two-meson production reactions near threshold with completely determined kinematics.

Calibration of the entire detector setup has been performed using the reactions $pd \rightarrow {}^3\mathrm{He}\ \pi^0$ and $pd \rightarrow {}^3\mathrm{H}\ \pi^+$. In a first data taking run at COSY at 1150 MeV/c proton momentum MOMO has then detected some 350 events from the reaction $pd \rightarrow {}^3\mathrm{He}\ \pi^+\pi^-$. Fig. 2 shows the double-hit pattern of these first two pion events at COSY in the meson vertex detector triggered by ^{3}He-identification in the focal plane detectors.

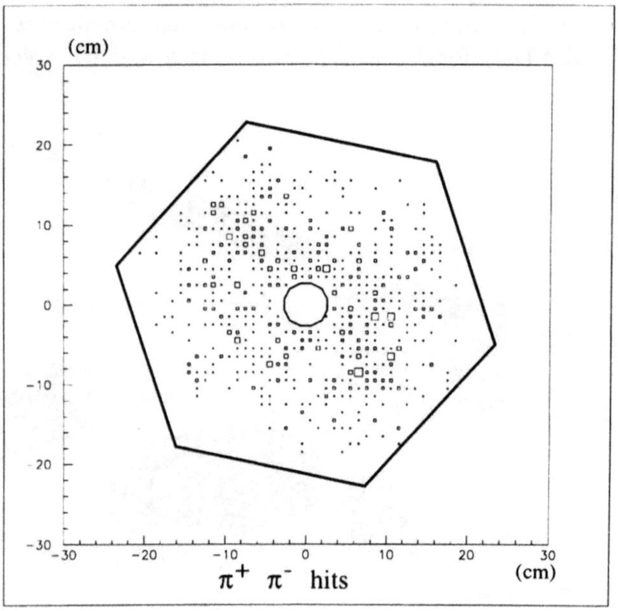

Figure 4. Meson hit pattern in vertexdetector

Their unambiguous assignment to the exclusive two pion reaction is proven by coplanarity of these double hits with respect to the total meson momentum axis as defined by the beam and ^{3}He-momenta. The events fit nicely into the expected phase space of this three particle reaction.

While this paper is written, MOMO has collected some other 5000 of these two pion events in a two weeks run with improved COSY performance. They cover a phase space with $\pi\pi$- relative energy ranging from 0 to 60 MeV. Evaluation of these data is in progress and will concentrate on the study of $\pi\pi$ and π-Nucleus interaction. These studies will be extended to the exclusive K^+K^- channel of the $pd \rightarrow {}^3$He + X reaction at higher energies of the extracted COSY beam.

With these first data from COSY-experiments, EDDA and MOMO as we believe have opened up a new field of precision experiments on hadron interaction at intermediate energies.

References

1. EDDA collaboration
   J. Bisplinghoff[a], R. Bollmann[b], P. Cloth[c], O. Diehl[a], F. Dohrmann[b], V. Drüke[c], S. Eisenhardt[b], H. P. Engelhardt[a], J. Ernst[a], P. D. Eversheim[a], D. Filges[c], M. Gasthuber[b], R. Gebel[a], J. Greif[b], A. Groß[b], R. Groß-Hardt[a], S. Heider[a], A. Heine[a], F. Hinterberger[a], R. Jahn[a], M. Jeske[a], U. Lahr[a], R. Langkau[b], R. Maier[c], R. Maschuw[a], T. Mayer-Kuckuk[a], G. Mertler[a], B. Metsch[d], F. Mosel[a], M. Müller[b], M. Münstermann[b], H. R. Petry[d], D. Prasuhn[c], H. Rohdjeß[a], D. Rosendaal[a], U. Roß[a], P. von Rossen[c], H. Scheid[a], H. Paetz gen. Schieck[e], N. Schirm[b], M. Schulz-Rojahn[a], F. Schwandt[a], V. Schwarz[a], W. Scobel[b], G. Sterzenbach[c], H. J. Trelle[a], A. Wellinghausen[b], W. Wiedmann[a], K. Woller[b], R. Ziegler[a]
   [a] Inst. f. Strahlen- und Kernphysik, Univ. Bonn, [b] I. Inst. f. Experimentalphysik, Univ. Hamburg, [c] Inst. f. Kernphysik, KFA Jülich, [d] Inst. f. Theor. Kernphysik, Univ. Bonn, [e] Inst. f. Kernphysik, Univ. Köln

2. MOMO collaboration
   F. Bellemann[a], A. Berg[a], J. Bisplinghoff[a], G. Bohlscheid[a], J. Ernst[a], R. Frascaria[f], C. Henrich[a], F. Hinterberger[a], R. Jahn[a], L. Jarczyk[b], R. Joosten[a], U. Kuhn[a], A. Kozela[b], H. Machner[c], R. Maschuw[a], T. Mayer-Kuckuk[a], G. Mertler[a], B. Metsch[d], P.v. Neumann-Cosel[e], T.v. Oepen[a], H.R. Petry[d], D. Rosendaal[a], P.v. Rossen[c], K. Scho[a], F. Schwandt[a], R. Siebert[f], J. Smyrski[b], A. Strzalkowski[b], R. Tölle[c], W. Wiedmann[a], R. Wurzinger[a], R. Ziegler[a]
   [a] Inst. f. Strahlen- und Kernphysik, Universität Bonn, [b] Inst. f. Physik, Jagellonian Universität Krakau, Polen, [c] Inst. f. Kernphysik, Forschungszentrum Jülich, [d] Inst. f. Theor. Kernphysik, Universität Bonn, [e] Inst. f. Kernphysik, Technische Hochschule Darmstadt, [f] IPN Orsay, Frankreich.

3. A.J. Simon et al.: Phys. Rev. **C48**, 662 (1993)

Few-Body Systems Suppl. 8, 156–160 (1995)

Few-Body Systems

Study of Proton–Neutron Correlations with the reaction $^{3,4}He(e, e'd)$

C.M. Spaltro [1,2] *, Th.S. Bauer [3], H.P. Blok [1,2], T. Botto [2], E. Cisbani [4], G.E. Dodge [1,2] †, S. Frullani [4], F. Garibaldi [4], E. Jans [2], M. Iodice [4], W.J. Kasdorp [2], C. Kormanyos [1,2], L. Lapikás [2], R. de Leo [5], A. Misiejuk [3], C.J.G. Onderwater [1,2], R. Perrino [5], M. van Sambeek [1,2], R. Starink [1,2], G. van der Steenhoven [2], M.A. van Uden [2], G.M. Urciuoli [4], H. de Vries [2], M. Yeomans [3]

[1] Department of Physics and Astronomy, Vrije Universiteit, de Boelelaan 1081, 1081 HV Amsterdam, The Netherlands

[2] National Institute for Nuclear Physics and High-Energy Physics (NIKHEF-K), P.O. Box 41882, 1009 DB Amsterdam, The Netherlands

[3] Subatomic Physics Department, R.J. Van de Graafflaboratorium, Utrecht University, P.O. Box 80.000, 3508 TA Utrecht, The Netherlands

[4] INFN Sezione Sanità, V.le Regina Elena 299, 00161 Roma, Italy

[5] INFN Via per Arnesano, 73100 Lecce, Italy

Abstract. A $^{3,4}He(e, e'd)$ experiment has been performed at NIKHEF. The data have been taken in both (q, ω)-constant kinematics and in parallel-kinematics for transferred momenta q of 410, 500 and 600 MeV/c using 576 MeV and 370 MeV electron beams. Preliminary results for ^3He at 576 MeV are reported here. The (q, ω) constant results allow to extract the contribution of the structure function W_{LT} from the cross section. The parallel kinematics results show an interesting q-dependence that cannot be described by a cluster-knockout model.

1 Introduction

Proton-neutron correlations in a nucleus can be studied with the reaction $(e, e'd)$ provided that the reaction mechanism is understood [1, 2]. The study of this reaction on ^3He has the virtue that exact calculations with the Faddeev technique are available both for the initial bound state and the final continuum

* *E-mail address:* chiara@nikhefk.nikhef.nl

† *Present address:* Department of Physics, Old Dominion University, Norfolk, VA 23529, USA

state [3, 4]. It is of interest to study the reaction as well on ^{4}He in view of its rather high density, which may result in an enhancement of correlation effects. Moreover, in addition to the ^{2}H final state one can also study transitions to the $p-n$ continuum final state. No exact calculations are available for the four-nucleon system, but good approximate methods have been developed [5]. The validity of these calculations can be checked in the ^{3}He case by comparing them with results obtained from exact calculations and with the data.

Another important aspect of the comparison between results on ^{3}He and ^{4}He is the different role of the isospin in both reactions. In the reaction ^{3}He$(e,e'd)p$ the initial and final nucleus have isospin $T = 1/2$. Hence both $\Delta T = 0$ and $\Delta T = 1$ transitions will contribute to the cross section. In the reaction ^{4}He$(e,e'd)$ one can study these separately, since the transition to the bound ^{2}H final state is purely $\Delta T = 0$, whereas the transition to the $p-n$ breakup state just above threshold is dominantly $\Delta T = 1$.

In a Distorted Wave Impulse Approximation (DWIA) description [2] the cross section for the reaction $(e,e'd)$ is expressed in the cluster knockout model as:

$$\sigma^{\mathrm{DWIA}} = K\ \sigma_{e,pn}(q)\ S_{pn}(\mathbf{p}_m, E_m, \mathbf{p}_d), \tag{1}$$

where K is a phase-space factor, $\sigma_{e,pn}$ represents the cross section describing the interaction of an electron with a correlated proton-neutron pair (not necessarily a deuteron) in the initial nucleus, q is the transferred momentum, S_{pn} is the (distorted) momentum distribution of the proton-neutron pair, $\mathbf{p}_m$ is the missing momentum, E_m is the missing energy and $\mathbf{p}_d$ is the momentum of the outgoing deuteron. The q-dependence of $\sigma_{e,pn}$ contains information on the relative motion of the correlated proton and neutron, while the p_m dependence of S_{pn} contains information on the center-of-mass motion of the pair within the nucleus. This approach has proven to be successful in the description of deuteron knockout from ^{6}Li and ^{12}C [2].

Here we present a first analysis of the data in terms of a simpler Plane Wave Impulse Approximation (PWIA) assuming a cluster model. This means that we assume that the virtual photon couples to a pre-formed deuteron, which is knocked out of the target nucleus, that the deuteron inside the nucleus has the same properties as a free deuteron and that the motion of the outgoing deuteron with respect to the residual nucleus is described by a plane wave. In this model the cross section for a transition to a bound final state factorises into:

$$\sigma^{\mathrm{PWIA}} = K\ \sigma_{e,d}(q)\ \rho_2(\mathbf{p}_m), \tag{2}$$

where K is a phase-space factor, $\sigma_{e,d}$ is the elastic electron-deuteron cross section, defined in full analogy to $\sigma_{e,p}$ used in $(e,e'p)$ experiments, and $\rho_2(\mathbf{p}_m)$ is the momentum distribution of the deuteron in the nucleus before being knocked out. We present the data in the form of reduced cross sections $\sigma_{\mathrm{reduced}}$, i.e. the measured cross section divided by $K\ \sigma_{e,d}(q)$. It is important to realize that in this simple model the reduced cross section yields a q-independent deuteron momentum distribution for data taken at different values of q.

Independent of any model we want to disentangle the different components that contribute to the cross section. We will do this by expressing the $(e,e'd)$ cross section in terms of structure functions W:

$$\frac{d^6\sigma}{dE_{e'}d\Omega_{e'}d\Omega_d dE_d} = C\,\sigma_{\rm Mott}\,(v_L W_L + v_T W_T + v_{LT} W_{LT}\cos\phi + v_{TT} W_{TT}\cos 2\phi), \tag{3}$$

where ϕ is the angle between the scattering and the reaction plane and the v's are the electron kinematical factors. Since the different structure functions result from different combinations of electro-magnetic current components, the separate structure functions are expected to be sensitive to different aspects of the reaction mechanism. In this way we will use the separation of the cross section into structure functions as a tool to extract effects that are not clearly visible on the cross section level.

2 Experiment

The experiment has been performed using the extracted beam from the pulse stretcher ring AmPS at NIKHEF [6] at beam energies of 576 MeV and 370 MeV. In order to study simultaneously the reactions ^{3}He$(e,e'd)p$ and ^{4}He$(e,e'd)$ a mixed ^{3}He/^{4}He cryogenic gas target was used, operating at 20 K and 15 bar. The ^{3}He/^{4}He atomic ratio was determined from elastic scattering, and was about 2:1. For both energies data have been taken in (q,ω) constant kinematics in order to extract the W_L+W_{TT}, W_T and W_{LT} structure functions. Also data in parallel kinematics have been collected, for which a Rosenbluth separation of W_L and W_T can be made, since W_{LT} and W_{TT} vanish under these kinematical conditions. Measurements have been carried out at q-values of 410, 500 and 600 MeV/c, and different $\mathbf{p}_m$-ranges between 0 and 200 MeV/c.

3 Preliminary Results

We will only discuss results on ^{3}He here. All data shown are preliminary with an overall systematic uncertainty of 10 %. In Fig. 1 the results are shown for the reduced cross section as measured in (q,ω)-constant kinematics at q=410 MeV/c. The two kinematics differ only in the value of ϕ, being 0 and 180 deg (see Eq. 3). In the PWIA description of Eq. 2 the value of W_{LT} (see Eq. 3) is completely due to the longitudinal-transverse interference part of $\sigma_{e,d}$, so the reduced cross section should show no asymmetry. This is clearly not the case, the largest deviation occurring at a missing momentum $\mathbf{p}_m$ between 0 and 50 MeV/c. Hence a PWIA cluster knockout description is within the present uncertainty already ruled out.

In Fig. 2 the results are shown for the reduced cross sections measured in parallel kinematics at q = 410, 500 and 600 MeV/c. The lowest q data agree within experimental uncertainties with previous NIKHEF data obtained at q=380 MeV/c [9]. The present data show a strong q-dependence invalidating the cluster-knockout description. The observed differences in the reduced cross

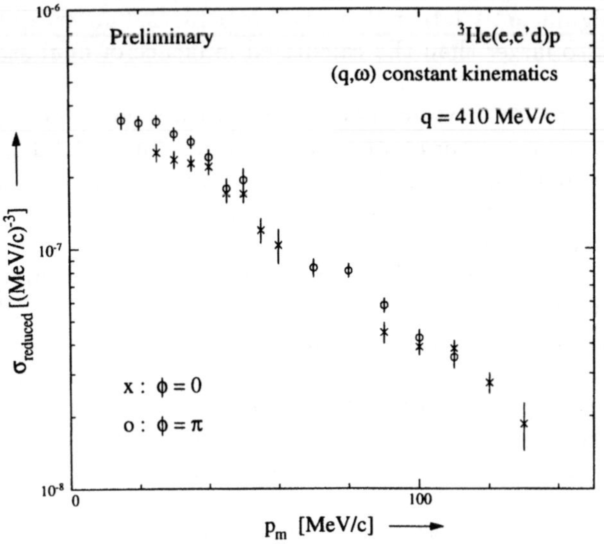

Figure 1. Reduced cross sections at $E_{\mathrm{beam}} = 576$ MeV for the reaction $^3\mathrm{He}(e, e'd)$ in (q,ω) constant kinematics. The overall uncertainty of these preliminary data is about 10 %.

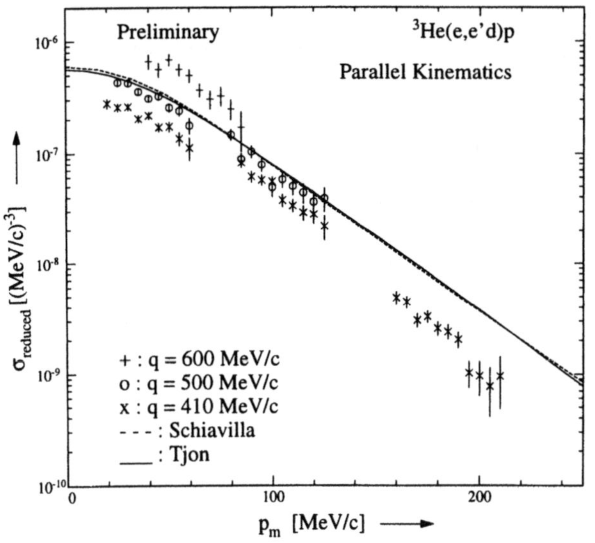

Figure 2. Reduced cross sections obtained at $E_{\mathrm{beam}} = 576$ MeV for the reaction $^3\mathrm{He}(e, e'd)$ in parallel kinematics. The overall uncertainty of these preliminary data is about 10 %. The solid and dashed curves are the deuteron momentum distributions as calculated by Tjon [7] and Schiavilla [8], respectively.

sections are significantly larger than the experimental uncertainties. The size of the effect is also larger than the calculated influence of final state interactions (FSI) [3].

From these preliminary results we can already conclude that the cluster-model description is too simplistic since the q-dependence of deuteron knockout from ^{3}He does not follow the elastic $e - d$ cross section. Various explanations can be given for this observation. First of all, the spatial extent of a $p - n$ pair in ^{3}He is expected to be different from that of a free deuteron. Secondly, elastic $e - d$ scattering has only a $\Delta T = 0$ part, which is almost purely longitudinal, whereas the reaction ^{3}He$(e, e'd)$ will also have $\Delta T = 1$ contributions, which are purely transverse. In a further analysis of the present data we will separate the longitudinal and transverse structure function in order to disentangle these different contributions. Similar information on the isospin dependence of the deuteron knockout reaction will emerge from the analysis of the two-body and three-body breakup channels in the reaction ^{4}He$(e, e'd)$.

Acknowledgement. This work is part of the research program of the National Institute for Nuclear Physics and High-Energy Physics (NIKHEF) made possible by the financial support from the Foundation for Fundamental Research of Matter (FOM) and the Netherlands' Organization for the Advancement of Research (NWO).

References

1. W. Leidemann et al.: Phys. Rev. **C50**, 630(1994)
2. R. Ent et al.: Nucl. Phys. **A578**, 93(1994)
3. S. Ishikawa et al.: Il Nuovo Cimento **107A**, 305-330(1994)
4. E. van Meijgaard, J.A. Tjon: Phys. Rev. **C42**, 96-107(1990)
5. S.I. Nagornyi et al.: Yad. Fiz. **57**, 1001(1994)
6. P.K.A. de Witt Huberts: Nucl. Phys. **A553**, 845(1993)
7. J.A. Tjon: Private Communication (1993)
8. R. Schiavilla et al.: Nucl. Phys. **A449**, 219-242(1986)
9. P.H.M. Keizer et al.: Phys. Lett. **157B**, 255(1985)

Few-Body Systems Suppl. 8, 161–165 (1995)

Few-Body Systems

Selected Topics of the Few–Nucleon Research Program at TUNL

W. Tornow[1], C.R. Howell[1], R.T. Braun[1], Q. Chen[1], D.E. Gonzalez Trotter[1], C.D. Roper[1], F. Salinas[1], H.R. Setze[1], R.L. Walter[1], H. Witała[2]

[1] Department of Physics, Duke University and Triangle Universities Nuclear Laboratory, Durham, NC 27708, USA

[2] Institute of Physics, Jagellonian University, PL-30059 Cracow, Poland

Abstract. Results of kinematically-complete cross-section measurements of the $n-d$ breakup reaction at $E_n = 13$ MeV are reported. The new data for the space-star configuration of the three outgoing nucleons are in good agreement with previous data but are considerably different from rigorous $n-d$ calculations. A status report on our $n-n$ and $n-p$ FSI measurements is given. New information about the importance of three-nucleon force effects in the $n-d$ reaction is expected from these data. New data for the analyzing power in $n-p$ scattering at $E_n = 12$ MeV are presented. The preliminary analysis of these accurate data does not support the small value obtained by the Nijmegen group for the charged πNN coupling constant.

1 Neutron-Deuteron Breakup Experiments and Rigorous $3N$ Calculations

The measurements were performed using a shielded neutron source at the Triangle Universities Nuclear Laboratory (TUNL). To obtain an absolute cross section for the $n-d$ breakup reaction, we normalized to $n-d$ elastic scattering, which was measured simultaneously. The relative efficiency of each neutron detector was measured by detecting fission neutrons emitted from a ^{252}Cf source. The absolute efficiency was determined by detecting neutrons from the ^{2}H$(d,n)^3$He reaction at discrete energies. In addition, the absolute detection efficiency was deduced from the measured light-response function obtained for our detectors at the Physikalisch-Technische Bundesanstalt in Braunschweig, Germany, for more than 30 energies in the neutron energy range from 1 to 15 MeV.

Our preliminary results for the space-star configuration are shown in Fig. 1 in comparison to previous $n-d$ data [1] and rigorous $3N$ calculations [2] using

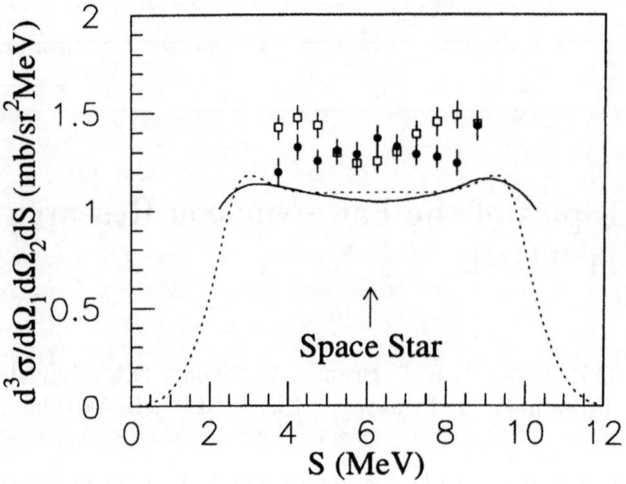

Figure 1. Cross section for $n-d$ space-star configuration at $E_n = 13$ MeV. Squares: data of Strate et al. [1]; dots: present data. Dotted curve: Rigorous $3N$ calculations for point geometry using the Bonn B NN potential [3]; solid curve: calculated finite geometry result using rigorous $3N$ calculations in a Monte Carlo simulation of the experimental setup.

the Bonn B nucleon-nucleon (NN) potential [3]. The normalization uncertainty in the present cross-section data is ±4.6% and is due to the combined uncertainties in the absolute neutron detection efficiency and the normalization factor determined from the $n-d$ elastic scattering yields. Since a unique procedure does not exist for extracting point-geometry cross sections from the present finite-geometry data, we smeared the calculations via Monte Carlo techniques over the finite geometry and energy of our experimental setup. The finite-geometry effects in the present data are small, as indicated by the difference between the dashed and solid curves. Our data are in good agreement with those of Strate et al. [1], but are roughly 20% higher than theoretical predictions that use pair-wise realistic NN forces. The disagreement between data and theory for the space-star configuration presents a legitimate reason for detailed experimental and theoretical studies of three-nucleon forces (3NF) in the $3N$ system.

2 Search for 3NF Effects via Measurements of the $n-n$ and $n-p$ FSI Cross Section in the $n-d$ Breakup Reaction at $E_n = 13$ MeV

In addition to the $3N$ bound state, the $3N$ continuum system can provide valuable information about 3NF effects. Here, the experimentalist has the freedom to position detectors at any angle in the laboratory system to study any $3N$ configuration of interest. The idea behind the TUNL approach is the following:

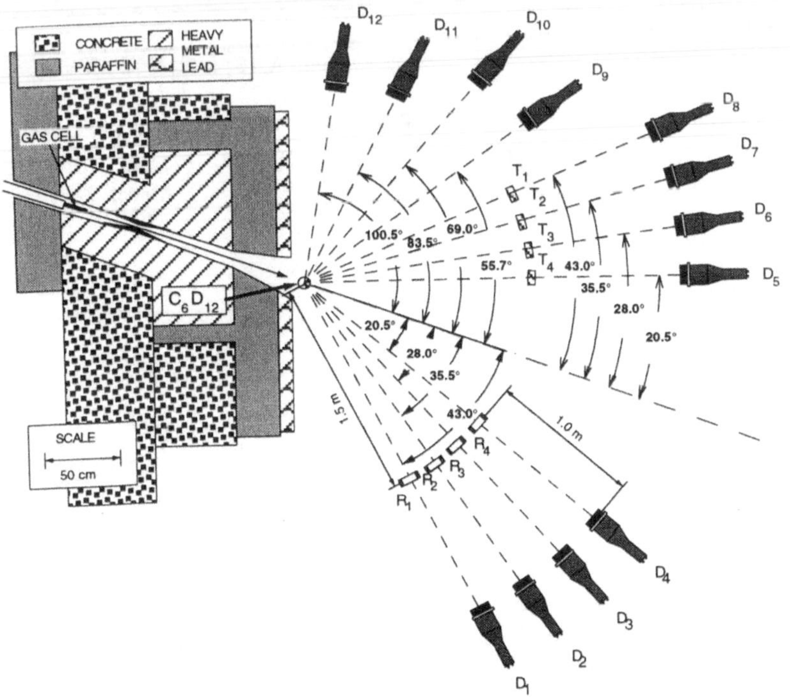

Figure 2. Experimental setup for TUNL a_{nn} and a_{np} experiment.

Suppose that the $^1S_0$ neutron-neutron scattering length a_{nn} is known accurately from the $\pi^- - d$ capture reaction ($a_{nn} = -18.6 \pm 0.3$ fm) [4] where, in contrast to the $n - d$ breakup reaction, only two strongly interacting particles exist in the exit channel. Then, any discrepancy between this value and results extracted from $n-n$ final-state-interaction (FSI) cross-section data of the $n-d$ breakup reaction and rigorous $3N$ calculations with 2N forces only is a strong indication of 3NF effects. Similarly, the $^1S_0$ neutron-proton scattering length a_{np} is very well known from free $n - p$ scattering ($a_{np} = -23.748 \pm 0.010$ fm) [4]. Again, any difference between this value and results obtained from $n - p$ FSI cross-section data must be attributed to 3NF effects. Since the 3NF effects are expected to be angle dependent, the $n - n$ and $n - p$ FSI cross sections are being measured simultaneously at different production angles of the $n - n$ and $n - p$ pairs (see Fig. 2). The four ring detectors, R_1 to R_4, and the four end detectors, D_1 to D_4, positioned on the right side of the incident neutron beam are used for the $n - n$ FSI measurements. In order to increase the statistical accuracy of these data, the detector positions are mirrored about the beam axis. Here, instead of ring detectors, four transmission detectors, T_1 to T_4, are used. All detectors on the right side of the beam axis are also used together with the four end detectors, D_9 to D_{12}, on the the left side of the beam axis to

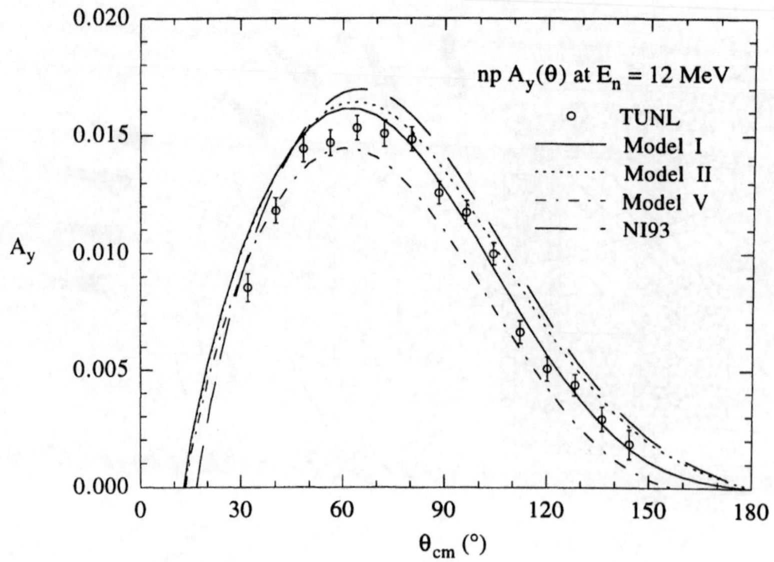

Figure 3. Preliminary results for the $n-p$ analyzing power $A_y(\theta)$ at $E_n = 12$ MeV in comparison to the NI93 $n-p$ phase-shift analysis prediction [6] and to model studies of Machleidt and Li [5] based on different values for the πNN coupling constants. Details are given in the text.

measure the $n-p$ FSI cross section. Almost 65% of the required 2000 hours of data taking have been completed already and first results are expected to be available in about a year.

3 The Low-Energy $n-p$ Analyzing Power and the Charged πNN Coupling Constant

Machleidt and Li [5] have recently shown that the new value for the charged πNN coupling constant of $g^2_{\pi\pm}/4\pi = 13.54 \pm 0.05$ obtained by the Nijmegen group [6] makes it impossible to reproduce accurately the quadrapole moment Q_d of the deuteron with existing realistic meson-exchange-based NN potential models, unless new tensor force generating mechanisms are introduced. The Nijmegen group has repeatedly pointed out that there exists no single NN observable that can be used to determine $g^2_{\pi\pm}/4\pi$: it is the bulk of the NN data that makes an accurate determination of the coupling constant possible. Nevertheless, in measurements at TUNL we focused on a single observable: the analyzing power $A_y(\theta)$ in $n-p$ scattering. At sufficiently low energies only s- and p-waves contribute to this observable. Since the p-waves are peripheral waves, they mainly probe the long-range part of the NN interaction, which is primarily governed by OPE. As is well known, $A_y(\theta)$ for $n-p$ scattering at low energies is determined almost completely by the ^{3}P NN interactions;

therefore, $A_y(\theta)$ is expected to be sensitive to $g^2_{\pi^\pm}/4\pi$. Recent studies performed by Machleidt and Li [5] support this intuitive picture.

We have measured the $n-p$ $A_y(\theta)$ at incident neutron energies of 7.6 and 12.0 MeV to a statistical accuracy of ± 0.0005. Our preliminary results at $E_n = 12$ MeV are presented in Fig. 3 and compared to the Nijmegen phase-shift analysis [6] and to NN potential models [5] based on different values for $g^2_{\pi^\pm}/4\pi$. The new data definitely rule out the proposed charge splitting [5] of the πNN coupling constants (see dashed-dotted curve), i. e., the new value $g^2_{\pi^0}/4\pi = 13.47 \pm 0.11$ and the old value $g^2_{\pi^\pm}/4\pi = 14.4$, a combination that reproduces both Q_d and p-p scattering data [5].

In the angular range $\theta_{cm} > 60°$ our data are in agreement with Model I (solid curve) which is based on $g^2_{\pi^0}/4\pi = g^2_{\pi^\pm}/4\pi = 14.4$. Model II (dotted curve) uses $g^2_{\pi^0}/4\pi = g^2_{\pi^\pm}/4\pi = 13.5$ and predicts $A_y(\theta)$ values which are slighly larger than the present data for $\theta_{cm} > 60°$. The model dependence of the study of Machleidt and Li [5] can be inferred from the comparison of the dotted curve (Model II) with the long-dashed curve, which was calculated from the Nijmegen $n-p$ phase-shift analysis NI93 [6]. Our data are still subject to small corrections associated with the polarization-dependent efficiency of our neutron detectors. The n-^{12}C $A_y(\theta)$ data needed for these corrections will be taken in the near future at TUNL. However, it can be stated already at the present time that our final data will not be in agreement with the NI93 prediction.

Acknowledgement. This work was supported in part by the U. S. Department of Energy, Office of High Energy and Nuclear Physics, under grant No. DEFG05-ER40619 and by the Maria Skłodowska-Curie II Fund under grant No. MEN/NSF-94-161. The calculations were performed on the Cray Y-MP of the North Carolina Supercomputing Center at Research Triangle Park, North Carolina.

References

1. J. Strate et al.: Nucl. Phys. **A51** (1989).

2. H. Witała, Th. Cornelius, W. Glöckle: Few-Body Syst. **3**, 1 (1988)

3. R. Machleidt: Adv. Nucl. Phys. **19**, 189 (1989)

4. G. A. Miller, B. M. K. Nefkens, I. Šlaus: Phys. Rep. **194**, 1 (1990)

5. R. Machleidt, G. O. Li: *Constraints on the πNN Coupling Constant from the NN system*, presented at the 5th International Symposium on Meson-Nucleon Physics and the Structure of the Nucleon, Boulder, Colorado (1993)

6. V. Stoks, R. Timmermans, J. J. de Swart: Phys. Rev. C**47**, 512 (1993)

Few-Body Systems Suppl. 8, 166–170 (1995)

Few-Body Systems

Measurement of the Spin–dependent Structure Function $g_1(x)$ of the Deuteron and the Proton

M.F. Litmaath, for the Spin Muon Collaboration at CERN

NIKHEF, postbus 41882, NL-1009 DB Amsterdam, The Netherlands

Abstract. In 1994 the Spin Muon Collaboration has measured the spin-dependent structure function g_1^d of the deuteron in deep inelastic scattering of 190 GeV polarised muons on polarised deuterons, in the kinematic range $0.003 < x < 0.7$ and $1 < Q^2 < 60\,\mathrm{GeV}^2$. The first moment $\Gamma_1^d = \int_0^1 g_1^d(x)\,dx$ obtained from all SMC deuteron data evaluated at $Q_0^2 = 10\,\mathrm{GeV}^2$ is $0.034 \pm 0.009\,(\mathrm{stat.}) \pm 0.006\,(\mathrm{syst.})$. This value is below the Ellis-Jaffe sum rule prediction by three standard deviations. Using our earlier determination of Γ_1^p we obtain $\Gamma_1^p - \Gamma_1^n = 0.199 \pm 0.038$ which agrees with the Bjorken sum rule prediction within one standard deviation.

1 Introduction

The structure function g_1^d and the asymmetry $A^d = (\sigma^{\uparrow\downarrow} - \sigma^{\uparrow\uparrow})/(\sigma^{\uparrow\downarrow} + \sigma^{\uparrow\uparrow})$ of the single-photon exchange cross sections for antiparallel ($\uparrow\downarrow$) and parallel ($\uparrow\uparrow$) longitudinal spins of the muon and deuteron are related to the virtual-photon deuteron asymmetries A_1^d and A_2^d by:

$$A^d = D(A_1^d + \eta A_2^d) \tag{1}$$

$$g_1^d = \frac{F_2^d}{2x(1+R)}(A_1^d + \gamma A_2^d), \tag{2}$$

where η, γ depend only on kinematic variables and are small in the domain of the SMC measurement. The virtual-photon deuteron asymmetry A_1^d is defined as $\frac{1}{2}(\sigma_0 - \sigma_2)/(\frac{1}{3}(\sigma_0 + \sigma_1 + \sigma_2))$, where σ_J is the virtual-photon deuteron absorption cross section for total spin projection J in the direction of the virtual photon. The upper bound of A_2^d has been included in the systematic error on A_1^d. The depolarisation factor D depends also on R, the longitudinal-transverse virtual-photon absorption cross section ratio. For the unpolarised structure functions F_2 and R we used the parametrisations by the NMC and SLAC, respectively.

2 The Experiment

The principal components of the experimental setup [1] are the polarised target, the forward spectrometer, and the beam polarimeter. The target previously used for the measurement of $g_1^{\rm p}$ [2, 4] was slightly enlarged and employed with fully deuterated butanol as target material. The target cells were polarised in opposite directions along the beam axis, by dynamic nuclear polarisation using frequency modulated microwaves. The polarisation was measured by NMR coils connected to series Q-meter circuits with an overall accuracy of $\Delta P_{\rm t}/P_{\rm t} = 5.4\%$. An average polarisation difference between upstream and downstream target cells of 0.97 was obtained throughout the data-taking. The spin direction of the polarised deuterons was reversed every five hours by a rotation of the magnetic field, with a polarisation loss of less than 0.003.

The μ^+ beam had an intensity of 4×10^7 per spill of 2.4 s duration with 14.4 s period, an average muon momentum of 190 GeV and a polarisation $P_\mu = -0.811 \pm 0.028\,(\text{stat.}) \pm 0.029\,(\text{syst.})$. The polarisation was determined using the shape of the energy spectrum of positrons from the decay $\mu^+ \to e^+\nu_e\bar{\nu}_\mu$. It is in agreement with our previous measurement at the same energy. The incident muon momentum was determined with an accuracy of $\Delta p/p = 0.5\%$ in a magnetic spectrometer upstream of the target. The incident muon track was reconstructed with an angular resolution of 0.1 mrad using scintillator hodoscopes and a multiwire proportional chamber.

The trajectory of the scattered muon was reconstructed from hits in about 150 planes of proportional chambers, drift chambers and streamer tubes located upstream, downstream and inside of the forward spectrometer magnet. The muon was identified as a track behind a hadron absorber made of 2 m thick iron. Due to the large number of detector planes the spectrometer acceptance was largely insensitive to any efficiency changes of the individual planes. The angle and the momentum of the scattered muon were determined with average accuracies of 0.4 mrad and 1.3 GeV, respectively. The interaction vertex was reconstructed with a resolution better than 30 mm (0.3 mm) parallel (perpendicular) to the beam direction.

Events were selected by applying cuts which ensured that the beam flux was the same for both target cells and that events from unwanted target material were suppressed. Further cuts removed muons from decays of hadrons and limited the size of kinematic smearing effects and radiative corrections. In total 6×10^6 reconstructed events were used in the analysis.

The asymmetry $A_1^{\rm d}$ was extracted from combinations of data sets taken before and after reversal of the spin directions. In the evaluation of the asymmetry it is assumed that in each combination the ratio $r = a_{\rm u}/a_{\rm d}$ is independent of time, where $a_{\rm u(d)}$ is the acceptance of the upstream (downstream) target cell. If this assumption is false, the evaluated asymmetry has a spurious contribution $\Delta A_1^{\rm d}$. For each data set we have determined the efficiencies of all tracking detectors. From the largest changes observed an upper limit of 0.003 has been estimated for $|\Delta A_1^{\rm d}|$. The radiative corrections were found to change $A_1^{\rm d}$ by less than 0.003 in the entire kinematic range.

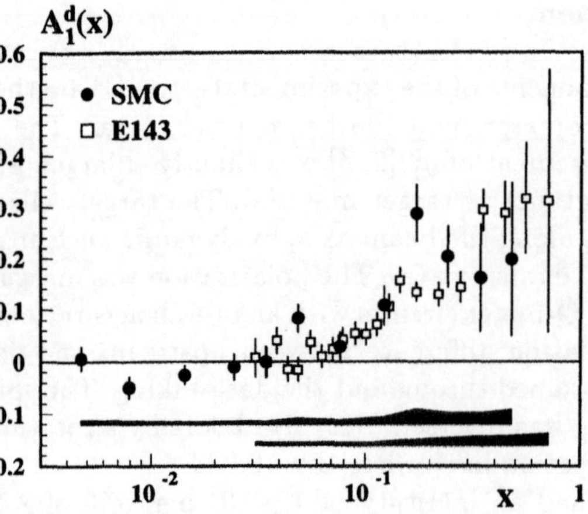

Figure 1. The virtual-photon deuteron cross section asymmetry A_1^d as a function of the scaling variable x at the average Q^2 of each x bin. Only statistical errors are shown on the data points. The size of the systematic errors is indicated by the shaded area. Results from SLAC E143 are shown for comparison.

3 Results

The results for $A_1^d(x)$ at the average Q^2 of each bin of x are shown in Fig. 1. In the analysis we have included data taken in 1992 with a beam of 100 GeV [3]. The results from both measurements were found to be in agreement. The dominant systematic errors are due to uncertainties in A_2^d, radiative corrections, time dependence of the acceptance ratio r, target and beam polarisations, and R. The total systematic error shown in Fig. 1 combines all contributions in quadrature.

The results for $g_1^d(x)$ at the average Q^2 of each x bin are shown in Fig. 2. Below $x \approx 0.03$ we find g_1^d to be negative. To obtain $g_1^d(x)$ at $Q_0^2 = 10$ GeV^2 it is assumed that A_1^d is independent of Q^2. The contribution to Γ_1^d from the measured x range then is:

$$\int_{0.003}^{0.7} g_1^d(x, Q_0^2)\,\mathrm{d}x = 0.0358 \pm 0.0087 \pm 0.0051 \quad (Q_0^2 = 10\ \mathrm{GeV}^2), \qquad (3)$$

where here and in the following the first error is statistical and the second one is systematic. The contribution from the unmeasured region at small x is $\int_0^{0.003} g_1^d(x)\,\mathrm{d}x = -0.0028 \pm 0.0016$. This contribution has been obtained assuming a Regge-type dependence $g_1^d(x) = \mathit{constant}$ that we fit to the two lowest x points. The error has been enlarged to account also for extrapolations obtained using the general form of Regge behaviour $g_1^d(x) \propto x^\alpha$, with $0 \leq \alpha \leq 0.5$. The contribution from the unmeasured region at large x is $\int_{0.7}^{1} g_1^d(x)\,\mathrm{d}x = 0.0006 \pm 0.0010$. For the extrapolation to $x = 1$ we used $A_1^d = 0.4 \pm 0.6$ which is consistent with the data and with the bound $|A_1^d| \leq 1$. The result for the first moment then is:

$$\Gamma_1^d = 0.034 \pm 0.009 \pm 0.006 \qquad (Q_0^2 = 10 \text{ GeV}^2) \tag{4}$$

The relation

$$\Gamma_1^p + \Gamma_1^n = 2\Gamma_1^d/(1 - \tfrac{3}{2}\omega_D) \tag{5}$$

can be used to determine the contributions to the nucleon spin from the sum of the quark spins ($\Delta\Sigma$) and from the strange quark spins (Δs), assuming $SU(3)_f$ relations for the axial vector coupling constants within the baryon octet. For the probability for the deuteron to be in the D state we have taken $\omega_D = 0.05 \pm 0.01$ which covers most of the values proposed. Off-mass-shell effects in g_1^d were calculated to be negligible in the region of our measurement.

The QCD corrections were calculated in second order using four quark flavours and $\alpha_s(Q_0^2 = 10 \text{ GeV}^2) = 0.24 \pm 0.03$, corresponding to $\alpha_s(m_Z^2) = 0.117 \pm 0.005$. For the $SU(3)_f$ coupling constants F and D we used $F + D = g_A/g_V = 1.2573 \pm 0.0028$ and $F/D = 0.575 \pm 0.016$. Neglecting contributions from polarised charm quarks and polarised gluons we obtain:

$$\Delta\Sigma = \Delta u + \Delta d + \Delta s = 0.20 \pm 0.11 \tag{6}$$

$$\Delta s = -0.12 \pm 0.04 \tag{7}$$

These results are inconsistent with the Ellis and Jaffe assumption of $\Delta s = 0$ and the resulting prediction $\Delta\Sigma = 3F - D = 0.579 \pm 0.025$. For the first moment the Ellis-Jaffe sum rule predicts $(\Gamma_1^d)_{EJ} = 0.070 \pm 0.004$ which is three standard deviations above the measured value. The conclusions on Δs and on the Ellis-Jaffe sum rule depend on the $SU(3)_f$ symmetry assumption. The value of $\Delta\Sigma$ appears rather insensitive to $SU(3)_f$ breaking effects.

Using Eq. 5 we combine our result for Γ_1^d (Eq. 4) with our result for $\Gamma_1^p = 0.136 \pm 0.011 \pm 0.011$ [2, 4], and find agreement with the Bjorken sum rule prediction at $Q_0^2 = 10 \text{ GeV}^2$, which includes third order α_s corrections in perturbative QCD for four quark flavours:

$$\Gamma_1^p - \Gamma_1^n = 0.199 \pm 0.038 \qquad (SMC,\ Q_0^2 = 10 \text{ GeV}^2) \tag{8}$$

$$\Gamma_1^p - \Gamma_1^n = 0.187 \pm 0.003 \qquad (Theory,\ Q_0^2 = 10 \text{ GeV}^2) \tag{9}$$

For the measurement the statistical and systematic errors were added in quadrature. The error in the prediction is due to the uncertainty in α_s. Our experimental results also give $\Gamma_1^n = -0.063 \pm 0.024 \pm 0.013$.

The spin-dependent structure function $g_1^n(x)$ of the neutron was determined by combining g_1^d and g_1^p [2, 4]. Results are are shown in Fig. 3, compared to those obtained similarly from the SLAC E143 data and to the results from the E142 experiment that used a ^{3}He target, all evolved to $Q_0^2 = 5 \text{ GeV}^2$. All measurements agree within the common x-range. However, our data at small x are more negative than expected from Regge-type extrapolations from the common region.

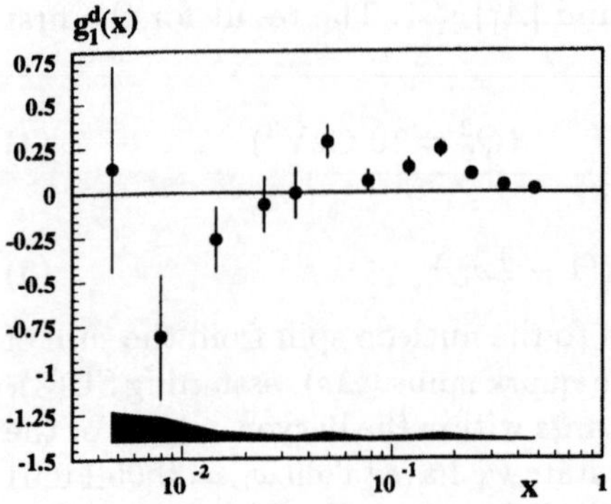

Figure 2. The spin-dependent structure function $g_1^d(x)$ as a function of the scaling variable x at the average Q^2 of each x bin. Only statistical errors are shown on the data points. The size of the systematic errors is indicated by the shaded area.

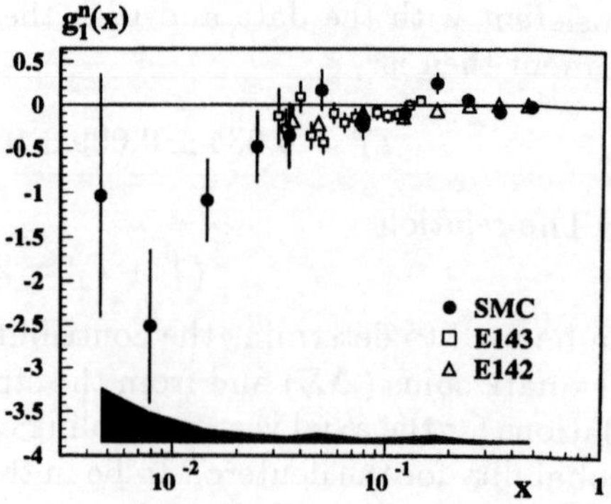

Figure 3. The spin-dependent structure function $g_1^n(x)$ as a function of the scaling variable x evaluated at a common $Q_0^2 = 5\,\mathrm{GeV}^2$. Only statistical errors are shown on the data points. The size of the systematic errors is indicated by the shaded area. Results from the SLAC E142 and E143 experiments are shown for comparison.

In summary, we have presented a new measurement of the spin-dependent structure function g_1^d extending down to $x = 0.003$. The result for the first moment Γ_1^d is found to disagree with the Ellis-Jaffe prediction by three standard deviations and gives $\Delta\Sigma = 0.20 \pm 0.11$ and $\Delta s = -0.12 \pm 0.04$ for the total spin fractions carried by all quarks and by the strange quarks, respectively. The SMC results for the proton and the deuteron confirm the Bjorken sum rule.

References

1. SMC, D. Adams et al.: Phys. Lett. **B357**, 248 (1995), and references therein.

2. SMC, D. Adams et al.: Phys. Lett. **B329**, 339 (1994);
Erratum, Phys. Lett. **B339**, 332 (1994)

3. S. Platchkov: Few-Body Systems Suppl. **7**, 449 (1994)

4. R. van Dantzig: AIP Conference Proceedings, *Few-Body Problems in Physics*, **334**, 343 (1995)

Few-Body Systems Suppl. 8, 171–175 (1995)

Few-Body Systems

Double Pion Photoproduction on the Proton up to 800 MeV

A. Braghieri[1]*, L.Y. Murphy[2], J. Ahrens[3], G. Audit[4], N. d'Hose[4], V. Isbert[4], S. Kerhoas[4], J.M. Laget[4], M. Mac Cormick[4], P. Pedroni[1], T. Pinelli[5], G. Tamas[4], A. Zabrodin[6]

1 INFN-Sezione di Pavia, via Bassi 6, 27100 Pavia, Italy

2 Physics Department, Rensselaer Polytechnic Institute, Troy NY 12180

3 Institut für Kernphysik, Universität Mainz, 55099 Mainz, Germany

4 CEA DAPNIA-SPhN, C.E. Saclay, 91191 Gif sur Yvette, France

5 Dipartimento di Fisica Nucleare e Teorica, Università degli Studi di Pavia, via Bassi 6, 27100 Pavia, Italy

6 INR, Russian Academy of Sciences, Moscow, Russia

Abstract. The total cross sections for the three $\gamma p \rightarrow N\pi\pi$ reactions have been measured for photon energies from 450 to 800 MeV. The $\gamma p \rightarrow p\pi^0\pi^0$ and $\gamma p \rightarrow n\pi^+\pi^0$ cross sections have never been measured before while the $\gamma p \rightarrow p\pi^+\pi^-$ results are much improved compared to earlier data. These measurements were performed with the large acceptance hadronic detector DAPHNE, at the tagged photon beam facility of MAMI microtron in Mainz.

1 Introduction

Of the three possible double pion photoproduction channels on hydrogen, only one, $\gamma p \rightarrow p\pi^+\pi^-$, has to date been measured with any accuracy [1]-[2]. Earlier experiments were for the most part performed with untagged bremsstrahlung photons incident on bubble chambers. In these instances, the $p\pi^+\pi^-$ channel was identified by the detection of at least two charged particles in coincidence. The two other possible reaction channels, each involving two neutral particles in the final state, were thus not measurable using such techniques.

In this paper, we present measurements of $\sigma_{\text{tot}}(p\pi^+\pi^-)$, $\sigma_{\text{tot}}(n\pi^+\pi^0)$ and $\sigma_{\text{tot}}(p\pi^0\pi^0)$, from 450 up to 800 MeV. In this energy range, the cross sections are sensitive not only to the Δ, but also to the higher baryon resonances. In

* *E-Mail address:* braghieri@pv.infn.it

particular the $p\pi^0\pi^0$ channel is the most interesting one because the Born terms are suppressed since the photon does not couple with neutral pions. Thereby the higher resonances contribution should be more clearly seen. As a consequence the simultaneous measure of the three channels gives a strong constraint to the theoretical studies of these resonances.

2 Experimental Set-Up

The experiment was performed with the large acceptance detector DAPHNE [3] installed on the tagged photon beam line [4] operated by Glasgow University at the continuous wave accelerator MAMI in Mainz [5].

The photon energy ranged from 350 up to 800 MeV with an average resolution of 2 MeV. The total tagged photon flux was $3 \cdot 10^6 \gamma\ s^{-1}$. The flux was continuously monitored throughout the experiment, such that the accuracy of the photon normalization has been evaluated to better than $\pm 1\%$.

We used the detector Daphne as described in ref. [3]. Its relevant characteristics are: i) high precision measurements of the angles of emission of the charged particles, ii) good p/π particle discrimination, iii) good proton momentum resolution ($\Delta p/p = 2.5 - 10\%$ for p=300-900 MeV/c), iv) modest π^0 detection efficiency, v) large solid angle acceptance (94% of 4π sr), vi) a large momentum acceptance (proton detection threshold: $p_p = 280$ MeV/c, and charged pion detection threshold: $p_\pi = 78$ MeV/c).

3 Data Analysis

There are five active channels in our energy range:

- (a) $\gamma p \to p\pi^0$; (b) $\gamma p \to n\pi^+$
- (c) $\gamma p \to n\pi^+\pi^0$; (d) $\gamma p \to p\pi^+\pi^-$; (e) $\gamma p \to p\pi^0\pi^0$

Given the characteristics of our detector, only channel (b) cannot be uniquely identified, as the detection of a single charged π is not a sufficient condition to distinguish it from either (c) or (d). In the majority of events, this π escapes from the detector so that its momentum cannot be directly measured, as discussed in ref. [6]. Kinematical constraints, consequently, may not be applied.

3.1 The channels $\gamma p \to p\pi^0\pi^0$, $\gamma p \to n\pi^+\pi^0$, and $\gamma p \to p\pi^0$

These channels, each with at least one π^0 in the final state, are unambiguously signed by a triple coincidence between a charged particle and the two γ's resulting from the decay of the π^0. The π^0 detection efficiency as a function of photon energy, ϵ_{π^0}, was evaluated using the GEANT code. All three reaction channels were simulated with the complete set of geometrical and electronic thresholds taken into account.

To achieve these results, we have used the $p\pi^0$ angular distribution found in ref. [7] and have assumed a three-body phase space distribution for the channels $p\pi^0\pi^0$ and $n\pi^+\pi^0$. The results from these calculations yield a global efficiency of $\epsilon_{\pi^0} \simeq 20\%$ for channels (a) and (c), and $\epsilon_{\pi^0} \simeq 38\%$ for (e).

In the second stage of the analysis, $n\pi^+\pi^0$ events are selected by p/π discrimination as in ref. [6]. Of the remaining events, distinction between $p\pi^0$ and $p\pi^0\pi^0$ events is based on the analysis of the missing mass spectrum $\gamma p \rightarrow pX$. The good proton energy resolution allows final states containing one and two π^0 to be distinguished.

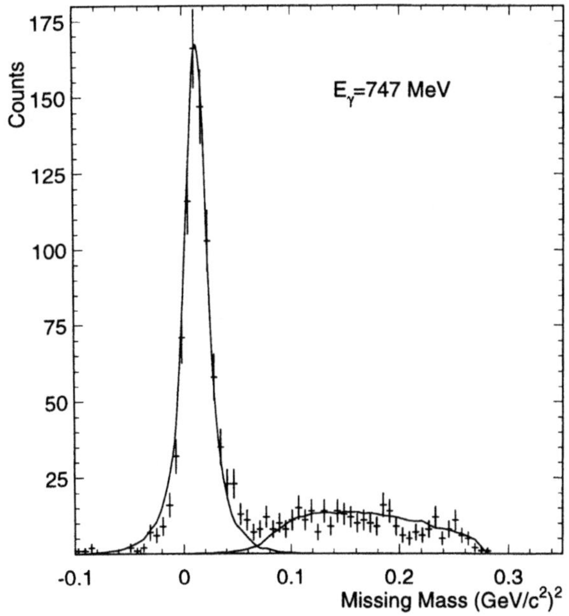

Figure 1. Missing mass distribution for protons in coincidence with a π^0.

As an example, we show, in fig. 1, the missing mass spectrum at $E_\gamma = 747$ MeV. The sharp peak, corresponding to the squared mass of the π^0, is due to the two body reaction (a). The curve through the figure is the result of a simulation which takes into account the various experimental resolutions as well as the known angular distribution associated with reaction (a). The distribution found at higher squared masses, corresponds clearly to the three body process (e). It has also been fitted with a three-body phase space angular distribution. The contributions from processes (a) and (e) are clearly separated.

3.2 The Channel $\gamma p \rightarrow p\pi^+\pi^-$

The measurement of this channel was the most straightforward: it is separated from the others by selecting events containing two or three charged particles in coincidence.

4 Total Cross Section Evaluation

Total cross sections were determined by extrapolation of the angular and momentum distributions in the dead angles and below the detection threshold by the help of a model recently developed by L.Y. Murphy and J.M. Laget [8].

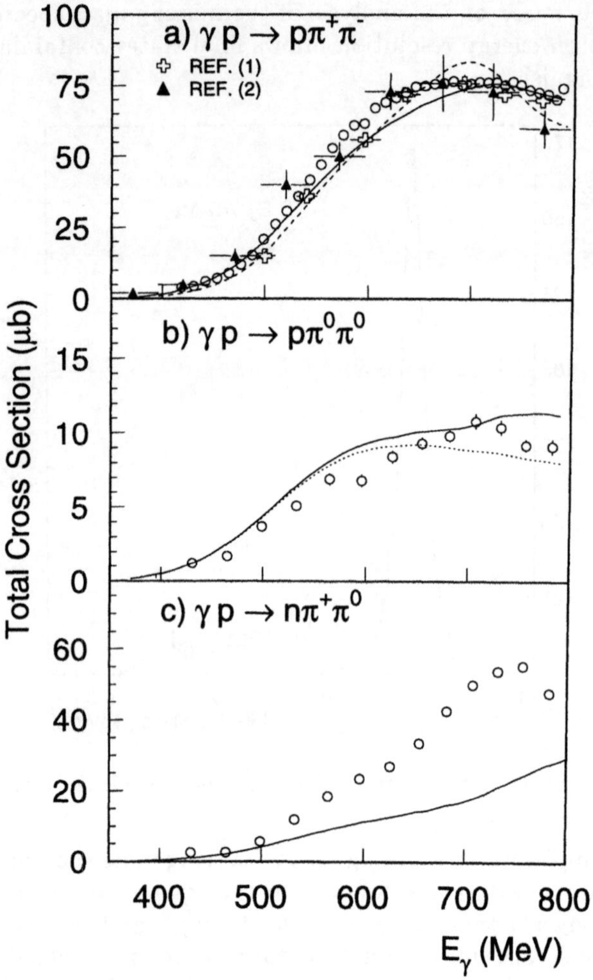

Figure 2. Total cross sections for the three double pion photoproduction channels.

For the $p\pi^+\pi^-$ and $n\pi^+\pi^0$ channels this extrapolation is mostly of the order of 20-25% whereas it is greater than 50% for $p\pi^0\pi^0$ channel for which the proton, mostly forward peaked, has to be detected.

The model takes into account both the contribution of the $\pi\Delta$ Born terms and the decay of N^* resonances (namely the $D_{13}(1520)$ and $P_{11}(1440)$) in the $\Delta\pi$ and $N\pi\pi$ channels. It reproduces well the cross section integrated over

DAPHNE acceptance for $p\pi^+\pi^-$ and $p\pi^0\pi^0$, but it fails for $n\pi^+\pi^0$. However in this case the extrapolation is quite small. The comparison with a 3-body phase-space extrapolation allow us to estimate to 2% the systematic error on the total cross section.

5 Results and Comments

Figs. 2a-b-c show the results for the three double pion photo-production channels. In complete agreement with previously published data, but with markedly improved precision, $\sigma_{\mathrm{tot}}(p\pi^+\pi^-)$ exhibits a rapid rise from 450 to 600 MeV, and then saturates. This is in contrast to $\sigma_{\mathrm{tot}}(n\pi^+\pi^0)$ and $\sigma_{\mathrm{tot}}(p\pi^0\pi^0)$ which continue to rise smoothly to about 720 MeV.

In the same figures our results are compared with the Murphy and Laget model (full lines) and in fig. 2a also with the Gómez Tejedor and Oset model [9] (dashed line).

The most relevant feature is that the model predicts a dominant contribution of the P_{11} resonance for the $p\pi^0\pi^0$ channel (dotted line in fig. 2b). This is an important result in view of understanding the nature of the P_{11} which, in the naive quark constituent model, is thought to be the radial excitation of the nucleon. The prediction of Murphy and Laget tends to be in conflict with this representation as real transverse photons may not excite such a mode.

References

1. ABBHHM Collaboration: Phys. Rev. **175**, 1669 (1968)
2. G. Gialanella et al.: Nuovo Cimento **63A**, 892 (1969)
3. G. Audit et al.: Nucl. Instr. and Meth. **A301**, 473 (1991)
4. I. Anthony et al.: Nucl. Instr. and Meth. **A301**, 230 (1991)
5. H. Herminghaus et al.: Nucl. Instr. and Meth. **A138**,1 (1976)
6. A. Braghieri et al.: Nucl. Instr. and Meth. **A343**, 623 (1994)
7. Landolt-Börnstein: *Photoproduction of Elementary Particles* (group 1, vol. 8). Berlin: Springer 1973
8. L.Y. Murphy, J.M. Laget: Phys. Lett. (to be published)
9. J.A. Gomez Tejedor, E. Oset: Nucl. Phys. **A571**, 667 (1994)

Few-Body Systems Suppl. 8, 176–180 (1995)

High Missing–Momentum Components in ^{4}He Studied with the $(e, e'p)$ Reaction

J.J. van Leeuwe[1]*, H.P. Blok[1,2], J.F.J. van den Brand[3], H.J. Bulten[3], G.E. Dodge[1,2]†, R. Ent[4], W.H.A. Hesselink[1,2], E. Jans[1], W.-J. Kasdorp[1], J.M. Laget[5], L. Lapikás[1], C.J.G. Onderwater[1,2], A. Pellegrino[1,2], C.M. Spaltro[1,2], J.J.M. Steijger[1], J.A. Templon[1,2], O. Unal[3]

[1] National Institute for Nuclear Physics and High-Energy Physics, section K, P.O. Box 41882, 1009 DB Amsterdam, The Netherlands

[2] Department of Physics and Astronomy, Free University, Amsterdam, The Netherlands

[3] Department of Physics, University of Wisconsin, Madison, Wisconsin, USA

[4] CEBAF, Newport News, Virginia, USA

[5] Service de Physique Nucléaire, Centre d'Etudes de Saclay, 91191 Gif-sur-Yvette Cedex, France

1 Introduction

The nucleon-nucleon (NN) interaction in the nuclear medium is of fundamental importance for the description of nuclear properties. In particular its short-range part is still poorly known. High-momentum components of the nuclear wave function are sensitive to short range correlations (SRC). Measurement of high momenta will therefore yield information on the NN-interaction at short internucleon distances [1]. Because of the availability of realistic calculations of the wave function of few-body systems, a light nucleus is preferred for such a study. Since the effects of short range correlations are expected to be enhanced in dense nuclei we choose for ^{4}He.

The $(e, e'p)$ reaction is suited to study high-momentum components, since in this way both the initial momentum and removal energy of the ejected proton can be determined. Calculations indicate that the high-momentum components are correlated with high removal energies [2]. In this kinematical domain the

* *email*: koos@nikhefk.nikhef.nl

† *present address*: Department of Physics, Old Dominion University, Norfolk, VA 23529, USA

reaction is sensitive to both the central part and the tensor part of the interaction, while the ^{4}He$(e,e'p)^3$H reaction is sensitive to its central part only. For this reason we have chosen to determine the proton momentum distribution of ^{4}He not only for high removal energies, but also for the channel in which the (A-1) system remains a triton.

2 Experimental Setup

The experiment has been performed at the electron scattering facility of NIKHEF with the extracted electron beam from the Amsterdam Pulse Stretcher (AmPS) [3]. As a target we used a cryogenic (15 K) high-pressure (4 MPa) ^{4}He gas target. The electrons were detected in the QDQ high-resolution magnetic spectrometer.

Protons were detected in the highly-segmented HADRON-4 scintillator detector (see ref. [4] for prototype). At the nominal distance of 80 cm from the target, the detector has an angular acceptance of $\pm 21°$, both in-plane and out-of-plane, resulting in a solid angle of 500 msr. The angular resolution is 1° and 2° for in-plane and out-of-plane, respectively. The total thickness of the scintillator material is 18 cm, which in combination with a 5.2 mm lead-absorber, results in a proton energy acceptance of 67-195 MeV.

The coincidence timing resolution, after applying all corrections, is 0.8 ns FWHM. The energy resolution determined with the ^{1}H$(e,e'p)$ reaction is 3.3 MeV (FWHM) for a proton energy of 122 MeV. This corresponds to 2.7 %, a value that turns out to be reasonably constant over the total energy acceptance.

3 Kinematics

The quantities missing momentum ($\mathbf{p}_{\mathrm{m}}$) and missing energy (E_{m}) are defined in the following way:

$$\boldsymbol{p}_m = \boldsymbol{q} - \boldsymbol{p}' \tag{1}$$

$$E_m = \omega - T_{p'} - T_{A-1} \tag{2}$$

in which $\mathbf{q}$ and ω are the momentum and energy transferred by the virtual photon, respectively. $\mathbf{p}'$ is the momentum and $\mathrm{T}_{\mathrm{p}'}$ the kinetic energy of the knocked out proton and $\mathrm{T}_{\mathrm{A-1}}$ the kinetic energy of the residual A-1 system. In the Plane Wave Impulse Approximation $\mathbf{p}_{\mathrm{m}}$ can be identified with the recoil momentum $\mathbf{p}_{\mathrm{A-1}}$ and is equal in size and opposite in direction to the initial proton momentum. $\mathrm{T}_{\mathrm{A-1}}$ can be calculated from the measured quantities.

For the present experiment we used an electron beam with an energy of 525 MeV and a duty factor of 50%. The scattered electron was detected at an angle of 49.6°. Since multi-nucleon processes are expected to be important in the dip region, between the quasi-elastic peak and the delta resonance region, we chose the energy transfer correspondingly: $\omega = 215$ MeV, leading to a momentum transfer of 400 MeV/c. The resulting virtual photon polarization ε was 0.63. At this fixed electron kinematics we used four settings of HADRON-4, covering a range of 8° to 101° for the angle between $\mathbf{q}$ and $\mathbf{p}'$.

4 Results

The kinematical settings mentioned in the previous section result in a coverage of the (E_m, p_m) plane up to 700 MeV/c in p_m and 150 MeV in E_m. Because of a dependence of the E_m-acceptance on p_m, the cutoff of the acceptance shifts to lower E_m when p_m increases. In Fig. 1 the E_m spectra for various values of p_m are shown. The data shown are preliminary, but the results are not ex-

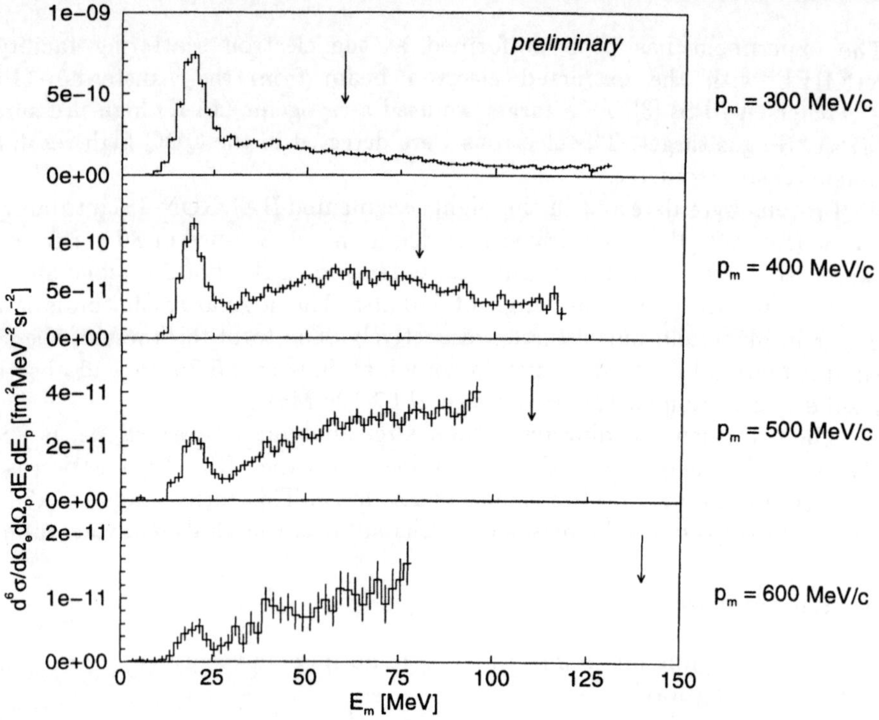

Figure 1. Preliminary six-fold differential cross section as a function of missing energy for 4 p_m intervals of 20 MeV/c wide. The arrows indicate where the centroid of the continuum is expected according to the 'scattering on a pair' model.

pected to change by more than 10 %. Radiative unfolding has been applied. The missing-energy resolution turns out to be 6 MeV at the highest proton energy, which corresponds to the lowest missing momentum. The peak around 19.8 MeV corresponds to the ^{4}He$(e,e'p)^3$H channel, i.e. the reaction in which the (A-1) system remains in its ground state. The thresholds for the ^{4}He$(e,e'p)dn$ and the ^{4}He$(e,e'p)ppn$ reactions are 26.1 and 28.3 MeV, respectively. This means that above 28.3 MeV both reactions, which are experimentally indistinguishable, play a role. This part of the E_m-spectrum will be referred to as the 'continuum'. From the figure it can be seen that the strength of the two-body breakup channel decreases rapidly with increasing p_m, whereas the strength in the continuum decreases less fast.

If a model is assumed in which the reaction takes place on a nucleon pair with thc other two nucleons as spectators, the undetected nucleon of the pair takes the recoil energy. The missing energy of such a process is:

$$E_m = E_{thr} - T_{A-1} + (\sqrt{p_m^2 + M_N^2} - M_N) \tag{3}$$

The center of mass motion of the pair will cause a broadening of the peak. The arrows in the figure indicate where the centroid is expected to show up.

To obtain the five-fold differential cross section, we integrated the two-body breakup peak for different values of p_m, by fitting the spectrum with a gaussian and a three-parameter function for the continuum. The result is shown in Fig. 2. From the figure it can be seen that the data do not show any

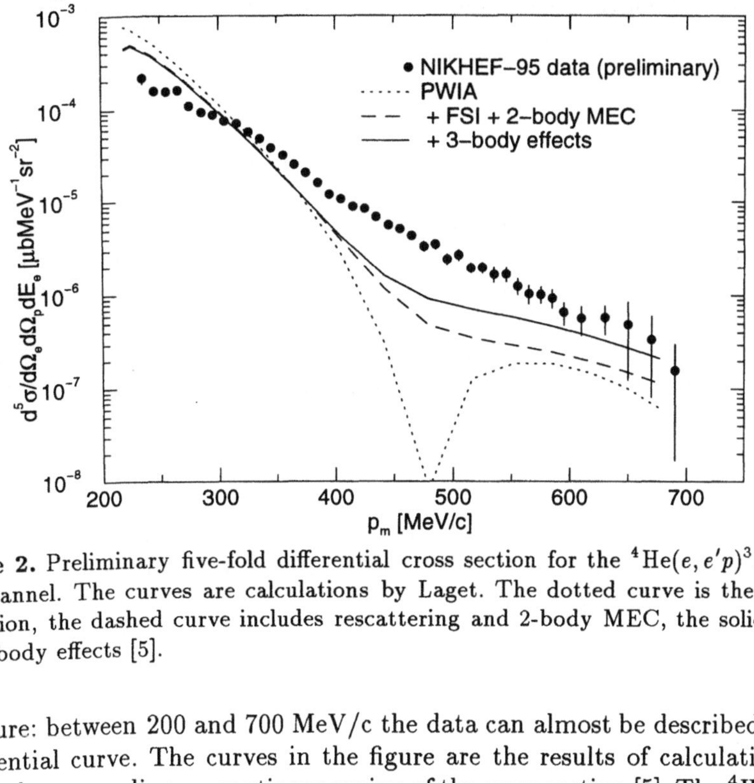

Figure 2. Preliminary five-fold differential cross section for the $^4He(e,e'p)^3H$ reaction channel. The curves are calculations by Laget. The dotted curve is the PWIA prediction, the dashed curve includes rescattering and 2-body MEC, the solid curve also 3-body effects [5].

structure: between 200 and 700 MeV/c the data can almost be described by an exponential curve. The curves in the figure are the results of calculations by Laget who uses a diagrammatic expansion of the cross section [5]. The ^{4}He wave function results from a variational calculation with the Urbana-V14 potential [6]. The minimum in the PWIA prediction is washed out when FSI and MEC are taken into account, but contrary to the data, still a change of slope can be seen at p_m=450 MeV/c in the full calculation. The data extend in a region where 3-body effects become important, but because of the discrepancy with the calculation between 250 and 450 MeV/c, the importance of 3-nucleon effects can not be quantified.

Dedicated calculations have also been performed by Laget for the continuum, the results of which are shown in Fig. 3. The spectra shown in this figure are six-fold differential cross sections for two values of the angle, γ_{pq}, between

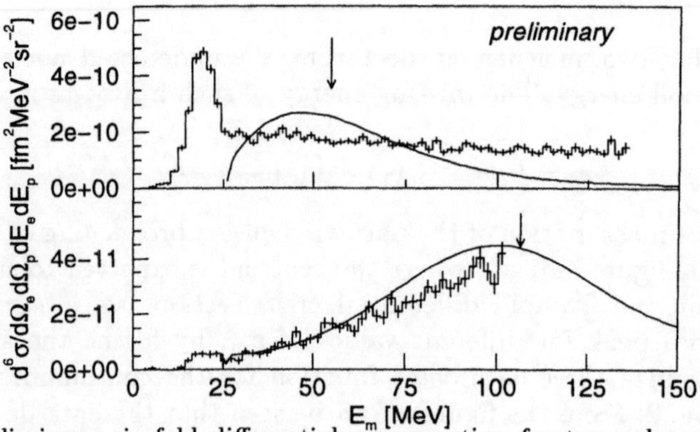

Figure 3. Preliminary six-fold differential cross section for two values of the angle between $\mathbf{p}'$ and $\mathbf{q}$, 30° (top) and 79° (bottom). The curves are the full calculations by Laget.

$\mathbf{q}$ and $\mathbf{p}'$. The arrows indicate where the bump is expected according to the 'scattering on a pair model' (Eq. 3). From the upper figure it can be concluded that the shape of the full calculation disagrees with that of the data, but that the strength at the peak of the calculation, which has the same position as in the naive 'scattering on a pair' model, is of roughly the right size. Apparently at higher missing energy other processes play a dominant role. In the lower figure it can be seen that both the shape and the strength of the data are well reproduced and that the position of the peak of the calculation agrees with the prediction of the model mentioned above. Hence it can be concluded that scattering on a pair is the dominant process at this value of γ_{pq}.

Acknowledgement. This work is part of the research program of the National Institute for Nuclear Physics and High-Energy Physics (NIKHEF) made possible by the financial support from the Foundation for Fundamental Research of Matter (FOM) and the Netherlands' Organization for the Advancement of Research (NWO). This work is supported in part by the National Science Foundation under Grant no. PHY-9314783 (Wisconsin).

References

1. J.M. Laget: Phys. Lett. **B199**, 493 (1987); J.M. Le Goff et al.: Phys. Rev **C50**, 2278 (1994)
2. H. Morita, T. Suzuki: Prog. Theor. Phys. **86**, 671 (1991)
3. P.K.A. de Witt Huberts: Nucl. Phys. **A553**, 845c (1993)
4. A. Zondervan et al.: Nucl. Instr. and Meth. **A342**, 436 (1994)
5. J.M. Laget: Nucl. Phys. **A579**, 333 (1994)
6. R. Schiavilla et al.: Nucl. Phys. **A449**, 219 (1986)

Few-Body Systems Suppl. 8, 181–185 (1995)

Electron–Deuteron Scattering with the New Generation of Nucleon–Nucleon Potentials

R.F. Wagenbrunn*, W. Plessas

Institute for Theoretical Physics, University of Graz, Universitätsplatz 5, A-8010 Graz, Austria

Abstract. We have employed the recent N-N interaction models developed by the Bonn and Nijmegen groups in elastic e-d scattering. In particular, we report here corresponding results for the electric and magnetic structure functions, the deuteron tensor polarization and the charge form factor. A comparison to presently available experimental data is given.

Recently the Bonn and Nijmegen groups have developed new versions of their meson-exchange models for the N-N interaction. The new potentials do not only provide a number of refinements over previous versions but also guarantee a better reproduction of the whole N-N data base. We consider here the meson-theoretical Bonn-B [1], FULLF as well as OBEPF [2], and Nijm93 [3] potentials. In addition, we include in our present analysis also the phenomenological Reid-like N-N potentials Nijm-I (non-local) as well as Nijm-II (local) of the Nijmegen group [3]. Our results provide a check on the (half-)off-shell behaviour of the new N-N potentials. This is important for their further applications in few-nucleon problems.

Our calculations of elastic e-d scattering observables (for more details see ref. [4]) include relativistic (spin-orbit, Darwin-Foldy, and nuclear-motion) effects (RC) and the most important meson-exchange-current (MEC) contributions (π-pair, $\rho\pi\gamma$, and π-retardation currents). For the meson-exchange N-N potentials the vertex form factors were chosen consistently with the specific underlying interaction. For the phenomenological N-N potentials a "natural" parametrization of the MEC was taken, relying on experimental coupling strengths and common cut-off parameters. For the nucleon form factors our calculation, whose results are demonstrated below, used Dirac-Pauli form factors and employed the best-fit parametrization of Lomon [5], usually called IJLG [6]; of course, we have also investigated the effects of different parametrizations

*Supported by the Paul-Urban Foundation

of the nucleon form factors but we cannot present nor discuss the corresponding results here.

We have found that neither of the above N-N models is able to describe the electromagnetic deuteron structure for momentum transfers $q^2 \lesssim 80\ \text{fm}^{-2}$ in a really satisfactory manner. With respect to the electric structure function $A(q^2)$ the difficulties are evident from Figs. 1 and 2. Especially at low q^2, all potentials fail to explain the newer Saclay data [7] within the experimental errors. The same kind of shortcomings were observed before for previous N-N potentials [8].

Another (related) problem consists in the reproduction of the zero of the charge form factor $F_C(q)$. The latest experimental analysis [10] reports it to occur at $q_0 = 4.39 \pm 0.16\ \text{fm}^{-1}$. Only three theoretical models (Bonn-B, FULLF, and Nijm-I) yield a q_0 perfectly within the experimental error, while the others are too low (see Table 1). In the same manner also the description of the tensor polarization $T_{20}(q)$ beyond the minimum is affected (Fig. 3). Only the potentials with a correct q_0 succeed in reproducing the three outmost data points.

Table 1. Positions of the zero of the charge form factor $F_C(q)$ as resulting from our IA + RC + MEC calculation.

Potential	q_0 [fm^{-1}]
Bonn-B	4.28
FULLF	4.26
OBEPF	4.08
Nijm93	4.15
Nijm-I	4.27
Nijm-II	4.15
Experiment [10]	4.39 ± 0.16

The magnetic structure functions $B(q^2)$ are shown in Fig. 4. The best descriptions are provided by FULLF, Bonn-B, Nijm-I, and Nijm-II. No diffractive minimum is observed. However, we consider our calculation reliable only up to $q = 6\ \text{fm}^{-1}$. Beyond that region any theory is afflicted with rather large uncertainties, especially with respect to $B(q^2)$.

Concentrating on the low-q^2 domain, we have carefully checked the dependence of our results on various ingredients of the calculation (type of relativistic corrections, meson-exchange currents, nucleon form factors, etc.). There the corresponding uncertainties cannot account for the differences between theory and experiment. At least we have found no acceptable (reasonable) choice of the debatable ingredients that would allow to reproduce all the experimental data simultaneously, even below $q \lesssim 6\ \text{fm}^{-1}$.

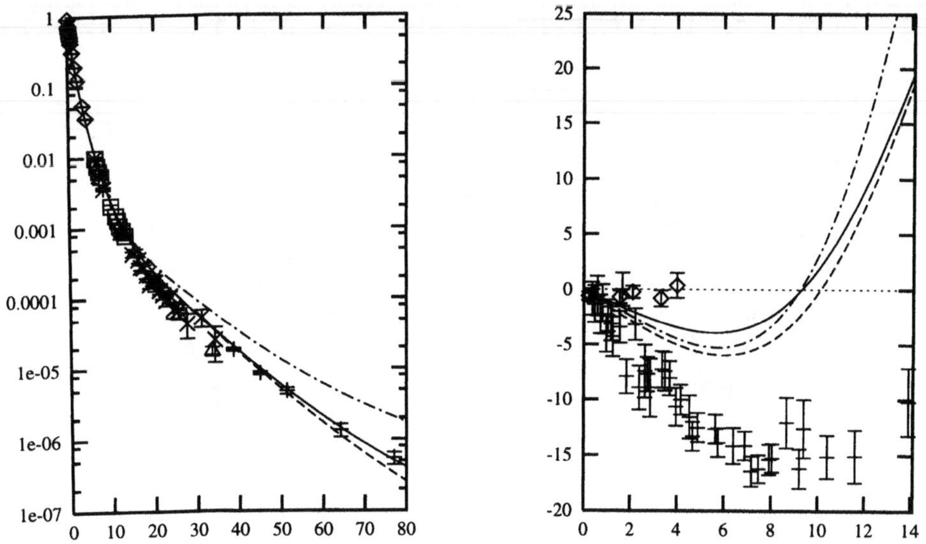

Figure 1. Electric structure functions $A(q^2)$ for the Bonn-B (solid), FULLF (dashed), and OBEPF (dash-dotted) potentials as resulting from our IA + RC + MEC calculation. For the relative percentage deviations shown in the right part the Paris-potential [9] IA calculation was taken as the reference. Experimental data as in ref. [8].

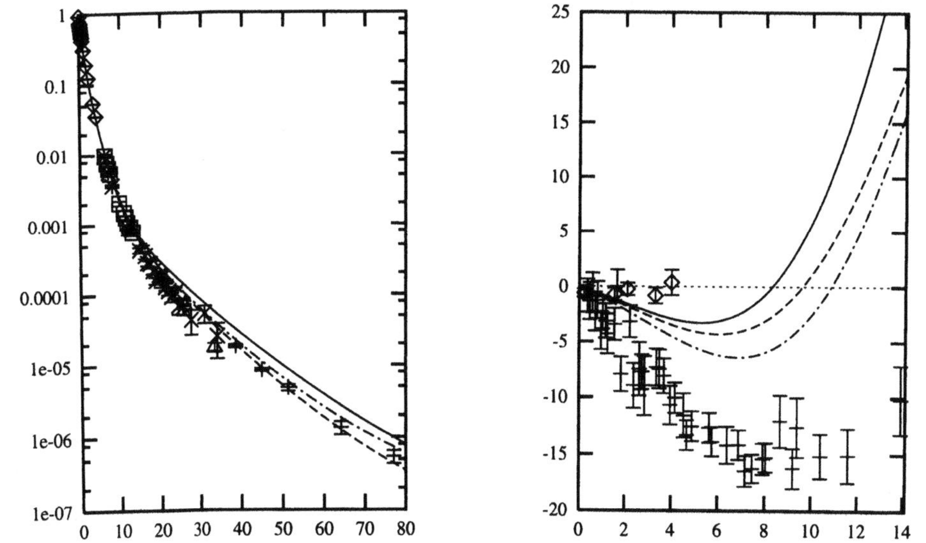

Figure 2. Same as in Fig. 1 for the Nijm93 (solid), Nijm-I (dashed), and Nijm-II (dash-dotted) potentials.

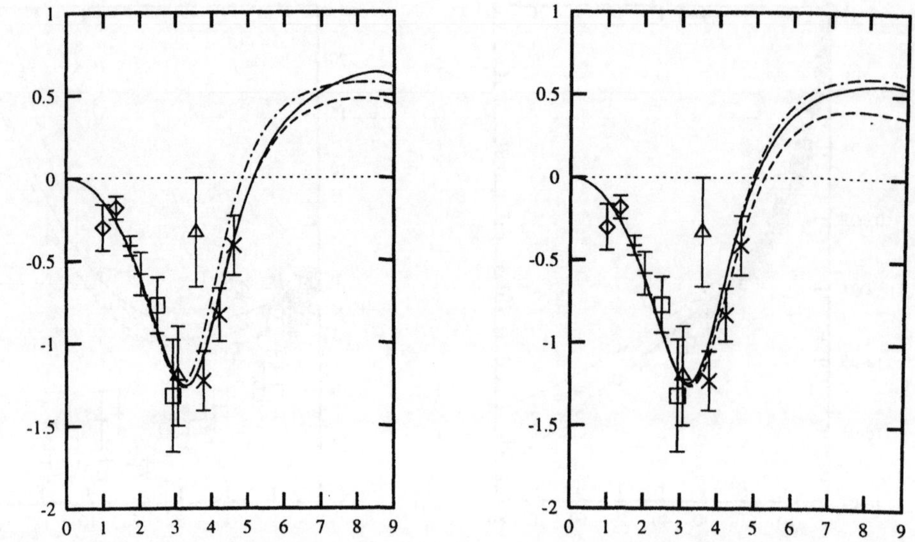

Figure 3. Deuteron tensor polarizations $T_{20}(q)$ for the Bonn potentials of Fig. 1 (left) and for the Nijmegen potentials of Fig. 2 (right). Experimental data from refs. [10, 11, 12, 13, 14, 15].

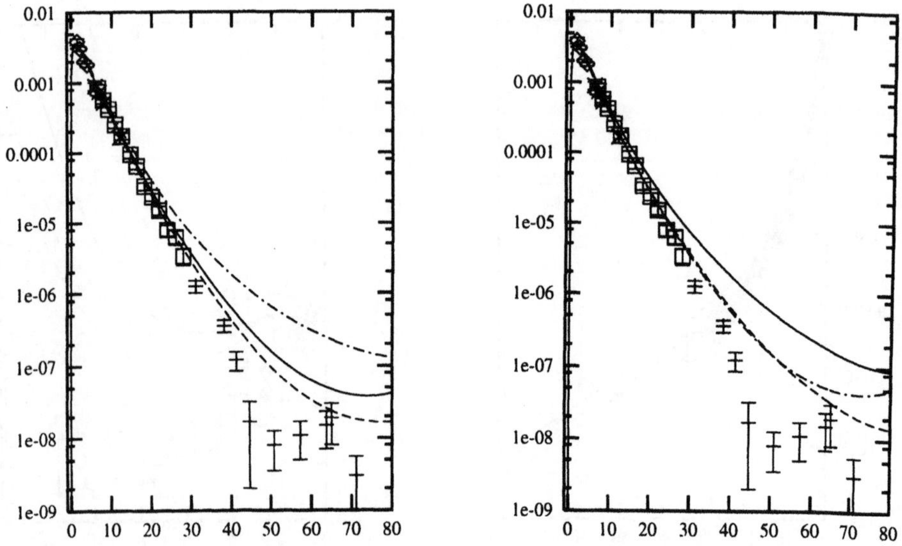

Figure 4. Magnetic structure functions $B(q^2)$ for the Bonn potentials of Fig. 1 (left) and for the Nijmegen potentials of Fig. 2 (right). Experimental data as in ref. [4].

References

1. R. Machleidt: Adv. Nucl. Phys. **19**, 189 (1989)
2. J. Haidenbauer, K. Holinde, and M.B. Johnson: Phys. Rev. C **45**, 2055 (1992)
3. V.G.J. Stoks, R.A.M. Klomp, C.P.F. Terheggen, J.J. de Swart: Phys. Rev. C **49**, 2950 (1994)
4. P. Obersteiner, W. Plessas, J. Pauschenwein: Few-Body Systems Suppl. **5**, 140 (1992)
5. E.L. Lomon: Ann. Phys. (N.Y.) **125**, 309 (1980)
6. F. Iachello, A.D. Jackson, A. Landé: Phys. Lett. **43B**, 191 (1973); S. Galster et al.: Nucl. Phys. **B32**, 221 (1971)
7. S. Platchkov, A. Amroun, S. Auffret, J.M. Cavedon, P. Dreux, J. Duclos, B. Frois, D. Goutte, H. Hachemi, J. Martino, X.H. Phan, I. Sick: Nucl. Phys. **A510**, 740 (1990)
8. W. Plessas, Ch. Brandstätter, S. Cvijetic, J. Haidenbauer, L. Mathelitsch, P. Obersteiner, J. Pauschenwein, and R. Wagenbrunn: Few-Body Systems Suppl. **7**, 251 (1994)
9. M. Lacombe, B. Loiseau, J.M. Richard, R. Vinh Mau, J. Côté, P. Pirès, R. de Tourreil: Phys. Rev. C **21**, 861 (1980)
10. M. Garçon, J. Arvieux, D.H. Beck, E.J. Beise, A. Boudard, E.B. Cairns, J.M. Cameron, G.W. Dodson, K.A. Dow, M. Farkhondeh, H.W. Fielding, J.B. Flanz, R. Goloskie, S. Høibråten, J. Jourdan, S. Kowalski, C. Lapointe, W.J. McDonald, B. Ni, L.D. Pham, R.P. Redwine, N.L. Rodning, G. Roy, M.E. Schulze, P.A. Souder, J. Soukup, I. The, W.E. Turchinetz, C.F. Williamson, K.E. Wilson, S.A. Wood, W. Ziegler: Phys. Rev. C **49**, 2516 (1994)
11. V.F. Dmitriev et al.: Phys. Lett. **157B**, 143 (1985)
12. B.B. Voitsekhovskii et al.: JETP Lett. **43**, 733 (1986)
13. M.E. Schulze et al.: Phys. Rev. Lett. **52**, 597 (1984)
14. R. Gilman et al.: Phys. Rev. Lett. **65**, 1733 (1990)
15. C.E. Jones, K.P. Coulter, R. Gilman, R.J. Holt, E.R. Kinney, R.S. Kowalczyk, D.H. Potterveld, L. Young, V.V. Frolov, S.I. Mishnev, D.M. Nikolenko, S.G. Popov, I.A. Rachek, D.K. Toporkov, E.P. Tsentalovich, B.B. Wojtsekhowski, C.W. de Jager, G. Retzlaff, J. Theunissen, H. de Vries, V.V. Nelyubin, V.V. Vikhrov, V.N. Stibunov, A.V. Osipov: Few-Body Systems Suppl. **7**, 112 (1994)

Few-Body Systems Suppl. 8, 186–190 (1995)

Few-Body Systems

© by Springer-Verlag 1995

Polarization Studies of Few Nucleon Systems at Laboratory for High Energies of Joint Institute for Nuclear Research

E.A. Strokovsky *

Laboratory for High Energies, JINR, 141980, Dubna, Moscow region, Russia

Abstract. The LHE facilities and beams used for polarization studies of few nucleon systems are characterized. Recent results of experiments on deuteron breakup at 0° and elastic backward *dp* scattering with unpolarized target bring to the conclusion that the available data display new features which hardly could be interpreted within the conventional picture of the deuteron as a 2–nucleon system with nucleons being in relative *S* and *D* orbital states.

New opportunities for studies of few nucleon systems were opened with installation of the Movable Polarized Target in the LHE JINR.

Two accelerators are now in operation in the Laboratory for High Energies of Joint Institute for Nuclear Research. The first machine is the well–known Synchrophasotron, which accelerates protons up to $T_{\rm kin} \simeq 9$ GeV and nuclei (including ^{28}Si) with $Z/A = 1/2$ up to $T_{\rm kin} \simeq 4$ GeV/A; the new machine is the superconductive synchrotron, the Nuclotron, which will accelerate protons up to $T_{\rm kin} \simeq 12.8$ GeV and nuclei (including ^{238}U) up to $T_{\rm kin} \simeq 6$ GeV/A (for $Z/A = 1/2$). The main characteristics on the complex are given in refs. [1].

The strategy taken in the LHE can be characterized as a **smooth transition** from the Synchrophasotron to the Nuclotron. The idea is to replace beams from the Synchrophasotron by beams from the Nuclotron so, that experiments can be carried out with that machine, which is able to provide requested beams.

At present, the Nuclotron reached half of its maximal energy and provides beams for experiments with internal targets [1]. Completion of the slow extraction system is among the 1-st priority tasks of the Laboratory [1].

Parameters of the beams of special interest for studies of few nucleon systems are listed in Table 1 (see also refs. [1]). Polarized proton and neutron beams are obtained from stripping of the vectorially polarized deuterons.

* *E-mail address:* strok@sunhe.jinr.dubna.su

Table 1. Beams: intensity (particles per cycle) and polarization

Beam	Synchro-phasotron	Nuclo-tron (I)	Nuclo-tron (II)	Polarization (Synhrophasotron)
p	$4 \cdot 10^{12}$	10^{11}	10^{13}	
n	10^{10}	$5 \cdot 10^{9}$	10^{13}	
d	10^{12}	$5 \cdot 10^{10}$	10^{13}	
d_{pol}	$(1–5) \cdot 10^{9}$	$3 \cdot 10^{8}$	10^{11}	$\mid P_{zz} \mid \sim 0.8$, $\mid P_z \mid \sim 0.5$
p_{pol}	$\sim 2 \cdot 10^{6}$			$\mid P_z \mid \sim 0.5$
n_{pol}	$\sim 10^{6}$			$\mid P_z \mid \sim 0.5$
^{3}He	$2 \cdot 10^{10}$			
^{4}He	$5 \cdot 10^{10}$	$5 \cdot 10^{9}$	$2 \cdot 10^{12}$	

Most of **experiments made at LHE with polarized deuterons** are devoted to the problem of deuteron structure at short distances. These are the experiments on the inclusive deuteron breakup [2] at 0°, backward dp elastic scattering (BES) [3], where the most detailed information was obtained, and exclusive deuteron breakup on protons in 4π geometry [4]. Studies of cumulative particle production by polarized deuterons are started.

In Fig. 1 all the significant observables for the **deuteron breakup** at zero degree with unpolarized targets are presented as functions of the light cone variable k (see refs. [6]) which has the meaning of the momentum of a nucleon in the deuteron. The data shown in the figure is a sample from the full data set obtained with hydrogen and nuclear targets at different energies [2, 5]. The data from inclusive $d(e, e')$ scattering [7] are shown for comparison.

The main results of these experiments can be summarized as follows:
1) the empirical momentum density (EMD), T_{20} and κ_0 are almost independent on the initial energy (T_{kin} from 1.5 to 7.4 GeV) [2, 5], on the type of a target (protons or nuclei) [2, 5] and on the type of a probe (compare with data [7]).
2) Dependence of all the observables on k disagrees drastically with calculations within existing models based on standard deuteron wave functions.
3) T_{20} tends to be a constant and κ_0 tends to be in a flat minimum at $k \geq 0.5$ GeV/c. All the existing models of this reaction, including those with final state interactions, predict that T_{20} should change its sign at k from 0.4 to 0.8 GeV/c; this qualitative theoretical result is not sensitive to the choice of the existing models for the deuteron wave function (DWF) with the S and D waves.

The data on **the elastic $p(d,p)d$ scattering at** $\vartheta_{\mathrm{cm}} = 180°$ (BES) with unpolarized target are summarized in Fig. 1 in the same way as the breakup data. The EMD was extracted from the cross sections according ref. [8].

The main results of these studies are summarized as follows:
1) the empirical momentum density (EMD) agrees surprisingly well with the EMD extracted from the breakup data.
2) T_{20} and κ_0 differ from those measured in the breakup experiments but display almost exactly the same trajectory on the $\kappa_0 - T_{20}$ plot [9, 10].

3) T_{20} has a new unexpected structure at $k \sim 0.65$ GeV/c; the structure at $k \sim 0.35$ GeV/c, first observed in ref. [11], is confirmed.

As in the case of breakup, these data put a real challenge to theory.

The analysis [9] of the data on the $\kappa_0 - T_{20}$ plot brings to the conclusion that something important is lost in existing models of the deuteron breakup and BES. It is possible that other components of the DWF exist, different from the S and D components (see also talks [12]). The other possibility, probably co-existing with the first, is that the mechanisms of both reactions have strong spin-flip contributions which change the spin state of the detected proton.

Installation of the **Movable Polarized Target** [13] (MPT) at the LHE opened new possibilities for studies of few nucleon systems. The first experiment with the MPT was started in 1995; the difference of the total np cross sections for longitudinally polarized initial particles in states with parallel and antipar-

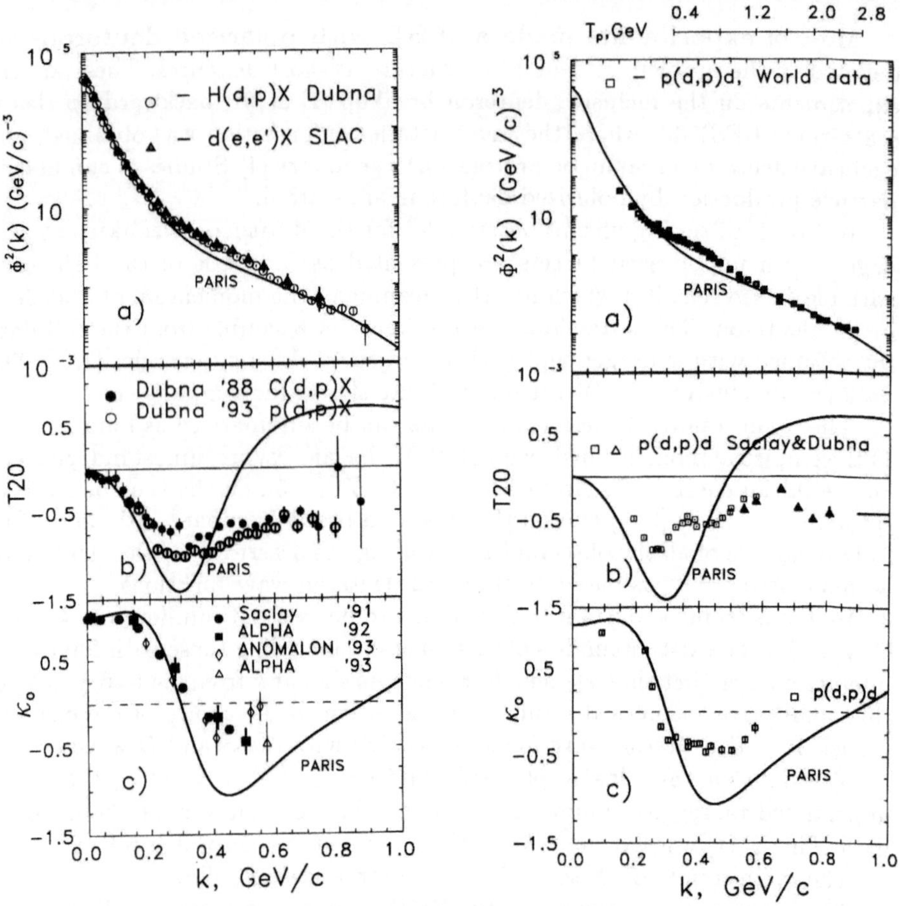

Figure 1. Summary of the 0° breakup data (left) and the data for the elastic $p(d,p)d$ scattering at $\vartheta_{cm} = 180°$ (right). Lines: calculations with the Paris DWF within the quasi-impulse approximation (left) and ONE-approximation [8] (right).

allel spins ($\Delta\sigma(np)_L$) was measured [13] up to the neutron energy $T_{\rm kin} \simeq 3.6$ GeV.

Several experiments which use the MPT are proposed [13]: (1) Further measurements of $\Delta\sigma(np)_{L,T}$; (2) Spin-spin correlations in BES; (3) $dp \to {}^3\text{He}+\phi$ study devoted to the OZI rule violation; (4) $dp \to {}^3\text{He}+(\pi,\eta)$ study, where spin-spin correlations should be measured [14]; (5) measurements of the spin transfer from target to proton-fragment [15] in the $p(d,p)X$ breakup at 0°.

The idea of the last proposal comes from the violation of the correlation [9] between T_{20} and κ_0, which originates from two main sources: the $S+D$ spin structure of the DWF and from the implicit assumption that the spin state of the detected proton–spectator is conserved during the breakup. If the latter assumption is not valid, there should be a significant correlation between the spin states of the target proton and the detected proton–fragment [15].

Measurements of T_{20} in inelastic (d,d') scattering at 0° are continuation of studies of charge exchange reactions with Δ excitations in nuclei [16].

The inelastic $p(d,d')$ reaction is equivalent to the $p(\alpha,\alpha')$ reaction in the sense that in both cases the isospin conservation results in the same selection rules for mechanisms of excitation of baryonic resonances [16, 17]. Some interference effects were expected [16] for these reactions and this is indeed the case [18]. Polarization observables might be sensitive to this interference and to a possible influence of the nuclear medium on properties of the excited resonances.

The characteristic features of the data obtained at deuteron momenta of 4.5 and 5.5 GeV/c are the following [19]:
1) T_{20} is negative and increases in absolute value when the transferred energy $Q = E_d - E_{d'}$ increases. 2) T_{20} is small and compatible with zero in the region of Δ excitation but it significantly differs from zero in the region of the $N(1440)$ resonance excitation and above. 3) $\mid T_{20} \mid$ becomes relatively big at large Q (about 0.4 to 0.6 at $Q \simeq 1$ GeV). 4) There is no visible difference between the $p(d,d')X$ and $C(d,d')X$ data at 0°.

Other polarization observables can be measured, but a theoretical input is necessary in order to see what one could learn from the polarization data about nucleonic resonances and mechanisms of their excitation.

Acknowledgement. This work was made possible in part by grants N° RFU000 from the International Science Foundation and N° RFU300 from the International Science Foundation and Russian Government.

References

1. A.M.Baldin: Proc. of the XI Intern. Sem. on High Energy Physics Problems, Ed. A.M.Baldin, V.V.Burov, p.672, Dubna, 1994; A.D.Kovalenko: Proc. of the 4th Eur. Particle Accelerator Conf., June 27 – July 1 1994, London, v.1 ed. V.Suller, Ch. Petit-Jean-Genaz, p.161, World Sci, 1994; E.V.Anoshina et al: INR MGU preprint 95-6/370, Moscow, 1995

2. V.G. Ableev et al: Pis'ma ZhETF **47**, 558 (1988); JINR Rapid Comm. **4[43]-90**, 5 (1990); B. Kuehn et al.: Phys. Lett. **B334**, 298 (1994); A.A. Nomofilov et al.: Phys. Lett. B325 (1994) 327 and references therein

3. L.S. Azhgirey et al: in *Few-Body Problems in Physics*, AIP Conference Proc. **334**, 415 Franz Gross ed.: AIP Press, New York (1995) and references therein

4. V.V. Glagolev et al: to be published

5. V. Punjabi et al.: Phys. Rev. **C39**, 608 (1989); E. Cheung et al.: Phys. Lett. **B284** (1992) 210 and references therein

6. B.L.G. Bakker et al.: Nucl. Phys. **B158**, 497 (1979); A.P. Kobushkin, V.P. Shelest, EChAYa **14**, 1146 (1983); V.A. Karmanov: *ibid* **19**, 525 (1988)

7. P. Bosted et al.: Phys. Rev. Lett. **49**, 1380 (1982)

8. A.P. Kobushkin: J. Phys. G: Nucl. Phys. **12**, 487 (1986)

9. B. Kuehn, C.F. Perdrisat, E.A. Strokovsky: JINR preprint E1-95-7, Dubna, 1995; will be published in Yad. Fiz. (1995)

10. V. Punjabi et al.: in *Few-Body Problems in Physics*, AIP Conference Proc. **334**, 419 Franz Gross ed.: AIP Press, New York (1995); *ibid.* : Phys. Lett. (in print)

11. J. Arvieux et al.: Phys. Rev. Lett. **50**, 19 (1983); Nucl. Phys. **A431**, 613 (1984)

12. F. Gross: this Conference; A.P. Kobushkin: this Conference

13. F. Lehar et al.: NIM in Ph. Res. **A356**, 58 (1995); F. Lehar et al.: to be published

14. V.P. Ladygin, N.B. Ladygina, JINR preprint E1-94-279, Dubna, 1994; *ibid.*: Yad. Fiz. **58** (1995) (in print)

15. E.A. Strokovsky,: JINR Rapid Comm. **2[70]-95** (1995) (in print)

16. E.A. Strokovsky, F.A. Gareev: JINR preprint E1-94-241 Dubna (1994); *ibid.*: Yad. Fiz. **58** (1995) (in print) and references therein

17. P. Fernández de Córdoba, E. Oset: Nucl. Phys. **A544**, 793 (1992); P. Fernández de Córdoba et al.: Phys. Lett. **B319**, 416 (1993), Nucl. Phys. **A586**, 586 (1995)

18. S. Hirenzaki et al.: this Conference

19. L.S. Azhgirey et al.: JINR preprint E1-95-263, Dubna (1995); submitted to Phys. Lett. **B**

Few-Body Systems Suppl. 8, 191–200 (1995)

Few-Body Systems

Achievements and Perspectives in the Field of Few-Body Systems Studied with Hadronic Probes

Jacques Arvieux

Laboratoire National Saturne, CEA-DSM/IN2P3-CNRS
F-91191 Gif-sur-Yvette Cedex, France

1 Introduction

The definition of a few-body system is wide ranging:

- it can be a system made of a few "real" sub-units like, for example, ^{3}He–nucleus as being made of three nucleons, or a nucleon considered as a composite system of three valence quarks

- it can also encompass the study of complicated many body systems in which one tries to extract underlying few-body mechanisms as in pion– or di–lepton production

- finally it can also be the interaction of a few *elementary* units as in $\pi - \pi$ scattering which, since targets made uniquely of pions do not exist, has to be studied in complex 2π-production reactions.

Therefore it is not easy to have a clear perspective of the achievements in the field as can be seen, for example, when looking at the many faces of the contribution to the poster session (Table 1). The problem may be more difficult for hadron probes where the scattering of topics is larger than for the electromagnetic probes, one reason being that, since the cross sections are larger, physicists have more freedom to explore rarer phenomena.

In the case of this conference, poster sessions are not fully representative of the whole field since they deal only with papers which have not been selected for oral presentation. This is particularly true in the case of anti-protons which are represented by four invited papers [1-4], none contribution appearing in the poster session. So a better way is to look at the entire program of the conference and then some trends begin to emerge.

Table 1. A selection of poster contributions showing the large spread of topics in the field of few-body studies with hadron probes.

Poster	E/T	Energy	Probe	Physics interest
W. Tornow et al	E	7.6 – 12 MeV	n	np
W. von Witsch et al	E	26 MeV	n	3-nucleon system
H. Anklin et al	E	45 MeV	p/d	3-nucleon system
M. Allet et al	E	65 MeV	p/d	3-nucleon system
M.A. Braun et al	T	180 MeV	π	^{3}He / T
E. Hernández et al	T	300 MeV	p	Off–shell pion production
J.P. Pinto et al	T	700 MeV	d	d-nucleus optical model potential
B.M. Abramov et al	E	600–1100 MeV	π	Inclusive double–charge exchange
A.V. Blinov et al	E	<1 GeV	^{3}He/^{4}He	Few-body breakup
R. Maschuw	E	2.5 GeV	p	Two–π production
S. Hirenzaki	T	4.2 GeV	^{4}He	Roper resonance
G.S. Averichev et al	E	Few GeV	p	Backward pion production
E. Bratkovskaya et al	T	Few GeV × A	H.I.	Dilepton production

2 Some Trends in Few-Body Studies

- In the low energy sector the Nucleon-Nucleon (NN) interaction still attracts a lot of interest. In the *pp* sector there are new data in the range of 2.1-2.5 GeV showing narrow structures in the analyzing power around $s^{1/2} = 2.1$ GeV which are related to rapid variations of some spin singlet amplitudes [5]. The underlying dynamics is not yet understood and it is still premature to talk of dibaryon resonances. More data are soon to be expected from COSY at KFA-Jülich (elastic scattering [6]) and AGOR at KVI-Groningen (*pp* bremstrahlung [7]).

- The *np* and *nd* interactions are studied at low energy with very high precision as emphasized by results obtained at TUNL [8]. At higher energies the *np* charge exchange reaction could produce the best $g_{\pi NN}$ determination as shown by T. Ericson [9] but the precision of the existing data is still inadequate.

- High precision data from Cooler rings (IUCF, COSY, CELSIUS) begin to appear as shown in Fig. 1 for $pp \longrightarrow pp\pi^0$ data from IUCF [10].

- The deuteron is still a fascinating object at all energy scales and there were many contributions on this topics at FBXV.

- Experiments with anti-protons have produced a lot of data of the highest interest in the elastic and charge-exchange channels, which have a strong impact on $N\bar{N}$ potentials.
- Strong violations of the OZI Rule have been found in $p\bar{p}$ annihilation and meson-production data from Saturne, which deserve more attention.
- Pion production still presents physics opportunities, a special case being 2π-production near threshold where it is related to Chiral Perturbation Theory (new experiments are in preparation at COSY [6], IUCF [11] and CERN [12]).

The last three topics will be developed in more details in what follows.

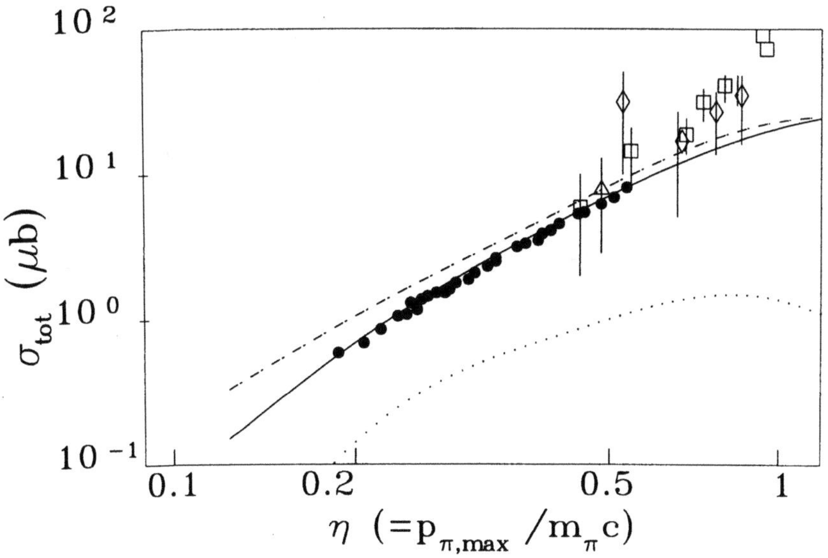

Figure 1. Total cross-section for pion production in pp interaction very near threshold, measured at the IUCF Cooler Ring (from ref. [10])

3 A Personal Selection of Advances in the Field

3.1 Deuteron Structure at Intermediate Range

In the One-Nucleon-Exchange (ONE) Impulse Approximation (IA) the deuteron tensor polarization (or analyzing power) T_{20} in $p(d,p)d$ backward scattering and the polarization transfer κ_0 from vector polarized deuterons to protons, are both related to the S- and D-momentum space wave-function of the deuteron $u(q)$ and $w(q)$ through

$$T_{20} = \frac{2\sqrt{2}uw - w^2}{\sqrt{2}(u^2 + w^2)}, \tag{1}$$

$$\kappa_0 = \frac{u^2 - w^2 - \frac{1}{\sqrt{2}}uw}{u^2 + w^2}. \tag{2}$$

From these formula one then deduces a circular relationship between T_{20} and κ_0

$$\kappa_0^2 + \left(T_{20}^2 + \frac{1}{\sqrt{2}}\right)^2 = \frac{9}{8}. \tag{3}$$

Data have been obtained recently [13] at Saturne and Dubna (Figs. 2 and 3) both in 180° elastic scattering and 0° deuteron break-up up to internal momenta of 0.5 GeV/c.

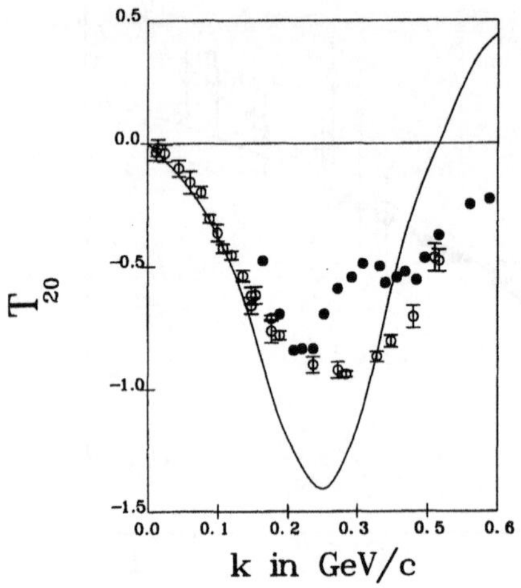

Figure 2. Tensor analyzing power T_{20} in $p(d,p)d$ backward elastic scattering (black dots) and $p(d,p)X$ break-up (open circles) (from ref. [14])

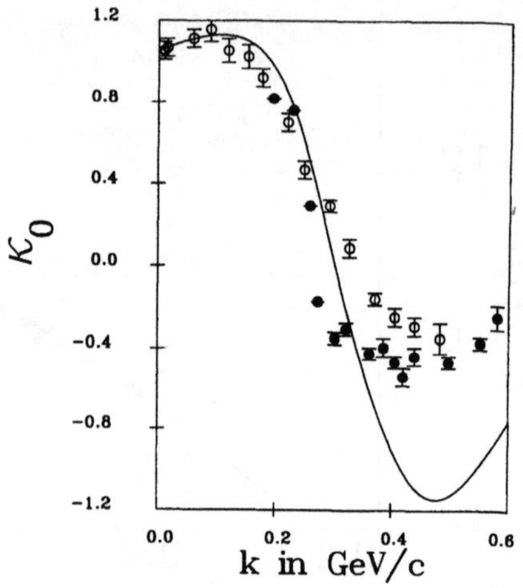

Figure 3. Same as Figure 2 for polarization transfer κ_0 from vector polarized deuteron to proton

Although the backward elastic scattering shows a characteristic dip between 0.3 and 0.5 GeV/c which does not exist in deuteron break-up, when T_{20} is plotted as a function of κ_0 the two reactions exhibit identical correlation plots as seen in Fig. 4.

The fact that the two reactions give an almost identical correlation up to $k = 0.5 - 0.6$ GeV suggests that it is produced by an intrinsic property of

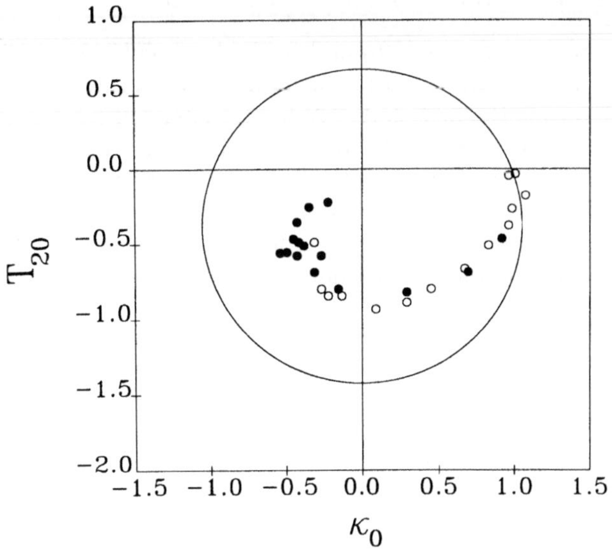

Figure 4. Same as Fig. 2 for the correlation plot of T_{20} versus κ_0

deuteron structure and not by the reaction mechanisms which may be completely different. Moreover one can see that the deviation from a circle starts at very low k. The circular correlation is a consequence of the ONE mechanism; it can only be destroyed if there is a helicity flip of the proton between the final state and the initial state. In particular FSI would not break this correlation except if the NN interaction has a strong spin-flip component. Other effects which may break this correlation are [14,15]:

- P-wave components in the deuteron wave-function,

- mechanisms in which the 3-nucleon could come at very low relative distance in the intermediate stage such as one would expect from a quark-quark direct interaction.

If one invokes an interaction mechanism more complex than I.A. or more complex deuteron wave-functions, then it would be extremely useful to measure other observables. Some of these show the same sensitivity as T_{20} and K_0 to the deuteron wave-function. They are the polarization transfer $K_{0,y}^{0,y}$ from a polarized proton target to the break-up proton of an unpolarized deuteron. The other ones are the vector to vector transfer $K_{y,0}^{y,0}$ and the vector to tensor transfer $K_{y,0}^{yy,0}$ in elastic backward deuteron scattering. Proposals to measure these observables at Saturne, Dubna and KEK are under way.

3.2 *The Nucleon-Antinucleon ($N\bar{N}$) Interaction*

At distances of the order of 1 fm and beyond the $N\bar{N}$ interaction is mediated, like the NN interaction, by boson exchanges. What makes the two interaction channels different is the fact that $N\bar{N}$ is characterized by a strong annihilation whose range is of the order of 0.5 to 0.8 fm. Below 0.5 fm quark and gluon degrees of freedom are supposed to dominate as shown in the picture of Fig. 5 (from ref. [16]). In the hybrid model the real part of the potential is deduced from the NN interaction through application of the G-parity rule. The imaginary part is obtained phenomenologically by fitting the $N\bar{N}$ data.

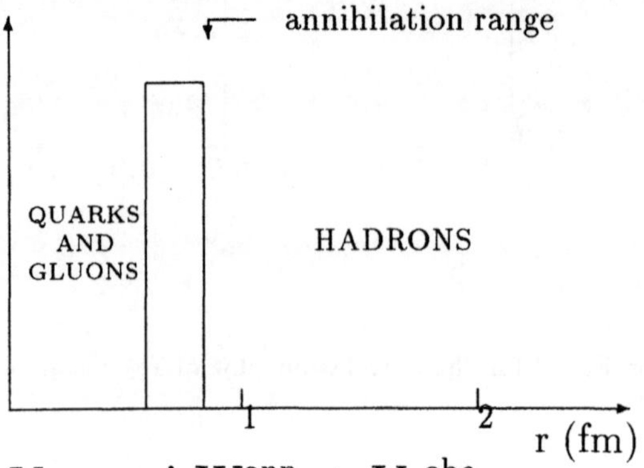

$$V_{N\bar{N}} = i\, W^{ann} + V^{obe}_{N\bar{N}}$$

Figure 5. Hybrid description of the $N\bar{N}$ potential (from ref. [16])

A lot of new data of high precision and/or large variety including sophisticated spin observables have been obtained at LEAR in the range of a few hundreds of MeV/c up to 1.3 GeV/c. In particular the charge-exchange channel $p\bar{p} \longrightarrow n\bar{n}$, for which existing data were crude and scattered, has been studied in two experiments. The first one (PS199) focused on the spin observables and the second one (PS206) on a high precision measurement of differential cross-sections.

The analyzing power A_{0n} which measures the asymmetry of the outgoing n or $\bar{n}$ after scattering of a beam of $\bar{p}$'s by a polarized target proton has been obtained at four momenta 545, 875, 1083 and 1287 MeV/c [17]. The gross features of the data can be reproduced by potential model calculations like the Paris potential but the quality is far from reaching the one achieved in NN scattering predictions.

Partial wave analysis based on the Nijmegen potential give better fits to the data although the agreement is far from being perfect at the largest energies where the importance of the annihilation channel is emphasized, due to a smaller distance of approach between p and $\bar{p}$.

In the same experiment the depolarization parameter D_{0n0n} which is obtained by scattering a $\bar{p}$ on a polarized $\bar{p}$ target and measuring the polarization of the outgoing neutron through a second scattering in a Carbon target has also been measured [18]. The data at 875 MeV/c are shown in Fig. 6 where they are compared to the Paris and Nijmegen calculations. The Paris potential reproduces remarkably well this parameter.

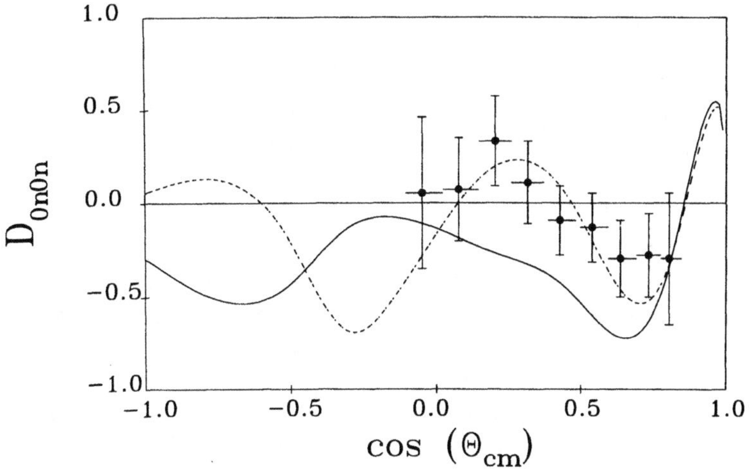

Figure 6. Depolarization parameter D_{0n0n} in the $p\bar{p} \to n\bar{n}$ charge exchange reaction measured at LEAR (from ref. [18]). The curves are theoretical predictions from the Nijmegen group (continuous line) and from the Paris model (dashed line).

The present situation of the $N\bar{N}$ interaction is not very satisfactory. There are new data which cannot be reproduced by existing models and other, more microscopic approaches are warranted. On the other hand the situation is urgent since in a short time (2-3 years) LEAR will be shut-down and there will be no way, for many years to come, to improve the present data or to obtain new ones in this energy range. Therefore and this was one of the topic of the panel session, theorists have to come forward and say if the present data base is good enough to allow significant advances in this field. If the answer is no, they must, as soon as possible say what data are needed to give a chance to experimentalists to measure them, especially since $N\bar{N}$ scattering studies are no more a priority at LEAR.

3.3 Meson Production and the Violation of the OZI Rule

The OZI rule states that disconnected quark diagrams as the one in Fig. 7a describing the reaction $p\bar{p} \longrightarrow \phi\pi^0$, are forbidden. In the production of ϕ and ω-mesons, it can be expressed quantitatively as an angle δ defined by a ratio

Table 2. The ratio of $R = \phi X/\omega X$ for production of the ϕ and ω mesons in the antinucleon annihilation at rest. The parameter Z of the OZI–rule violation is calculated for $\delta = \Theta - \Theta_1 = 3.7°$, assuming equal phases of the ϕ and ω production amplitudes

Final state	$BR \times 10^4$	$R = \phi X/\omega X \times 10^3$	$\|Z\|(\%)$	Comments
$\phi\pi^0$	4.0 ± 0.8	76.9 ± 17.1	21 ± 3	$^3S_1$
$\phi\eta$	0.3 ± 0.39	29 ± 3.8	1 ± 4	$^3S_1$
$\phi\rho$	3.4 ± 1.0	17.8 ± 6.3	7 ± 2	$^1S_0$
$\phi\omega$	5.3 ± 2.2	18.9 ± 11.3	7 ± 4	$^1S_0$
$\phi\pi\pi$	4.7 ± 1.1	7.2 ± 1.8	2 ± 1	S-wave
$\phi\pi$	0.0 ± 0.3			$^1P_1$
$\phi\rho$	3.7 ± 0.9	5.8 ± 1.8	1 ± 1	^3P_J
$\phi\pi\pi$	6.6 ± 1.5	9.3 ± 2.5	3 ± 1	P-wave
$\phi\gamma$	0.22 ± 0.07	234 ± 134	49 ± 11	liquid
$\phi\pi^0$	7.8 ± 1.6	137 ± 31	30 ± 4	liquid
$\phi\eta$	0.9 ± 0.3	6.0 ± 2.0	1.3 ± 1.2	liquid
$\phi\pi^0\pi^0$	1.2 ± 0.6	6.0 ± 3.0	1.3 ± 2.0	liquid
$\phi\pi^-$	14.8 ± 1.1	133 ± 26	29 ± 3	$\bar{p}d,\ p < 200\mathrm{MeV/c}$
$\phi\pi^-$		113 ± 30	27 ± 4	$\bar{p}d,\ p > 400\mathrm{MeV/c}$
$\phi\pi^+$		110 ± 15	26 ± 2	$\bar{n}p$
$\phi\pi^-$	9.0 ± 1.1	83 ± 25	22 ± 4	bubble chambers
$\phi\pi^-\pi^+$	4.6 ± 0.9	7.0 ± 1.4	1.9 ± 0.8	bubble chambers
THEORY		0.15 – 4.2		Linear–quadratic mass formula

of branching ratios

$$tg^2\delta = R = \frac{B.R.(AB \longrightarrow \phi X)}{B.R.(AB \longrightarrow \omega X)} \tag{4}$$

where $\delta = \theta - \theta_i$; θ_i being the ideal mixing angle between ϕ and ω ($\theta_i = 35.3°$). If the ϕ is a pure $s\bar{s}$-state and the ω a pure $(u\bar{u}, d\bar{d})$ state, then $\delta = 0$.

Using Gell-Mann-Okubo relationships one finds that $\theta = 35.3°$ or $38.4°$ very close to θ_i depending if one uses the linear or quadratic mass-formula. Therefore one expects a very small value of R, of the order of $0.15 - 4.2 \times 10^{-3}$. Alternatively one may characterize the OZI-violation by a parameter Z defined as

$$R^{1/2} = \frac{Z + \tan(\theta - \theta_i)}{1 + Z\tan(\theta - \theta_i)}, \tag{5}$$

Z being 0 if the OZI rule holds perfectly. In most experiments involving ϕ and ω-production one finds a *small* violation of the OZI-rule ($R \sim 10 - 20 \times 10^{-3}$ corresponding to $Z \sim 0 - 10$ %) [19]. It can be easily explained by Final State Interaction (FSI) effects [20].

What is new, is that, in some special reactions involving the production

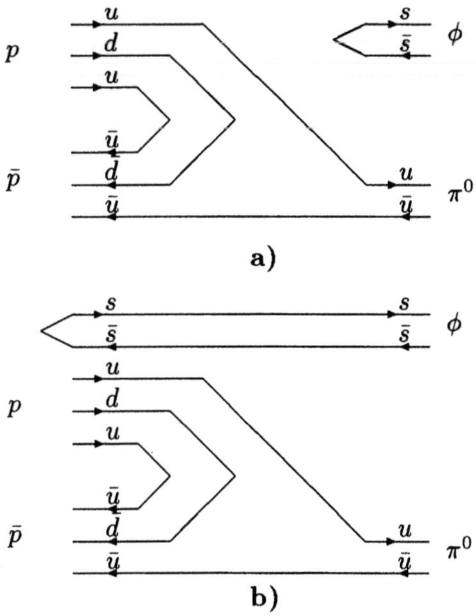

Figure 7. ϕ–production mechanism in $p\bar{p}$ annihilation (from ref. [21]). a) OZI–forbidden disconnected diagram b) ϕ–production from $s\bar{s}$ component in the nucleon

of a π° or a γ in the $p\bar{p}$ annihilation channel, very large violations have been observed up to $R \approx 100 - 300 \times 10^{-3}$ or $Z = 20 - 50$ % as shown in Table 2.

Violations of the same order have also been found in the reaction $pd \rightarrow {}^3\mathrm{He}\phi/\omega$ at Saturne [21].

The OZI-violation in $p\bar{p}$ annihilation seems to be related to the initial spin state : it is much stronger in the S-state for which the initial spin is S=1 than in the P-state for which S=0. This fact is very interesting since it allows to connect the violation of the OZI-rule to intrinsic strangeness components in the nucleon (Fig. 7b) which have been deduced from Deep Inelastic Scattering on (DIS) polarized nucleon. Many further tests of this hypothesis in which polarization effects play an important role have been proposed [19]. In particular, a full understanding of the origin of the OZI-rule violation could come from the use of fully polarized initial states as in the $pp \rightarrow pp\phi$ and $pd \rightarrow^3 \mathrm{He}\phi$ with polarized beam and target.

4 Conclusions

Even at the time when new electron machines with high duty cycle are coming into operation at Amsterdam (AmPS), Bonn (ELSA), Mainz (MAMI B) and CEBAF, which will bring precise data on few nucleon systems, the hadronic probes, with their large cross-section and their possibilities to study quark rearrangement mechanisms, still keep a strong potential for discovery as demonstrated at FBXV. The future of this field lies now with the Cooler Rings of protons and heavier nuclei which will come into operation at Bloomington (IUCF Cooler Ring), Julich (COSY) and Upsala (CELSIUS).

References

1. E. Klempt: Contribution to this Conference
2. Yu. Kuperin: Contribution to this Conference
3. J. Carbonell: Contribution to this Conference
4. O. Dalkarov: Contribution to this Conference
5. J. Ball et al.: Phys. Lett. **B320**, 206 (1994)
6. R. Maschuw: Contribution to this Conference
7. N. Kalantar-Nayestanaki: Contribution to this Conference
8. W. Tornow: Contribution to this Conference
9. T.E.O. Ericson: Contribution to this Conference
10. H.O. Meyer et al.: Nucl. Phys. **A539** 633 (1992)
11. S.E. Vigdor et al.: IUCF letter of Intent (1992)
12. B. Adeva et al.: Proposal SPSLC 95-1, CERN (1995)
13. V. Punjabi et al.: to be published
14. E.A. Strokovsky: JINR Rapid Communication **2[70]-95** (1995)
15. A.P. Kobushkin: Proceedings of the Int. Symposium *Dubna-Deuterons 93* and contribution to this Conference
16. F. Bradamante: Conf. Proc. Vol. 39, *Flavour and Spin in Hadronic and Electromagnetic Interactions*, Eds F. Balestra, R. Bertini and R. Gargagnini, SIF, Bologna (1993) p.295
17. R. Birsa et al.: Nucl. Phys. **B403**, 25 (1993)
18. A. Ahmidouch et al.: to be published
19. J. Ellis et al.: preprint CERN (1994)
20. M.P. Locher, Y. Lu, B.S. Zou: Z. Phys. **A347** 281 (1994)
21. R. Wurzinger: Preprint LNS/Ph/94-21 (1994)

Few-Body Systems Suppl. 8, 201–212 (1995)

Few-Body Systems

Few–Body Systems Probed in Several Ways

J.F.J. van den Brand * †

NIKHEF, Kruislaan 411, 1098 SJ Amsterdam, The Netherlands

Abstract. New developments in nuclear few-body systems are discussed. Results will be reported for the nucleon using virtual-compton scattering as a probe. For the deuteron, the high Q^2 (up to 6.8 $(\mathrm{GeV}/c)^2$) data from SLAC will be discussed, while also new data for the $(e, e'p)$ reaction using tensor polarized deuterium from NIKHEF will be presented. Finally, the results from quasi-free knockout experiments using polarized protons incident on polarized ^{3}He performed at IUCF will be shown. Plans for similar experiments in electronuclear physics will be outlined.

1 Introduction

A comprehensive program of measurements of the electromagnetic response of the few-body systems is in progress. The scientific program is focussed on the study of these systems in terms of nucleon structure, the ground-state few-body structure built from the nucleon-nucleon (NN) interaction and the nature of the interaction of the virtual photon. These aspects of the electromagnetic response are interrelated and can only be unambiguously separated by a broad study of the few-body systems. The choice of few-body systems and the electromagnetic probe, especially at relatively low momentum transfers, should allow accurate comparison with theoretical calculations.

A central question in subatomic physics is the description of the nucleon in terms of quarks and gluons. Traditionally, the electromagnetic structure of the nucleon has been investigated with elastic electron scattering, deep-inelastic scattering, and real Compton scattering.

In section 2, first evidence [1] is shown for the Virtual-Compton (VC) Scattering process $\gamma^* N \rightarrow \gamma N$, where a photon with virtuality Q^2 is absorbed on a nucleon which subsequently radiates a real photon with energy (momentum) ω' $(\boldsymbol{q}')$. This process is intrinsically a rich probe of nucleon dynamics, since one

* *On leave from:* Department of Physics, University of Wisconsin-Madison, Madison, WI 50263, USA

† *E-mail address:* jvdb@dxhrb3.desy.de

can independently map out the dependence of the VC process on Q^2, the virtual photon's polarization, and the invariant mass of the hadronic intermediate state W. In addition, one can experimentally select the polarization of incident beam, target, or recoiling proton. In the limit $Q^2 \rightarrow 0$ the VC reduces to the real Compton process, while in the limit $W^2 \rightarrow M^2$ it becomes equivalent to elastic electron-proton scattering.

In section 3, a high momentum transfer quasi-elastic electron scattering experiment [2] on deuterium is discussed. Until recently, such experiments were performed at relatively moderate Q^2, typically $Q^2 \approx 0.2$ (GeV/c)^2. The analysis of quasi-elastic $(e, e'p)$ experiments has generally been performed in a non-relativistic framework, to facilitate the calculation of nuclear wave functions in the initial and final state. For the current operator often a non-relativistic reduction is used, truncated at lowest order in p/M, where p (M) is the proton momentum (mass). At high Q^2 this approximation is invalid since the proton momentum is not small compared to its mass.

Apart from relativistic ingredients in the $(e, e'p)$ reaction mechanism, new phenomena may occur due to the fact that the wavelength of the photon becomes small ($1/Q \approx 0.08$ fm at $Q^2 = 6.8$ (GeV/c)^2) with respect to the size of the nucleon ($< r >_{RMS} \approx 0.8$ fm). It is thus not clear a priori that quasi-elastic scattering still can be described in terms of the impulse approximation using nucleon-meson fields.

Finally, in section 4 the measurement of spin-dependent electron [3] and proton [4] scattering from polarized few-body systems is discussed. Experimentalists will utilize the latest technology available in the form of electron scattering from pure, polarized internal gas targets. It has been widely recognized that this arrangement can provide the lowest combined statistical and systematic error.

Measurements of the spin dependence of the electromagnetic response are particularly important because they enhance our knowledge of the small components of the nucleon and nuclear wave functions. These small amplitudes enter the spin observables through their interference with larger, and in general better known, amplitudes.

The study of the small components of the two and three-body ground states is an important aspect of the scientific effort. In addition to being necessary for extraction of information on neutron charge structure at low Q^2, this is also crucial for unambiguous determination of neutron spin structure from high-energy measurements at CERN, SLAC, and DESY.

2 Virtual Compton Scattering as a Probe of Nucleon Structure

The VC process [1] provides a particularly interesting probe of nucleon structure, since it represents a fully electromagnetic *exclusive* reaction, where significant initial- and final-state interactions are absent. In the non-perturbative QCD regime, below the π^0-production threshold, the VC process allows the determination of electromagnetic polarizabilities through the measurement of

q'-dependent contributions to the LET. These observables are determined by the virtual excitation spectrum of the nucleon. In the region above the π^0-production threshold VC scattering allows the study of s-channel effects of baryon resonances. Here W^2 can be chosen to optimize the sensitivity to the electromagnetic decay of a particular resonance (e.g. $W^2 \approx M_\Delta^2$). In the hard-scattering region, $Q^2 > 1$ (GeV/c)^2, VC scattering can be employed to investigate the perturbative QCD (PQCD) description of the nucleon wave function. However, up to now no data has been obtained for VC scattering.

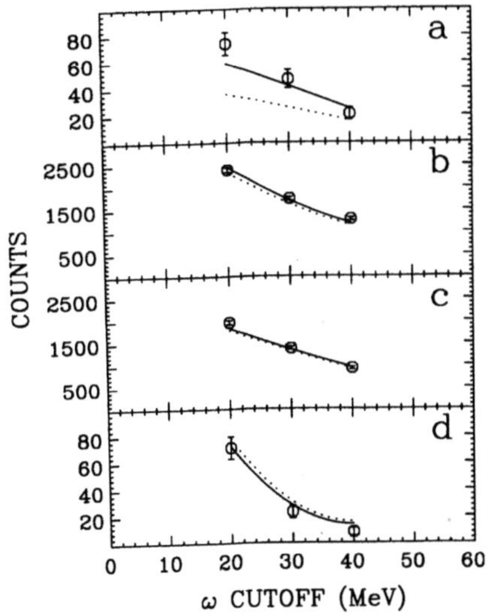

Figure 1. Number of radiated events versus the cut-off energy ω', in the intervals $-60° < \theta_\gamma < -20°$ (radiation along the proton direction, Fig. 1a), $-20° < \theta_\gamma < 20°$ (radiation along the incident electron direction, Fig. 1b), $20° < \theta_\gamma < 50°$ (radiation along the scattered electron direction, Fig. 1c), and the region $50° < \theta_\gamma < 90°$ (Fig. 1d). The solid (dotted) curve shows the prediction in the soft-photon limit for Bethe-Heitler and VC contributions (Bethe-Heitler only).

The reaction $eN \rightarrow eN\gamma$ is dominated by contributions from the photons radiated by the incoming and outgoing electrons. The corresponding lowest-order Feynman diagrams can be calculated exactly using QED and are termed Bethe-Heitler contributions. The VC contributions are due to photon radiation by the nucleons and can only be unambiguously calculated in the soft-photon limit, where the cross section is independent of nucleon-structure ambiguities. In general they involve a sum over all possible intermediate hadronic states.

Experimentally, the reaction $eN \rightarrow eN\gamma$ can be kinematically determined by detecting a scattered electron in coincidence with a proton using high-resolution magnetic spectrometers. The NE-18 experiment at the Stanford Linear Accelerator Center (SLAC) measured the A$(e, e'p)$ reaction for several target nuclei, using the 1.6 GeV/c and 8 GeV/c spectrometers in coincidence.

In Fig. 1 the number of radiated events are shown versus the cut-off energy ω', in various intervals of the reconstructed real photon angle, $\Delta\theta_\gamma$. When only Bethe-Heitler contributions are taken into account, the radiation along the incident and scattered electron is well described. This is also the case for the radiated photons in the angular range $50° < \theta_\gamma < 90°$ (Fig. 1d), which is especially sensitive to the effects of experimental resolutions. However, the calculation with Bethe-Heitler contributions only shows a deficiency in the description for radiation along the proton direction. The solid curves represent the prediction in the soft-photon limit for both Bethe-Heitler and VC contributions. The good agreement between data and prediction of the number of counts when proton radiation is included clearly demonstrates the experimental viability of the technique even with the 2×10^{-4} duty factor available at the SLAC end-station. In the near future the high-luminosity beams at Mainz and CEBAF in combination with their high-resolution spectrometers, will allow for a meaningful study of the nucleon structure [5].

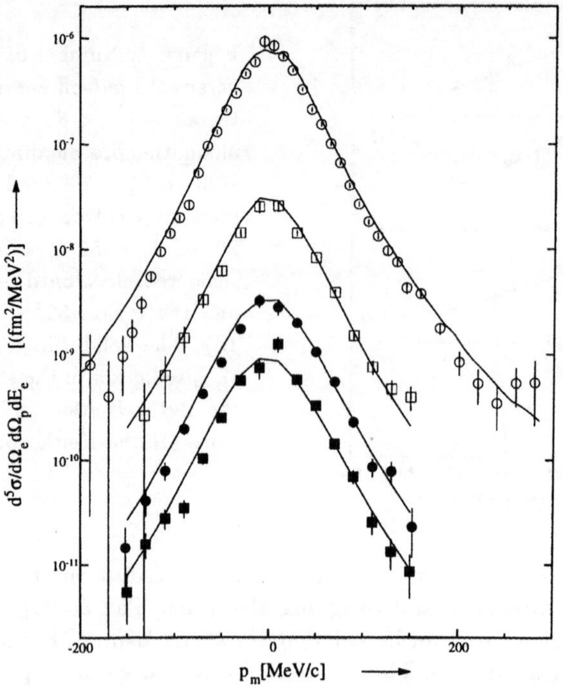

Figure 2. Cross section for the reaction $^2\text{H}(e, e'p)n$ as a function of the missing momentum. The open circles, open squares, filled circles and filled squares represent the data at four-momentum transfer squared of 1.2, 3.0, 5.1 and 6.8 $(\text{GeV}/c)^2$, respectively.

3 Scattering from Deuterium at High Momentum Transfer

Exact calculations for realistic NN potentials can be performed for the two-body system. This allows for an important benchmark study of the $(e, e'p)$ quasi-elastic reaction mechanism using deuterium. The dependence of the cross section on Q^2 is used to investigate the validity of the impulse approximation for quasi-elastic scattering and to determine the importance of relativistic effects in the cross section.

Fig. 2 shows the coincidence cross section measured at SLAC as a function of missing momentum [2] for $Q^2 = 1.2, 3.1, 5.0$, and 6.8 $(\text{GeV}/c)^2$. The curves represent the results of a PWIA calculation, based on the prescription σ_{cc1} of de Forest for the off-shell electron-proton cross section and the Paris spectral function. The curves exceed the data on average by about 10 %. Note, that a reduction of the cross section is to be expected due to final-state interaction. Reasonable agreement is obtained for all Q^2 and the data are consistent with a description in terms of the quasi-free scattering mechanism.

The reaction mechanism was studied in more detail by comparing the data at $Q^2 = 1.2$ (GeV/c)^2 to several modern calculations. In Tjon's relativistic field theoretical model [6] for the reaction ^{2}H$(e, e'p)n$, the Bethe-Salpeter equation is solved for a one-boson exchange potential. The non-relativistic calculations of Arenhövel et al. [7] start from the Schrödinger equation. Their relativistic calculations include third and fourth-order terms in (p/M_p) in the current operator and contributions from the boost operator that translates the initial-state wave function from the laboratory frame to the center-of-mass frame.

The comparison reveals that the non-relativistic calculation of Arenhövel significantly exceeds the data. The relativistic calculations of Arenhövel and Tjon give a good description of the cross section, except for $p_m <$ -125 MeV/c, where the data have limited statistical accuracy.

Priority should be given to future ^{2}H$(e, e'p)$ experiments with complete structure function separation which may achieve sufficient precision to reveal genuine relativistic contributions to the reaction mechanism.

4 Spin-Dependent Scattering from Polarized Few-Body Systems

4.1 *Measurement of Tensor Analyzing Powers for Electron Scattering from a Polarized ^{2}H Target Internal to a Storage Ring.*

Measurements of spin-dependent electron scattering have the potential to greatly enhance our understanding of nucleon and nuclear structure. For example, spin observables in elastic, quasi-elastic, and deep-inelastic scattering from polarized deuterium are predicted to provide important information on the effects of D-wave components in the ground state of ^{2}H, the largely unknown charge form factor of the neutron, and the neutron spin structure functions. This has prompted development of both polarized ^{2}H targets for use with external beams and polarimeters for measuring the polarization of recoiling hadrons. Indeed a first round of measurements of spin-dependent $e - {}^2$H scattering have been carried out at Novosibirsk [8], Bonn, MIT-Bates and SLAC.

The measurement of analyzing powers and spin-correlation parameters in spin-dependent electron scattering from polarized nuclei is optimally performed by scattering electrons from a pure undiluted target. Polarized internal gas targets in electron storage rings have the advantage that spin-dependent scattering from chemically and isotopically pure atomic species of high polarization can be realized. They offer rapid polarization reversal and flexible orientation of the nuclear spin direction using low magnetic holding fields, a low thickness at high luminosity which allows for detection of low-energy recoiling hadrons, and access to a broad kinematic range by using large acceptance detectors. For polarized deuterium one has the additional ability to reverse tensor polarization, P_{zz}, at fixed vector polarization, P_z, and vice versa. Subsequently, low systematic errors can be expected.

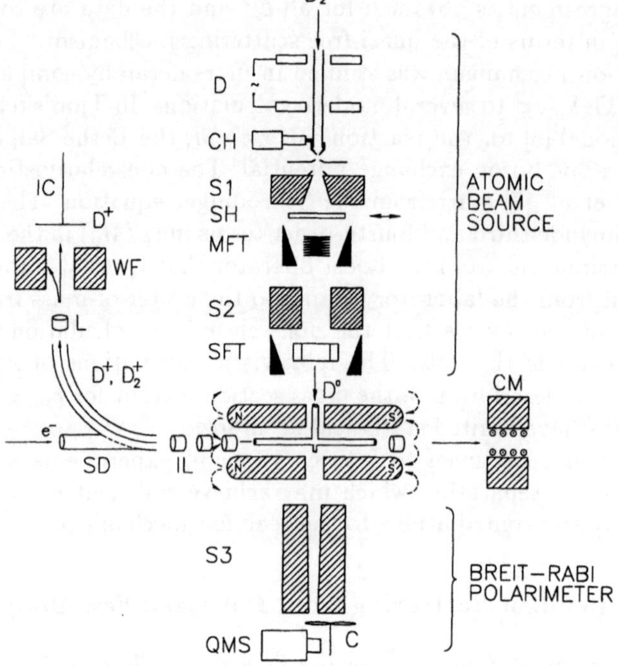

Figure 3. A schematic outline of the atomic beam source, Breit-Rabi polarimeter, internal target, and ion extraction system. All components, except the target holding field magnets, are inside the vacuum system. D: RF dissociator; S1, S2, S3: sextupole magnets; MFT, SFT: medium and strong field transition units; SH: shutter; C: chopper; QMS: quadrupole mass spectrometer; CM: correction magnet; IL: triplet of ion extraction lenses; SD: spherical detector; L: electrostatic lens; WF: Wien filter; IC: ion collector.

The first pioneering measurements [8] with a polarized deuterium internal target have been carried out at VEPP-3 in Novosibirsk. Here, we discuss an experiment [3] measuring tensor analyzing powers in electron-deuteron scattering, where special attention was paid to determine the average polarization in the target cell, using techniques that allow for high systematic accuracy. A tensor polarized deuterium internal target was used at the Amsterdam Pulse Stretcher, AmPS, at NIKHEF. The experiment was equipped with several diagnostic tools and polarimeters to enable a *direct* measurement of tensor analyzing powers. This experiment constitutes a major step towards a future program of spin-dependent electron scattering from polarized internal targets, and represents an important bench test for the polarized internal target technique in an electron scattering environment. Electrons were accelerated to an energy of 565

MeV and injected into a storage ring. Several beam bunches were stacked into the ring, yielding currents up to 120 mA. The life time of the electron current in the storage ring exceeded 15 minutes.

Fig. 3 shows a schematic of the experiment. An Atomic Beam Source (ABS), consisting of an RF dissociator, a cooled nozzle, collimators, sextupole magnets, and RF transition units, provides a flux of 1.3×10^{16} deuterium atoms/s (for 2 hyperfine states), injected into a T-shaped storage cell. A medium-field and a strong field transition unit enable flipping the tensor polarization of the deuterium gas from -2 to +1, with a corresponding vector polarization of zero. The target thickness obtained with this ABS, in a cryogenic storage cell cooled down to 80 K, amounted to 2×10^{13} atoms$\cdot$cm^{-2}. This constitutes an order of magnitude increase over previous experiments [8].

The polarized deuterium atoms were injected into the storage cell, while a small fraction of these atoms was sampled and analyzed in a Breit-Rabi Polarimeter (BRP). This polarimeter consists of a sextupole magnet, a chopper and a quadrupole mass analyzer. The BRP proved useful to optimize the intensity of atoms injected into the storage cell as function of e.g. nozzle temperature, deuterium flow into the RF dissociator, and sextupole fields. Furthermore it was used to test the performance of the field transition units. A drop of 1/3 in amount of atoms detected by the quadrupole mass analyzer with field transition units on indicates a 100% efficiency of the transition. These kind of measurements also prove that the polarization of the injected atoms is not lost in passing through the target holding field magnets, where the field components have zero crossings. In addition, no beam-induced depolarization was observed at currents as high as 120 mA.

Ions, produced by the electron beam, were extracted and analyzed in an ion-extraction channel, using a combination of electrostatic lenses, a sperical deflector, a Wien filter, and an ion collector (see Fig. 3). These ions were prevented from reaching the walls of the storage cell by a longitudinal holding field. They were on one side of the cell reflected by an electrostatic repeller lens, on the other side extracted using a triplet of lenses and a deflector. A Wien filter separated the atoms from molecules, where each fraction was measured in a faraday cup. Using the sextupole magnets, a distinction between the molecular fraction caused by molecules originating from the ABS and by recombination in the storage cell could be made. A measurement of the molecular and atomic fractions with this ion-extraction channel is equivalent to an actual polarization measurement, if recombination is the sole source of depolarization *inside* the storage cell. This was verified independently in a separate measurement where ions were produced by a 1 kV e-gun, and were accelerated onto a tritium foil. The reaction $^3\text{H}(d,n)^4\text{He}$ was then used to measure the target tensor polarization directly. Thus, combining the results of the Breit-Rabi polarimeter and the ion-extraction system, the nuclear tensor polarization of the target has been determined with the same weighting over target-gas and electron beam density as in the actual experiment.

The reconstructed vertex distribution for $^2\text{H}(e,e'd)$ events showed the expected triangular density distribution. The experiment established an essentially background-free measurement of the ^{2}H$(e, e'd)$ reaction. Analyzing powers were measured for the target spin vector coaligned with the momentum transfer ($\boldsymbol{S} \parallel \boldsymbol{q}$) and for the target spin vector directed perpendicular to the momentum transfer ($\boldsymbol{S} \perp \boldsymbol{q}$). A change of sign in the analyzing powers was observed for these two spin orientations, with A($\boldsymbol{S} \parallel \boldsymbol{q}$) $\approx$ -2 $\times$ A($\boldsymbol{S} \perp \boldsymbol{q}$). The measured analyzing powers agree well with the results from several state-of-the-art models [9].

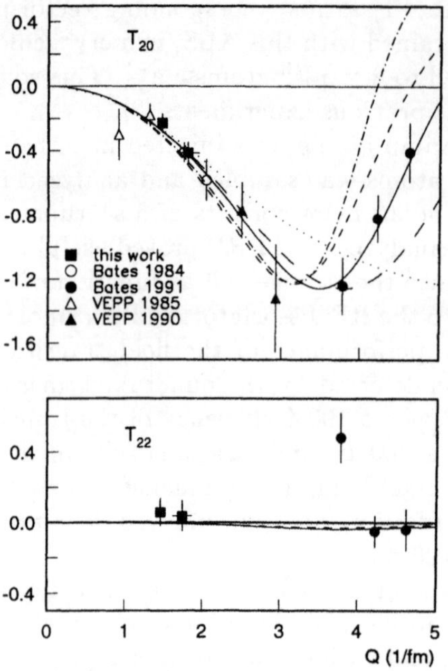

Figure 4. Experimental results and theoretical predictions for the tensor analyzing powers T_{20} (a) and T_{22} (b), as a function of momentum transfer. The solid circles from the present experiment while the other data are from ref. [10]. The curves are results of theoretical calculations; see ref. [9].

A combination of the measurements for these two spin orientations allows a decomposition into T_{20} and T_{22}. In Fig. 4a (4b) the derived result for T_{20} (T_{22}) are compared with the results of other measurements and with various theoretical models. Both in Fig. 4a and Fig. 4b the results are included of experiments measuring T_{20} and T_{22} through the recoil polarization technique [10]. Here one uses an unpolarized deuterium target and measures the tensor polarization of the deuteron, recoiling into a magnetic spectrometer equipped with a polarimeter, instead. Note the small uncertainty of the new data points. The variation in theoretical predictions for T_{20} shows that precise measurements up to q = 4 fm^{-1} can distinguish between the various models shown. The techniques established in this experiment are sufficient to obtain such measurements, with low statistic and systematic uncertainties, within a reasonable time frame.

4.2 Ground-State Spin Structure of ^{3}He

Exact non-relativistic calculations predict the following components to dominate the ^{3}He ground-state wave function: (a) the S-state is spatially symmetric, accounts for ~90% of the spin-averaged wave function, has the ^{3}He spin entirely due to the neutron and has the two protons in a spin singlet state; (b) the D-state is due to the tensor force, accounts for ~ 8% of the spin-averaged wave function and has the three nucleon spins dominantly oriented opposite to the ^{3}He nuclear spin; (c) the S'-state is a mixed-symmetry configuration of the nucleons which arises from spin-momentum correlations and accounts for ~2% of the spin-averaged wave function. All other components are predicted to be negligibly small.

The nuclear-structure information is contained in the spin-dependent spectral function $S^N_{\hat{\sigma}_3}(E, \mathbf{p}, \hat{\sigma}_\mathrm{N})$ which is the probability to find a nucleon N with spin $\hat{\sigma}_\mathrm{N}$ with binding energy E and momentum $\mathbf{p}$ in the ^{3}He nucleus with spin $\hat{\sigma}_3$. $S^N_{\hat{\sigma}_3}(E, \mathbf{p}, \hat{\sigma}_\mathrm{N})$ is calculated exactly in non-relativistic approximation using the Faddeev technique If a neutron is removed from ^{3}He the remaining nucleons are unbound so S^n contains only 3-body final states. Whereas, if a proton is removed the remaining nucleons can be bound as a deuteron or unbound. In this sense S^p is the sum over 2-body and 3-body final-state components. These components of S^p are large and of opposite sign and at low $\mathbf{p}$ the difference in their magnitudes comes from the S'-state. This primarily reflects the fact that the neutron and proton spins must be parallel to bind as a deuteron. The S and S' state contributions to the spectral function are a maximum for zero momentum while the D-state contribution is greatest for about 400 MeV/c.

Quasi-elastic spin-dependent knockout of the constituent nucleons of ^{3}He offers the most direct approach to experimentally constrain the ground state spin structure. In PWIA the observables $A^{3,N}_{00ij}$ for ^{3}He(p, pN) scattering (N is a proton p or neutron n) are related to the observables A^N_{00ij} for $N(p,p)$ elastic scattering by

$$A^{3,N}_{00ij} = A^N_{00ij} \cdot \frac{S^N_{+\hat{\sigma}_3} - S^N_{-\hat{\sigma}_3}}{S^N_{+\hat{\sigma}_3} + S^N_{-\hat{\sigma}_3}} , \tag{2}$$

where the observables A^N_{00ij} are obtained from global fits to spin-dependent $N(p,p)$ elastic scattering. The PWIA model (2) is expected to be a reasonable approximation at low missing momentum or p_m where comparison of $(p, 2p)$ and $(e, e'p)$ data indicates that scattering effects of initial and final-state nucleons are not large.

A series of previous measurements of spin-dependent ^{3}He$(p, 2p)$ and ^{3}He(p, pn) quasi-elastic scattering at 220 and 290 MeV [11] incident polarized proton energies disagreed significantly with the PWIA model of (2). In particular, the observable $A^{3,n}_{000n}$ was observed to be strongly suppressed. These previous experiments were carried out with relatively small acceptance detectors in limited kinematics at low momentum transfers. In contrast the experiment described here employed a large-acceptance detector to extend the previous measurements over a large kinematic range. In addition, data were taken at several incident proton energies and at each energy all asymmetries were measured to high momentum transfer and over an extended range in p_m.

The experiment was performed at the Indiana University Cyclotron Facility (IUCF) Cooler Ring [4] using the new technique of a polarized internal gas target in a storage ring [12]. The target consisted solely of pure vertically polarized ^{3}He nuclei. Quasi-elastic $^3\mathbf{He}(p, 2p)$ and $^3\mathrm{He}(p, pn)$ data were acquired with vertically polarized beam at energies of 197, 300 and 415 MeV.

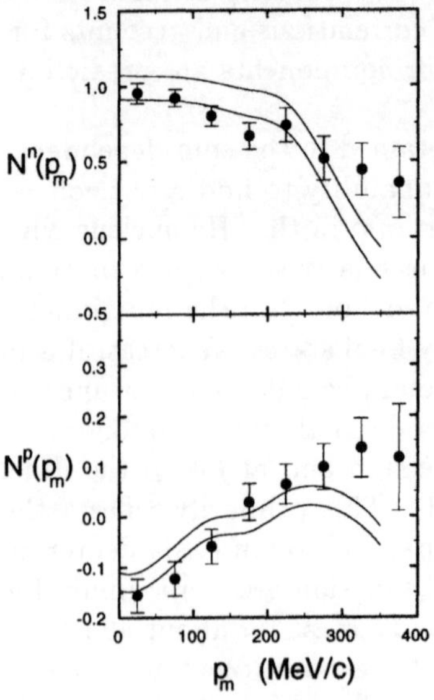

Figure 5. At 197 MeV incident energy the missing momentum distribution of the following observables: (a) A_{00nn} in ^{3}He(p, pn); (b) A_{00nn} in ^{3}He$(p, 2p)$. The curves correspond to the upper and lower bounds of the PWIA prediction allowed by various phase shift solutions for the free observables.

The spin observables $A^{3,N}_{00ij}$ were extracted as a function of p_m for both ^{3}He$(p, 2p)$ and ^{3}He(p, pn) reactions at 197 MeV incident proton energy. The data at low $|\mathbf{q}|$, where $\mathbf{q}$ is the 3-momentum transfer to the recoiling nucleon, were compared to the previous data at similar kinematics and were in good agreement. Fig. 5 shows (a) the $A^{3,n}_{00nn}$ and (b) $A^{3,p}_{00nn}$ observables as a function of p_m. The solid point at $p_m = 0$ is the PWIA prediction. In 5(b) we see that the data at low p_m are in good agreement with the PWIA prediction; here the asymmetry is sensitive to the S' state in the ^{3}He ground-state wave function.

Using the PWIA model of (2) above we have extracted the probability $P_N(p_m = 0)$ to find the nucleon N at $p_m = 0$ polarized parallel to the ^{3}He nuclear spin $\hat{\sigma_3}$. From 5(a) and 5(b) we find $P_n(p_m = 0) = 0.98 \pm 0.11$ and $P_p(p_m = 0) = -0.16 \pm 0.04$, respectively. The Faddeev calculations both predict $P_n(p_m = 0) = 1.00$ and $P_p(p_m = 0) = $ -0.12, in good agreement with the data.

The above described IUCF experiment has given significant impetus to similar experiments in electron scattering. The polarized internal target program is unique and complementary to those envisaged at other laboratories. Spin-dependent scattering experiments with polarized electrons in combination with recoil polarimetry or an external polarized ^{3}He target have been carried out at MIT-Bates. In addition, they have ambitious plans for a dedicated internal target set-up, the so-called BLAST project [13]. At CEBAF measurements of G_E^n are approved with a recoil neutron polarimeter and with polarized deuterium and ^{3}He targets. The Mainz facility has an ongoing program of measurements [14] to determine G_E^n both by the recoil neutron polarimeter technique and by measurement of $(e, e'n)$ scattering from polarized ^{3}He.

At NIKHEF, experiments are planned in the quasi-elastic region where the polarized internal target technique allows to make measurements of the asymmetries in $(e, e'n)$ scattering from deuterium and ^{3}He. These data will allow determination of $G_E^n(Q^2)$ for the neutron in two quite different but complementary nuclear systems. The experimental configuration allows the measurements to be made on pure nuclear targets in a common apparatus with high statistical and systematic precision. Completely simultaneously, data will be acquired on quasi-elastic $(e, e'p)$ spin dependent scattering as well as inclusive scattering. These experiments allow an unambiguous determination of a relatively small quantity like $G_E^n(Q^2)$ which then can be firmly based on a solid understanding of the ground-state spin structure and the details of the reaction mechanism at quasi-elastic kinematics.

5 Conclusion

In summary, a comprehensive program of measurements of electron scattering from polarized few-body systems is in progress. The measurement of nucleon form-factors and the spin structure of few-body systems are of central importance to nuclear physics. To address these important problems our field can now take advantage of the best techniques available. Especially, the polarized internal target technique places experimentalists in the unique position to provide the answers to important questions concerning the spin structure of few-body systems.

Acknowledgement. I want to thank my wife for her permanent interest, vivid encouragement, and mental support of my work.

References

1. J.F.J. van den Brand, R. Ent, P.L. Anthony, R.G. Arnold, J. Arrington, E.J. Beise, J.E. Belz, P.E. Bosted, H.-J. Bulten, M.S. Chapman, K.P. Coulter, F.S. Dietrich, M. Epstein, B.W. Filippone, H. Gao, R.A. Gearhart, D.F. Geesaman, J.-O. Hansen, R.J. Holt, H.E. Jackson, C.E. Jones, C.E. Keppel, E.R. Kinney, S. Kuhn, K. Lee, W. Lorenzon, A. Lung, N.C.R. Makins, D.J. Margaziotis, R.D. McKeown, R.G. Milner, B. Mueller, J. Napolitano, J. Nelson, T.G. O'Neill, V. Papavassiliou, G.G. Petratos, D.H. Potterveld, S.E. Rock, M. Spengos, Z.M. Szalata, L.H. Tao, K. van Bibber, D.A. Wasson, J.L. White, B. Zeidman: Accepted for Publ. in Phys. Rev. **D**, (1995)
2. H.J. Bulten, P.L. Anthony, R.G. Arnold, J. Arrington, E.J. Beise, E. Belz, K. van Bibber, P.E. Bosted, J.F.J. van den Brand, M.S. Chapman, K.P.

Coulter, F.S. Dietrich, R. Ent, M. Epstein, B.W. Filippone, H. Gao, R. A. Gearhart, D.F. Geesaman, J.-O. Hansen, R.J. Holt, H.E. Jackson, C.E. Jones, C.E. Keppel, E. Kinney, S.E. Kuhn, K. Lee, W. Lorenzon, A. Lung, N.C.R. Makins, D.J. Margaziotis, R.D. McKeown, R.G. Milner, B. Mueller, J. Napolitano, J. Nelson, T.G. O'Neill, V. Papavassiliou, G.G. Petratos, D.H. Potterveld, S. E. Rock, M. Spengos, Z.M. Szalata, L.H. Tao, J.L. White, B. Zeidman: Phys. Rev. Lett. **74**, 4775 (1995)

3. M. Ferro-Luzzi, M. Bouwhuis, E. Passchier, Z-L. Zhou, R. Alarcon, R. van Bommel, J.F.J. van den Brand, M. Buchholz, H.-J. Bulten, S. Choi, J. Comfort, D. Dimitroyannis, S. Dolfini, R. Ent, C. Gaulard, T. Henkes, C.W. de Jager, C.E. Jones, J. Lang, D.J. de Lange, M. Miller, N. Papadakis, I. Passchier, H.R. Poolman, S.G. Popov, I. Rachek, E. Six, J.J.M. Steijger, O. Unal, N. Vadakos, H. de Vries: Submitted to Phys. Rev. Lett. (1986)
4. M. A. Miller, K. Lee, A. Smith, J.-O. Hansen, C. Bloch, J.F.J. van den Brand, H.J. Bulten, D. DeSchepper, R. Ent, C.D. Goodman, W.W. Jacobs, C.E. Jones, W. Korsch, L.H. Kramer, M. Leuschner, W. Lorenzon, N.C.R. Makins, D. Marchlenski, H.O. Meyer, R.G. Milner, J.S. Neal, P.V. Pancella, S.F. Pate, W.K. Pitts, B. von Przewoski, T. Rinckel, G. Savopulos, J. Sowinski, F. Sperisen, E.R. Sugarbaker, C. Tschalär, O. Unal, T.P. Welch, Z-L. Zhou: Phys. Rev. Lett. **74**, 502 (1995)
5. CEBAF experiment 94-011, J.F.J. van den Brand, H.J. Bulten, and R. Ent, spokespersons; CEBAF experiment 93-050, P.Y. Bertin, G. Fournier, and C.E. Hyde-Wright, spokespersons.
6. J.A. Tjon: Few-Body Systems Suppl. **5**, 17 (1992); E. Hummel, J.A. Tjon: Phys. Rev. **C42**, 423 (1990); E. Hummel: Ph. D. thesis, Rijks Universiteit Utrecht (1991), unpublished
7. G. Beck, H. Arenhövel: Few-Body Systems **13**, 165 (1992)
8. V.F. Dimitrev et al.: Phys. Lett. 157**B**, 143 (1985); B.B. Woitsekhovskii et al.: JETP Lett. **43**, 733 (1985); R. Gilman et al.: Phys. Rev. Lett. **65**, 1733 (1990)
9. E. Hummel, J.A. Tjon: Phys. Rev. C**42**, 423 (1990); R. Schiavilla, D.O. Riska: Phys. Rev. C**43**, 437 (1990); P.G. Blunden, W.R. Greenberg, E.L. Lomon: Phys. Rev. C**40**, 1541 (1989); C.E. Carlson: Nucl. Phys. **A508**, 481c (1990)
10. M.E. Schulze et al.: Phys. Rev. Lett. **52**, 597 (1984); I. The et al.: Phys. Rev. Lett. **67**, 173 (1991); M. Garçon et al.: Phys. Rev. C**49**, 2516 (1994)
11. E.J. Brash et al.: Phys. Rev. C **47**, 2064 (1993); A. Rahav et al.: Phys. Lett. **275**, 259 (1992); A. Rahav et al.: Phys. Rev. C **46**, 1167 (1992)
12. K.Lee et al.: Phys. Rev. Lett. **70**, 738 (1993); K.Lee et al.: Nucl. Instr. and Meth. A **333**, 294 (1993)
13. S. Kowalski: see elsewhere in these proceedings
14. T. Walcher: invited talk at this Conference

Few-Body Systems Suppl. 8, 213–224 (1995)

Theoretical Description of Helium Clusters

E. Krotscheck[1,2] *, S.A. Chin[1]†

[1] Department of Physics, Texas A&M University, College Station, TX 77843, USA

[2] Institute for Theoretical Physics, Johannes Kepler Universität Linz, A4040 Linz, Austria

Abstract. Helium droplets are unique bosonic finite systems that can be studied from a few-body bound state, through droplets of thousands of particles, up to the bulk limit. They therefore provide an excellent laboratory for the study of finite-size dependence of many physical properties. Moreover, due to the simplicity of the helium-helium interaction, exact Monte Carlo simulations [1] of the ground state can be performed to critically assess the validity of approximate theoretical techniques and (semi)-phenomenological models.

In this work, we present some studies on the structural features of small liquid droplets, with particle numbers ranging from 10 to 70. We pay particular attention to excited state properties that are currently beyond the reach of Monte Carlo methods. It is shown that in many respects even very small clusters retain surprisingly bulk-like properties.

1 Introduction

The reasons for the current widespread interest in helium clusters are manifold: The experimentalist is generally interested in finite size effects, the measurement of surface profiles, the possible detection of Bose-Einstein condensate in a finite system, and the characteristic spectral changes of an imbedded impurity [2-8]. For the theorist, helium clusters are of interest [1,9-16] because they are the simplest self-bound, non-uniform many-particle systems. They thus provide an ideal physical system for understanding how various bulk properties are modified in restricted geometries, or conversely, how various bulk properties emerge from finite systems. Because of the relative simplicity of the helium-helium interaction, helium clusters can be accurately solved by Monte Carlo method or by variational theory without recourse to phenomenology or

*Supported by NSF grant PHY-9108066

†Supported by NSF grant PHY-9213502

ad hoc modeling. One is therefore in an excellent position to concentrate on the physics involved rather than be distracted by discussions of approximate methods or modeling adequacy. The recent suggestion that the Bose condensate in droplets may be detectable *via* atom-cluster collision [14, 17, 18] has given further impetus to the study of helium clusters.

Calculations of the ground state structure of quantum liquid droplets have been done using variational and diffusion Monte Carlo algorithms [11, 12, 13, 1, 19], density functional theory [20, 22], and the optimized integral-equation hypernetted chain (HNC/EL) method [16]. One may have the impression that, in view of the exactness of the Green's Functions and Diffusion Monte Carlo methods, there is no further use for approximate techniques. This may be true to some limited extent with regard to *ground state* properties. For excited states, our experience in ref. [1] has shown that such impression is simply unwarranted. In ref. [1], low angular momentum excitations were extracted from the partial wave expansion of the two-body density via a two dimensional binning of approximately 4000 numbers. Very long running time is required to produce adequate statistics for these 4000 numbers, roughly one to two orders of magnitude longer than that for computing the ground state energy. Furthermore, the extraction of higher partial wave amplitudes is nearly impossible because they can only be projected out via highly oscillatory Legendre polynomials.

Severe convergence problems can also arise in a Monte Carlo simulation where the importance sampling function, through providing a reasonable estimate of the energy, may nevertheless poorly model the exact wavefunction in details. This is particularly relevant for the study of excitations: While the energy of the system comes mostly from the *interior* of the droplet, most low-lying *excitations* live on the surface. It is therefore extremely difficult to determine low energy surface excitations by direct simulation. In contrast, the optimized variational theory presented here is free from statistical or convergence uncertainties which are crucial for deciding such fine structures.

On the other hand, the existence of *some* exact simulation results for non-uniform quantum liquids, clusters, in particular, have re-defined the level of acceptance for doing analytic many-body physics: There is no longer any excuse for not comparing results of an approximate theoretical description with all of the exact information that is currently available. In the case of HNC/EL, extensive and detailed comparison [16] have shown that, in all cases studied so far, this theory is predictive to within a few percent of Diffusion or Green's Functions Monte Carlo results. These comparisons, which require minimal phenomenological input, form the basis by which we can be confident of the theoretical predictions presented below.

By the use of our optimized variational theory, we shall address here the question *what is the meaning of "many" in "many-body physics"* or, more appropriately for this meeting, how "small" must a physical system be in order to reveal truely few-body effects. For this purpose, we have computed the ground state energy, chemical potential, collective monopole, dipole, quadrupole and higher multipole excitations, one and two-body densities, and the dynamic structure function for helium droplets with mass numbers ranging from $N = 10$ to N=1000. This allows us to address specific questions left unresolved in our earlier DMC calculations [1] and enable us to discern systematic trends of physical observables as a function of the particle number.

2 Ground State of Pure ^{4}He Clusters

The optimized variational theory begins with the hamiltonian

$$H = -\frac{\hbar^2}{2m}\sum_i \nabla_i^2 + \sum_{i<j} v(|\mathbf{r}_i - \mathbf{r}_j|), \tag{1}$$

where one normally uses the Aziz potential [23] for the two-body interaction. The method is *approximate* since it uses a Jastrow-Feenberg wave function

$$\Psi_0(\mathbf{r}_1,\ldots,\mathbf{r}_N) = \exp\frac{1}{2}\left[\sum_i u_1(\mathbf{r}_i) + \sum_{i<j} u_2(\mathbf{r}_i,\mathbf{r}_j) + \sum_{i<j<k} u_3(\mathbf{r}_i,\mathbf{r}_j,\mathbf{r}_k)\right] \tag{2}$$

and calculates quantities of physical interest by integral equation techniques. The method has been described repeatedly in the literature [24, 25]. The only new aspect for finite systems is that we must consider the "internal energy"

$$E_{\mathrm{int}} = E - T_{\mathrm{cm}}, \tag{3}$$

where T_{cm} is the kinetic energy connected with the center-of-mass motion. Quantities of interest are the ground state energy, one- and the two-body densities, and the coordinate-space representation of the static structure function

$$\begin{aligned} \rho_1(\mathbf{r}) &= \frac{\langle\Psi_0 \mid \sum_i \delta(\mathbf{r}_i - \mathbf{r}) \mid \Psi_0\rangle}{\langle\Psi_0 \mid \Psi_0\rangle}, \\ \rho_2(\mathbf{r},\mathbf{r}') &= \frac{\langle\Psi_0 | \sum_{i\neq j} \delta(\mathbf{r}_i - \mathbf{r})\delta(\mathbf{r}_j - \mathbf{r}')|\Psi_0\rangle}{\langle\Psi_0 \mid \Psi_0\rangle}, \\ S(\mathbf{r},\mathbf{r}') &= \delta(\mathbf{r} - \mathbf{r}') + \frac{\rho_2(\mathbf{r},\mathbf{r}') - \rho_1(\mathbf{r})\rho_1(\mathbf{r}')}{\sqrt{\rho_1(\mathbf{r})\rho_1(\mathbf{r}')}}. \end{aligned} \tag{4}$$

These densities are necessary for the calculation of excitation energies and the dynamic structure function.

Space does not permit us to display the large array of comparisons that we have carried out; an extensive discussion can be found in a forthcoming publication [16]. We shall only show here the energy (as the best) and the one-body density (as the worst) case of agreement. The results of our HNC/EL calculations for the ground state energy and the unit radius $r_0 = (5/3)^{1/2}\langle r^2\rangle^{1/2}N^{-1/3}$ agree with our DMC calculations better than expected. The best global fit to the energy per particle for clusters with 70 or more particles has been obtained by the form

$$\frac{E(N)}{N} = -7.21 + 17.71N^{-1/3} - 5.95N^{-2/3}\ \mathrm{K}. \tag{5}$$

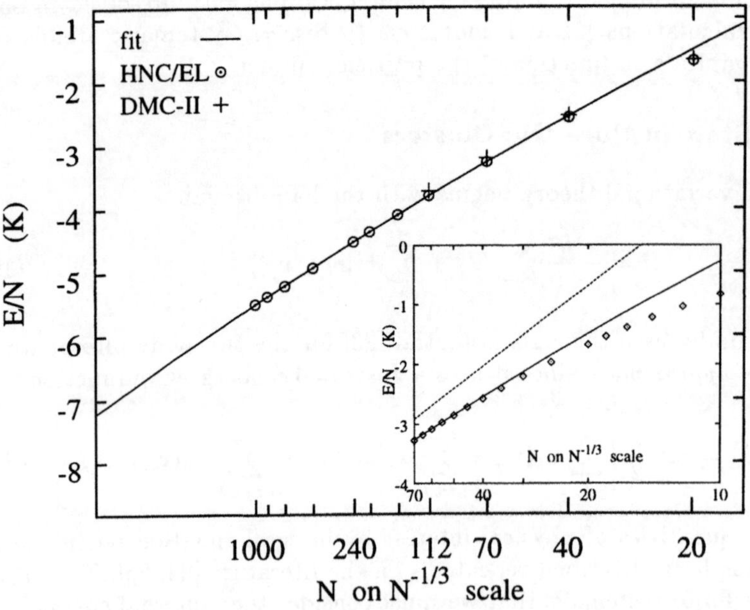

Figure 1. The HNC/EL energies of helium clusters (circles) are compared with the DMC results of ref. [1] (crosses). Also shown is the fit (5) (solid line), recall that only the energies for particle numbers $N \geq 70$ have been fitted. The insert shows the energies for small droplets along with the fit (5) (solid line) and the same fit with the curvature term omitted (dashed line).

Clusters with less than 40 particles are surface dominated, their central densities has not yet reached the bulk value. They were therefore not used for extrapolating to the bulk limit. In Fig. 1 we show our HNC/EL results, together with the above fit and the DMC results of ref. [1]. Eq. (5) suggests an asymptotic energy per particle of -7.21 K and a surface energy of $\sigma \approx 0.272$ K Å^{-2}. The asymptotic energy has the expected accuracy from the bulk theory, The surface energy also compares quite favorably with the experimental value [26] of 0.256 K Å^{-2}. To assess the extent to which the energetics of *smaller* clusters can be well represented by the fit (5), the insert in Fig. 1 shows the energies for *small* droplets with with N = 10 to 70. Also included is the mass formula fit (5) with and without the "curvature term"(i.e., the $N^{-2/3}$ term). For the ordinary liquid drop model, retaining only the volume and surface terms is usually adequate and keeping the curvature term is sufficient for all practical purposes. It needs to be pointed out here that our theory is a *many-body* theory and its present implementation is subject to errors on the order of $1/N$. We have, however, tested the theory for particle numbers N = 10 in atoms [27] and for N = 16 in a model nucleus [28] and have found it quite satisfactory.

The agreement between our HNC/EL calculated density profile and the

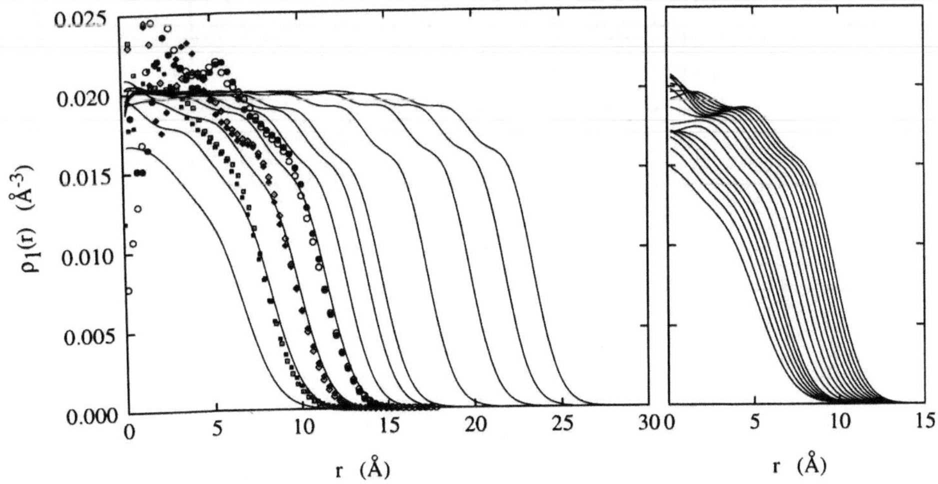

Figure 2. The ground state density profiles of ^{4}He droplets with particle numbers $N = 20, 40, 70, 112, 150, 200, 240, 400, 600, 800$, and 1000 as obtained from the HNC/EL theory. The boxes, diamonds and circles are DMC results from ref. [1] from different runs for N = 40, 70, and 112 particles. The right figure shows density profiles of small droplets with particle numbers $N = 10, 12, \ldots, 20, 25, \ldots, 70$.

DMC results is less quantitative (cf. Fig. 2). They are systematically too low in comparison and appear to converge towards a central density slightly larger than 0.02 $Å^{-2}$. We attribute this to discretization errors.

In our DMC calculation [1], we have found persistent ground state density oscillation in smaller clusters attributable to a geometric "shell structures". Our present calculations clearly demonstrate that the one-body densities are indeed *not* smooth. Moreover, Fig. 2 shows how the shell structure in smaller clusters develops into a broader shoulder in the density profile of larger clusters, suggesting a two-dimensional "skin" of ^{4}He surrounding each cluster. These density oscillations from our HNC calculations are somewhat smaller in amplitude and out of phase with those found in DMC, but they are undeniably present and are rather similar.

3 Collective Excitations

Excited states can be estimated by assuming a variational *ansatz* for the wave function of the excited state of Feynman form

$$\Psi_F(\mathbf{r}_1, \ldots \mathbf{r}_N) = F(\mathbf{r}_1, \ldots \mathbf{r}_N)\Psi_0(\mathbf{r}_1, \ldots \mathbf{r}_N) = \sum_{i=1}^{N} f(\mathbf{r}_i)\Psi_0(\mathbf{r}_1, \ldots \mathbf{r}_N), \qquad (6)$$

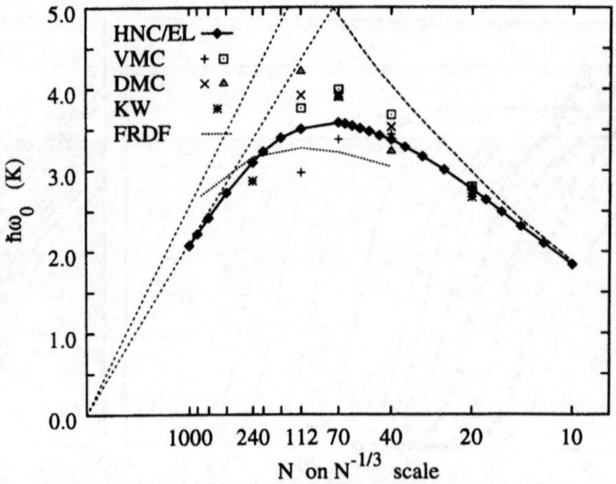

Figure 3. The monopole excitation energies of droplets from 10 to 1000 particles (heavy solid line with diamonds) are compared with the results of our VMC and DMC calculations [1], those of Krishna and Whaley (ref. [11]) and the density functional calculation by Barranco and Hernández (ref. [22]). The long-dashed line is the continuum limit $-\mu(N)$. The upper short-dashed line is the liquid drop approximation as calculated from the *bulk* equation of state, and the lower dashed line is an "extrapolated" liquid drop limit from the microscopic calculation of the excitation energies.

where the *excitation function* $f(\mathbf{r}_i)$ is determined by minimizing the excitation energy

$$\hbar\omega \equiv E_F - E_0 = \frac{1}{2}\frac{\langle\Psi_0\,|[F,[T,F]]\,|\Psi_0\rangle}{\langle\Psi_0\,|F^2\,|\Psi_0\rangle}. \tag{7}$$

This minimization leads [29] to the generalized eigenvalue problem

$$H_1(\mathbf{r})\psi_\omega(\mathbf{r}) = \hbar\omega \int d^3r' S(\mathbf{r},\mathbf{r}')\psi_\omega(\mathbf{r}') \tag{8}$$

where

$$H_1(\mathbf{r}) = -\frac{\hbar^2}{2m}\frac{1}{\sqrt{\rho_1(\mathbf{r})}}\nabla\rho_1(\mathbf{r})\cdot\nabla\frac{1}{\sqrt{\rho_1(\mathbf{r})}}. \tag{9}$$

For each energy $\hbar\omega$, the corresponding excitation function $\psi_\omega(\mathbf{r}) = \sqrt{\rho(\mathbf{r})}f(\mathbf{r})$ is related to the time-dependent part of the density $\delta\rho_\omega(\mathbf{r})$ through

$$\sqrt{\rho_1(\mathbf{r})}H_1\psi_\omega(\mathbf{r}) = \delta\rho_\omega(\mathbf{r}). \tag{10}$$

Our results for monopole and quadrupole excitation energies are compared with our previous DMC energies and a few other recent calculations in Figs. 3 and 4. Overall, we find satisfactory agreement between our HNC/EL calculations and our Monte Carlo results. Both exhibit the same trend as a function

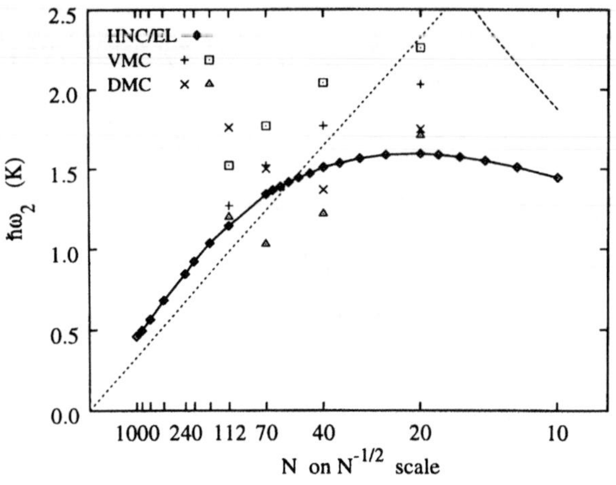

Figure 4. Same as Fig. 3 for the quadrupole excitations.

of the cluster size. At higher particle numbers, Monte Carlo results are less reliable due to extrapolation uncertainties in the one and two-body densities. This is demonstrated by the large scatter between different calculations. Density functional calculations approach the liquid drop limit too rapidly.

In the case of the monopole mode, the excitation energy appears to have approached such a limit quite well, albeit with a different slope. In the limit of large particle numbers, one would expect the monopole energy to approach the liquid drop model result of

$$\hbar\omega_0 = \frac{\hbar c \pi}{R} = 25.6 N^{-1/3}\,\mathrm{K}. \tag{11}$$

Fig. 3, however, suggests a coefficient of only 21. This number agrees actually quite well with the prediction from the *calculated* equation of state [30], which yields, at the *calculated* central density, a coefficient of 20.

For the quadrupole excitations, the DMC calculations suffered from significant extrapolation uncertainties. It is again interesting to know to what extent a "liquid drop limit" has been reached for cluster sizes considered here. Fig. 4 seems to indicate that one is still far from such a limit. However, one can extract further information on "how close" one is to the liquid drop limit by considering higher multipole excitation modes. As we shall see, for somewhat shorter wavelengths, the agreement between the dispersion relation of the surface modes calculated from the Feynman theory and the one calculated form the surface energy extracted from the mass formula (5) is indeed excellent.

We have computed all *discrete* excitation energies for angular momenta $\ell \geq 2$. Since the excitation function is generally given by $\sim r^\ell$, these high ℓ excitations are essentially surface vibrational modes. In general, one expects the droplet excitation energies to depended on both N, the droplet size, and ℓ, the multipolarity or angular symmetry. However, if a surface mode ℓ were to

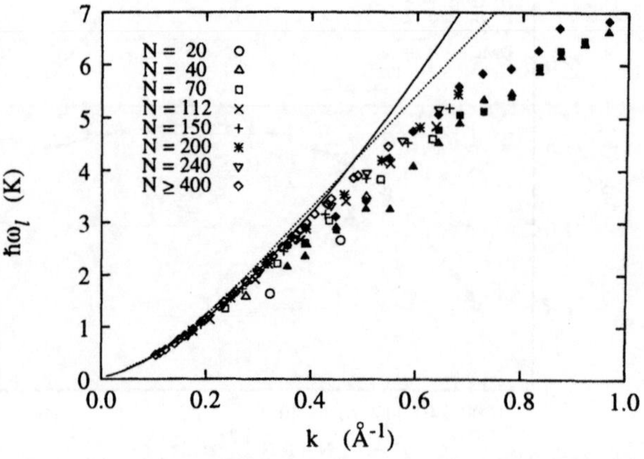

Figure 5. The lowest excitation energies with an angular momentum up to $\ell = 9$ are shown, for all pure ^{4}He droplets considered here, as a function of an effective wave number $k = \sqrt{\ell(\ell+1)}/R$, where we have taken $R \equiv \sqrt{5/3} r_{\mathrm{rms}}$. The droplets with $N \geq 150$ are marked with diamonds, $N = 112$ with asterikses, $N = 70$,with $N = 40$ with triangles and $N = 20$ with squares. Only energies corresponding to discrete modes are shown. Also shown is ripplon dispersion relation (13) (solid line) the lowest excitation energy of a *helium film* (dashed line, from ref. [25]).

be excited by a projectile with momentum $\hbar k$, then k must be of the order

$$k = \sqrt{\ell(\ell+1)}/R, \tag{12}$$

where we can take the droplet radius R to be the equivalent hard sphere radius $R = \sqrt{5/3} r_{\mathrm{rms}}$. This suggests that one may view the excitation energies in terms of a single variable, k. The validity of this picture is demonstrated in Fig. 5. The excitation energies for all droplets fall very closely together to form an universal curve. The only two noticeable deviations are the lowest $\ell = 2$, 3 excitation energies of cluster size $N = 20$. Moreover, the surface excitation energies are in excellent agreement with the *experimentally observed* surface excitations of helium films [31], they follow a curve which can, for small k (<0.3 Å^{-1}), be fitted by a characteristic "ripplon" dispersion relation

$$\hbar\omega = \sqrt{\frac{\hbar^2 \sigma}{m \rho_\infty}}\, k^{3/2} = 12.399\, k^{3/2} \ \ \mathrm{K}. \tag{13}$$

The fit (13) is plotted as a solid line in Fig. 5. This allows us to extract from the dispersion relation a surface energy of $\sigma = 0.279 \mathrm{K\AA}^{-2}$, in excellent agreement with the one obtained from our mass formula (5). Both the droplet excitations and the ripplon dispersion relation (13) agree well with that of a film of three atomic layers of helium atoms adsorbed onto a *plane* substrate [25]. Data from neutron scattering experiments [31] on such films are shown as solid markers in Fig. 5. The considerable overlap between theoretical droplet excitation energies and experimental film data suggests, again, that certain features of even very small droplets can be quite well obtained from bulk properties.

4 Dynamic Structure Function

Up to now, we have focused attention on the low-lying, discrete, collective excitations which have revealed much about the microscopic structure of helium droplets. However, experimentally one is interested in scattering cross sections and the extent to which these cross sections are sensitive to the excitation spectrum of the droplet. For sufficiently high energy scattering where the Born approximation is valid, the cross section is given in terms of the dynamic structure function, $S(k,\omega)$. In our theory of excitations, the $S(k,\omega)$ can be obtained directly from the Fourier transform of the transition densities (10),

$$S(k,\omega) = |\delta\tilde{\rho}_\omega(\mathbf{k})|^2 \tag{14}$$

with

$$\delta\tilde{\rho}_\omega(\mathbf{k}) = \int d^3k e^{i\mathbf{k}\cdot\mathbf{r}} \delta\rho_\omega(\mathbf{r}). \tag{15}$$

In practice, we obtain $S(k,\omega)$ by summing over all partial wave contributions

$$S(k,\omega) = \sum_\ell S_\ell(k,\omega), \tag{16}$$

where both the individual components and the total sum are of interest. For this propose one must compute, not just the bound, collective excitations, but also high-energy resonances in the continuum. In ref. [1] we have considered only excitations with $\ell \leq 2$, and have demonstrated that they contribute relatively little to the total scattering cross section of a plane wave. To compute the full $S(k,\omega)$, one must therefore include partial waves of higher multipolarity.

For large droplets one expects that the excitation of the droplet is dominated by a bulk phonon-roton mode and a surface ripplon mode. However, neither the size dependence nor the relative strength of these two modes are known. It is also not clear how each partial wave would (16) contributes to each of these two modes. Fig. 6 shows the dynamic structure function for a large droplet. We have broadened the discrete modes discussed above by a Gaussian of 0.5 K width. On the left, where only the $\ell = 0$ partial wave contribution is shown, the phonon-roton spectrum is clearly outlined with no hint of any surface excitations. This is as expected since the monopole mode is a bulk excitation and cannot be sensitive to any surface structures. When the first 50 angular momentum components are summed according (16) to get the full dynamic structure function, two distinct dispersion branches emerge. One branch again overlaids the phonon/roton spectrum, but the other now traces out the *surface excitation* spectrum as discussed above. The appearance of the surface spectrum is due entirely to high ℓ partial wave contributions. While

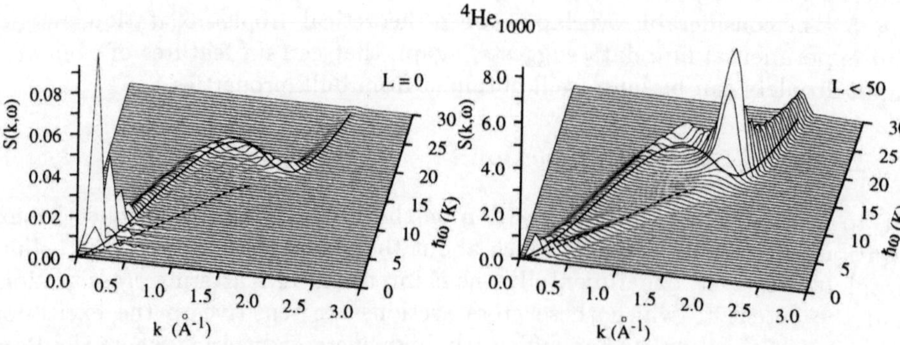

Figure 6. The dynamic structure function of a helium droplet with $N = 1000$ particles in Feynman approximation. The *left* figure shows the result when only monopole excitations are included, and the right figure shows a converged result, which retains all spherical harmonimcs with $L < 50$. The solid curve in the $k - \hbar\omega$-plane of the figure is the Feynman phonon-roton spectrum for the bulk liquid, and the dashed curve the surface excitation calculated in ref. [25].

the *general* features of the spectrum are as expected, the quantitative details of the dynamic structure function cannot be determined without an actual calculation. For example, the importance of the roton is clearly revealed by its dominant strength. Even more interesting is that, for $N = 1000$, up to a wave vector of approximately 1.5 Å^{-1}, the phonon-roton and the ripplon excitation have comparable strength. This suggests an intriguing possibility that both excitations may be observed cleanly by scatter off of a "mist" or "fog" of helium droplets.

This picture of the dynamic structure function for $N = 1000$ remains qualitatively correct down to cluster sizes of about 100 particles. The strength of the surface excitation relative to that of the bulk increases slowly and all resonances gradually broadens. What happens at even smaller particle numbers? How many of these excitation features will persist for really *small* clusters? We have already seen that *some* droplet properties, like energies or the high multipole surface excitations, are quite well described by simple features of the liquid drop model. One expects that at $N \leq 20$, when the central density is well below that the bulk liquid, all traces of bulk-like behaviors should disappear and no extrapolation from liquid drop data is possible. Fig. 3 clearly shows that the monopole excitation energies at these particle numbers are far from a "liquid drop" limit.

Fig. 7 shows four contour plots of the dynamic structure function $S(k, \omega)$ for clusters of very small to medium size. A remarkable feature, common to all cases, is that the strength remain piled up at the vicinity of the "roton minimum". The peak moves, with increasing particle number, ever closer to the bulk roton minimum with increasing strength, and persists even in a system of only 10 particles. This supports the view that actually nothing is "rotating"

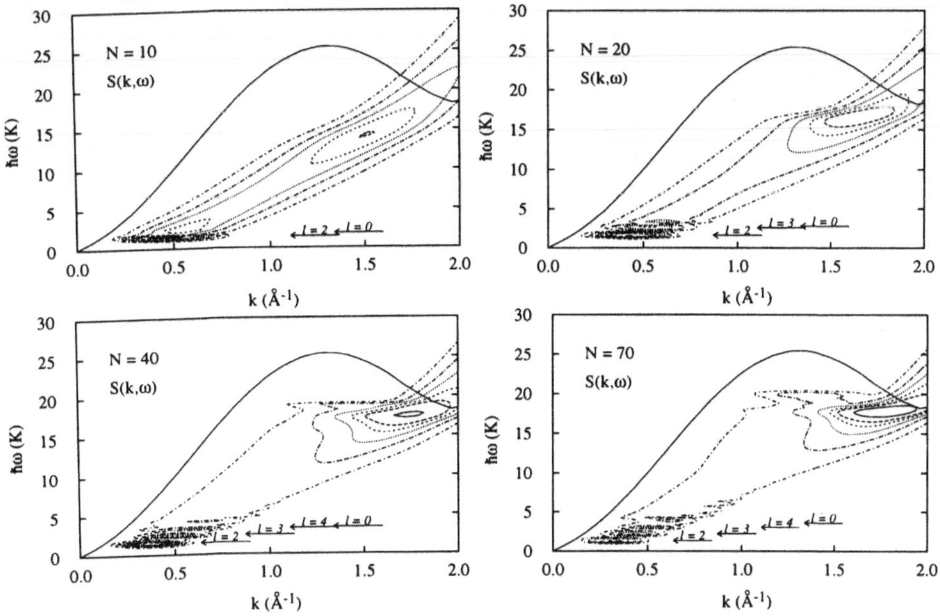

Figure 7. Contour plots of the dynamic structure function of small droplets. The contour levels are drawn at relative height 0.2, 0.4, 0.6, 0.8, 1.0, and 1.5. The solid line is the Feynman dispersion relation, $\hbar\omega(k) = \hbar^2 k^2 / 2mS(k)$ for the *bulk* helium liquid. The arrows point to the energies of the *discrete* modes.

at the roton minimum, but rather that this resonance is an effect of the short-range repulsive nature of the helium interaction.

In all cases, there is still a certain amount of "discreteness" visible. It is also apparent that the strength of these modes follows the surface-mode dispersion law discussed earlier. At $N = 70$, the "roton minimum", which is a generic many-body effect, begins to assert itself as the strongest feature in the spectrum.

References

1. S.A. Chin, E. Krotscheck: Phys. Rev. B **45**, 852 (1992)
2. E. Syskakis, F. Pobell, H. Ullmaier: Phys. Rev. Lett. **55**, 2964 (1985)
3. H. Buchenau et al.: J. Chem. Phys **91**, 6875 (1990)
4. J. Gspann: Physica Status Solidi B **169**, 519 (1991)
5. T. Jiang, J.A. Northby: Phys. Rev. Lett. **68**, 2620 (1992)
6. S. Goyal, D.L. Schutt, G. Scoles: Phys. Rev. Lett. **69**, 933 (1992)

7. S. Goyal, D.L. Schutt, G. Scoles: J. Phys. Chem. **97**, 2236 (1993)

8. R. Fröchtenicht, J.P. Toennies, A. Viselov: Chem. Phys. Lett **229**, 1 (1994)

9. U. Helmbrecht, J.G. Zabolitzky: in *Monte Carlo Methods in Quantum Problems*, edited by M. Kalos (D. Reidel, Holland, 1984)

10. P. Sindzingre, M.J. Klein, D.M. Ceperley: Phys. Rev. Lett. **63**, 1601 (1989)

11. M.V.R. Krishna, K.B. Whaley: Phys. Rev. Lett. **64**, 1126 (1990)

12. M.V.R. Krishna, K.B. Whaley: J. Chem. Phys. **93**, 6738 (1990)

13. S.A. Chin, E. Krotscheck: Phys. Rev. Lett. **65**, 2658 (1990)

14. J.W. Halley, C.E. Campbell, C.F. Giese, K. Goetz: Phys. Rev. Lett. **71**, 2429 (1993)

15. S.A. Chin, E. Krotscheck: Phys. Rev. Lett. **74**, 1143 (1995)

16. S.A. Chin, E. Krotscheck: Phys. Rev. B (1994), (submitted)

17. C.E. Campbell: J. Low Temp. Phys. **93**, 907 (1993)

18. S.A. Chin: J. Low Temp. Phys. **93**, 921 (1993)

19. R.N. Barnett, K.B. Whaley: J. Chem. Phys. **99**, 9730 (1993)

20. M. Casas, S. Stringari: J. Low. Temp. Phys. **79**, 135 (1990)

21. F. Dalfovo: Z. Physik D **29**, 61 (1994)

22. M. Barranco, E.S. Hernandez: Phys. Rev. B **49**, 12078 (1994)

23. R.A. Aziz et al.: J. Chem. Phys. **70**, 4330 (1979)

24. B.E. Clements, J.L. Epstein, E. Krotscheck, M. Saarela: Phys. Rev. B **48**, 7450 (1993)

25. B.E. Clements et al.: Phys. Rev. B **50**, 6958 (1994)

26. M. Iino, M. Suzuki, A.J. Ikhushima: J. Low Temp. Phys **61**, 155 (1985)

27. C.E. Campbell, E. Krotscheck, T. Pang: Physics Reports **223**, 1 (1992)

28. S.A. Chin, E. Krotscheck: Nucl. Phys. A **560**, 151 (1993)

29. C.C. Chang, M. Cohen: Phys. Rev. B **11**, 1059 (1975)

30. E. Krotscheck, M. Saarela: Phys. Rep. **232**, 1 (1993)

31. H.J. Lauter, H. Godfrin, V.L.P. Frank, P. Leiderer: Phys. Rev. Lett. **68**, 2484 (1992)

Few-Body Systems Suppl. 8, 225–234 (1995)

Few-Body Systems

Antiprotonic Helium Atoms

O.I. Kartavtsev*

Bogoliubov Laboratory of Theoretical Physics, Joint Institute for Nuclear Research, 141980 Dubna, Russia

Abstract. Metastable antiprotonic helium atoms $^{3,4}\mathrm{He}\,\bar{p}e$ have been discovered recently in experiments of the delayed annihilation of antiprotons in helium media. These exotic atoms survive for an enormous time (about tens of microseconds) and carry the extremely large total angular momentum $L \sim 30-40$. The theoretical treatment of the intrinsic properties of antiprotonic helium atoms, their formation and collisions with atoms and molecules is discussed.

1 Introduction

Metastable antiprotonic helium atoms $^{3,4}\mathrm{He}\,\bar{p}e$ have been discovered recently in experiments on the delayed annihilation of antiprotons in helium media [1–3]. A few percents of antiprotons are captured in these exotic systems, which survive for an enormous time (about tens of microseconds in comparison with the mean lifetime of antiprotons in media $\sim 10^{-12}$s) and carry the extremely large total angular momentum $L \sim 30-40$. Analogous long-lived systems were observed in experiments with negative kaons [4] and pions [5]. This new class of three-body systems was first predicted in [6] and theoretically studied in [7]. As far as the experimental data concern mainly the antiprotonic helium atom, only this system is considered in this report. One can mention that described theoretical considerations can be easily applied also to other hadronic helium atoms.

The observation of the resonant laser-induced annihilation [8,9] has initiated thorough investigations of these unusual systems. Initial populations, level lifetimes and very precise values of the transition energies (with the relative accuracy $< 10^{-5}$) have been obtained in these experiments.

The antiprotonic helium can be considered as a usual helium atom with one electron replaced by an antiproton. A large angular momentum $L \sim (M/m_e)^{1/2}$ (M – $\bar{p}$–nucleus reduced mass) provides that the antiproton-nucleus and electron-nucleus distances are approximately equal. Besides, this

* *e-mail:* oik@thsun1.jinr.dubna.su

system can be considered as exotic diatomic molecule, where one nucleus has charge -1.

Main reasons for the enormous lifetime of antiprotonic helium atoms are as follows. The annihilation of an antiproton is inhibited due to the extremely large total angular momentum $L \sim 35$. The Auger decay is inhibited due to sufficiently large values ($l_e \geq 4$) of the angular momentum of an outgoing electron. Usual mechanism of the de-excitation by the Stark mixing is not appropriate to the three-body system due to the lack of degeneracy. The collisional de-excitation by surrounding He atoms is suppressed due to the screening of antiproton by the electron in an antiprotonic helium. The only remaining de-excitation mechanism is multistep radiative transitions, whose rates are of order μs^{-1}.

For clear understanding of the antiproton fate in the helium medium one needs to study the slowing down and capture of antiprotons, intrinsic properties of antiprotonic helium atoms, their decay processes and interactions with surrounding atoms and molecules. The purpose of this report is to present the results of existing theoretical calculations on antiprotonic helium atoms and related topics with the reference to the modern experimental situation.

2 Energy Levels

Main problems in the calculation of energy levels arise from very large values of the total angular momentum $L \sim 30 - 40$ of antiprotonic helium atoms and instability of these systems. However, very small decay rates allow one to regard in calculations states of these systems as true bound states.

The first calculations of energy levels were made in the pioneering Russel papers [7]. Using a simple two-parameter variational wave function, energy levels were calculated with the relative accuracy about 3 percents. The Born-Oppenheimer approximation will be quite reliable in solving this problem due to very small mass ratios $m_e/m_{\bar{p}}, m_e/m_{\rm He}$ and the next step has been made in the framework of this approach [10]. Only ground states were considered in this paper and more systematic study of antiprotonic helium eigenenergies within the same approach was carried out in papers [11,12] after the discovery of the delayed antiproton annihilation [1]. The relative accuracy of eigenenergies obtained in these calculations can be estimated as 10^{-4}.

The configuration interaction calculation [13] provides a comparable accuracy of eigenenergies without the reference to the small mass ratio. Systematic variational calculations [14], with simple trial functions of the form

$$\chi^{LM}_{nml\lambda i}(\mathbf{r}, \mathbf{r}_1) = \mathcal{Y}^{LM}_{l\lambda}(\hat{\mathbf{r}}, \hat{\mathbf{r}}_1) r^{l+i} r_1^{\lambda} exp(-a_n r - b_m r_1), \tag{1}$$

where $\mathbf{r}, \mathbf{r_1}$ are the electron-nucleus and antiproton-nucleus radius-vectors, also provide accuracy of the same order of magnitude. Very recently, V. I. Korobov has obtained more precise spectra of an antiprotonic helium, using the correlated trial functions in variational calculations [15].

By using the method of laser-induced resonant annihilation, two transition wavelengths 597.259 ± 0.002nm [8] and 470.724 ± 0.002nm [9] were mea-

Table 1. Experimental and theoretical transition wavelengths (nm) of the $^4\mathrm{He}\,\bar{p}e$ system.

Assignment	Experim.	[16]	[11]	[14]	[15]
$35,4 \to 34,4$	597.259	597.341	598.010	597.544	597.229
$34,3 \to 33,3$	470.724	470.594	471.351	470.871	470.705

sured in the $^4\mathrm{He}\,\bar{p}e$ system and assigned to the transitions $(35,4 \to 34,4)$ and $(34,3 \to 33,3)$, respectively. The most important for proving the existence of antiprotonic helium atoms is to compare experimental and theoretical transition wavelengths given in Table 1. Here and below in this report the L, N notation of states, where L is the total angular momentum and N enumerates the states of the same L value, will be used.

While the agreement of experimental and theoretical transition wavelengths approves the formation of antiprotonic helium atoms unambiguously, it is worthwhile to mention a possible ambiguity in the assignment of the 470.724nm transition. In fact, the calculated wavelength of the transition $(35,2 \to 34,2)$ is also very close to the experimental value.

In conclusion, a more precise description of energy spectra requires to take into account the effect of relativistic and QED corrections, spin-dependent interactions and the coupling with continuous spectrum.

3 Radiative Transitions

Due to large lifetimes of metastable states against Auger decay and collisional de-excitation the radiative transitions become most important in the description of the system. As far as only dipole transitions are significant, the total angular momentum changes by unity in each transition. Thus, the system looses the angular momentum and energy step-by-step and finally reaches the state with large Auger decay rate. As a result, the total lifetime is determined by the number of radiative transitions necessary to reach this state and rates of each transition.

The dipole transition rates were calculated in papers [11,13,14,17], whose results are fairly close to each other. It was found that the most probable are transitions between states of the same N, i. e. the excitation number is approximately conserved in the radiative transitions. At the same time, the rates of transitions between states with different N are smaller at least by 1–2 orders of magnitude. Thus, radiative cascades in an antiprotonic helium proceed almost independently along the chains of states of fixed N.

Up to now, lifetimes of the states $(L,N) = (35,4)$ and $(34,3)$ of $^4\mathrm{He}\,\bar{p}e$ system are determined experimentally by the method of resonant laser-induced annihilation [9,18]. Table 2 contains the experimental decay rates and theoretical radiative transition rates of these states. The difference of experimental

Table 2. Experimental decay rate and theoretical radiative transition rates ($10^6 s^{-1}$) for two states of the $^4He\bar{p}e$ system.

L, N	Experiment	[17]	[11]	[14]
35, 4	0.72	0.614	0.619	0.597
34, 3	1.18	0.734	0.754	0.713

and theoretical values is not still understood and may be caused by additional nonradiative decay channels.

One can conclude that the existing calculations of radiative transitions are reasonable accuracy and a detailed consideration of other decay processes is needed.

4 Auger Decay

The main feature of the Auger decay rates of antiprotonic helium atoms is their essential dependence on the multipolarity, i. e. the angular momentum of the outgoing electron l_e. This feature was supposed already in [6] and any simple estimate gives the Auger transition rate of order $10^5 s^{-1}$ for the multipolarity $l_e = 4$ and $10^8 s^{-1}$ for $l_e = 3$. This important result was initially obtained in [7]. Bearing in mind the radiative transition rates of order ($10^6 s^{-1}$), one can consider antiprotonic helium states to be metastable due to the multipolarity of the Auger decay $l_e \geq 4$. Calculation of eigenenergies unambiguously determines multipolarities of the Auger decay and it was performed firstly in [7] and more exactly in subsequent papers [10,11,13,14].

Up to now the only progress in this direction is due to calculations of the Auger decay rates of the $^4He\,\bar{p}e$ system [16,17]. In these calculations, an increase of the multipolarity from $l_e = 3$ to $l_e = 4$ as a rule gives a decrease of the Auger rate by three orders of magnitude. Besides, for some states the calculated Auger rates are much smaller than the typical values prescribed by the multipolarity. This interesting peculiarity is a result of the interference of different parts of the initial and final state wave functions.

It is worthwhile to mention that these results are based on the approximation of the continuous spectrum final state wave function. Namely, the effective two–body model was used to describe the interaction of the outgoing electron and the hydrogenlike $\bar{p}He^{++}$ system.

As it was mentioned in the preceding section, the experimental data call for the intensive theoretical study of decay processes and precise few–body calculations are desirable for understanding the cascade processes in an antiprotonic helium.

5 Energy Levels Splitting and Relativistic Corrections

As it was discussed above, the precise measurement of transition energies of antiprotonic helium atoms in recent experiments on the laser-induced resonant annihilation invokes the calculations of comparable accuracy. For this reason, relativistic corrections of order α^2 (in a.u.) to the pure Coulomb interaction should be taken into account in the precise calculations of energy spectra of antiprotonic helium atoms. Since the precision of experiments can be improved significantly, also QED corrections to energies of higher orders on α will be calculated. Precise calculations and measurements of the energy spectra can be also used for the determining the antiproton properties, e.g. the magnetic moment. Experimentally, the most simple is the measurement of the energy-level splitting due to the spin-dependent part of the relativistic interaction.

Recently, the energy-level splitting of antiprotonic helium atoms has been calculated with the wave functions [14]. The details of this calculation and analysis of the level structure will be presented in [19]. Short results of this calculation are given below.

The splitting of levels arise due to the spin-dependent part of the Breit interaction in each pair of particles in antiprotonic helium atoms. Since the magnetic moments and velocities of particles scale inversely proportional to the particle mass and very small ratios m_e/m_{He}, $m_e/m_{\bar{p}}$, the largest contribution to the energy splitting comes from the interaction with the electron spin. This part of relativistic interaction can be written as follows:

$$H_s = \alpha^2 \sum_{i=\bar{p},\mathrm{He}} \frac{Z_i}{r_{ei}^3} \mathbf{s}_e \cdot \mathbf{r}_{ei} \times \left(\tfrac{1}{2}\mathbf{v}_e - \mathbf{v}_i\right), \tag{2}$$

where α is the fine structure constant, $\mathbf{r}_i$, $\mathbf{v}_i$, $\mathbf{s}_i$, Z_i are the radius-vector, velocity, spin and charge of particle i, $\mathbf{r}_{ei} = \mathbf{r}_e - \mathbf{r}_i$.

This interaction conserves the sum $\mathbf{j} = \mathbf{L}+\mathbf{s_e}$ of the total angular momentum $\mathbf{L}$ and electron spin $\mathbf{s}_e$ and splits each level into two sublevels, corresponding to the eigenvalues $j = L \pm 1/2$.

Splitting values, defined as $\Delta E_L = E(j = L + 1/2) - E(j = L - 1/2)$, have been calculated in the first order of perturbation theory over H_s by using variational nonrelativistic wave functions [14]. These values for a number of states of $^4\mathrm{He}\,\bar{p}e$ systems in the range $33 \leq L \leq 37$ are presented in Table 3.

Experimentally, only the difference in splitting of two levels can be observed. Since the dependence of calculated splitting values on L is rather slow, it is not possible to resolve such a small difference in splittings for the favoured transitions, i. e. transitions between states of the same N. For this reason, the experimental proposal for near future [20] is aimed at searching the splitting in the unfavoured transitions $(L, N) \rightarrow (L-1, N+2)$.

The part of interaction depending on heavy particle spins removes the remaining degeneracy and split each $j = L \pm 1/2$ sublevel further into two or four levels for $^4\mathrm{He}\,\bar{p}e$ and $^3\mathrm{He}\,\bar{p}e$ systems, respectively. Values of this secondary splitting are much smaller in comparison with the initial splitting.

Table 3. Splitting values $\Delta E_L(10^{-4}\text{eV})$ of the ${}^4\text{He}\bar{p}e$ system.

$N \backslash L$	33	34	35	36	37
1	-.311	-.313	-.311	-.305	-.298
2	-.296	-.294	-.289	-.282	-.272
3	-.278	-.273	-.266	-.256	-.245
4	-.256	-.255	-.246	-.234	-.223
5	-.277	-.252	-.245	-.228	-.226

6 Formation Probability and Initial Populations

The calculation of the formation probability and initial populations of antiprotonic helium atoms is important for the understanding of cascade processes and the interpretation of experimental data. Antiprotonic helium atoms are produced in the reaction of the antiproton capture by a helium atom with an electron emission. As antiprotons mainly loose energy in inelastic collisions, one can assume that their initial kinetic energy does not exceed the ionization energy of helium atoms. Calculations of the antiproton capture and slowing down in helium [21], made in the framework of semiclassical model, confirm this assumption.

The formation probability of 0.22 for one stopped antiproton, calculated in this paper, is far from the experimental value 0.03 – 0.04. The possible reason for this difference is due to the model used in the calculation. One can mention that calculated energies of the considerable part of final states are smaller in comparison with results of precise calculations.

The important qualitative result, obtained in this calculation, is a strict connection on the initial kinetic energy of an antiproton and the total angular momentum of the formed antiprotonic helium atom. As a consequence of this connection the formation of antiprotonic helium atoms with the fixed angular momentum is due to a capture of an antiproton with the fixed energy.

Another approach, used quasi-classical conceptions, have been applied to the calculation of initial populations of antiprotonic helium atoms [22]. Lower bound for the populations of antiprotonic helium atoms have been obtained. Thus, only states of the large enough L and N values can be formed. Full formation probability in this calculation is about 0.3 per one stopped antiproton and also differs from the experimental value.

Up to now, the experimental information on populations of states for $N = 3, 4$ near the metastability boundary have been obtained [9,18]. One can mention the necessity of the reliable calculations of the formation probability and initial populations by using few-body methods for the precise solution of this problem.

7 Interaction with Helium Medium

The interaction of antiprotonic helium atoms with the surrounding medium plays a crucial role in its survival. The angular momentum and energy transfer is hindered due to the neutrality of antiprotonic helium, screening of the antiproton by the electron and large excitation energy of the ordinary helium atom. This qualitative arguments are supported by the experiment [3,23], where only a small, although significant, density dependence of the delayed annihilation time spectra have been found.

At the same time an average antiproton lifetime, transition lifetimes and a fraction of the so-called fast component in the delayed annihilation time spectrum depend on density. An average antiproton lifetime depends also on the phase of the helium medium and its value in solid helium is 20 percent shorter than in liquid helium of the same density. One more problem is the observed isotope dependence of the decay rate of the fast component [23]. One should obtain the quantitative description of the existing density and phase dependence.

Only calculations of the ordinary helium atom – antiprotonic helium interaction were obtained [22] in the framework of multi-channel approach. The quenching of the delayed annihilation due to processes

$$\bar{p}\mathrm{He}^+(L_i N_i) + \mathrm{He} \rightarrow \bar{p}\mathrm{He}^+(L_f N_f) + \mathrm{He}$$

in collisions with surrounding helium atoms has been considered. Main conclusions are as follows:
The states with $L + N \leq 40$ are stable against collisional quenching at all densities up to the density of liquid helium. The states with $L + N = 41$ are intermediate: they are destroyed by collisions at high densities. The states with $L + N \geq 42$ are fully destroyed at $p \geq 1$ bar, and can survive at very low densities ($p \sim 1$ mbar).

One can conclude that the influence of medium on antiprotonic helium atoms, including the external Auger effect, collisional de-excitation, collisional broadening of radiation spectral lines and solid phase effect are still awaiting for their investigation.

8 Quenching by Impurities

The effect of impurities on the delayed annihilation was discovered in experiment [2,3,24]. The admixtures of noble gases (Ne, Ar, Kr) of order 10 percent shorten the average $\bar{p}$ lifetime slightly. The effect of molecular gases (H_2, N_2, O_2) is noticeable at impurity concentrations as low as 20ppm. For the most investigated case of the molecular hydrogen the following estimate for the quenching cross-section σ_q was found: $\sigma_q v = 6 \cdot 10^{-11}\mathrm{cm}^3/\mathrm{s}$. Thus the quenching cross-section is close to the geometrical cross-section, i.e. metastable states are destroyed in a single collision with a hydrogen molecule.

The qualitative understanding of these facts is based on the easy possibility to transfer the angular momentum to a diatomic molecule. On the contrary, such a possibility does not exist for noble gases due to large excitation energies.

Up to now there are no calculations of the interaction of antiprotonic helium atoms and another atoms or molecules. At the same time, delayed annihilation time spectra reveal some peculiarities in the dependence on the impurity concentration. As a result, the theoretical study is necessary to provide the quantitative description of observed quenching phenomena.

9 Other Exotic Systems and Reactions

The discovery of antiprotonic helium atoms caused searching of other long-lived systems with an antiproton. The most probable candidates are systems consisted of an antiproton captured by the noble gas atom. Another possibility is the antiprotonic lithium and energies of this system were calculated in [10]. However, no delayed annihilation was detected experimentally in gaseous Ne [3], metallic Li and LiH [23]. Calculations of formation and quenching rates in these media are necessary to clear up the situation.

Besides of neutral antiprotonic helium atoms, $\bar{p}$He systems of charge -1 can be formed [25] due to the attractive polarization $\bar{p}$ – He potential. If these systems survive for sufficiently long time, they can contribute to the kinetics of antiprotons in media.

New exotic systems can be produced in reactions of long-lived antiprotonic helium atoms with other atoms and molecules. In particular, the reaction

$$\bar{p}\mathrm{He}^+ + \mathrm{Ps} \to \bar{H} + \mathrm{He}$$

was proposed [26] for the antihydrogen production. One can assume also the possibility of exotic molecules, which include as a part the antiprotonic helium atom.

10 Conclusion

Nowadays new three-body systems are discovered and intensively investigated both theoretically and experimentally. The description of these exotic systems meet some difficulties due to metastability and the extremely large total angular momentum $L \sim 30 - 40$.

There is the principal possibility to study antiproton properties by using measurements of antiprotonic helium atoms.

A number of problems arise in the theoretical treatment of intrinsic properties of antiprotonic helium atoms, their formation and collisions with atoms and molecules. These problems represent a wide area for the application of few-body methods of calculations.

Acknowledgement. Fruitful discussions with V. Belyaev, J. Eades, R. Hayano, G. Korenman, V. Korobov, K. Ohtsuki, I. Shimamura, E. Widmann and T. Yamazaki were of great importance in the preparation of this report. The author is grateful to the University of Valencia and Russian Academy of Sciences for financial support in the participation in Few–Body XV.

References

1. M. Iwasaki et al.: Phys. Rev. Lett. **67**, 1246 (1991)
2. T. Yamazaki et al.: Nature **361**, 238 (1993)
3. S.N. Nakamura et al.: Phys. Rev. **A49**, 4457 (1994)
4. T. Yamazaki et al.: Phys. Rev. Lett. **63**, 1590 (1989)
5. S.N. Nakamura et al.: Phys. Rev. **A45**, 6202 (1992)
6. G.T. Condo: Phys. Lett.**9**, 65 (1964)
7. J.E. Russell: Phys. Rev. **A1**, 721 (1970); **A1**, 735 (1970); **A1**, 742 (1970)
8. N. Morita et al.: Phys. Rev. Lett.**72**, 1180 (1994)
9. F. Maas et al.: Phys. Rev. Lett. (in print)
10. R. Ahlrichs et al.: Z. Phys. **A306**, 297 (1982)
11. I. Shimamura: Phys. Rev. **A46**, 3776 (1992)
12. P.T. Greenland, R. Thürwächter: Hyperfine Int. **76**, 355 (1993)
13. T. Yamazaki, K. Ohtsuki: Phys. Rev. **A45**, 7782 (1992)
14. O.I. Kartavtsev: Proc. of the 3rd Intern. Symp. on Muon and Pion Interactions with Media, Dubna, 1995., p. 138
15. V.I. Korobov: Private communication
16. K. Ohtsuki: Private communication
17. N. Morita, K. Ohtsuki, T. Yamazaki: Nucl. Instr. Meth. **A330**, 439 (1993)
18. R.S. Hayano et al.: Phys. Rev. Lett., **73**, 1485 (1994); **73**, 3181 (1994)
19. O.I. Kartavtsev: Talk given at the Int. Symp. on Exotic Atoms and Nuclei, Hakone, Japan, 1995
20. PS205 Collaboration: Preprint CERN SPSLC 95–12/SPSLC I 201.
21. W.A. Beck, L. Wilets, M.A. Alberg: Phys. Rev. **A48**, 2779 (1993)
22. G.Ya. Korenman: Talk given at the Int. Symp. on Exotic Atoms and Nuclei, Hakone, Japan, 1995
23. E. Widmann et al.: Phys. Rev. **A51**, 2870 (1995)

24. E. Widmann et al.: Nucl. Phys. **A558**, 679c (1993)

25. J. Carbonell et al.: see elsewhere in these Proceedings

26. Ya. Ito, E. Widmann, T. Yamazaki: Hyp. Int. **76**, 163 (1993)

Few-Body Systems Suppl. 8, 235–247 (1995)

Few-Body Systems

Muon Catalyzed Fusion

Claude Petitjean

Paul Scherrer Institut, CH-5232 Villigen PSI, Switzerland

Abstract. Reviewing the discoveries of muon catalyzed fusion and resonant mesomolecule formation, the data on $d\mu d$ and $d\mu t$ fusion are presented and the fast $d\mu t$ fusion cycle is discussed in detail. The newest developments in $\mu\alpha$ sticking, muon transfer and resonant $d\mu t$ formation are shown. Epithermal μt atoms and large resonances at eV energies play the dominant role for the $d\mu t$ kinetic cycle. These phenomena are enhanced in triple mixtures due to the Ramsauer-Townsend effect. In solid targets, μd and μt beams can be produced in vacuum allowing direct studies of μCF reactions.

1 Introduction

The process of Muon Catalyzed Fusion (μCF) was discovered in 1956, almost 40 years ago, during the euphoric times at the Berkeley Bevatron, when the team of Alvarez set up the new Hydrogen Bubble Chamber making every month several discoveries, among them the event displayed in Fig. 1. It shows a muon track entering and stopping in the chamber; the track then reappears twice with a length characteristic for 5.3-MeV muons. Finally, an electron track is visible signifying the muon decay into electron + two (invisible) neutrinos. Alvarez et al. [1] interpreted these events correctly as muon catalyzed fusion processes of the type

$$\mu d + p \rightarrow p\mu d \rightarrow^3 \mathrm{He} + \mu + 5.5 \ \mathrm{MeV} \ , \qquad (1)$$

where most of the reaction energy is transferred to a conversion muon. Only a few weeks later, Jackson [2] published in 1957 already a complete theory about the *"Catalysis by μ-Mesons"*. He estimated in his work that, while the *pd* reaction might often be a dead end process due to "sticking" muons in the output channel $p\mu d \rightarrow \mu\ ^3\mathrm{He} + \gamma$, the deuterium-tritium fusion cycle

$$\mu t + d \rightarrow d\mu t \rightarrow \alpha + n + \mu + 17.6 \ \mathrm{MeV} \qquad (2)$$

should be very fast and the sticking probability only ~1%. He concluded: *"For the d-t reaction, the μ meson can catalyse ~100 reactions in its lifetime."*

Figure 1. Bubble chamber picture (Berkeley 1956) showing two $p\mu d$ fusion events

Alas, nobody followed Jacksons suggestion, because the known (non resonant) formation process of mesic molecules with Auger emission

$$\mu t + \mathrm{D}_2 \rightarrow [(d\mu t)\ d\ e]^+ + e_{\mathrm{Auger}}\ , \tag{3}$$

was (correctly) considered to be too slow on the scale of the muon lifetime $\tau_o = 2.2 \cdot 10^{-6}$ s. So, this idea was forgotten for the following 20 years!

In 1964–66 the Russian researchers at the Dubna SC, led by Dzhelepov, investigated muon catalyzed reactions with the help of cloud chambers [3]. They made the crucial discovery of a much larger formation rate $\lambda_{d\mu d}$ in

$$\mu d + d \rightarrow d\mu d \tag{4}$$

at the temperature of cloud chambers as compared to the low temperature results (20K) of bubble chambers. Soon thereafter, in 1967 Vesman [4] suggested the resonance mechanism that correctly explained the increase of $\lambda_{d\mu d}$ with temperature: The $d\mu d$ mesic molecule has at $\epsilon = -2$ eV a weakly bound state. Thus, the released binding energy needs not to be radiated by an Auger electron, but is directly transferred to the D_2 molecule. Vesmans mechanism leads to the famous resonance curve of temperature-dependent meso-molecule formation [5] shown in Fig. 2 for the case of "steady-state" $d\mu d$ formation. The results of many extensive measurements of $d\mu d$ fusion, performed at the four laboratories Dubna [6], PSI [7], Gatchina [8] and LAMPF [9] are also presented in Fig. 2 demonstrating the almost perfect agreement with theory [10].

In 1977, the technique of solving the three-body Coulomb problem of the system "muon + two hydrogen nuclei" was well enough advanced by the Dubna theory group, that all energy levels of the μ-mesic molecules were precisely known. A weakly bound state was also found in the $d\mu t$ molecule [5] being the same analog state as in $d\mu d$ with quantum numbers (J,v) = (1,1), and Gershtein and Ponomarev [11] predicted that resonant formation

$$\mu t + \mathrm{D}_2 \rightarrow [(d\mu t)_{1,1}\ d\ e\ e]^*_{K,\nu} \tag{5}$$

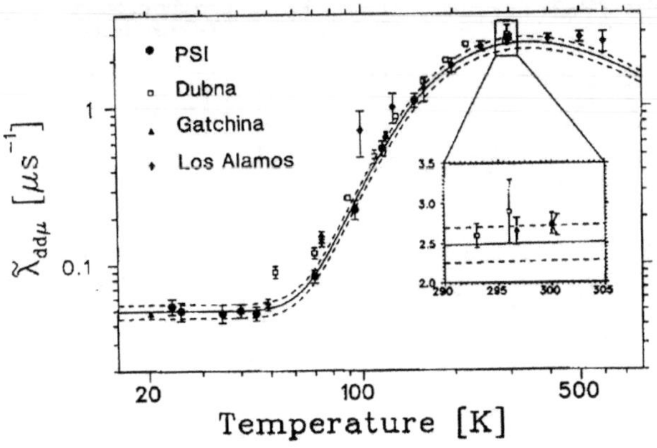

Figure 2. Resonance curve [10] of steady-state $d\mu d$ formation vs D_2 temperature

will work at even much larger rates than $\mu d + D_2$, i.e. $\lambda_{d\mu t} \sim 10^8 s^{-1}$ resulting in an ultra-fast $d\mu t$ cycle. So, in 1977 the theory was back at Jacksons prediction [2] two decades ago, but this time it was based on solid grounds!

In the eighties, after an exploratory experiment at Dubna [12], extensive measurements of high yield $d\mu t$ fusion followed at the two meson factories LAMPF [9, 13] and PSI [14, 15, 16]. Fig. 3 presents the cycle rates λ_c measured in low temperature D-T mixtures versus tritium concentration c_t and density ϕ. The rates are normalized to the density of liquid hydrogen $\phi = 1$ ($4.25 \cdot 10^{22}$ atoms/cm^3). Effective rates up to $2 \cdot 10^8 s^{-1}$ [15], 450 times faster than the muon decay rate λ_o, were observed. The total fusion yield, however, was mainly restricted by the effect of $\mu\alpha$-sticking ω_s, defined as ratio

$$\omega_s = \frac{d\mu t \rightarrow \mu\alpha + n}{d\mu t \rightarrow \text{all}}, \tag{6}$$

removing the muon from the cycle. Maximum yields of 150 ± 20 (LAMPF [9]) and 124 ± 10 (PSI [15]) dt-fusions per muon were reported. The muon loss terms, $w \geq 0.57\%$ per cycle, were mainly limiting the yields.

While $d\mu d$ fusion is well understood (Fig. 2), the $d\mu t$ cycle is more difficult to interprete, because at high yields only the average cycle rates and *not* the underlying kinetics are accessible to direct observation. Therefore, open questions remained on a number of problems:

i) There is a strong *non linear density dependence* of the cycle rate λ_c at large tritium concentrations c_t, where *$d\mu t$ formation* dominates.

ii) There is a strong *enhancement* of λ_c in *non-equilibrated mixtures* $D_2 + T_2$, indicating larger formation rates for D_2 than for DT molecules.

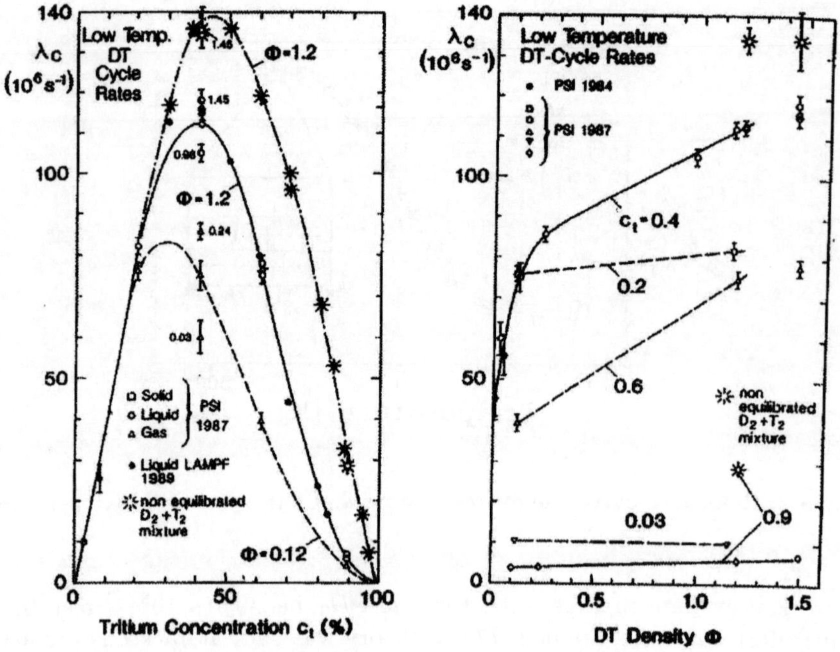

Figure 3. Overview of normalized $d\mu t$ cycle rates vs c_t (left) and ϕ (right)

iii) There is on the other hand *no density dependence at small* c_t, where the *transfer* $\mu d \to \mu t$ dominates. From the excited states transfer, however, a significant density effect is predicted (chapter 4).

iv) The extraction of $\mu\alpha$ *sticking* ω_s from slope data has led to *disagreements among experiments and theory* [17].

2 The Fast $d\mu t$ Cycle

The fast $d\mu t$ kinetic cycle in dense mixtures of deuterium and tritium is presented in Fig. 4. The basic processes develop as follows:

(1) Atomic capture with rate $\Lambda_a = \phi \cdot \lambda_a$ in the outer orbits (n $\sim$ 14) of μd or μt with the ratio according to the atomic abundances.

(2) For $(\mu d)_n$: Deexcitation to the 1s groundstate, possibly interrupted by transfer from excited state $(\mu d)_n \to (\mu t)_n$ with release of 48 eV$/n^2$ energy. The probability to reach the $(\mu d)_{1s}$ state is denoted as $q_{1s}(0 \leq q_{1s} \leq 1)$.

(3) For $(\mu d)_{1s}$: Groundstate transfer $\mu d \to \mu t$ with rate $\Lambda_{dt} = \phi \cdot c_t \cdot \lambda_{dt}$.

(4) Mesic molecule formation with rate $\Lambda_{d\mu t} = \phi \cdot c_d \cdot \lambda_{d\mu t}$.

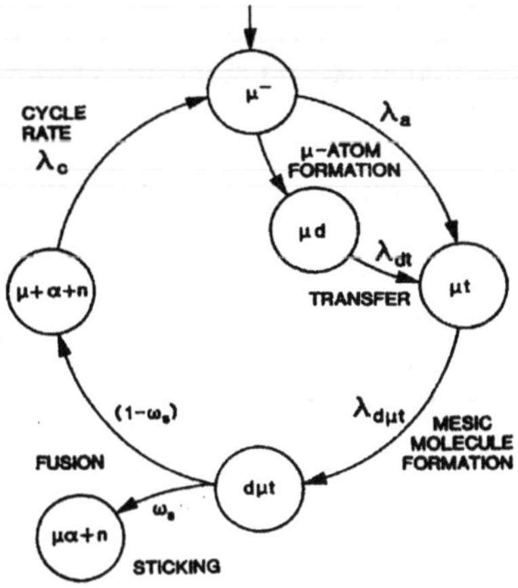

Figure 4. Schematic presentation of the fast $d\mu t$ kinetic cycle

(5) Nuclear fusion $d\mu t \rightarrow \alpha + n + \mu$ with rate λ_f^{dt} .

(6a) Start of the next cycle at (1) with probability (1-ω_s).

(6b) $\mu\alpha$ sticking with probability ω_s. The original sticking factor ω_s^o is larger ($\sim$ 0.9%). During the $\mu\alpha$ slowdown from 3.5 MeV, the μ can be shaken off with probability R (slightly ϕ dependent). Thus, $\omega_s = \omega_s^o \cdot$ (1 - R) [17].

At large densities, i.e. in the liquid phase ($\phi \sim 1.2$), the kinetic system reaches quickly the steady state, and the only observables are contained in the fusion time distribution

$$\frac{d}{dt} N_f(t) = \Lambda_c \cdot \exp[-(\lambda_o + w \cdot \Lambda_c) \cdot t], \qquad (7)$$

where $\Lambda_c = \phi\ \lambda_c$ = effective (observed) cycle rate,
$\lambda_o = 0.455\ 10^6 \mathrm{s}^{-1}$ = free muon decay constant,
$w = \omega_s$ + other loss terms per cycle = raw sticking factor.

The cycle rate Λ_c can be written in terms of the kinetic system in Fig. 4 as

$$\frac{1}{\Lambda_c} = \frac{1}{\Lambda_a} + \frac{q_{1s} \cdot c_d}{\Lambda_{dt}} + \frac{1}{\Lambda_{d\mu t}} + \frac{1}{\lambda_f^{dt}}. \qquad (8)$$

Of the 4 terms in Eq. (8), two constitute very fast rates, namely $\Lambda_a \sim \phi \cdot 4 \cdot 10^{12}\ \mathrm{s}^{-1}$ and $\lambda_f^{dt} \sim 10^{12}\ \mathrm{s}^{-1}$. The two other (slower) components, i.e. the transfer term and molecular formation, will be discussed in chapters 4 and 5.

3 $\mu\alpha$-Sticking ω_s

The dt-fusion yield per muon Y_f is obtained by integrating Eq. (7):

$$Y_f = \frac{\Lambda_c}{\lambda_o + w \cdot \Lambda_c}. \qquad (9)$$

The muon loss factor w (containing ω_s) sets the ultimate limit, since $Y_f \rightarrow 1/w$ for $w \cdot \Lambda_c \gg \lambda_o$. At large densities the slope measurements gave rather low w values and considerably smaller sticking factors than theory:

$$\omega_s^{\text{exp}} = 0.35\% - 0.48\% \text{ , while } \omega_s^{th} = (0.61 \pm 0.03)\%, \tag{10}$$

raising some doubts about the understanding of $\mu\alpha$ sticking [18]. So, new experiments were proposed aiming at the direct observation of sticking. Such an experiment using the "Gatchina" method of charge collection by a high pressure ionization chamber [8, 19] was recently completed at PSI. Fig. 5(a) shows the energy spectrum of (non sticking) α events and a small peak of $\mu\alpha$'s which is nicely separated due to a recombination effect. In the spectrum Fig. 5(b) the sticking peak is absent, because a second fusion was requested ("survived muon method"). The final result of this experiment (at $\phi = 17\%$) is [17, 20, 21]

$$\omega_s^{\text{exp}} = (0.56 \pm 0.04)\% \text{ , while } \omega_s^{th} = (0.65 \pm 0.03)\%. \tag{11}$$

The slight difference of the theory values between Eqs. (10) and (11) is the predicted density effect in muon reactivation R.

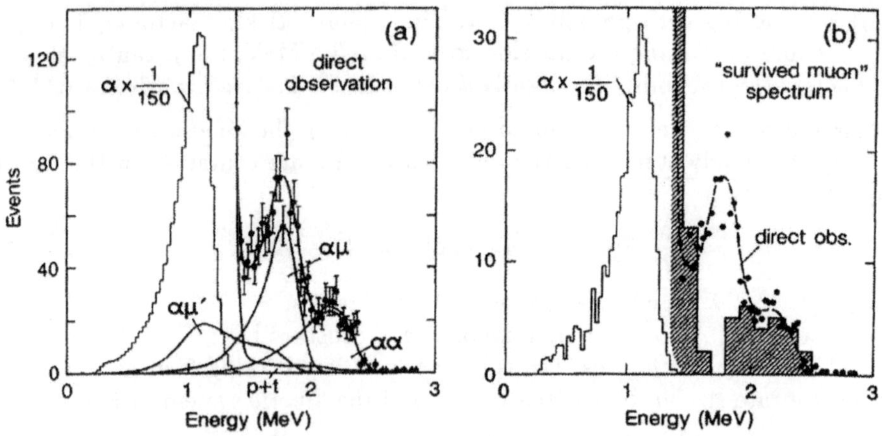

Figure 5. Final data of direct $\mu\alpha$-sticking observation: (a) full spectrum of dt-fusion events, (b) "survived muon" spectrum of events followed by another fusion

4 Muon Transfer $\mu d \rightarrow \mu t$

The phenomenon of muon transfer by collisions $(\mu d)_n + t \rightarrow (\mu t)_n + d$ was originally studied by Men'shikov and Ponomarev [22]. The excited states transfer (rates $\lambda_{\text{ex}}^n \sim 10^{11} - 10^{12}\text{s}^{-1}$ with $n \geq 2$) was found to be significant with respect to radiative cascade transitions leading to very small q_{1s} probabilities.

The transfer from the groundstate (rate λ_{dt}) on the other hand is much smaller, $\sim 3 \cdot 10^8$ s^{-1}. Experimentally, however, only a very weak q_{1s} effect and ϕ dependence was observed [9, 14, 16]. New theoretical attempts to explain the experimental observations were made in refs. [23, 24] assuming the presence of epithermal μd atoms ($\epsilon_{\mu d} \sim$ 1-6 eV) where λ^n_{ex} drops by factors 2 - 3. Good agreement was only obtained in the cascade calculations of Markushin [25] by including a strong Coulomb deexcitation mechanism

$$(\mu d)_n + d(or\ t) \longrightarrow (\mu d)_{n-1} + d(or\ t) + \Delta\epsilon_{n,n-1} \tag{12}$$

with $\Delta\epsilon_{n,n-1} \sim 2\text{Ry} \cdot (1/n^3)$, Ry = 2.6 keV being the Rydberg constant for μd atoms. This mechanism *accelerates the μd atoms* by several eV to 100 eV, depending on n. Fig. 6(a) shows the q_{1s} dependence obtained in a recent kinetic analysis of the PSI *dt*-data by Ackerbauer et al. [16] in comparison with theory [24] and with calculation [25]. The problem of the atomic cascade was recently treated again by Markushin [26] for the very similar case of μp atoms, see Fig. 6(b). For n $\leq$ 6, Coulomb deexcitation seems indeed to be one of the important cascade processes which was probably underestimated in the past.

A new experimental technique using Charged Coupled Devices (CCD) was recently introduced by Egger et al. [27]. It is a planar silicon pixel detector with pixel sizes of 22 μ acting like a photoplate. With energy resolution 120 eV at 2 keV, this detector is ideal for measurements of soft X-rays in the energy range from 1 to 10 keV. Very efficient background suppression of charged particle tracks is achieved by analyzing the cluster size of events. An example of the X-ray spectra is given in Fig. 6(c), where due to the isotopic shifts K_α and K_β lines can be identified for each isotope allowing the direct determination of q_{1s}. The CCD technique will be employed for the study of atomic cascades and transfer reactions of exotic hydrogen and helium atoms. Measurements with muons in H/D and H/D-helium mixtures are underway at PSI [28].

More evidence for acceleration mechanisms in muonic hydrogen atoms was observed in the μ diffusion experiments by Siegel et al. [29]. Muons are stopped in H_2 or D_2 gas ($\sim$ 0.1 bar) in gaps between thin goldcoated foils. After formation of atoms, they drift to the foils, transfer to gold and radiate characteristic γ rays. The time between μ stop and γ arrival is used to determine the initial energy distribution of the atoms. The preliminary results are $\epsilon_{\mu p} \simeq 2.5$ eV , $\epsilon_{\mu d} \simeq 1.4$ eV.

Acceleration effects were also observed in a pion experiment by Crawford et al. [30] studying the pion charge exchange reaction $\pi^- p \rightarrow \pi^o$. This reaction takes place when the pion is still in an outer orbit n $\sim$ 3 - 5 and is sensitive to the kinetic energy distribution of the πp atoms. About 43% of the atoms were found to be epithermal with energies between 5 eV and 100 eV.

5 Resonant $d\mu t$ Formation

Fig. 7 presents the scheme of resonant $d\mu t$ formation [31] demonstrating todays very accurate knowledge about energies and levels of the molecular complex

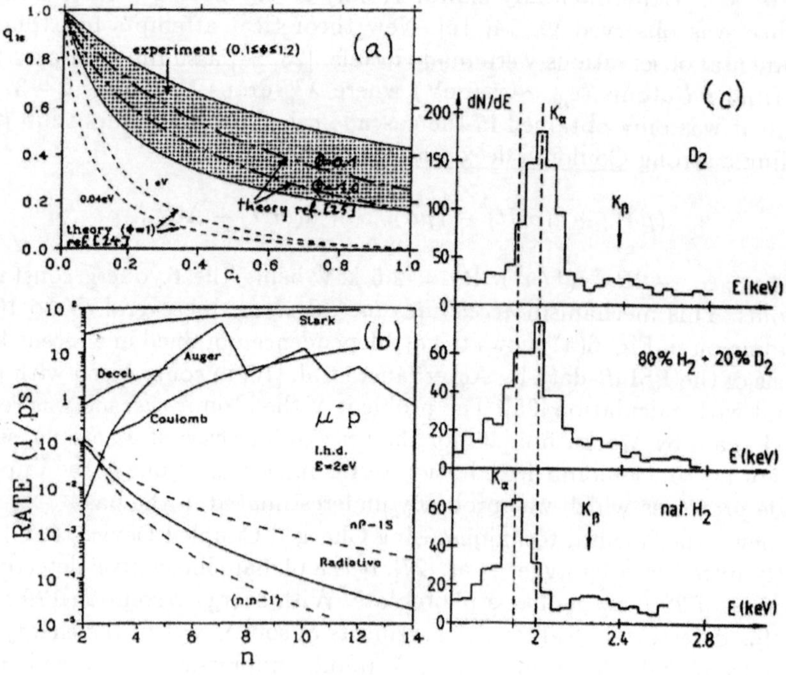

Figure 6. (a) q_{1s} kinetic analysis [16] of the PSI dt-data compared with theories [24] and [25], (b) rates of cascade processes in μp atoms [26], (c) CCD spectra of muonic K_α and K_β lines in mixtures of liquid H_2 and D_2.

$[(d\mu t)$ d e e$]$. The strongest resonances of mesomolecule formation ($\nu_f = 2$, $K_f = 0, 1, 2$) are at negative energies. As a consequence, the $d\mu t$ formation on D_2 becomes at low temperatures strongly density dependent because of tripel collisions and level broadening effects. Since there are no low energetic resonances of the DT molecule, the thermal $d\mu t$ formation goes almost entirely via collisions $\mu t + D_2$. At epithermal energies ($\epsilon_{\mu t} \geq 0.1$ eV), however, the resonances with vibrational quantum number $\nu_f = 3$ start to dominate. Detailed calculations of all resonance rates were made by Faifman et al. [32]. Fig. 8 shows as an example the behaviour of $\lambda_{d\mu t}^{F=0}$ in D_2 molecules as function of $\epsilon_{\mu t}$ and D_2 temperature. The other molecular species DT and HD and the F = 1 hfs levels show also a very similar behaviour with strong epithermal resonance structures and rates $\lambda_{d\mu t} \sim 10^9 - 10^{10}\mathrm{s}^{-1}$.

At low temperatures, $\lambda_{d\mu t}$ has sharp structures. Thus, the evolution of $\epsilon_{\mu t}$ with time is crucial for the $d\mu t$ cycle. At high temperatures (T $\geq$ 1000 K), the resonances are washed out due to the Maxwell distribution of kinetic energies, and the formation rates become everywhere large ($\lambda_{d\mu t} \geq 10^9\mathrm{s}^{-1}$), but exhibit

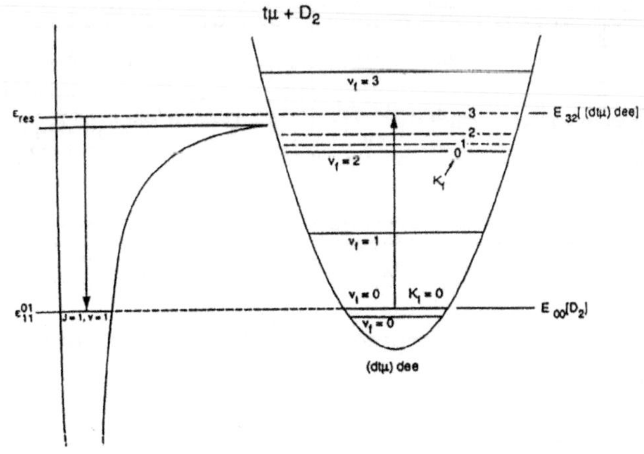

Figure 7. level scheme of resonant $d\mu t$ formation (from ref. [31]): The energies $\epsilon_{res} + \epsilon_{1,1}$ of the μt +d system (left) are exciting the $(d\mu t)dee$ molecular complex (right)

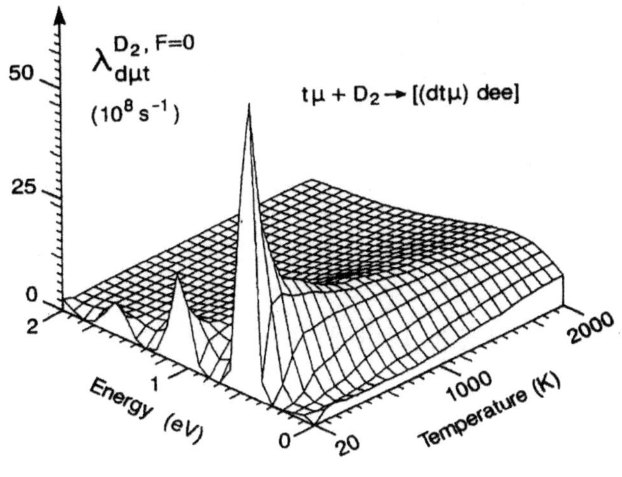

Figure 8. Resonant mesomolecule formation rates for μt(F=0) + D_2 as function of μt energy and D_2-temperature (from ref. [32])

less pronounced maxima. This general picture of resonant $d\mu t$ formation is now confirmed by many experiments:

(1) At LAMPF, $d\mu t$ fusion was investigated in D-T mixtures at densities $\phi = 0.1 - 1.3$ in the temperature range from 70K to 800K [9, 13]. $\lambda_{d\mu t}^{D_2}$ was found to increase from $4 \cdot 10^8\ s^{-1}$ to $11 \cdot 10^8\ s^{-1}$, while $\lambda_{d\mu t}^{DT}$ was

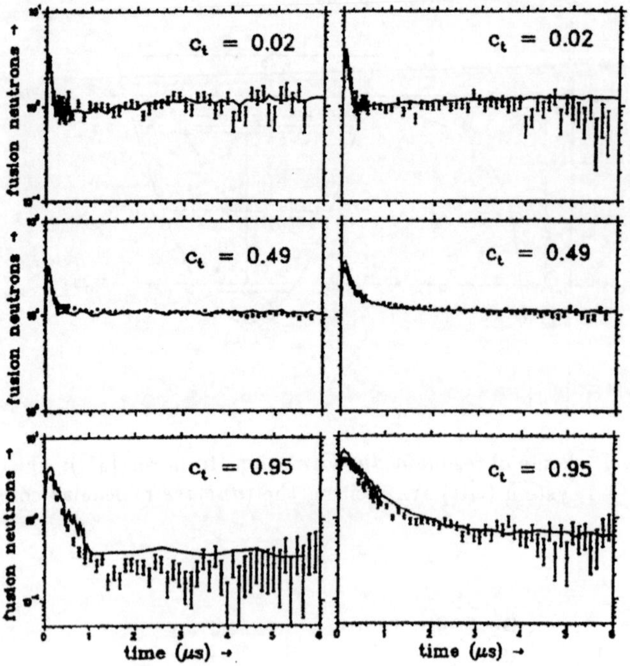

Figure 9. Time spectra of $d\mu t$ fusion in D-T gas at $\phi = 1\%$, T = 30K (left), T = 300K (right): experimental points from [14], Monte Carlo reproductions (full lines) from [33] (the muon life time $\exp[-\lambda_o \cdot t]$ is removed)

up to 300K small ($0.3 \cdot 10^8$ s^{-1}), but increasing from 300K to 800K to $5 \cdot 10^8$ s^{-1}.

(2) At PSI, $d\mu t$ fusion was first measured in low density D-T gas ($\phi = 0.01$) from 30K to 300K [14]. Fig. 9 shows a selection of the observed time distributions at various conditions. In the initial part ($t \leq 1\mu s$) large rates are visible which are interpreted as epithermal μt atoms passing through the big $d\mu t$ resonances. The thermalisation is slow because of the low density. This allows detailed studies of the energetical development of the μt atoms with time. The curves in Fig. 9 show a Monte Carlo reproduction of the measured time structures by Jeitler et al. [33] using the $d\mu t$ formation rates from [32] and the new data collections of theoretical scattering crossections from refs. [34]. Good agreement with the experimental data was reached after some adjustments of $\lambda_{d\mu t}^{D_2}$ at low temperature.

(3) The experiments on $\mu\alpha$ sticking (chapter 3) were performed in tripel mixtures with c_p = 90.5% - 97%, c_d = 3% - 9.5%, c_t = 0.036% - 0.05%

and density $\phi = 17\%$. Very similar peaks of the initial time distribution of fusion events were observed indicating a much slower thermalisation than in D-T mixtures. Markushin et al. [35] have analyzed these data by Monte Carlo simulations and obtained also a good reproduction of the data after a general reduction of the resonance rates [32] by a factor 2-4. The slow thermalisation in tripel mixtures is explained by the Ramsauer-Townsend effect, i.e. a quantum mechanical minimum of the elastic scattering crossections $\sigma_{el}^{\mu d+p}$ and $\sigma_{el}^{\mu t+p}$ in the energy range $\epsilon_{\mu d}$, $\epsilon_{\mu t} \cong 0.1 - 10$ eV.

6 Physics with Atomic μd and μt Beams

The "re-discovery" of the Ramsauer-Townsend effect (known already from the gaps seen in bubble chambers pictures) has led to a new atomic beam technique, which is explored by Marshall et al. [36] at TRIUMF: Muons are stopped in a solid layer of H_2 at 3K, doped with about 10^{-3} parts deuterium or, alternatively, tritium. The muons form μp atoms and are within ~ 100 ns transferred to μd (or μt). Due do the Ramsauer-Townsend effect, these atoms remain at epithermal energies, penetrating the solid layer by up to 1mm and exiting into the vacuum. After some distance (~ 1 cm) another layer of solid material (e.g. D_2 can be placed for direct studies of reaction rates. The time between muon stop and the secundary reaction determines the energy of the colliding atom. Several studies with μd and μt beams are in progress, and it will be possible to directly observe the epithermal resonances [32].

7 Conclusions

μCF has reached a mature state where most phenomena are qualitatively and many even quantitatively understood. Perhaps the most important progress of knowledge was made in recognizing that muonic atoms of the hydrogen isotopes are produced at epithermal energies ($\epsilon_{\mu t} \geq 1$ eV) due the existence of acceleration mechanisms in the atomic cascade, in addition to the already known transfer reactions. This phenomenon plays for the $d\mu t$ cycle a crucial role, since the largest formation rates are definitively at epithermal energies ($\sim$ 0.5 eV). The developments of Monte Carlo codes to simulate the μCF processes has increased our detailed knowledge of the kinetics and rates, and will do so even more in the future. In connection with studies of triple mixtures and the use of atomic beams, we can reach the full understanding of all fusion cycles. However, $\mu\alpha$ sticking is now established to be near 0.5%. So, it will be difficult to exceed dt-fusion yields bejond the limit of 200 per muon.

References

1. L.W. Alvarez et al.: Phys. Rev. **105**, 1127 (1957)

2. J.D. Jackson: Phys. Rev. **106**, 330 (1957)

3. V.P. Dzhelepov et al: Sov. Phys. JETP **19**,1376 (1964) and **23**, 820 (1966)

4. E.A. Vesman: Sov. Phys. JETP Lett. **5**, 9 (1967)

5. S.I. Vinitsky et al.: Sov. Phys. JETP **47**, 444 (1978)

6. V.M. Bystritsky et al.: Sov. Phys. JETP **49**, 232 (1979)

7. P. Kammel et al.: Phys. Rev. **A28**, 2611 (1983); J. Zmeskal et al.: Phys. Rev. **A42**, 1165 (1990)

8. D.V. Balin et al.: Phys. Lett. **141B**, 173 (1984) and Muon Cat. Fusion **5/6**, 163 (1990/91); A.A. Vorobyov: Muon Cat. Fusion **2**, 17 (1988)

9. S.E. Jones et al.: Phys. Rev. Lett. **56**, 588 (1986); A.J. Caffrey et al.: Muon Cat. Fusion **1**, 53 (1987)

10. M.P. Faifman: Muon Cat. Fusion **2**, 247 (1988); A. Scrinzi et al.: Phys. Rev. **A47**, 4691 (1993)

11. S.S. Gershtein, L.I. Ponomarev: Phys. Lett. **72B**, 80 (1977)

12. V.M. Bystritsky et al.: Phys. Lett. **94B**, 476 (1980)

13. S.E. Jones et al.: Phys. Rev. Lett. **51**, 1757 (1983); S.E. Jones: Muon Cat. Fusion **1**, 21 (1987)

14. W.H. Breunlich et al.: Phys. Rev. Lett. **53**, 1137 (1984) and **58**, 329 (1987); *ibid.*: Muon Cat. Fusion **1**, 67 (1987)

15. C. Petitjean et al.: Muon Cat. Fusion **2**, 37 (1988)

16. P. Ackerbauer et al.: Hyperfine Int. **82**, 357 (1993)

17. C. Petitjean et al.: Hyperfine Int. **82**, 273 (1993)

18. C. Petitjean, Nucl. Phys. **A543**, 79c (1992)

19. D.V. Balin et al.: Muon Cat. Fusion **5/6**, 481 (1990/91) and **7**, 1 (1992)

20. T. Case: Thesis (U.C. Berkeley, 1993) unpublished; T. Case et al.: Hyperfine Int. **82**, 295 (1993)

21. K. Lou: Thesis (ETH Zurich, 1993) unpublished; K. Lou et al.: Hyperfine Int. **82**, 313 (1993)

22. L.I. Men'shikov, L.I. Ponomarev: JETP Lett. **39**, 663 (1984) and **42**, 13 (1985); *ibid.*: Z.Phys. **D2**,1 (1986)

23. A.V. Kravtsov et al.: Phys. Lett. **A132**, 124 (1988); *ibid.*: Muon Cat. Fusion **2**, 183 (1988)

24. W. Czaplinski et al.: Phys. Rev. **A50**, 518, 525 (1994)

25. V. Markushin: in *Electromagnetic Cascade and Chemistry of Exotic Atoms*: Plenum Press N.Y. (1990) p.73, eds. L.M. Simons et al.

26. V. Markushin: Phys. Rev. **A50**, 1137 (1994)

27. J.P. Egger et al.: in *Muonic Atoms and Molecules*, Birkhäuser Basel (1993) p. 331, eds. L.A. Schaller and C. Petitjean.

28. B. Lauss et al.: LEMS'93 workshop (Santa Fe, N.M.), Los Alamos rep. **LA-12698C**, 300 (1994)

29. J.B. Kraiman et al.: Phys. Rev. Lett. **63**, 1942 (1989); *ibid.*: Muon Cat. Fusion **5/6**, 43 (1990/91)

30. J.F. Crawford et al.: Phys. Rev. **D43**, 46 (1991)

31. J.S. Cohen, LEMS'93 workshop (Santa Fe, N.M.), Los Alamos rep. **LA-12698C**, 228 (1994)

32. M.P. Faifman et al.: Muon Cat. Fusion **4**, 1 (1989); *ibid.*: Phys. Lett. **B265**, 201 (1991)

33. M. Jeitler et al.: Phys. Rev. **A51**, 2881 (1995)

34. L. Bracci et al.: Muon Cat. Fusion **5/6**, 21 (1990/91); C. Chiccoli et al.: Muon Cat. Fusion **7**, 87 (1992); A. Adamczak et al.: Hyperfine Int. **82**, 91 (1993)

35. V.E. Markushin et al.: Hyperfine Int. **82**, 373 (1993)

36. B.M. Forster et al.: Hyperfine Int. **65**, 1007 (1990); P.E. Knowles et al.: Hyperfine Int. **82**, 521 (1993); G.M. Marshall et al.: Hyperfine Int. **82**, 529 (1993)

Few-Body Systems Suppl. 8, 248–253 (1995)

Few-Body Systems

Precision Measurement of Nuclear Muon Capture by ^{3}He

D.V. Balin[1], V.N. Baturin[1], Yu.S. Grigoriev[1], E.M. Maev[1], G.E. Petrov[1], G.G. Semenchuk[1], Yu.V. Smirenin[1], A.A. Vorobyov[1], N.I. Voropaev[1], J. Deutsch[2], J. Govaerts[2], R. Prieels[2], P. Ackerbauer[3], W.H. Breunlich[3], P. Kammel[3], B. Lauss[3], J. Marton[3], W. Prymas[3], J. Egger[4], C. Petitjean[4], Th. Petitjean[4], B. van den Brandt[4], H. Daniel[5], F.J. Hartmann[5], M. Mühlbauer[5], W. Schott[5], T. von Egidy[5], P. Wojciechowski[5], T. Case[6], K.M. Crowe[6], V.E. Markushin[7], G.A. Beer[8]

[1] St.-Petersburg Nuclear Physics Institute, Gatchina 188'350, Russia

[2] Université Catholique de Louvain, 1348 Louvain-la-Neuve, Belgium

[3] Austrian Academy of Sciences, 1090 Vienna, Austria

[4] Paul Scherrer Institut, 5232 Villigen PSI, Switzerland

[5] Technische Universität München, 85747 Garching, Germany

[6] University of California Berkeley and LBL, Berkeley CA 94720, USA

[7] Russian National Center I.V. Kurchatov Institute, Moscow 123'182, Russia

[8] University of Victoria, Victoria B.C. V8W 2Y2, Canada

We report here on the preliminary results of the experiment performed in 1993 at PSI with the goal to determine the absolute rate of the nuclear muon capture reaction

$$\mu^- + {}^3\text{He} \longrightarrow {}^3\text{H} + \nu_\mu \tag{1}$$

Nuclear muon capture was discovered more than forty years ago, but we still lack a precise experimental verification of its understanding. This is so because muon capture on the proton is hampered by considerable molecular difficulties, and in muon capture on medium A nuclei it is generally the reliability of the theoretical expectations which can be questioned. A unique opportunity to improve these tests is offered by the ^{3}He $\rightarrow$ ^{3}H partial muon capture transition. The initial state of the reaction, the ^{3}He-muon system, does not suffer from the molecular problems which handicap the neutral proton-muon system, and the $A = 3$ nuclei are certainly the best studied and understood ones. Moreover, the ^{3}He $\rightarrow$ ^{3}H nuclear muon capture transition is the only one connecting members of an isospin doublet, which is especially well suited for a reliable application of the so-called "elementary particle model" (EPM) pioneered by C.Kim and H. Primakoff [1]. This approach fully profits from the fact that the electromagnetic properties of the $A = 3$ system are well studied.

Up to now there were only three measurements of the absolute rate of reaction (1), all performed more than 30 years ago. For the first time the muon capture by ^{3}He was observed in 1963 by Falomkin et al. [2] in a diffusion cloud chamber filled with ^{3}He. They identified about 200 events and determined the muon capture rate as $\Lambda_C = (1410 \pm 140)\mathrm{s}^{-1}$. The next result was reported by Clay et al. [3] from an experiment where the muons were stopped inside a gas scintillation counter filled with a ^{3}He/Xe mixture. They found $\Lambda_C = (1465 \pm 67)\mathrm{s}^{-1}$. A similar technique was used in the experiment of Auerbach et al. [4] (gas scintillation counter, pure ^{3}He). They reported $\Lambda_C = (1505 \pm 46)\mathrm{s}^{-1}$. However, there is an error in the analysis of the data from that experiment. When evaluating the wall effect, the authors used a wrong value for the range of the triton (1.75 mm instead of 6.3 mm). As a consequence, the error on Λ_C reported in this experiment should be considered as underestimated.

In the present experiment we used a new technique recently developed at Gatchina. The muons were stopped in a high pressure gridded ionization chamber IC filled with 120 bar of isotopically ultra-pure ^{3}He gas (Fig. 1).

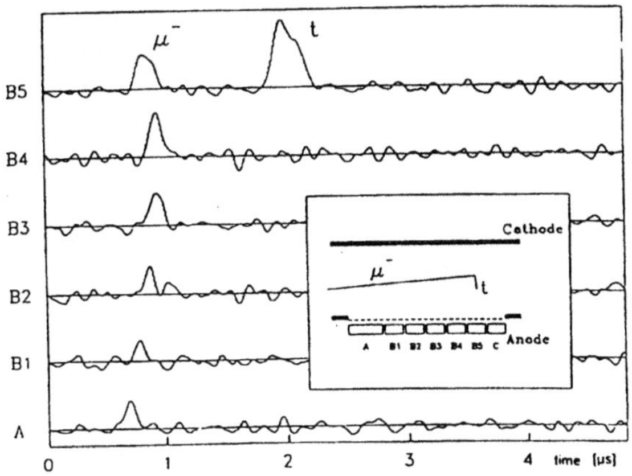

Figure 1. Display of the flash ADCs showing a capture event. Inserted is a cross view of the Ionization Chamber.

The sensitive volume is $15\,\mathrm{cm}^3$ with 1.5 cm vertical drift distance (cathode to grid) and 10 cm^2 anode area ($B1 - B5$). The anodes A are to define the incoming muon direction. The anodes C serve as an anticoincidence ring. The IC operated at -40 kV on the cathode and -5 kV on the grid. It allowed to detect both the stopped muons and the charged reaction products, i.e. tritons from reaction (1) and protons and deuterons from the break-up reactions

$$\mu^3\mathrm{He} \rightarrow p + 2n + \nu_\mu, d + n + \nu_\mu.$$

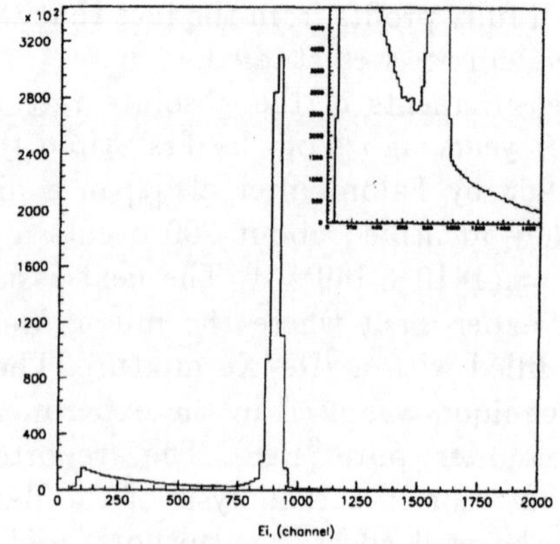

Figure 2. Amplitude spectrum of triton signals separated in time from muon signals

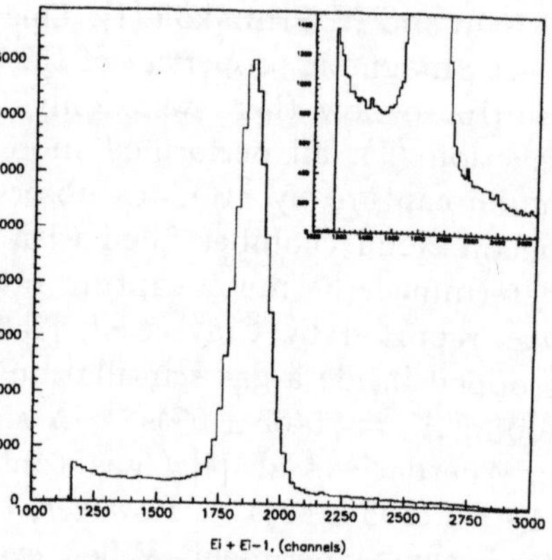

Figure 3. Distribution of the first signals as a function of the sum of the amplitudes from two neighboring anodes B_i and B_{i-1}

Each anode has a separate read-out, the signals being digitized with fast 8-bit flash ADCs during $10\mu s$. The amplitude E, the arrival time T, and the duration of each signal were measured. The strategy of the measurements was to select a very clean sample of muon stops, N_μ, in the zone of B-anodes well isolated from the cathode and the grid. This procedure should guarantee the 100% efficiency in detection of the tritons without any corrections for wall effects. Then the ratio N_t/N_μ will be a direct measure of the muon capture rate.

During 4 weeks of data taking 374028500 muon stops not accompanied by muon capture have been accumulated that satisfy the above requirements. The muons were registered with prescaling factors from 500 to 2000.

At the same time 1229240 muon stops followed by triton signals have been detected.

The majority (78.3%) of the triton signals proved to be separated in time with the total energy of the tritons measured by one anode. The amplitude spectrum of such signals is shown in Fig. 2. The energy resolution is 30 keV (σ). The background (3.7%) is easily subtracted contributing to the total error on a level of 0.08%. In about 7% of the events, the triton energy was divided between two neighboring anodes. In this case the energy was determined by the sum of the amplitudes from these anodes. Some of the triton signals (14.7%) pile up with muon signals. The selection of tritons in this case was done using the $(E_i + 2 \cdot E_{i-1})$ plot (Fig. 3).

The main correction (6.45%) to the measured number N_t was a trivial one taking into account the time beyond the analyzed time interval of 6 μs. Other corrections take into account some small losses of the triton events. The total value of these corrections is 0.73%. Finally we obtained

$$\Lambda_C = 1496 \pm 3(\text{stat}) \pm 3(\text{syst})\ \text{s}^{-1}. \qquad \text{(preliminary result)}$$

The systematic error was evaluated by varying cuts in wide ranges and comparing results from different anodes and under different run conditions.

The measured value of Λ_C corresponds to statistical population of the singlet and triplet states of $^3He\mu$ atoms. The $s \to t$ transition with $\Lambda_{st} \geq 0.01\mu s^{-1}$ would change the observed Λ_C by 0.17%. Fortunately, we can control this process in our experiment by measuring the exponential decay constant from the time distribution of the triton events (Fig. 4). This decay constant proved to be exactly equal to the sum of $\lambda_0 + \Lambda_{Ctotal}$ thus eliminating the $s \to t$ transition on a level $\Lambda_{st} < 0.01\mu s^{-1}$. ($\lambda_0$ is the free muon decay rate).

Another uncertainty might arise if the transitions from the metastable $2S$ states would take place during the muon life-time. According to the present understanding, however, such transitions should occur in much shorter time and not influence our result. But again we can control this effect with our own data by comparing the area extrapolated to $\Delta T = T_t - T_\mu = 0$ (Fig. 4) with the observed number of pile-ups. From this comparison we obtain: $\tau_{2S} \leq 50$ns. So, the effect of this transition on our results is negligible.

The obtained result can be analysed in the frame of the EPM following the work of J.Congleton and W.Fearing [5]. In this model the weak current of the $^3He \to {}^3H$ transition is parametrized by four form-factors $F_{V,M,A,P}$ evaluated at the relevant value of the four-momentum transfer $q^2 = -0.954m_\mu^2$. Two of these parameters, F_V and F_M, are derived from the isotriplet vector current hypothesis (CVC) and the results of elastic electron scattering by ^3He and ^3H:

$$F_V(-0.954m_\mu^2) = 0.834 \pm 0.011, \; F_M(-0.954m_\mu^2) = -13.969 \pm 0.052.$$

Taking these data as an input, our measurement determines the correlation

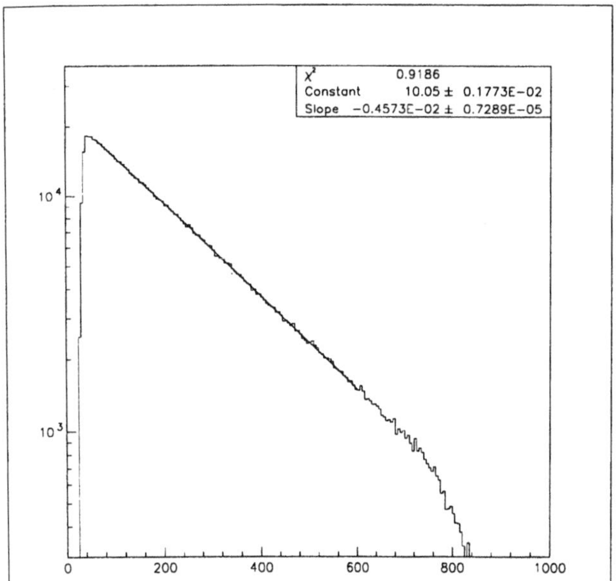

Figure 4. Delay time spectrum of the triton signals

between F_P and F_A as shown in Fig. 5. The width of the allowed region is determined now mostly by the error in F_V. The next step can be done if we use $F_A(-0.954m_\mu^2) = 1.052 \pm 0.011$ extrapolated from $F_A(0) = 1.212 \pm 0.004$ as measured by the ^{3}H beta decay (vertical lines in Fig. 5). It allows to determine $F_P = 21 \pm 3$. A disadvantage of this method is to involve some theoretical considerations based on the Impulse Approximation in the extrapolation from $q^2 = 0$ to $q^2 = 0.954m_\mu^2$ based on electron scattering that resulted in an increase of the error in $F_A(q^2)$. A more precise determination of F_P would be possible if one succeeds to measure the triton asymmetry parameter A_ν for polarized μ^3He capture to 5% precision or better. Such an experiment is going on at TRIUMF. Figure 5 displays also the relation between F_P and F_A based on the Partially Conserved Axial Current (PCAC) hypothesis:

$$F_P(q^2) = \frac{2m_\mu M_3}{m_\pi^2 - q^2} \cdot F_A(q^2)$$

One can see from Fig. 5 that the central values of F_A and F_P obtained in our analysis are in surprisingly good agreement with the PCAC relation.

It is well known that the induced pseudo-scalar coupling F_P describes the capture of the muon on a virtual pion in the nucleus which therefore constitutes

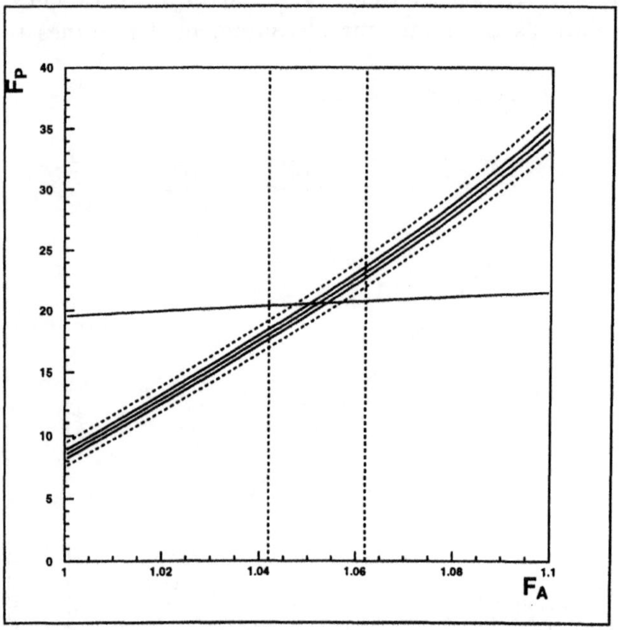

Figure 5. Constraints on the axial F_A and pseudo-scalar F_P form-factors. Solid lines from Λ_C with its errors only. Dashed lines from Λ_C with errors in F_V and F_M added. Vertical dashed lines from F_A extrapolated from ^{3}H beta decay. Horizontal solid line the PCAC relation

a sensitive probe of the pionic degrees of freedom in nuclei and their possible modification by nuclear matter. That is why it is important to compare the experimental data on muon capture with the nuclear microscopic theory. The first comparison [5] showed that the Impulse Approximation leads to a much lower μ^3He decay rate: $\Lambda_C = 1304\text{s}^{-1}$. It means that the theory should explicitely take into account the meson exchange effects. Such work is in progress now by J. Congleton.

References

1. C.W. Kim, H. Primakoff: Phys. Rev. **B140**, 556 (1965)
2. I.V. Falomkin et al.: Phys. Lett. **3**, 229 (1963)
3. D.B. Clay et al.: Phys. Rev. **B140**, 586 (1965)
4. L.B. Auerbach et al.: Phys. Rev. **B138**, 127 (1965)
5. J.G. Congleton, W.H. Fearing: Nucl. Phys. **A552**, 534 (1993)

Few-Body Systems Suppl. 8, 254–258 (1995)

Few-Body Systems

The πNN Coupling from High Precision np Charge Exchange Data

T.E.O. Ericson[1]*, B. Loiseau[2], J. Nilsson[3], N. Olsson[3]

[1] CERN, CH-1211 Geneva 23, Switzerland, and Department of Theoretical Physics, Uppsala University, Box 803, S-75108 Uppsala, Sweden

[2] Division de Physique Théorique, IPN, 91406 Orsay Cedex, and LPTPE Université P. & M. Curie, 4 Place Jussieu 75252 Paris Cedex 05, France

[3] Department of Neutron Research, Uppsala University, Box 535, S-75121 Uppsala, Sweden

Abstract. Differential np cross sections for unpolarized neutrons of 162 MeV have been measured to high precision with particular attention to the absolute normalization. These data can be extrapolated precisely and model-independently to the pion pole and give a πNN coupling constant $g^2 = 14.62 \pm 0.06 \pm 0.29$ or $f^2 = 0.0808 \pm 0.0003 \pm 0.0017$. This is higher than recently suggested values.

At the Few Body Conference in Adelaide three years ago the panel discussion on the πNN coupling constant concluded that there was an urgent need for a direct determination of this quantity [1]. The textbook value is $g^2 = 14.3$ to 14.5 deduced from πN scattering data in the resonance region and from forward dispersion relations in pp scattering as the principal sources of information. In the last few years the Nijmegen group [2] has proposed a substantially smaller value for the charged pseudoscalar coupling constant, $g^2 = 13.58(5)$ (or $f^2 = 0.0748(3)$) on the basis of their extensive and global analysis of NN scattering data. It has become evident that the basis for the standard value in the literature is weaker than previously thought.

Neutron-proton charge exchange reactions at intermediate energies are in principle very suitable for extrapolation to the charged pion pole. In this case both the CM unpolarized cross section as well as the polarization transfer one are incoherent combinations of 5 amplitudes:

$$\frac{d\sigma}{d\Omega}(q^2) = 1/2(|a|^2+|b|^2+|c|^2+|d|^2+|e|^2); \; \frac{d\sigma}{d\Omega}(q^2)\,(1-K_{n00n}) = |b|^2+|d|^2, \quad (1)$$

* *E-mail address:* Ericson@cernvm.cern.ch

where q^2 is the squared momentum transfer from the proton to the neutron. Only the amplitudes b and d contain the pion pole at $q^2 + m_\pi^2 = 0$ and these terms are the dominant ones also in the physical region. The characteristic shape and magnitude of these terms are shown for the Paris model in Fig. 1a. In the Born approximation the amplitude b corresponds mainly to a spin-spin interaction and d to a pure tensor one. These amplitudes are crucial to the extrapolation. The model independent determination of the coupling constant g^2 requires accurate single-energy data with absolute normalization N of the unpolarized differential cross section. Since g^2 is always proportional to $\sqrt{N}$, the polarization information is of no particular use in the present context.

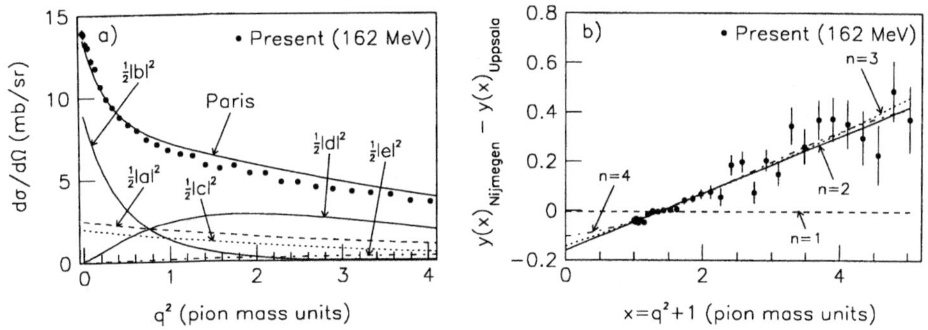

Figure 1. a) Contributions of the a, b, c, d and e amplitudes to the differential np charge exchange cross section using the Paris model. b) Extrapolation to the pion pole of the difference between the $y(x)$ of Eq. (2) for the Nijmegen potential at 162 MeV and the $y(x)$ for our data. The model coupling constant is $g_R^2 = 13.58$.

The history of physics teaches us that a high precision determination of a constant of nature should ideally be based on single experiments, so that sources of uncertainties can be analyzed in great detail and improved systematically. The procedure of analysis should be as transparent and direct as possible with particular attention to sources of systematical errors. In other words: the devil is in the details. In addition to the method used here, only the forward πN dispersion relations, the forward pp dispersion relation and, with improved data, the πN Hüper-plot appear to fulfill this condition at present.

We exemplify our approach using the precision data from 162 MeV np scattering from the recent Uppsala experiment [3], which represents a new generation of high precision np scattering measurements, with higher relative and absolute precision than was previously possible. The angular shape differs noticeably from older and less accurate data [4], but it has been carefully verified using different tests. The main problem is to obtain a reliable experimental normalization. In addition, the systematic error and the stability of the extrapolation must be controlled on the % level. The extrapolated value at the pion pole will then determine the coupling constant model-independently.

Absolute np scattering cross sections were obtained by normalizing to the total np cross section of Lisowski *et al.* [5] exploiting the phenomenological phase-shift solution VZ40 in the angular range not covered by the experiment. This source of uncertainty will be diminished in the next phase of the experiment. The total uncertainty in the cross section scale is estimated to be less than 4%, based on the differences obtained using different phase-shift solutions. The final data are shown as filled circles in Fig. 1a. The errors given are the combined statistical and systematic uncertainties in the relative cross sections.

We now extract a value for the charged πNN coupling constant from these data by extrapolation to the pion pole using both the standard Chew method [6] as well as a more efficient difference method.

The Chew method [6] extrapolates the function y(x) to the pion pole,

$$y(x) = \frac{sx^2}{m_\pi^4 g_R^4}\frac{d\sigma}{d\Omega}(x) = \sum_{i=0}^{n-1} a_i x^i \; ; \qquad y(0) \equiv a_0 \equiv g^4/g_R^4. \tag{2}$$

Here, $x = \mathbf{q}^2 + m_\pi^2$ (at the pole $x = 0$), s is the square of the total energy, m_π the charged pion mass and g_R^2 is a reference scale for the coupling.

Since the pure pion Born term fails to give a qualitatively correct angular distribution, we find it convenient to use the regularized pion Born amplitudes, a_π, b_π, c_π and d_π, as given in Eqs. (12-16) of ref. [7], to ensure a non-zero cross section at 180°. In these amplitudes, the r-space δ-function has been subtracted. This does not in any way affect the model-independence of the procedure.

The difference method is based on the Chew function, but it recognizes that a major part in the cross section behavior is described by models with exactly known values for the coupling constant. It therefore applies the Chew method to the difference between the value of the function $y(x)$ for a model and the experimental values with g_R replaced by the model value g_M:

$$y_M(x) - y_{\exp}(x) = \sum_{i=0}^{n-1} d_i x^i \; ; \qquad y_M(0) - y_{\exp}(0) \equiv d_0 \equiv \frac{g_M^4 - g^4}{g_M^4}. \tag{3}$$

Thus one now determines the *difference* between the model coupling constant and experimental value, which should diminish systematic extrapolation errors.

In Table 1 we give the results for the different methods applied to the present data. We have also generated pseudodata with uncertainties corresponding to the present experiment from 10000 computer simulations using exact data points from the Nijmegen [2] and the Paris [8] potentials with a Gaussian, random error distribution. The coupling constant was determined for each of these 'experiments' using the two methods. The average g^2, its r.m.s. uncertainty and the associated average χ^2 per Degrees of Freedom (χ^2/DoF) are given in the table. We have thus a check on extrapolation errors and can control the systematic corrections. For the Chew method a perfect fit is achieved with a fourth order polynomial in x, but with a small systematic downward shift of 0.21 (Nijmegen) and 0.22 (Paris) as compared to the original model values. With a third order polynomial fit the error becomes smaller, but the systematic shift is considerably larger (2.08 for Nijmegen and 2.39 for Paris).

The Chew method gives a large extrapolation uncertainty (0.93), which indicates that this method is not ideal for a precision determination of the coupling constant, although it gives reliable results with perfect data.

Table 1. Results for our 31 data points and for the corresponding Nijmegen and Paris potential pseudo-data (see text) using the model coupling constants $g^2_{Nijmegen} = 13.58$ and $g^2_{Paris} = 14.43$. The pseudodata are indicated by quotation marks. Optimal values are given in bold font.

n	χ^2/DoF	g^2	χ^2/DoF	g^2	χ^2/DoF	g^2
			Chew method			
	Uppsala		'Nijmegen'		'Paris'	
4	1.314	12.31 ± 0.43	1.129	11.50 ± 0.46	1.203	12.04 ± 0.43
5	1.074	**14.88±0.93**	1.000	**13.37±1.03**	1.001	**14.21±0.94**
			Difference method			
	Nijmegen-Uppsala		Paris-Uppsala		Paris-'Nijmegen'	
2	1.023	**14.62±0.06**	1.286	15.56 ± 0.06	1.136	14.58 ± 0.06
3	1.038	14.52 ± 0.15	1.126	15.26 ± 0.14	1.049	14.37 ± 0.15
4	1.054	14.26 ± 0.37	1.036	**14.65±0.36**	1.006	**13.99±0.37**
5	1.065	15.02 ± 0.92	1.068	15.04 ± 0.92	0.999	13.56 ± 1.02

The difference method requires only a few terms in the polynomial expansion in favorable cases, which gives a small extrapolation error. The similarity between the angular distributions from models and the experimental data are exploited, particularly for larger $\mathbf{q}^2$. This incorporates substantial additional physical information without introducing any model dependence. For the Nijmegen-Uppsala difference (Fig. 1b) the very good χ^2/DoF minimum is obtained with a linear ($n = 2$) expansion with $g^2 = 14.62 \pm 0.06$. The difference method using the Paris potential requires at least four terms and has then an upward systematic shift of 0.41 for the Nijmegen potential and an error of 0.36, corresponding to $g^2 = 14.21 \pm 0.36$, consistent, but smaller than the value from the difference with the Nijmegen potential. The Paris-Uppsala results for $n = 2$, 3 and 5 give also values for g^2 close to those of the Nijmegen-Uppsala difference, when we apply the systematic shifts obtained from the Paris-'Nijmegen' differences. The major uncertainty (0.29) in the analysis of the present experiment comes from the 4% experimental uncertainty in the overall normalization, which can be improved. The extrapolation error is very small and does not reflect systematics.

As a check on our procedure we have also applied our method to the recent precision data on antiproton charge exchange analyzed using the Chew method [9]. This analysis finds with $n = 6$ a value $g^2 = 12.80(36)$ for the charged coupling constant well below the Nijmegen values ($g^2 = 13.58$). We can reproduce this analysis ($g^2 = 12.76$), but we also find that systematic errors favor a higher value, closer to the Nijmegen one, but well below our results for the np data. We do not know the origin of this difference. However, the pronounced minimum in the antiproton cross section makes it nearly mandatory to use a relatively large number of parameters for the extrapolation, even in the case of the difference method. This as well as the shape of the extrapolation functions tends to enhance extrapolation uncertainties in the antiproton case.

Our analysis of the present precise experiment on np charge exchange demonstrates that such data can be used for a direct and precise determination of the πNN coupling constant. We reproduce the original coupling constant using the present procedures for pseudodata from the Nijmegen and Paris potentials. The extrapolation error varies considerably, however, and it goes up with the number of parameters. The most accurate determination from the present data results from the comparison with the Nijmegen potential using the difference method. This gives $g^2 = \sqrt{N} g^2_{\text{exp}} \simeq 14.62(6)$ or $f^2 = \sqrt{N} f^2_{\text{exp}} \simeq 0.0808(3)$, where the normalization introduces an additional 2% uncertainty. The reason for this precision is probably that the recent Nijmegen potential has a good description of the tensor interaction strength. Our value is considerably larger than the Nijmegen result $g^2 = 13.58(5)$, but it is consistent with values given in earlier data compilations based on the analysis of πN and NN scattering data. We have also analyzed our scattering data at 96 MeV [3] with a consistent, although slightly lower, value for the coupling constant. Improvements in the present determination would result from experiments with higher statistics and better systematics, which would be further clarified by similar experiments at other energies. We will in a later version of the present experiment have a more complete range of angles, which will decrease the uncertainty in the overall normalization and give an improved value for the coupling constant. A detailed and more extensive analysis will be published elsewhere.

References

1. T.E.O. Ericson: Nucl. Phys. **A543**, 409c (1993)
2. V. Stoks et al.: Phys. Rev. **C47**, 512 (1993); V.G. Stoks et al.: Phys. Rev. **C49**, 2950 (1994) as given by SAID
3. T.E.O. Ericson et al.: Phys. Rev. Lett. (accepted); T. Rönnqvist et al.: Phys. Rev. **C45**, R496 (1992)
4. B.E. Bonner et al.: Phys. Rev. Lett. **41**, 1200 (1978)
5. P.W. Lisowski et al.: Phys. Rev. Lett. **49**, 255 (1982)
6. G.F. Chew: Phys. Rev. **112**, 1380 (1958)
7. W.R. Gibbs and B. Loiseau: Phys. Rev. **C50**, 2742 (1994)
8. M. Lacombe et al.: Phys. Rev. **C21**, 861 (1980)
9. F. Bradamante et al.: Phys. Lett. **B343**, 427 and 431 (1995)

Few-Body Systems Suppl. 8, 259–263 (1995)

Few-Body Systems

Two–Body Components in the Nucleon Wave Function and their Consequences in Various Phenomena

A. Szczurek [1,2], H. Holtmann [2], J. Speth [2]

[1] Institute of Nuclear Physics, ul.Radzikowskiego 152, PL-31-342 Kraków, Poland

[2] Institut für Kernphysik(Theorie), Forschungszentrum Jülich GmbH., 52425 Jülich, Germany

Abstract. The nucleon light-cone wave function is expanded in terms of two-body meson-(baryonic core) Fock components. The parameters of this expansion are found from both low and high-energy physics. We find many interesting and surprising results for the flavour, spin and electromagnetic structure of the nucleon. The model leads to asymmetry of the light sea quarks which has important consequences for nucleon-nucleon Drell-Yan processes. The experimental Q^2 dependence of the neutron electric form factor is reproduced.

This presentation is an extension of our earlier works on the role of the meson cloud in the nucleon for deep inelastic lepton-nucleon scattering [1, 2], Drell-Yan processes [3] and electromagnetic form factors [4]. At present we put special emphasis on the light-cone formulation of the problem. We expand the nucleon light-cone wave function in terms of two-body meson-(baryonic core) Fock components. Restricting to only the most important component (πN), the Fock state decomposition of the light-cone wave function of the proton reads:

$$|p\rangle = \sqrt{Z}\Big[|3q\rangle + \int dy d^2\boldsymbol{k}_\perp \phi(y,\boldsymbol{k}_\perp)\Big(\sqrt{\frac{1}{3}}|p\pi^0; y, \boldsymbol{k}_\perp\rangle + \sqrt{\frac{2}{3}}|n\pi^+; y, \boldsymbol{k}_\perp\rangle\Big) + \ldots\Big], \tag{1}$$

with Z being the wave function renormalization constant which can be calculated by imposing the normalization condition $\langle p|p\rangle = 1$. $\phi(y,\boldsymbol{k}_\perp)$ is the light-cone wave function of the πN Fock state, where y is the longitudinal momentum fraction of the bare nucleon and $\boldsymbol{k}_\perp$ its transverse momentum. We include pseudoscalar and vector mesons as well as octet and decuplet baryons, i.e. particles important for the description of low-energy proton-proton scattering. Our present approach ensures charge and momentum conservation by construction [5].

The parameters of this expansion are found from both low and high-energy physics [6]. The radii of the meson-baryon Fock components of the nucleon are obtained from high-energy reactions: $pp \to nX$, $pp \to \Delta^{++}X$, $pp \to \Lambda X$ [6]. We find an universal radius for all states involving octet baryons. This allows us to study the role of the two-body components in the nucleon wave functions for different phenomena. We find many interesting and surprising results for the flavour, spin and electromagnetic structure of the nucleon.

As a first surprise we find that the structure of the nucleon obtained from the analysis leads to the violation of the Gottfried Sum Rule [7]

$$S_G = \int_0^1 [F_2^p(x) - F_2^n(x)] \; \frac{dx}{x} \; = \; \frac{1}{3}. \tag{2}$$

Our result ($S_G = 0.224$) is consistent with the experimental result of the New Muon Collaboration $S_G = 0.24 \pm 0.016$ at $Q^2 = 5$ GeV^2 [8]

Our model gives an unique predictions for the $\bar{d} - \bar{u}$ asymmetry. The asymmetry obtained from our model almost coincides with the asymmetry found from the fits of quark distributions to the world data on deep-inelastic and Drell-Yan processes (Fig. 1).

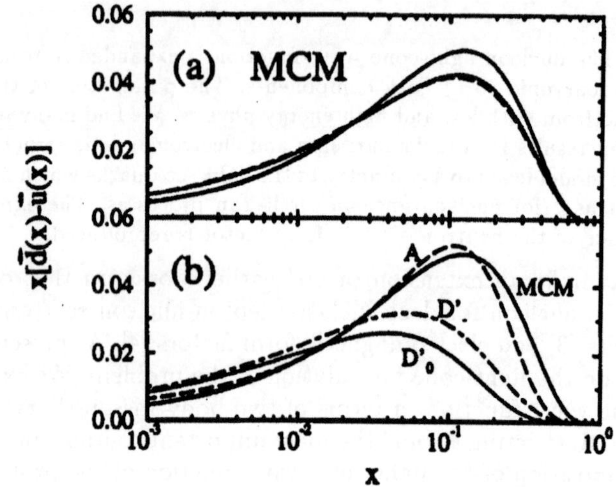

Figure 1. The $x(\bar{d} - \bar{u})$ difference as a function of the Bjorken x calculated in our model for different quark distributions in the pion (panel (a)). In panel (b) we compare our result (solid line) with the newest MSR(A) [12] parameterization of the world data on DIS and Drell-Yan processes (dashed line).

The asymmetry of the sea quark distributions leads to a very good description of the asymmetry of the cross sections

$$A_{DY} = \frac{\sigma_{pp} - \sigma_{pn}}{\sigma_{pp} + \sigma_{pn}} \tag{3}$$

for the dilepton production in proton-proton and proton-neutron collisons observed in the recent NA51 experiment at CERN [11] as seen in Fig. 2. We get

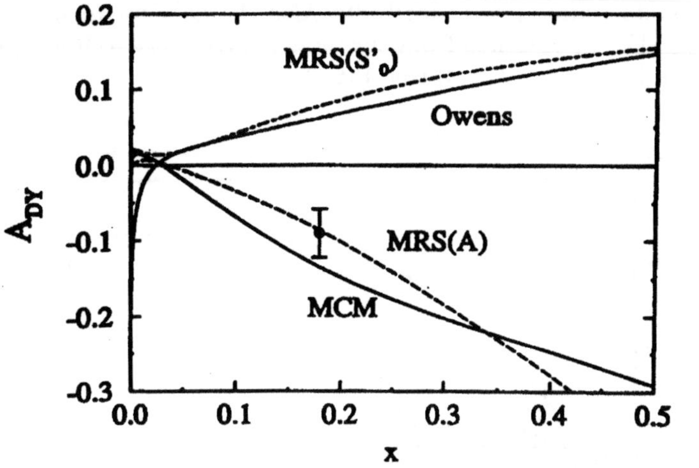

Figure 2. The Drell-Yan asymmetry along the $x_1 = x_2$ diagonal. The solid line is the result of our calculation. The experimental data point is taken from the recent result of the NA51 Collaboration at CERN [11].

similar good description of the nuclear Drell-Yan data [10].

Our model can also be applied to calculate mesonic corrections to the nucleon electromagnetic properties consistently in the light-cone approach. The Fock expansion (1) implies that in our model the Dirac/Pauli form factors can be expressed as

$$
\begin{aligned}
F_{1/2}^{p/n}(Q^2) = Z\big[F_{1/2}^{p'/n'}(Q^2) &+ (F_{1/2}^{p'/n'}(Q^2) + 2F_{1/2}^{n'/p'}(Q^2))I_{1/2,1/2}(Q^2) \\
&+ (F_{2/1}^{p'/n'}(Q^2) + 2F_{2/1}^{n'/p'}(Q^2))I_{1/2,2/1}(Q^2) \\
&+ (G^{\pi^0}(Q^2) + 2G^{\pi^+/\pi^-}(Q^2))I_{1/2,0}(Q^2)\,\big] \qquad (4)
\end{aligned}
$$

The primed quantities are for the bare nucleon. The explicit formulas for $I_{11}, I_{12}, I_{10}, I_{22}, I_{21}, I_{20}$ can be found in refs. [4, 9]. We find a somewhat different pattern of the corrections than in the commonly used static approximation. In distinction to many calculations in the literature we do include the form factor of the virtual pions ($G^{\pi}(Q^2)$) which we find rather important.

The electromagnetic form factors of the core are not known *a priori* and have to be parameterized. Inspired by the success of the standard dipole parameterization for physical nucleons we assume the same functional form for the bare nucleons. The cut-off parameter of the dipole form factor has been fitted to the existing data for G_E^p, G_M^p and G_M^n [9].

We discuss here only predictions for the neutron electric form factor; other details can be found in [9]. For the corresponding radius we get $\sqrt{|\langle r_E^2 \rangle|} = 0.310\ fm$ compared to $0.346 \pm 0.003\ fm$ obtained from the experimental data [13]. Here the whole(!) effect comes from the meson cloud (the traditional dipole parameterization gives exactly 0). Also the Q^2 dependence at very low ($Q^2 < 0.5\,(GeV/c)^2$) obtained from our model agrees very well with the Saclay experimental data from ref. [14] (see Fig. 3).

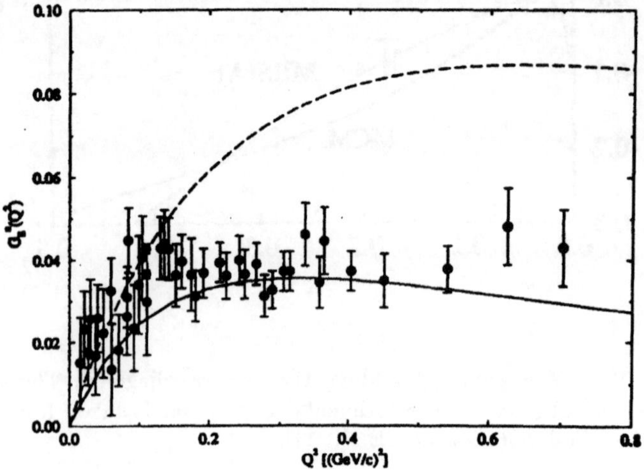

Figure 3. The Q^2 dependence of the electric form factor of the neutron. The experimental data are from ref. [14]; the Gari-Krümpelmann [15] parameterization is shown by the dotted line.

We wish to point out that the model of the pion induced sea discussed here gives also unique predictions for the semi-inclusive deep inelastic scattering $e + p \to e' + n + X$ [16]. The spectrum of neutrons in this reaction directly measures the y-dependence of the flux $f_{\pi/N}(y)$. Such a measurement could not have been performed in the fixed target experiments. The HERA kinematics is ideally suited for measuring the fragmentation of the proton beam into fast neutrons. Such a measurement is being prepared by the ZEUS collaboration at HERA, which is installing the forward neutron calorimeter. A pilot experiment on detecting neutrons from the beam-gas interactions has already been performed with the encouraging result [17] that the observed spectrum of neutrons is in perfect agreement with the pion exchange predictions.

In future lattice QCD calculation should address quantitatively the problem of the pion (meson) cloud. At present the lattice QCD calculation gave only some qualitative evidence for the importance of pion loop effects for nucleon properties [18]. Some authors discuss the pionic dressing of the constituent quarks in the nucleon [19], which is also capable of generating the $\bar{u}$-$\bar{d}$ asymmetry. In contrast to our approach these models are, however, unable to make any link to high-energy production of baryons.

Acknowledgments. We wish to thank M. Ericson and N.N. Nikolaev for collaboration on some points presented here.

References

1. A. Szczurek and J. Speth: Nucl. Phys. **A555** 249 (1993).
2. A. Szczurek and H. Holtmann: Acta Phys. Pol. **B24** 1833 (1993).
3. A. Szczurek, J. Speth and G.T. Garvey: Nucl. Phys.**A570** 765 (1994).
4. N.N. Nikolaev, A.Szczurek, J.Speth and V. Zoller: Z. Phys.**A349** 59 (1994).
5. H. Holtmann, A. Szczurek and J. Speth: *Preprint KFA-IKP(TH)-1993-33.*
6. H. Holtmann, A. Szczurek and J. Speth: *Preprint KFA-IKP(TH)-1994-25.*
7. K. Gottfried: Phys. Rev. Lett.**18** 1174 (1967).
8. P. Amaudruz et al.: Phys. Rev. Lett. **66** 2712 (1991).
9. H. Holtmann, A. Szczurek and J. Speth: *Preprint KFA-IKP(TH)-1994-33.*
10. A. Szczurek, M. Ericson, H. Holtmann and J. Speth: *Preprint KFA-IKP(TH)-1995-05.*
11. A.Baldit et al.: Phys. Lett. **B332** 244 (1994).
12. A.D.Martin, W.J.Stirling and R.G.Roberts: Phys. Rev. **D50** 6734 (1994).
13. L. Koester et al.: Phys. Rev. Lett. **36** 1021 (1976).
14. S. Platchkov et al.: Nucl. Phys. **A510** 740 (1990).
15. M. Gari and W. Krümpelmann: Z. Phys. **A322** 689 (1985).
16. H. Holtmann, G.Levman, N.N. Nikolaev, A. Szczurek and J. Speth: Phys. Lett. **B338** 363 (1994).
17. S.Bhadra et al.: DESY preprint, DESY 94-084.
18. T.D. Cohen and D.B. Leinweber: Comments Nucl. Part. Phys. **21** 137 (1993).
19. E. Eichten, I. Hinchliffe and C. Quigg: Phys. Rev. **D45** 2269 (1992).

Few-Body Systems Suppl. 8, 264–268 (1995)

Few-Body Systems

Hidden Local Symmetries and Meson Exchange Currents

J. Smejkal[1], S. Platchkov[2], E. Truhlík[1]*

[1] Institute of Nuclear Physics, Czech Academy of Sciences, CZ-250 68 Řež n. Prague, Czech Republic

[2] Service de Physique Nucléaire, CEN Saclay, F-91191 Gif-sur-Yvette, France

Abstract.
The operator of the electromagnetic isovector one-pion meson exchange current is constructed using the concept of the hidden local symmetries. Its comparison with the analogous operator, obtained from the hard pion Lagrangian of the Yang-Mills type is presented. It is shown that the observables depend non-negligibly on the choice of the current operator.

1 Introduction

The theory of the isovector electroweak meson exchange currents (MECs) developed in [1]-[3] is based on a hard pion Lagrangian of the $NA_1\rho\pi$ system. The full Lagrangian consists of two parts. One part forms $\mathcal{L}^{YM}_{NA_1\rho\pi}$ containing all the contributions from the nucleon-meson interaction. Another part is formed by $\mathcal{L}^{YM}_{A_1\rho\pi}$ of the subsystem $A_1\rho\pi$, constructed in [4] within the framework of the Yang-Mills (YM) gauge fields. This theory of MECs respects Lorentz and global chiral $SU(2)\times SU(2)$ invariances. The local chiral invariance is broken only by the heavy mesons mass.

Extensive calculations [3], [5] of the cross sections for the backward deuteron electrodisintegration,

$$e + d \longrightarrow e' + np \tag{1}$$

show that such a theory of MECs describe the data [6], [7] satisfactorily. Its specific feature is reflected in the existence of a transverse MEC $\jmath^{YM}_{A_1\pi,\mu}$, the spatial component of which is in the non-relativistic approximation of the form,

$$\jmath^{YM}_{A_1\pi} = -i\frac{g}{2M}\frac{g_A}{2f_\pi}(\boldsymbol{\tau}_1\times\boldsymbol{\tau}_2)^3\, F_1^V(k)\Delta_F^A(q_1^2)\Delta_F^\pi(q_2^2)\,\cdot$$

*Supported by grants Nos. 148410 (Grant Agency of the Czech Academy of Sciences) and 202/94/0370 (Grant Agency of the Czech Republic).

$$\boldsymbol{k} \times (\boldsymbol{\sigma}_1 \times (\boldsymbol{q}_2 + \boldsymbol{k})) + (1 \leftrightarrow 2) . \tag{2}$$

We refer for details to Sect.2.5 of [3]. This current contains a part $\sim \boldsymbol{k}^2$ (k_μ is the momentum transfer), the origin of which can be traced to the corresponding part in the photo-production amplitude, which makes it compatible with the soft pion theorem [3], [8]. This piece of the current Eq. (2) should be present in any correct theory of MECs.

A part of the current Eq. (2) which is $\sim \boldsymbol{q}_2$ is not fixed by low energy theorems and the question arises how it depends on the model of MECs. Here we report shortly on our study of this problem. It is based on a comparison of the prediction from the hard pion theory of MECs discussed above with the results deduced from the hidden local symmetries (HLS) approach to the problem as developed in [9].

In Sect.2, we discuss the Lagrangians. In Sect.3, we pay attention to the structure of currents. Our results and conclusion are presented in Sect.4.

2 Lagrangians

In the HLS approach, the symmetry group of the studied model is extended to some other (hidden) local symmetries. This procedure introduces some redundant non-physical degrees of freedom (compensators) into the model. In the next step, the HLS are spontaneously broken and the non-physical compensators give rise to the mass terms of the HLS gauge fields.

In other words, the remarkable difference between the discussed YM theory and the HLS approach is that in the last one, the heavy mesons masses are generated by the standard Higgs mechanism and the local chiral symmetry of a Lagrangian is conserved. However, it is broken spontaneously.

Generally we found that the class of HLS Lagrangians comprises the YM Lagrangians as a subclass. Both models comprise a set of constants which can be fixed using additional principles and relations between constants such as vector meson dominance, ρ–universality, KSFR relation and Weinberg sum rules [4], [9].

First we constructed a HLS Lagrangian $\mathcal{L}^{HLS}_{A_1\rho\pi}$ of the $A_1\rho\pi$ system, an analogue of $\mathcal{L}^{YM}_{A_1\rho\pi}$. We found that the main difference between these two Lagrangians is observed for the A_1-ρ-π vertex. In terms of the π- ($\boldsymbol{\pi}$), ρ- ($\boldsymbol{\rho}_\mu$) and A_1-meson ($\boldsymbol{a}_\mu$) fields we have explicitly,

$$\mathcal{L}^{YM}_{A_1\rho\pi} = -\frac{1}{f_\pi}\, \boldsymbol{\rho}_{\mu\nu} \cdot (\boldsymbol{a}_\mu \times \partial_\nu \boldsymbol{\pi} + \frac{1}{4}\, \boldsymbol{\pi} \times \boldsymbol{a}_{\mu\nu}) , \tag{3}$$

$$\mathcal{L}^{HLS}_{A_1\rho\pi} = \frac{1}{f_\pi} \left[\boldsymbol{\rho}_{\mu\nu} \cdot \boldsymbol{a}_\mu \times \partial_\nu \boldsymbol{\pi} + \frac{1}{2} \boldsymbol{\rho}_\mu \cdot \partial_\nu \boldsymbol{\pi} \times \boldsymbol{a}_{\mu\nu} \right] . \tag{4}$$

The difference in the A_1-ρ-π vertex depends essentially on how the non-physical A_1-π coupling is removed. There are 4 principal points that influence the process of removing:

1. In [4], the Lagrangian is demanded to contain no more than two field derivatives. In [9], too a strong momentum dependence of the coupling $g_{\rho\pi\pi}$ is eliminated and the ρ meson dominance of the decay $A_1 \longrightarrow \gamma\pi$ is demanded.
2. Consequently, different higher order correction Lagrangians in [4] and [9] are chosen so that p. 1 were satisfied.
3. *The parametrization of the gauge fields:* In the parametrization used in [4], the gauge fields transform according to the linear realization of the chiral symmetry, whereas in the parametrization of [9], they transform in accordance with a non-linear one.
4. A different redefinition of the A_1 meson field is taken in order to suppress the A_1-π coupling.

Next we analyzed the difference between the Lagrangian $\mathcal{L}^{YM}_{NA_1\rho\pi}$ and its HLS analogue, $\mathcal{L}^{HLS}_{NA_1\rho\pi}$. We found that they satisfy equation,

$$\mathcal{L}^{HLS}_{NA_1\rho\pi} = \mathcal{L}^{YM}_{NA_1\rho\pi} - \Delta\mathcal{L}_{NA_1\rho\pi}\,, \tag{5}$$

where

$$\begin{aligned}\mathcal{L}^{YM}_{NA_1\rho\pi} \;=\; & -\bar{N}\gamma_\mu\partial_\mu N - M\bar{N}N - i\frac{g_A}{2f_\pi}\bar{N}\gamma_\mu\gamma_5\,\partial_\mu\boldsymbol{\pi}\cdot\boldsymbol{\tau}N \\ & + ig_\rho\frac{g_A}{2f_\pi}\,\bar{N}\gamma_\mu\gamma_5(\boldsymbol{\tau}\cdot\boldsymbol{\rho}_\mu\times\boldsymbol{\pi})N \\ & -\frac{i}{2}\frac{g_\rho}{f_\pi}\,\bar{N}\gamma_\mu(\boldsymbol{\tau}\cdot\boldsymbol{\pi}\times\boldsymbol{a}_\mu)N - ig_A g_\rho\bar{N}\gamma_\mu\gamma_5(\boldsymbol{\tau}\cdot\boldsymbol{a}_\mu)N \\ & -\frac{1}{2}ig_\rho\bar{N}\left(\gamma_\mu\boldsymbol{\rho}_\mu - \frac{i}{2}\frac{\kappa^V_\rho}{2M}\sigma_{\mu\nu}\widetilde{\boldsymbol{\rho}}'_{\mu\nu}\right)\cdot\boldsymbol{\tau}N + \mathcal{O}(\pi^2)\,,\end{aligned} \tag{6}$$

$$\widetilde{\boldsymbol{\rho}}'_{\mu\nu} = \boldsymbol{\rho}_{\mu\nu} - g_\rho\boldsymbol{\rho}_\mu\times\boldsymbol{\rho}_\nu + \mathcal{O}(\pi)\,, \qquad \boldsymbol{\rho}_{\mu\nu} = \partial_\mu\boldsymbol{\rho}_\nu - \partial_\nu\boldsymbol{\rho}_\mu\,, \tag{7}$$

and

$$\Delta\mathcal{L}_{NA_1\rho\pi} = ig_\rho\frac{g_A}{2f_\pi}\,\bar{N}\gamma_\mu\gamma_5\boldsymbol{\tau}\cdot(\boldsymbol{\rho}_\mu + \frac{1}{g_A}\boldsymbol{a}_\mu)\times\boldsymbol{\pi}N\,. \tag{8}$$

We can see from Eqs. (5)-(8) that there are no $NN\pi\rho$ and $NN\pi A_1$ contact terms present in $\mathcal{L}^{HLS}_{NA_1\rho\pi}$. This is due to the fact that in the HLS approach [9] all fields correspond to the non-linear realization.

3 Currents

The difference between both concepts appears also in the form of the currents. The vector current is,

$$\boldsymbol{J}^{HLS}_{V,\mu} = \boldsymbol{J}^{YM}_{V,\mu} + 2f_\pi g_\rho\,\boldsymbol{\pi}\times\boldsymbol{a}_\mu + \mathcal{O}(\pi^2), \qquad \boldsymbol{J}^{YM}_{V,\mu} = -\frac{m_\rho^2}{g_\rho}\,\boldsymbol{\rho}_\mu\,. \tag{9}$$

The existence of a contact electromagnetic (e.m.) vertex can be deduced in the HLS approach from the form of the current $\boldsymbol{J}^{HLS}_{V,\mu}$ in Eq. (9),

$$\mathcal{L}_{A_1\gamma\pi} = -e\frac{m_\rho^2}{f_\pi g_\rho} A_\mu \left(\boldsymbol{a}_\mu \times \boldsymbol{\pi}\right)^{(3)} . \qquad (10)$$

Here A_μ is the four-vector of the e.m. potential.

In order to construct the nuclear MECs within the framework of the extended S-matrix method [10], one should first have the two-nucleon Feynman amplitudes. The total conserved YM Feynman amplitude is,

$$J^{YM}_\mu(k) = J^B_\mu(k) + J^{m.t.}_\mu(k) + J^{c.t.}_\mu(k) + J^{YM}_{A_1\rho\pi,\mu}(k) . \qquad (11)$$

Here the amplitudes with labels 'B', 'm.t.' and 'c.t.' are the standard Born, mesonic and contact terms, respectively. They are well known from the chiral model with the pseudovector πNN coupling incorporated in the Lagrangian Eq. (6). The last amplitude corresponds to the vertex Eq. (3) and provides the current of Eq. (2).

The conserved HLS Feynman amplitude is obtained from Eqs. (4)-(10),

$$J^{HLS}_\mu(k) = J^B_\mu(k) + J^{m.t.}_\mu(k) + J^{HLS}_{A_1\rho\pi,\mu}(k) + J_{\gamma,\mu}(k) . \qquad (12)$$

It can be shown that the current analogous to the one of Eq. (2) is,

$$\begin{aligned} \boldsymbol{\jmath}^{HLS}_{A_1\pi} &= -i\frac{g}{2M}\frac{g_A}{2f_\pi}(\boldsymbol{\tau}_1 \times \boldsymbol{\tau}_2)^3\, F^V_1(k)\Delta^A_F(q_1^2)\Delta^\pi_F(q_2^2)\,\cdot \\ &\quad \boldsymbol{k} \times (\boldsymbol{\sigma}_1 \times (-\boldsymbol{q}_2 + \boldsymbol{k})) + (1 \leftrightarrow 2) . \end{aligned} \qquad (13)$$

Comparing Eqs. (2) and (13) we can see that the model dependent part changed the sign.

4 Results and conclusion

The impact of the change of the sign of model dependent part in Eq. (2) on observables can be deduced from Fig. 1, where we show calculations of cross sections for reaction Eq. (1). The deuteron and $^1S_0$ final state nuclear wavefunctions correspond to the Bonn OBEPQB potential improved by including the A_1 meson exchange [11]. The cross sections were calculated for two different sets of the e.m. nucleon form factors. The curves labelled 'H8.2' correspond to the form factors of the fit 8.2 of ref. [12], while those labelled 'P' were calculated with the e.m. nucleon form factors modelled according to ref. [13], using the two-parameter formula for the neutron electric form factor and the deuteron wavefunction obtained from our improved potential OBEPQB. The data is taken from ref. [6].

For the values $t \sim 1\,\mathrm{GeV}^2$, the change in the spectra caused by the choice of the current model forms $30\% - 40\%$ of the change due to the model dependence of the e.m.nucleon form factors. We conclude that the studied effect is not negligible.

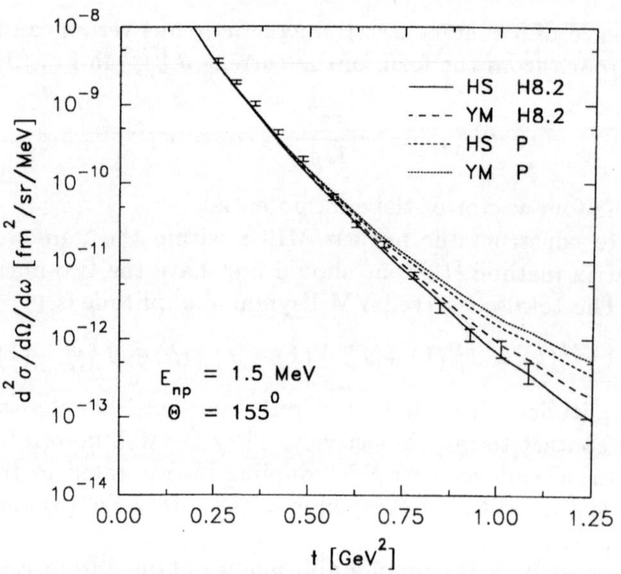

Figure 1. The cross sections for reaction (1).

References

1. E. Ivanov, E. Truhlík: Nucl. Phys. **A316**, 437 (1979)
2. J. Adam, Jr., E. Truhlík: Czech. J. Phys. **B34**, 1157 (1984)
3. E. Truhlík, J. Adam, Jr.: Nucl. Phys. **A492**, 529 (1989)
4. V.I. Ogievetsky, B.M. Zupnik: Nucl. Phys. **B24**, 612 (1970)
5. E. Truhlík, K.M. Schmitt: Few-Body Systems **11**, 155(1992)
6. S. Auffret et al.: Phys. Rev. Lett. **55**, 1362(1987)
7. M. Frodyma et al.: Phys. Rev. **C47**, 1599(1993)
8. E. Truhlík: Czech. J. Phys. **43**, 467(1993)
9. M. Bando, T. Kugo, K. Yamawaki: Phys. Reports **164**, 217 (1988)
10. J. Adam, Jr., E. Truhlík, D. Adamová: Nucl. Phys. **A492**, 556(1989)
11. P. Obersteiner, W. Plessas, E. Truhlík: In:*XIII International Conference Particles and Nuclei, Perugia, 28.6.-2.7. 1993*, p.430. Singapore: World Scientific 1994
12. G. Höhler et al.: Nucl. Phys. **B114**, 505(1976)
13. S. Platchkov et al.: Nucl. Phys. **A510**, 740(1990)

Few-Body Systems Suppl. 8, 269–274 (1995)

Few-Body Systems

Relativistic Calculations of the Deuteron Form Factors and Triton Binding Energy

Franz Gross[1,2], Alfred Stadler[1], J.W. Van Orden[2,3], N. Devine[2]

[1] Department of Physics, College of William and Mary, Williamsburg VA, 23185, USA

[2] Physics Division, MS12H2, CEBAF, Newport News, VA 23606, USA

[3] Department of Physics, Old Dominion University, Norfolk VA, 23529, USA

Abstract. Recent results of relativistic calculations of the deuteron form factors and the triton binding energy are described. The results are *preliminary*; several of them were only obtained a week before the conference, and are based on NN interaction models which are still under development.

1 Deuteron Form Factors

In collaboration with J. W. Van Orden and Neal Devine, a new relativistic calculation of the deuteron form factors is being completed [1, 2]. This calculation is based on an updated version of our NN one boson exchange (OBE) Model IIB [3]. The updated model was fitted to the Nijmegen '93 np phase shifts [4] using a revised NN code (with some minor errors corrected). The model fits the NN data very well, but has a D/S ratio and a $^3S_1$ scattering length and effective range which are too small, and we are currently looking for models with better $^3S_1$ low energy parameters.

A novel feature of our NN interaction model is the appearance of a strong form factor for the nucleon, $h(p^2)$, which is a function of the mass (squared) of the nucleon. The form factor is defined to be unity when the nucleon is on-shell ($h(m^2) = 1$) and approaches zero as $p^2 \to \pm\infty$. Gauge invariance [5] requires that the electromagnetic current be modified in the presence of this form factor, and a general form which satisfies the necessary Ward–Takahashi constraint is [2, 6]

$$J^\mu(p',p) = F_1(Q^2)f_0\,\gamma^\mu + \frac{F_2(Q^2)}{2m}h_0\,i\sigma^{\mu\nu}q_\nu + F_3(Q^2)g_0\,\frac{\not p' - m}{2m}\,\gamma^\mu\,\frac{\not p - m}{2m} \quad (1)$$

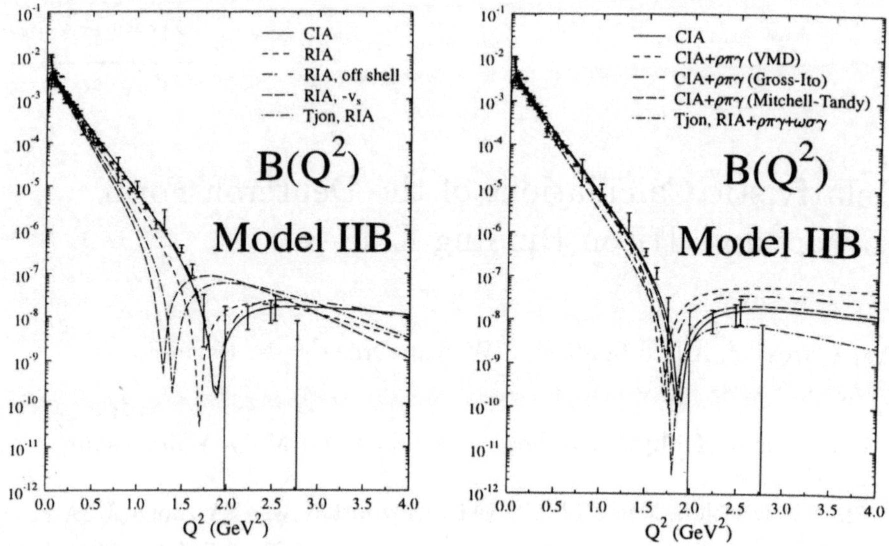

Figure 1. The B structure function in impulse approximation (RIA and CIA), and with exchange currents.

where

$$f_0 = f_0(p'^2, p^2) = \frac{h(p^2)}{h(p'^2)} \frac{m^2 - p'^2}{p^2 - p'^2} + \frac{h(p'^2)}{h(p^2)} \frac{m^2 - p^2}{p'^2 - p^2}, \tag{2}$$

$$g_0 = g_0(p'^2, p^2) = \left(\frac{h(p^2)}{h(p'^2)} - \frac{h(p'^2)}{h(p^2)} \right) \frac{4m^2}{p'^2 - p^2} \tag{3}$$

and $F_3(Q^2)$ and $h_0 = h_0(p'^2, p^2)$ are arbitrary functions subject only to the constraints $F_3(0) = 1$ and $h_0(m^2, m^2) = 1$. In the calculations presented here, we chose $h_0(p'^2, p^2) = f_0(p'^2, p^2)$, and $F_3(Q^2) = G_E(Q^2)$ for simplicity.

Our results for the B structure function are shown in Fig. 1. The left hand figure compares the relativistic impulse approximation (RIA), in which the spectator nucleon is on-shell, with the complete impulse approximation (CIA), which also includes terms in which the struck nucleon is on shell (see refs. [1, 2, 5]). The RIA, as first introduced [7], included a current operator of the form (1) with $F_3 = 0$ and $f_0 = h_0 = 1$. This result is the widely spaced dashed line. The introduction of the off shell current (i.e. $F_3 = G_E$ and $h_0 = f_0 \neq 1$) has a significant impact, moving the result very close to the data. The addition of the small extra terms in the CIA (the solid line) suppresses the second maxima slightly, in excellent agreement with the data. Note that our CIA differs significantly from the RIA obtained by Hummel and Tjon [8].

The good agreement between the CIA and the data seems to be due in part to the very small v_s component of the deuteron wave function (with a probablity of only 0.0009%). If the sign of this small component is changed, the minimum shifts down by almost 0.5 GeV^2, spoiling the agreement. This is a surprising result showing that the B structure function is very sensitive to relativistic details of the deuteron structure.

We have also calculated the contributions from the $\rho\pi\gamma$ exchange currents (see the right pannel in Fig. 1). This current is very sensitive to the model of the $\rho\pi\gamma$ from factor which is assumed; results for the vector dominance model used by Hummel and Tjon [8] and quark loop calculations [9, 10] are shown for comparison.

2 The Triton Binding Energy

In collaboration with Alfred Stadler, progress has been made on a relativistic calculation of the triton binding energy using the covariant spectator (or Gross) equation, and we are now able to report preliminary results for two models which appear to illustrate the essential features of the problem.

The three body spectator equation [11, 12, 13] has now been solved for a variety of NN interaction models based on OBE kernels. We realized at the beginning that the triton binding would be sensitive to the low energy behavior of the interaction model, and began looking for NN models with more accurate low energy properties (good D/S ratio, scattering lengths, and effective ranges). The original work [3] used a nucleon form factor of the form

$$f_N(p^2) = \frac{(\Lambda_N^2 - m^2)^2 + (m_0^2 - m^2)^2}{(\Lambda_N^2 - m^2)^2 + (m_0^2 - p^2)^2} \tag{4}$$

with $m_0 = \Lambda_N$, but subsequent experience [14] has suggested that a form factor with $m_0 = m$ is a more suitable choice. Making this change we refit the parameters of Model IIA (defined in ref. [3]). This model (referred to here as Model X) gives a binding energy of -6.185 MeV, compared to the experimental value of -8.48. When we first obtained this result, we were surprised at the extent to which this model underbinds the triton.

We now believe that we have a first understanding of the reason for this underbinding. In our approach, one of the interacting nucleons is always off-shell, and hence it is not possible to use the Dirac equation, $(\not{p}-m)\,u(p,s) = 0$, to simplify the meson-nucleon couplings. Our couplings allow for the presence of terms like $(\not{p} - m)$, and when evaluated for off-shell nucleons these terms give rise to nonlocal, energy dependent factors of the form $(W - 2E_p)$, where W is the energy of the NN system and $E_p = \sqrt{m^2 + p^2}$ is the energy of an on-shell nucleon. The underbinding shown by Model X seems to be, in large part, due to the presence of such terms.

This suggests looking for NN interaction models which include off-shell interactions which increase the binding, but still fit the scattering data and low energy parameters. One such model (Model Y) includes off-shell couplings in the effective σNN and δNN interactions which were previously ignored. The

Table 1. Convergence of the binding energy for two preliminary relativistic models

j_{max}, parity	ρ spin	channels	model X	Model Y	Paris
1 +	+	5	−6.059	−6.735	−7.30
1 ±	+	10	−5.901	−6.390	−7.11
2 ±	+	18	−6.118	−6.604	−7.38
1 +	±	10	−6.098	−6.786	
1 ±	±	20	−5.969	−6.469	
2 ±	±	36	−6.185	−6.679	

new interactions modify the scalar couplings as follows:

$$g_s \to g_s \left\{ 1 + \frac{\lambda_s}{2m} \left[\not{p} - m + \not{k} - m \right] \right\}, \tag{5}$$

where k and p are the four-momenta of the initial and final nucleons. A good fit to the NN data was obtained with the values $\lambda_\sigma = 0.134$ and $\lambda_\delta = 5.373$ (and changes in the other parameters also). Not only does this model give more binding (−6.679 MeV), but it also gives a better fit to the NN data and might have been discovered even without consideration of triton binding.

These results raise the intriguing possibility that a good NN model might be found which also gives a good value for the triton binding energy.

Table 1 shows how these two models converge. The upper part of the table shows the convergence if all negative energy channels (i.e. those with negative "ρ" spin) are removed from the calculation. The addition of negative ρ spin to the relativistic models doubles the number of channels, but only increases the binding in each case by about 50-70 keV. We conclude that negative ρ spin channels make only a small, attractive contribution to the triton binding.

The way in which Model X approaches a nonrelativistic limit is shown in Table 2. The first 4 steps, shown in the upper box of the table, display what happens when the *two body amplitude is changed without altering the three body equation*. In each step, *the new two body amplitude is obtained from a fit to the deuteron and to the two body data*. When the negative ρ spin channels are switched off [step (1)], the binding is increased by about 130 keV. If the energy dependent, off-shell effects are then removed from the meson nucleon couplings [step (2)], the binding energy increases by a rather large 1200 keV, leading to the conclusion that off-shell effects are important. To obtain a good fit to the NN data in this case it was necessary to allow the cutoff mass in the πNN form factor, Λ_π, to vary independently of the other meson form factor masses, Λ_m. The next step (3) removes some m/E factors, and the retardation terms from the meson propagators and form factors, giving another 500 keV binding. Finally, the nonrelativistic limit of the NN amplitude is arrived at by replacing the relativistic propagator by the nonrelativistic propagator, and setting

Table 2. Approach to the nonrelativistic limit. The four steps in the upper box are sequential operations on the *two body amplitude*, leaving the three body Faddeev equations in fully relativistic form. The last step (5) shows the effect of substituting the nonrelativistic amplitude (4) into nonrelativistic Faddeev equations.

Step	Short description	Energy	Comments
	full relativistic result	−6.059	$\Lambda_\pi = \Lambda_m$
(1)	remove − channels	−6.183	$\Lambda_\pi = \Lambda_m$
(2)	(1) and remove off-shell effects	−7.363	$\Lambda_\pi \neq \Lambda_m$
(3)	(2) and remove retardation and m/E factors	−7.824	$\Lambda_\pi \neq \Lambda_m$
(4)	nonrelativistic limit: (3) and change to nonrelativistic propagator and set the nucleon form factor to unity	−7.851	$\Lambda_\pi \neq \Lambda_m$
(5)	nonrelativistic two body amplitude in nonrelativistic Faddeev equations	−8.144	

the nucleon form factors to unity. This last step does not change the binding energy significantly. The fully nonrelativistic three body result is obtained by inserting the nonrelativistic two body amplitude obtained in step (4) into the nonrelativistic Faddeev equations, which increases the binding by about 300 keV, giving a final result of −8.144 MeV. A brief discussion of the origin of these last effects was already reported by A. Stadler at the 1994 Williamsburg conference [12].

3 Conclusions

Bearing in mind that the results presented in this report are *preliminary*, we draw the following conclusions:

1. Relativistic OBE models can be found which give excellent fits to the NN data.
2. The B structure function of the deuteron is very sensitive to the deuteron P states. This is unexpected.
3. Model IIB fits $B(Q^2)$ in the complete impulse approximation (CIA) very well. No exchange currents are needed.
4. The $\rho\pi\gamma$ exchange current improves $A(Q^2)$ (not discussed here) but moves $B(Q^2)$ away from the data. It is very sensitive to the $\rho\pi\gamma$ form factor, and might be very small.
5. The three body binding energy is sensitive to to the *energy dependence and nonlocality* inherent in relativistic OBE models. These may be the most important relativistic effects.

Acknowledgement. This work was supported in part by the U. S. Department of Energy under Grant Nos. DE-FG05-88ER40435, DE-FG05-94ER40832, and DE-AC05-84ER40140.

References

1. J.W. Van Orden, N. Devine, F. Gross: in preparation
2. J.W. Van Orden, N. Devine, F. Gross: talk given by J.W. Van Orden to the *Conference on Mesons and Light Nuclei*, Czech Republic, July 3–7, 1995, to be published in Few–Body Systems Suppl. (Springer-Verlag Wien)
3. F. Gross, J.W. Van Orden, K. Holinde: Phys. Rev. C **41**, R1909 (1990); **45**, 2094 (1992)
4. V.G.J. Stoks, R.A.M. Klomp, M.C.M. Rentmeester, J.J. de Swart: Phys. Rev. C **48**, 792 (1993)
5. F. Gross, D.O. Riska: Phys. Rev. C **36**, 1928 (1987)
6. Y. Surya, F. Gross: preprint # WM-95-101, CEBAF-TH-95-04
7. R.E. Arnold, C.E. Carlson, F. Gross: Phys. Rev. C **21**, 2361 (1980)
8. E. Hummel, J.A. Tjon: Phys. Rev. Lett. **63**, 1788 (1989); Phys. Rev. C **42**, 423 (1990)
9. H. Ito, F. Gross: Phys. Rev. Lett. **71**, 2555 (1993)
10. K.L. Mitchell: Ph.D. Thesis, Kent State University (1995) (unpublished); K.L.Mitchell, P.C.Tandy: to be published
11. F. Gross: Phys. Rev. C **26**, 2226 (1982)
12. A. Stadler, F. Gross: *Few–Body Problems in Physics*, AIP Conference Proceedings #334 (1995), F. Gross, ed., p. 867
13. F. Gross, A. Stadler, M. Frank: in preparation
14. F. Gross, Y. Surya: Phys. Rev. C **47**, 703 (1993)

Few-Body Systems Suppl. 8, 275–286 (1995)

Few-Body Systems

Glueballs, η and η' Mesons and Antiproton–Proton Annihilation Dynamics

E. Klempt

Institut für Strahlen– und Kernphysik, Nußallee 14 - 16, D-53115 Bonn, Germany

Abstract. A new scalar isoscalar state, the $f_0(1500)$, has been discovered in a series of recent experiments, in $\bar{p}p$ annihilation at rest and in flight, in charge–exchange, Pomeron–Pomeron scattering, and in radiative J/ψ decays. The state has large decay branching ratios to $\pi\pi$, $\sigma\sigma$, $\eta\eta$, and to $\eta\eta'$. The production characteristics and the decay modes suggest an interpretation of the $f_0(1500)$ as ground–state scalar glueball even though other interpretations cannot yet be excluded. The experimental situation is reviewed with particular emphasis on the results obtained by the Crystal Barrel Collaboration at LEAR in $\bar{p}p$ annihilation into 3 pseudoscalar mesons.

In $\bar{p}p$ annihilation at rest into two final–state mesons dynamical selection rules have been observed, the best–known example being the $\rho\pi$ puzzle. Even more striking examples are found in strange–meson production. These dynamical selection rules require an interpretation in terms of symmetries of the underlying dynamical amplitudes. In it assumed that this symmetry is SU(3). However, SU(3) can be formulated in different coupling schemes; in terms of the underlying quark dynamics, of s–channel resonances or in terms of baryon exchanges in the t or u channel. The different (squared) amplitudes are compared with experimental data.

1 Introduction

Antiproton–proton annihilation is a rich source of mesons. The phase space available in $\bar{p}p$ annihilation at LEAR is 2.40 GeV/c^2 at the maximum LEAR momentum; in annihilation at rest it is still 1.87 GeV/c^2. Hence a large number of final states is produced and new mesons can be searched for. Indeed, several mesons have been discovered in $\bar{p}p$ annihilation, from the very beginning of experiments with antiprotons when the ω meson was found until recent discoveries of scalar states [1, 2].

In the mass range accessible at LEAR glueballs and hybrides are predicted to exist which are a direct consequence of the confinement properties of quantum chromodynamics. The search for these states was the prime motivation for building the Crystal Barrel detector which is taking data on $\bar{p}p$ annihilation at rest and in flight since 1989. Recently, the Crystal Barrel collaboration found two new scalar states, one of which has properties fully compatible with those predicted for the lowest–mass scalar glueball. The state has large couplings to η and η' mesons; this is the "excuse" to present evidence and a discussion of its role in meson spectroscopy in a session on η production dynamics.

Scalar mesons are difficult to identify: in scattering experiments high angular momenta are dominant and scalar mesons hidden under large contributions from higher partial waves. Furthermore, scalar mesons often have large widths since no centrifugal barrier prevents them to fall apart into two pseudoscalar mesons. These facts have, for a long time, obscured the ground–state scalar meson nonet. On the other hand, scalar mesons are particularly interesting. The gluon gluon self–interactions introduce a new degree of freedom into hadron spectroscopy: not only $(q\bar{q})$ mesons should exist but also bound states without any constituent quarks [3]. This is certainly one of the most striking predictions of Quantum Chromodynamics. The lowest–mass glueball has likely scalar quantum numbers and a mass between 1 and 2 GeV/c^2. This is also the mass range of the 1^3P_0 and possibly of the 2^3P_0 light–quark nonets. The identification of the scalar glueball against $q\bar{q}$ mesons in the same mass range requires therefore a good understanding of the spectrum of scalar mesons.

Here, $\bar{p}p$ annihilation at rest offers one distinctive advantage: while peripheral processes like charge exchange are dominated by large angular momenta, $\bar{p}p$ annihilation at rest proceeds dominantly via S–states, i.e. via states with no orbital angular momenta between proton and antiproton. The production of scalar mesons is therefore enhanced, facilitating the search for scalar resonances. And $\bar{p}p$ annihilation offers a further distinctive advantage: very large data sets can and have been accumulated at rest and in flight. It is this enormous gain in statistics which has led to new and exciting results.

The second part of this paper addresses a different question: in $\bar{p}p$ annihilation into two mesons three quarks and three antiquarks annihilate and interact thus forming two outgoing $q\bar{q}$ mesons. How can we describe this process? Do we need quarks and gluons to understand the processes or are baryons and mesons still the fundamental particles? Recent data at LEAR show very remarkable features, in particular the presence of *dynamical selection rules* which suggest that symmetries play a decisive role. The selection rules have no generally accepted explanation and strong arguments were put forward at this conference by K. Holinde and A. Faessler against the interpretation given below. Inspite of this criticism I shall present here different symmetry schemes and discuss which one is best suited to reproduce the observed selection rules.

2 Search for Glueballs

Before we discuss the possible existence of glueballs, I would like to recall their distinctive properties which might help to identify them.

- Glueballs do not fit into any $q\bar{q}$ meson nonet.
- Glueballs are predicted to have a scalar ground state and a tensor and a pseudoscalar state as first excitations. The scalar ground state should have a mass of about 1.5 GeV/c^2; the tensor of about 2.4 GeV/c^2 [4].
- The widths of glueballs are rather uncertain: lattice gauge calculations neglect the couplings to the hadronic world. This 'quenched approximation' results also in uncertainties of the calculated glueball masses.
- Glueballs contain no constituent quarks; hence their decays should respect flavour symmetry, e.g. their couplings to $\pi^0\pi^0$, to $\eta\eta$, or to $K\overline{K}$ should be equal, the coupling to $\eta\eta'$ should vanish. However, glueballs may prefer to decay into η and η' mesons [5], and also mixing with $q\bar{q}$ states having the same quantum numbers will lead flavour symmetry breaking.
- Glueballs should be produced abundantly in 'glue–rich' processes like radiative J/ψ decays where the J/ψ radiates off one photon and two gluons; they then interact to form a glueball. Glueballs may also be formed in central production experiments via Pomeron–Pomeron scattering. Finally, $\bar{p}p$ annihilation might be a rich source of hybrids and glueballs since quarks and antiquarks are contained in a small volume and annihilate into gluons interacting among themselves and with $q\bar{q}$ pairs.

The Crystal Barrel experiment has recorded high–statistics data on 17 million events with only $\bar{p}p \rightarrow \pi^0\pi^0\pi^0$, $\pi^0\pi^0\eta$, $\pi^0\eta\eta$, $\pi^0\eta\eta'$ and $5\pi^0$. A special trigger has been set up in order to enhance the statistics on $\pi^0\eta\eta$ and $\pi^0\eta\eta'$. Fig. 1a shows the $3\pi^0$ Dalitz plot. Due to the symmetry in the final state there are six entries per event. The highest intensities reflect the interference of two respective amplitudes for the reaction $\bar{p}p \rightarrow f_2(1270)\pi^0$. The $f_2(1270)$ decay angular distribution is peaked in the forward and backward directions characteristic for a spin–2 particle. The $\pi\pi$ S–wave is also an important contribution in this mass range. In the three corners, a band and an additional blob–like structure are observed. The band shows up as $f_0(1500)$ in the analysis. Finally we notice a faint dip approximately at the $K\overline{K}$ threshold.

The reaction $\bar{p}p \rightarrow \pi^0\eta\eta$ is dominated by production of $a_0(980)$ mesons and of scalar $\eta\eta$ resonances (Fig. 1b). Two $\eta\eta$ bands are observed in the Dalitz plot: a striking high–mass band and some extra intensity in the upper right corner. The high–mass band corresponds to a state at 1500 MeV/c^2, the band at lower masses to 1300 MeV/c^2. Important for the understanding of the Dalitz plot is also the scalar $\pi\eta$ interaction. It can be constrained by analysing in parallel the reaction $\bar{p}p \rightarrow \pi^0\pi^0\eta$ (Fig. 1c). Apart from the well known $a_0(980)$, a new scalar resonance, $a_0(1450)$, is needed to describe this data. The strong enhancement in the lower right corner of the $\pi^0\eta\eta'$ Dalitz plot (Fig. 1d) can be identified as $\eta\eta'$ decay mode of the $f_0(1500)$. A very recent analysis of the reaction $\bar{p}p \rightarrow 5\pi^0$revealed that the $f_0(1500)$ has a large 4π decay branching ratio into two pion pairs in relative S–wave (into two 'σ' mesons [6]).

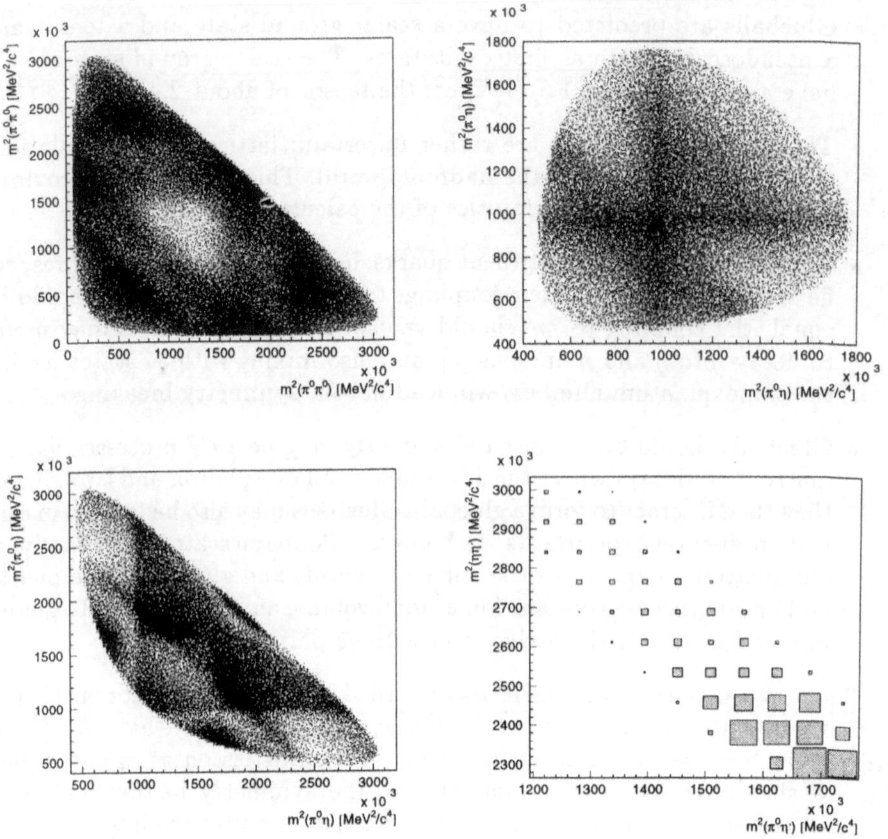

Figure 1. Dalitz plot for the reactions $\bar{p}p \rightarrow \pi^0\pi^0\pi^0$ (a), $\pi^0\eta\eta$ (b), $\pi^0\pi^0\eta$ (c), and $\pi^0\eta\eta'$ (d). The data are from the Crystal Barrel experiment. See text for a discussion.

A scalar isoscalar resonance at 1590 MeV/c^2 coupling to $\eta\eta$, $\eta\eta'$, and to $4\pi^0$ had been observed in a series of pioneering experiments at CERN and at Serpukhov by the GAMS collaboration. WA91 reported a scalar resonance at 1450 MeV/c^2 produced in central collisions [7]. Experiment E760 at Fermilab has collected huge statistics on the reactions $\bar{p}p \rightarrow \pi^0\pi^0\pi^0$ and $\pi^0\eta\eta$ at antiproton–proton invariant masses of about 3 and 3.5 GeV/c^2. Clear peaks at about 1500 MeV are observed.

The apparent absence of the $f_0(1500)$ in radiative J/ψ decays was a serious counter–argument against the glueball interpretation of the $f_0(1500)$. A recent reanalysis of the process J/$\psi \rightarrow \gamma 4\pi$ [8] reports, however, the presence of a scalar state at 1500MeV decaying into $\sigma\sigma$. This results suggests that the $f_0(1500)$ is also produced in radiative J/ψ decays.

Masses, widths and decay branching ratios of the $f_0(1500)$ as found in the different experiments differ more than allowed by statistics. Nevertheless I assume that these observations reflect one scalar resonance with mass and width of 1500 MeV/c^2 and 130 MeV/c^2, respectively, and with coupling to 4π, 2π, $\eta\eta$, $\eta\eta'$, and at most weak coupling to $K\overline{K}$.

The $f_0(1500)$ seems not to fit into the scalar $q\bar{q}$ nonet: there are too many isoscalar states. The $K\overline{K}$ mode is unobserved so that an $s\bar{s}$ assignment is untenable. The large couplings to $\eta\eta$ and to $\eta\eta'$ and the absence of the $K\overline{K}$ decay mode are unexpected for a $q\bar{q}$ state. The $f_0(1500)$ is observed in $\bar{p}p$ annihilation at rest and in flight, in charge exchange, in Pomeron–Pomeron scattering and in radiative J/ψ decays. Also, its mass agrees nicely with lattice gauge calculations [4]. In view of its production and decay characteristics, this state is presently considered as prime glueball candidate.

Nevertheless I have to point out that another interpretation of the $f_0(1500)$ exists [9] in which the $f_0(1500)$ is the flavour–octet state while the flavour–singlet state is at about 1000 MeV/c^2 (likely the broad $f_0(1000)$ advocated by Morgan and Pennington [10]). The mass inversion of the two isoscalar states arises naturally from the same instanton interaction introduced to describe the pseudoscalar mass splittings.

3 Annihilation Dynamics

Interactions of protons and antiprotons at low energies are very well described by meson exchange potentials derived from nucleon–nucleon scattering data. In potential models, baryons and mesons are considered as elementary particles; no quarks or gluons are needed to find an impressive agreement with data on elastic and charge exchange reactions. This success may lead to the hypothesis that baryon exchange should be the appropriate 'language' to describe the annihilation process and seems to exclude any necessity to introduce quark degrees of freedom.

It is, however, not obvious that baryon and meson exchanges can also be used to describe $\bar{p}p$ annihilation. New particles are created in this process and they may take advantage of the quarks which already exist in the initial state. From this point of view it seems more natural to describe annihilation in terms of effective quarks. Indeed, the first attempt to understand annihilation dynamics was a quark model [11].

The $\bar{p}p$ atom has a defined mass. Thus annihilation into two mesons could proceed via mixing of the initial six–quark configuration with four–quark states which then decay. This mixing should then be selective to four–quark states close to the mass of the $\bar{p}p$ system. Therefore one might expect annihilation to favor a set of s–channel resonances with specific flavour symmetry.

Here, we address the question whether the basic annihilation mechanism can be identified by use of symmetry arguments. We use SU(3) to derive three different coupling schemes describing transitions from the initial $\bar{p}p$ system to two mesons [12]. Two experimental observations are particularly important for the interpretation of the data within SU(3): annihilation into η and η' mesons give new insight into the role of phase space and angular momentum barrier for $\bar{p}p$ annihilation; the so-called *dynamical selection rules* [13, 14, 15] provide strong evidence that symmetry arguments should be useful in uncovering the nature of the annihilation mechanism.

3.1 Production of η and η' Mesons

The Crystal Barrel collaboration determined a large number of branching ratios for reactions $\bar{p}p \to \eta X$ and $\eta' X$ [16, 17]. Interesting in this context are the double ratios:

$$\frac{\bar{p}p \to \pi^0\eta'}{\bar{p}p \to \pi^0\eta} = 0.55 \pm 0.05 \quad \frac{\bar{p}p \to \eta\eta'}{\bar{p}p \to 2\eta\eta} = 0.65 \pm 0.07 \quad \frac{\bar{p}p \to \omega\eta'}{\bar{p}p \to \omega\eta} = 0.52 \pm 0.03$$

$$\frac{\bar{p}p \to \rho^0\eta'}{\bar{p}p \to \rho^0\eta} = 0.40 \pm 0.10 \qquad \frac{\bar{p}p \to (\pi^0\pi^0)_S\eta'}{\bar{p}p \to (\pi^0\pi^0)_S\eta} = 1.0 \pm 0.4$$

Note that these ratios are compatible with one unique ratio of 0.54 ± 0.03. This is very surprising since the decay momenta for production of η and η' mesons differ by up to a factor of two, and different decay angular momenta are involved (l=0,2 for two pseudoscalar mesons and l=1 for one pseudoscalar and one vector meson).

From these double ratios the pseudoscalar mixing angle can be determined by use of the relation [16]:

$$\frac{(\bar{p}p \to \eta' X)}{(\bar{p}p \to \eta X)} \cdot \frac{f_{\text{dyn}}\,(\eta X)}{f_{\text{dyn}}\,(\eta' X)} = ctg^2\,\{\Theta_{\text{PS}} - \Theta_{\text{ideal}}\}$$

The ratios are corrected for various dynamical functions f_{dyn}. It can be shown that only two dynamical functions yield compatible results for Θ_{PS} for all five double ratios, $f_{\text{dyn}} = 1$ [18] and $f_{\text{dyn}} = q \cdot \exp\{A(s - (m_a + m_b)^2)\}$ [19]. The latter expression gives a 10-20% correction to the double ratios. From these ratios and using these dynamical functions the pseudoscalar mixing angle is determined to $-(17.3 \pm 1.8)^\circ$. This value is fully compatible with our present knowledge.

This is an important observation: annihilation dynamics is incompatible with the expected phase space dependence or suppression due to angular momentum barriers. It is therefore very dangerous to draw conclusions on primary $q\bar{q}$ vertices from global comparisons of measured and calculated branching ratios without explicit control of dynamical aspects in the calculations.

3.2 Dynamical Selection Rules in $\bar{p}p$ Annihilation at Rest

The best known *dynamical selection rule*, the *$\rho\pi$ puzzle*, has attracted considerable interest; annihilation into $\rho\pi$ is observed strongly from the $^3S_1$ initial state of the $\bar{p}p$ atom but only weakly from the $^1S_0$ state. At least four different explanations have been offered: In [20], the $\rho\pi$ *puzzle* is related to the $\bar{q}q$ annihilation vertex which is supposed to carry the quantum numbers of a single gluon. Maruyama et al. [21] assume that the $\bar{q}q$ annihilation vertex carries the quantum numbers of the vacuum; ${}^3S_1$–${}^3D_1$ mixing in the $\bar{p}p$ initial–state wave function plays a decisive role to reproduce its large $\rho\pi$ branching ratio. Mull et al. [22] find the reason for the $\rho\pi$ *puzzle* in final–state interactions between the outgoing π and ρ mesons. Of course, the different interpretations of the $\rho\pi$ *puzzle* are mutually exclusive. No dynamical explanation of the effect is offered here; instead, the selection rule is related to other double branching ratios into two isovector mesons.

Annihilation into two Isovector Mesons.

- ${}^3P_{0,2}/{}^3S_1$: $\quad 14 \times \frac{\bar{p}p \to \pi^+\pi^-(I=1)}{\bar{p}p \to \pi^+\pi^-(I=0)} = 0.16 \pm 0.04$
- ${}^{3,1}S_0/{}^{1,1}S_0$: $\quad 3 \times \frac{\bar{p}p \to \rho^\pm\pi^\mp(I=1)}{\bar{p}p \to \rho^\pm\pi^\mp(I=0)} = 0.08 \pm 0.04$
- ${}^{3,3}S_1/{}^{1,1}S_0$: $\quad 1/3 \times \frac{\bar{p}p \to \pi^\pm a_2^\mp(I=1)}{\bar{p}p \to \pi^\pm a_2^\mp(I=0)} = 0.10 \pm 0.04$

The normalization factors 14, 3, 1/3 are independently determined weight factors. The symbols on the left characterize the $\bar{p}p$ atomic state from which annihilation is considered.

Even more striking are the selection rules observed in the production of strange mesons. Most models predict both isospins to contribute to $\bar{p}p$ annihilation leading to a small rate for production of two neutral kaons compared to two charged kaons. However, this is not seen in experiments.

Annihilation into $K\bar{K}$ and into $K^*\bar{K}$+c.c.

- ${}^3S_1$: $\quad \frac{\bar{p}p \to K\bar{K}(I=0)}{\bar{p}p \to K\bar{K}(I=1)} = 0.23 \pm 0.20$
- ${}^3S_1$: $\quad \frac{\bar{p}p \to K^*\bar{K}(I=0)}{\bar{p}p \to K^*\bar{K}(I=1)} = 0.03 \pm 0.02$
- ${}^3P_{0,2}$: $\quad \frac{\bar{p}p \to K\bar{K}(I=0)}{\bar{p}p \to K\bar{K}(I=1)} = 0.05 \pm 0.04$
- ${}^1S_0$: $\quad \frac{\bar{p}p \to K^*\bar{K}(I=1)}{\bar{p}p \to K^*\bar{K}(I=0)} = 0.23 \pm 0.07$

3.3 *Data on $\bar{p}p$ Annihilation at Rest*

The branching ratios of table 1 are fitted to SU(3) matrix elements by use of

$$BR(\bar{N}N \to M_a M_b) = \frac{1}{2^{\delta_{ab}}} \cdot F_l \cdot F_{\mathrm{HFS}} \cdot F_D \cdot f_{\mathrm{dyn}} \cdot |SU(3)|^2 \qquad (1)$$

The factor $2^{\delta_{ab}}$ takes care of Bose symmetry in case of two identical mesons. $F_{l=1} = 0.12$ represents the fraction of P–wave annihilation when antiprotons stop in liquid H_2. F_{HFS} counts the number of hyperfine structure states which contribute to the reaction (e.g. $F_{\mathrm{HFS}} = 6/12$ for $\bar{p}p \to \pi^0\pi^0$ allowed from five $^3P_2$ states and one $^3P_0$ state). The P–wave contribution in liquid D_2 is enhanced compared to liquid H_2 by a factor of $F_D = 1.73$. For the present analyis $f_{\mathrm{dyn}} = 1$ is assumed. The SU(3) matrix element is calculated for the different coupling schemes. It is assumed to have an error of ±20% due to SU(3) breaking effects.

Table 1. Branching ratios for $\bar{p}$N annihilation at rest into two pseudoscalar mesons and into one pseudoscalar and one vector mesons (in units of 10^{-4} [23])

Initial state		Branching ratio		Branching ratio
$\bar{p}p(^3P_{0,2})$	$\to \pi^0\pi^0$	(6.93 ± 0.43)	$\to \pi^0\eta$	(2.12 ± 0.12)
$\bar{p}p(^3P_{0,2})$	$\to \pi^0\eta'$	(1.23 ± 0.13)	$\to \eta\eta$	(1.64 ± 0.10)
$\bar{p}p(^3P_{0,2})$	$\to \eta\eta'$	(2.16 ± 0.25)	$\to K^+K^-$	(0.83 ± 0.05)
$\bar{p}p(^3P_{0,2})$	$\to K^0\overline{K}^0$	(0.26 ± 0.10)		
$\bar{p}p(^3S_1)$	$\to \pi^+\pi^-$	(30.7 ± 1.3)	$\to K^+K^-$	(9.9 ± 0.5)
$\bar{p}p(^3S_1)$	$\to K^0\overline{K}^0$	(9.0 ± 0.6)		
$\bar{p}n(^3P_{0,2})$	$\to \pi^-\eta$	(3.9 ± 0.6)	$\to \pi^-\pi^0$	(1.9 ± 0.7)
$\bar{p}n(^3S_1)$	$\to \pi^-\pi^0$	(4.0 ± 0.5)	$\to K^-K^0$	(8.4 ± 1.6)
$\bar{p}p(^3S_1)$	$\to \pi^0\rho^0$	(165 ± 8)	$\to \eta\omega$	(151 ± 12)
$\bar{p}p(^3S_1)$	$\to \eta'\omega$	(78 ± 8)	$\to \pi^0\omega$	(57.3 ± 4.7)
$\bar{p}p(^3S_1)$	$\to \eta\rho^0$	(46 ± 5)	$\to \eta'\rho^0$	(18.1 ± 4.4)
$\bar{p}p(^3S_1)$	$\to \pi^0\phi$	(5.5 ± 0.7)	$\to \eta\phi$	(0 ± 0)
$\bar{p}p(^{3,3}S_1)$	$\to K^*\overline{K}$+cc	(24.6 ± 2.4)		
$\bar{p}p(^{1,3}S_1)$	$\to K^*\overline{K}$+cc	(0.8 ± 0.4)		
$\bar{p}p(^1S_0)$	$\to \rho^\pm\pi^\mp$	(9.2 ± 4.0)		
$\bar{p}p(^{3,1}S_0)$	$\to K^*\overline{K}$+cc	(2.0 ± 0.4)		
$\bar{p}p(^{1,1}S_0)$	$\to K^*\overline{K}$+cc	(8.6 ± 1.7)		
$\bar{p}n(^{3,3}S_1)$	$\to \pi^-\omega$	(59 ± 10)		

3.4 Fits with Different Coupling Schemes

The three coupling schemes are fully equivalent when no further assumptions are introduced. In case of two pseudoscalar mesons there are 14 data points, which are used to fit 7 SU(3) amplitudes, one mixing angle (Θ_{PS}), and F_D. There are several solutions all resulting in the same χ^2 of 1.0 for 5 degrees of freedom (N_F). The 13 data points on annihilation into one vector and one pseudoscalar meson ($\phi\pi$ is left out!) yield a χ^2 of 0.42 for $N_F = 3$. Again, several solutions are found with the same χ^2. There are now eight SU(3) amplitudes to be determined and two mixing angles. The mixing angles are compatible with our present-day knowledge. In both cases, any interpretation of the results is ruled out because of the variety of the solutions. Next we discuss fits with reduced numbers of amplitudes in the different coupling schemes.

The Quark Coupling Scheme

The quarks line diagrams of this scheme are presented in Fig. 2. Diagrams which would correspond to OZI violating amplitudes are not shown; their expansion coefficients are set to zero. The validity of the OZI rule in $\overline{N}N$ annihilation is presently under debate (see, e.g. [25]) but the maximum OZI violation observed is at the 10% level; neglecting OZI violation is thus tolerable. Since the different coupling schemes are related via transformation matrices, the OZI rule can be imposed in a straightforward way to all three coupling schemes.

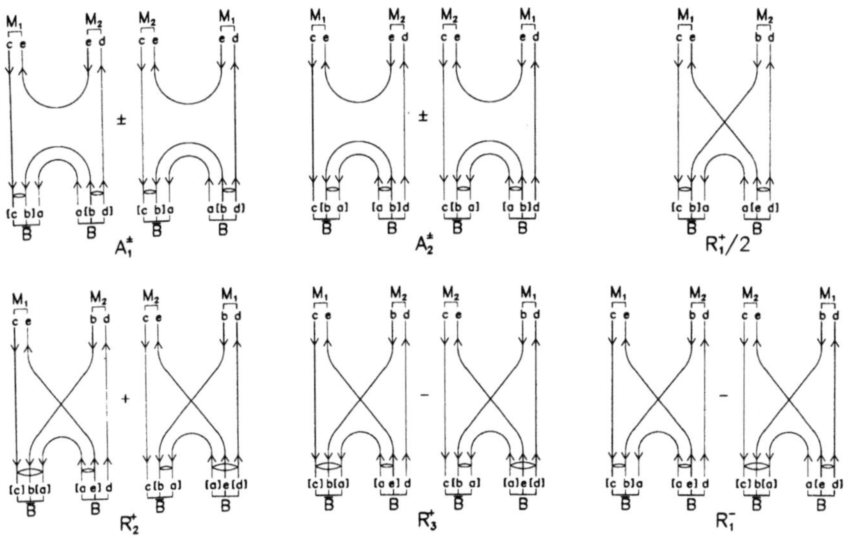

Figure 2. Quark-line-diagrams in the quark coupling scheme

There are 8 independent couplings, 4 of annihilation- and 4 of rearrangement-type. The data are compatible with the assumption that

$$A_1^{\pm} + A_2^{\pm} = 0 \qquad A_1^{+} \sim A_1^{-} \qquad R_2^{+} = R_3^{+} = R_1^{-} \ll R_1^{+}$$
$$A_1^{+}/R_1^{+} \sim 0.55(\mathrm{PSPS}) \text{ and } 0.35(\mathrm{PSV})$$

Imposing these equations leads to a χ^2 of 10 with now $N_F = 14$.

The s-Channel Coupling Scheme

The s-channel coupling scheme can be visualized by deforming the quark lines of the quark coupling scheme as shown in Fig. 3. The 8 SU(3) amplitudes now describe formation of $q\bar{q}$ and of $q\bar{q}q\bar{q}$ systems with defined flavour symmetry. The amplitudes are:

$$S_1^+, S_{ss}^+, S_{sa}^+, S_{as}^+, S_{10-\overline{10}}^+, S_{27}^+, S_{as}^-, S_{aa}^-, S_{10+\overline{10}}^- \qquad (2)$$

where $s(a)$ represent (anti–) symmetric couplings at the octet+octet→octet vertices. As mentioned above this coupling scheme gives the same χ^2 as above as long as all amplitudes are allowed.

This scheme permits to set to zero possible contributions from the "exotic" exchanges $S_{10-\overline{10}}^+, S_{27}^+$ and $S_{10+\overline{10}}^-$. Suppressing these amplitudes yields a χ^2 of 19.5 for $N_F = 13$. The data suggest no symmetry relations between the different octet couplings.

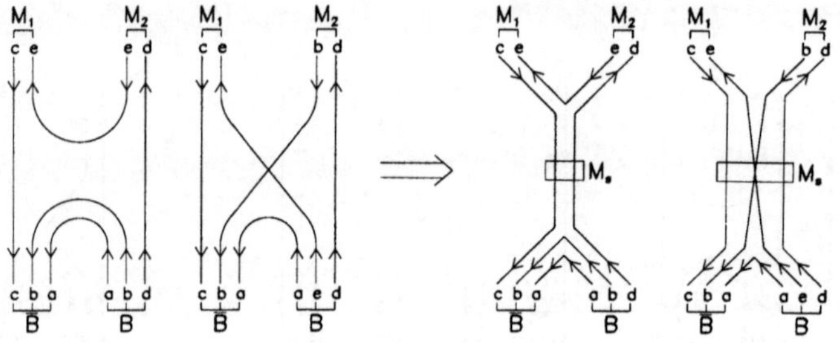

Figure 3. Connection between diagrams in the quark– and the s–channel coupling schemes

The Baryon–Exchange Coupling Scheme

The t– and u–channel coupling scheme (Fig. 4) is closely related to the concept of baryon exchange. The amplitudes can be written down in exactly the same form as in case the s–channel coupling scheme, with S_1^+ replaced by B_1, etc. Note that C–parity is no longer a good quantum number. The full scheme results, of course, again in identical χ^2's. We may however impose the absence of contributions from $B_{\overline{10}}, B_{27}$ and require constant F/D ratios at the baryon–baryon–meson vertices. This results in an unacceptable χ^2 of 190 for $N_F = 15$.

Conclusions

Antiproton–proton annihilation is an extremely rich field. In meson spectroscopy new mesons were discovered, the $f_0(1500)$ being particularly interesting since this state could be the long awaited ground–state scalar glueball. The excellent antiproton–beam quality and high rates permit to stop antiprotons in small targets and to trigger on rare annihilation modes likes $\bar{p}p \to \pi^0\eta\eta'$. Thus decay modes of new mesons can be searched for which are important for the interpretation of these states.

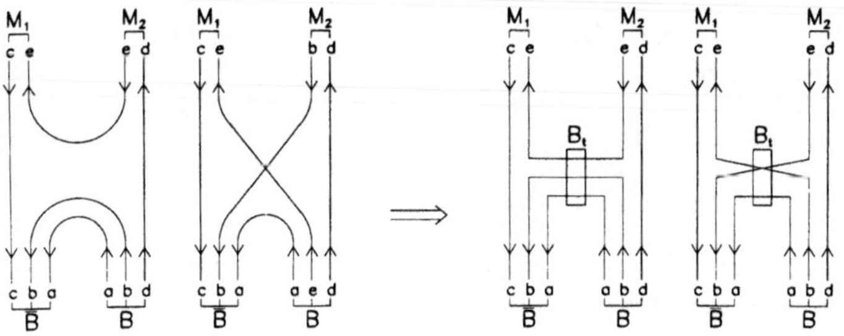

Figure 4. Connection between diagrams in the quark– and the baryon–exchange coupling schemes

I could not cover in my talk results on tests of chiral dynamics by precision measurements of η decay rates [26], on searches for new gauge bosons [27] and on systematic investigations on the role of the OZI rule for baryons [25]. Antiproton annihilation in deuterium provides a unique chance to search for new baryon resonances not coupling to the πN system. Data on $\bar{p}d \rightarrow \pi\pi N$ show clear evidence for production of the $\Delta(1232)$ and of higher nucleon resonances [28]; the statistics does however not yet allow a spin–parity analyis. High–statistics studies on reactions like $\bar{p}d \rightarrow \pi\omega N$ might reveal new baryon resonances decaying into $N\omega$. The search for these states is one of the topical issues of new electron accelerators like CEBAF. At Lear, these studies will not be pursued because of its premature closure.

Last not least, the $\bar{p}N$ annihilation process might give us unexpected insights into the dynamics of quarks and gluons in the confinement region. Most theoreticians refuse the possibility that quark effects might show up at these low energies. However, one of the founders of SU(3), Y. Ne'eman, used data on $\bar{p}p$ annihilation at rest to demonstrate that the Sakata model was wrong and the "eightfold way" was correct [29]. Personally I am still convinced that Ne'eman, was right.

References

1. C. Amsler et al.: *Coupled channel analysis of $\bar{p}p$ annihilation into $\pi^0\pi^0\pi^0$, $\pi^0\eta\eta$ and $\pi^0\pi^0\eta$.* Phys. Lett. **B** and references therein, in print

2. C. Amsler et al.: Phys. Lett. **B333**, 227 (1994)

3. H. Fritzsch, M. Gell-Mann, XVI Int. Conf. on High Energy Physics: Chicago - Batavia, Vol 2, 135 (1972)

4. C. Michael, M. Teper: Nucl. Phys **B314**, 347 (1989)

5. S.S. Gershtein et al.: Z. Phys. **C24**, 305 (1984)

6. St. Resag (Crystal Barrel Collaboration), contribution to Hadron95, Manchester 1995

7. S. Abatzis et al.: Phys. Lett. **B324**, 509 (1994)

8. D.V. Bugg: private communication

9. S. Hainzl et al.: *Scalar mesons in a relativistic quark model with instanton-induced forces*, in preparation

10. D. Morgan, M.R. Pennington: Phys. Rev. **D48**, 1185 (1993)

11. H. R. Rubinstein, H. Stern: Phys. Lett. **21**, 447 (1966)

12. E. Klempt, J.G. Körner, F. Walter, *Antiproton–proton annihilation into two mesons: SU(3) coupling schemes for quark-line rule, s-channel dominance and baryon exchange* submitted to Z. Physik **A**

13. A. Rotondi, in: Proc. LEAP'92, Courmayeur 1992, eds. C. Guaraldo, F. Iazzi, A. Zemoni: North Holland, 1993

14. E. Klempt: Phys. Lett. **B308**, 174 (1993)

15. E. Klempt: Phys. Lett. **B308**, 179 (1993)

16. C. Amsler et al.: Phys. Lett. **B294**, 451 (1992)

17. C. Amsler et al.: Z. Physik **C58**, 175 (1993)

18. E. Klempt: Phys. Lett. **B244**, 122 (1990)

19. J. Vandermeulen: Z. Phys. **C37**, 563 (1988)

20. J.A. Niskanen, F. Myhrer: Phys. Lett. **B157**, 247 (1985)

21. M. Maruyama et al.: Phys. Lett. **B215**, 233 (1988)

22. V. Mull et al.: Phys. Lett. **B347**, 193 (1995)

23. F. Walter: Dissertation, Mainz, unpublished

24. C. Amsler et al.: Phys. Lett. **B297**, 214 (1992)

25. J. Ellis, M. Karliner, D.E. Kharzeev, M.G. Sapozhnikov: CERN-TH.7326/94

26. C. Amsler et al.: Phys. Lett. **B346**, 203 (1995)

27. C. Amsler et al.: Phys. Lett. **B333**, 271 (1994)

28. C. Amsler et al.: Phys. Lett. **B352**, 187 (1995)

29. Y. Ne'eman: Nucl. Phys. **26**, 22 (1961) and private communication

Few-Body Systems Suppl. 8, 287–298 (1995)

η Production via the Strong Interaction

A. Boudard*
SPhN-DAPNIA-CEA, CEN Saclay, 91 191 Gif-sur-Yvette, FRANCE

Abstract. The η production is characterized by a strong cross section close to the threshold. The S_{11} nucleon resonance has a specific role in this domain with important consequences below the elementary threshold. The mass of the η, well above the π mass, provides interesting informations on the sharing of the momentum transfered in reactions involving a small number of nucleons. The η production as a possible source of e^+e^- is also studied in the context of dileptons produced by heavy ion collisions of a few GeV per nucleon. As a practical consequence, the production of η from baryonic interaction is now used to study the physics of η decays.

A short review of the present knowledge and activities on the η production from hadronic interaction in the energy range of 1-2 GeV is discussed with both experimental and theoretical aspects.

1 Introduction

The interest for the production of η mesons does not come from its importance in nucleus. In ordinary nuclear matter it can be neglected due to its weak coupling with the nucleon. It affects rather little the NN interaction. However it is with pions and kaons, a member of the octet of the light pseudo-scalar mesons, fundamental fields of the Chiral Perturbation Theory.

Its production in the reaction $pd \to^3 \mathrm{He}\eta$ at backward angles was found extremely structured as a function of the proton energies and sometime larger than the corresponding production of π^0 [1]. Due to its isospin the η is only coupled with N^* resonances and not with Δ as it is the case for the π. So the η production will provide information on the $N - N^*$ interaction.

It will also provide new tests of the isobaric models. The mass of the η, well above the pion mass gives large differences in the dynamics of the exclusive reactions. For the η production, a larger fraction of the incoming energy is converted in mass. This induces generally a larger momentum transfer (Table 1) than for the pion production and favors multi-step processes. So the

* *E-mail address:* boudard@phnx7.saclay.cea.fr

comparison between η and π production provides informations on the reaction mechanism and on the relative excitation of N^* and Δ in nuclei.

Table 1. Total cross section in the vicinity of the threshold for various two body reactions. The initial momentum in the center of mass (k_i^{cm}) is also, at threshold, the momentum transfer.

Reaction	Kinetic threshold energy in the lab. (MeV)	k_i^{cm} (fm^{-1})	$\sigma_{Tot.}$
$\pi^- p \to \eta n$	559.3	2.16	1 - 2 mb
$np \to d\eta$	1249.6	3.83	$\sim$ 100 μb
$pp \to pp\eta$	1254.6	3.83	≤ 1 μb
π^{-3}He $\to \eta$H^3	458.4	2.43	$\sim$ 80 μb
$pd \to^3$ Heη	891.4	4.39	0.4 μb
$pd \to^3$ Heπ^0	198.8	2.0	
$dd \to^4$ Heη	1120.3	5.12	$\sim$ 10 nb

For this reason the η production is considered in the collision of heavy ions of a few GeV per nucleon. Below the elementary threshold, it could give informations on the equation of state [2] and on the fraction of N* excited during the collision [3]. Through its Dalitz decay it is a large fraction of the dilepton spectra, one of the main hope to extract a clean information from the high nuclear matter density produced during the collision.

From the experimental point of view, the narrow width of the η makes it easy to identify in missing mass or invariant mass spectra. However, in spite of its interest [4], a beam of η is very difficult to handle.

2 The Primary Process $\pi N \to \eta N$.

This is the basic process to produce the η within the strong interaction. If the π is not the beam, it comes from the NN interaction. Then the two step process ($NN \to NN\pi + \pi N \to \eta N$) compete with the direct one ($NN \to NN\pi$) to produce the final η.

The reverse process is also a large fraction of the η absorption. It should be complemented at least by the process $\eta N \to N^*(S11) \to \pi\pi N$).

In the frame work of an isobaric model coupling the channels πN, ηN and $\pi\Delta$, Bhalerao and Liu [5] have adjusted the masses, the coupling constants and the cutoff parameters for the S_{11}, P_{11}, D_{13} and P_{33} resonances on the πN scattering phase shifts. The ηN interaction is found dominated, close to threshold, by the S_{11} resonance and higher in energy by the P_{11} (1710 MeV) (A compilation of experimental results for this reaction can be found in [9]). The D_{13} while weakly coupled to the η is found essential to reproduce the asymmetry of the cross section close to the threshold ($\sqrt{s} = 1542$MeV). In this study, it was deduced that the ηN interaction is attractive in the S wave with a scattering length

$$a_{\eta N} = (0.28 + i0.19)\text{fm}$$

This opened the possibility of η-nucleus bound states. Some predictions on bound η-nucleus states were done for $A > 12$ [6] with a potential possibility to study the ηN interaction in the nucleus. A binding of a few to 22 MeV in the S state was found but with the possibility of rather large width which will make those states very difficult to observe [7] [8].

More recently, Sauermann et al. [10] in view of the analysis of η inclusive production induced by p and heavy ion collisions, have parameterized the elementary $\pi N \rightarrow \eta N$ production in the S_{11} channel. In a T-matrix formalism, they consider two resonances (S_{11} (1535) and S_{11} (1650)) with there coupling to π, η and 2π. The parameters (coupling constants and masses) are adjusted on the elastic πN T-matrix amplitude up to $\sqrt{s} = 1.75\text{GeV}$. The parameters found are slightly model dependent (one ore two resonance considered) but the agreement on $\sigma_{Tot}(\pi^- p \rightarrow \eta n)$ is very good up to $p_{\pi lab} = 0.9\text{GeV/c}$. Above, the $P_{11}(1710)$ is needed. They found a larger scattering length

$$a_{\eta N} = (0.51 + i0.21)\text{fm}$$

than Bhalerao et al, in favor of a stronger ηN interaction.

The experimental search for the η nucleus bound states have been done with controversial results in $^{16}\text{O}(\pi^+, p)^{15}\text{O}$ at LAMPF [11] and in $^{12}\text{C}(\pi^+, p)^{11}\text{C}$ at BNL [12]. A test in $^{12}\text{C}(p, 2p)^{11}\text{B}$ at SATURNE [13] was also negative. More recently, a possible signal is found in the double charge exchange process $^{18}\text{O}(\pi^+, \pi^-)^{18}\text{Ne}$ [14]. The cross-section to the double isobaric analog states integrated on 7 MeV and for no momentum transfer has a resonance like picture for an energy of the pion beam of 420 MeV (width of 25 MeV). This interesting effect is still to be confirmed with at least a better statistics.

As a conclusion of this process we keep that very close to the threshold, the S_{11} (1535) resonance is crucial. Note that while its branching ratio to ηN is around 50%, through its width, 30% of this resonance remains below the threshold. It means that for an off shell η, the effect of the resonance is even larger. The ηN interaction is attractive in the S wave, but there is also a strong absorption of the η (at least all the other decaying channels of the S_{11}).

3 Production from NN Interaction.

3.1 $pp \rightarrow pp\eta$

Rather few was known about this interaction near threshold before the works done at SATURNE by the SPES 3 collaboration and the PINOT collaboration. The magnetic spectrometer SPES 3 has a large solid angle ($10^{-2}\text{sr}, \Delta\theta = 5.7^0$, $.6\text{GeV/c} < \Delta p < 1.4\text{GeV/c}$). Close to threshold, the two protons are detected without cut in θ. The identification of the η is done by the missing mass. After a careful treatment of the acceptance (including the distortion of the phase space due to the pp and the pη interaction [15]), the total cross section is deduced at three energies (1260, 1265 and 1300 MeV) close to the threshold (1254.6 MeV) Table 2. Note that in [16], a pure phase space was used to evaluate the acceptance.

Table 2. Total cross section for $pp \to pp\eta$ measured at SPES 3 [15]

Energy (MeV)	Q_{cm} (MeV)	True energy (MeV)	$\sigma_{pp\to pp\eta}$ (nb)
1260	0.64 ± 0.25	1257.3 ± 0.7	87 ± 14
1265	2.65 ± 0.25	1262.6 ± 0.7	742 ± 113
1300	16.0 ± 0.60	1297.7 ± 1.6	2060 ± 415

The PINOT detection is made of two arms rotating independently around the target. Each of them measure the energy of a γ from the $\eta \to \gamma\gamma$ decay and its direction. The η is identified by the invariant mass of the two γ. Then with various positions of the two arms, it is possible to detect the η at various angles (up to 40^0 to 50^0) and from 0 to 400 MeV. The agreement with the SPES 3 data at 1265 and 1300 MeV is excellent. In addition, they have measured the total cross section at 1400 and 1500 MeV [17]. The measurements at 1600 and 2000 MeV are still preliminary, but will bridge the gap from threshold to the existing data at higher energy [18] (Fig. 1).

At least three different calculations have been tried to describe this reaction in an isobaric approach.

Very close to threshold, Germond and Wilkin [19] include π, ρ, ω and η exchange to excite only the S_{11} resonance. They evaluate initial and final state interaction but don't use form factors at the vertex. The coupling constant ρNN^* is taken from vector meson dominance. It is a strong coupling and the ρ exchange is found to be the dominant contribution. Due to the different approximations this calculation is only valid very close to threshold. With a more sophisticated treatment of the final state interaction (interference of the pp and pη re-scattering), Moalem et al. [20] could get with basically the same model a good fit up to 15 MeV above threshold.

A more ambitious calculation [21] treats the $\pi N \to \eta N$ subprocess through its partial wave analysis up to the F wave. Form factors are included at the vertex. The ρ coupling is also strong. This calculation should be valid up to higher energies. Actually it gives a rather correct description of the total cross section up to 2 GeV. It is a common finding of [19], [21] and [20] that the interference between the π and the ρ exchange (not fixed by the theory) should be destructive in the case of $pp \to pp\eta$ and constructive for the $np \to d\eta$ process. They obtain also the correct magnitude with the unique measurement of $np \to d\eta$ [22].

The model of Vetter et al [23] consider the exchange of π, ρ, ω and σ with form factors, but with only the excitation of the S_{11} resonance. The ρNN^*

coupling constant is there taken directly from the decay of the S_{11} in the $N\rho$ channel. It is weaker by about a factor 2.7 than in the above mentioned works. Close to threshold this calculation is much below the data, but it is surprising that it gives a rather good overall agreement up to very high energies ($\sqrt{s} = 3.2$GeV). The absence of final state interaction could also be a reason for the poor agreement close to the threshold. Another problem for this model is that it gives a ratio of the order of 2 or 3 between the $pn \rightarrow pn\eta$ and the $pp \rightarrow pp\eta$ total cross section at $T_p = 1.5$GeV. We will see that the experimental value is around 5.

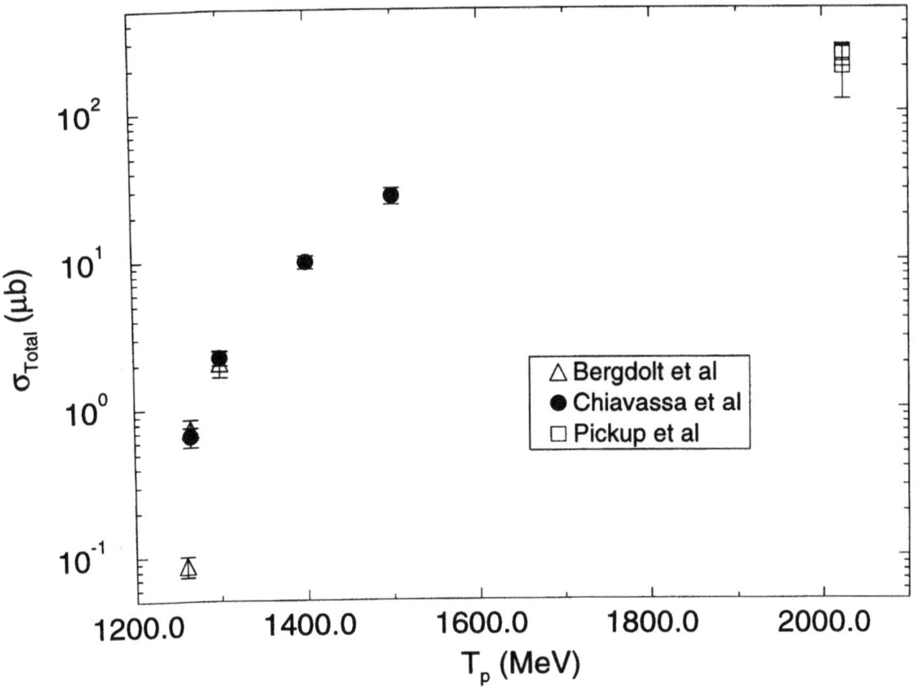

Figure 1. Total cross section of $pp \rightarrow pp\eta$ from the PINOT [17] and SPES 3 [16] measurements. The measurement of Pickup [18] is also shown.

3.2 $np \rightarrow d\eta$

We have still mentioned the unique measurement of $np \rightarrow d\eta$ ($110 \pm 20\mu b$) for a deuteron center of mass momentum of 35 MeV/c. It should not be forgotten that this value (surprisingly high) has been obtained from old data [24] with a central value of the neutron beam momentum (1.9 GeV/c) below the threshold (1.973 GeV/c). The neutrons were obtained from the breakup of a deuteron beam. It is only through the width of the distribution that the η could be produced at very small angle. Physical hypothesis on the various contributions to the spectra where done to extract a value for this cross section. It should certainly be measured directly, first to confirm the large value and also to check the strong energy dependence predicted by the theory [21].

3.3 $pn \to pn\eta$

To obtain information on this total cross section, the PINOT [25] collaboration has measured the η produced from a hydrogen and a deuterium target in the same conditions ($0^0 < \theta_\eta < 14^0, 100 < T_\eta < 220\text{MeV}$), and for a proton beam of 1.3 and 1.5 GeV. Normalized to the same number of incident protons and target nucleons, the production is clearly much higher on the deuterium target. Using a crude folding model on the elementary pp and pn cross sections, the ratio of $\sigma(pn)$ to $\sigma(pp)$ is found to be 5 ± 1 at 1.5 Gev and 10 ± 2 at 1.3 GeV. This non constant ratio is indicative of some dynamics possibly due to the $np \to d\eta$ channel or to final state interaction. An isoscalar exchange gives a ratio equal to 1, an isovector exchange, a factor 5.

4 Production from Light Nuclei in Exclusive Processes

4.1 $pd \to {}^3\text{He}\eta$, $\pi^- {}^3\text{He} \to \eta {}^3\text{H}$

The reaction $pd \to^3 \text{He}\eta$ is especially interesting for the reaction mechanism. At threshold, the momentum transfered is 4.4 fm^{-1}, more than two time larger than for the π^0 production at his threshold. At backward angles, the cross section was found rather large and the excitation function extremely structured [1]. Opposite to the $pd \to^3 \text{He}\pi^0$ at his threshold, the angular distribution remains isotropic while changing in magnitude with the beam energy much faster than a pure S-wave phase space. An isobaric model gives the gross structure of the excitation function at backward angles [26], stressing the importance of a three nucleon mechanism to share the large momentum transfer. However the precise experimental values close to threshold was poorly reproduced by this model. It was noticed [27] that at this energy, the two step process ($pp \to d\pi^+$ followed by $\pi^+ n \to \eta n$) involves an intermediate pion almost on shell. But to reproduce the very fast energy dependence of the cross section at threshold it was necessary to consider the final state interaction, and to treat unperturbatively the $\eta^3\text{He}$ interaction [28] [29]. With a scattering length

$$a_{\eta N} = (0.55 + i0.30)\text{fm}$$

adjusted on the $\pi^- p \to \eta n$ cross section (larger than in [5] but close to the one found recently by [10]), Wilkin gets through an optical potential a scattering length

$$a_{\eta^3\text{He}} = (-2.31 + i2.57)\text{fm}$$

The negative sign is indicative of a possible bound state. A pure fit to the last data analysis of the SPES 2 data [30] gives:

$$a_{\eta^3\text{He}} = (|3.77 \pm 0.49| + i(1.62 \pm 0.88))\text{fm}$$

related to thc cross section by:

$$\sigma = \frac{k_\eta}{k_p} \left| \frac{cte}{1 - ik_\eta a_{\eta^3\mathrm{He}}} \right|^2$$

It should be noted that the experiment does not state on the sign of the real part, but the calculation goes as well as the fit through the experimental values (look at Fig. 2). Several theoretical works on this question of quasi bound $\eta -^3$ He and $\eta -^4$ He are presented at this conference [31].

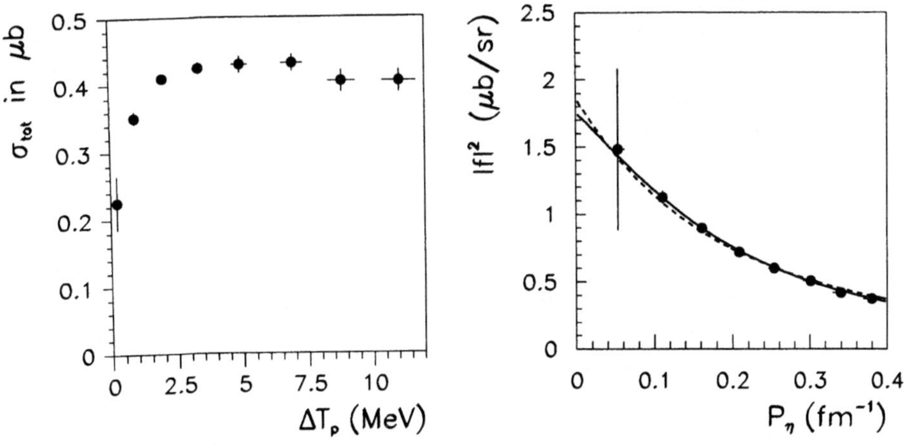

Figure 2. Total cross section of $pd \rightarrow^3$ Heη from the SPES 2 measurements [30] as a function of the beam kinetic energy above threshold. The nuclear matrix element $|f|^2$ is plotted on the right as a function of the outgoing momentum in the center of mass and compared to the fit (continuous line) and to the calculation [28] (dashed line).

It will be very interesting to confirm the possibility of a bound state for such a light nuclei. unfortunately, the $\pi^{-3}\mathrm{He} \rightarrow \eta^3\mathrm{H}$ [32] has not yet been measured close enough to threshold. For this reaction, the cross section computed at 0^0 close to threshold in a sophisticated coupled channel calculation with Faddeev wave functions underestimate the data by a factor $\sim$ 4, while being rather good at 40 MeV above threshold [33]. A better agreement was obtained [34] in a calculation stressing the importance of a coherent process of production on two nucleons (coupled to $J = 0, T = 1$) and a negligible contribution for a 3 nucleon process. But the momentum transfer in this reaction is two time smaller than for the $pd \rightarrow {}^3\mathrm{He}\eta$.

4.2 $dd \rightarrow \alpha\eta$

This reaction has been measured at four energies close to threshold by the detection of the α particle in the spectrometer SPES 4 [35]. The cross sections of the order of 10 nb have too large error bars on a too small range of energies for a conclusion about the η^4He interaction. They are compatible with a constant nuclear amplitude as well as with the predictions [28] of a scattering length

$$a_{\eta^4\mathrm{He}} = (-2. + i1.)\mathrm{fm}$$

based on the same model as for the $pd \rightarrow {}^3\mathrm{He}\eta$ reaction.

Another measurement at SPES 3 [36] not yet fully analysed, will give more informations. The cross section is there measured with a much larger solid angle (10 to 100 more). This allowed to have a larger range in energy above threshold. The total cross section has been measured at eight energies from 1120.4 to 1200 MeV, corresponding to a final center of mass momentum from 6 to 185 MeV/c. In this experiment the incident beam is polarized. This will bring with analyzing powers more constraints for the models. The mean tensor analyzing power A_{yy} is found rather large (-0.5 at 1123 MeV).

5 Inclusive η Production from Proton-Nucleus Interaction

The PINOT collaboration has provided useful data [38] [39] [40] [41] for the production of η meson from the proton-nucleus interaction around 1 GeV. At 1 GeV, below the elementary threshold (1.25 GeV), the A dependence of η produced in the range 30-100 MeV and 0^0-40^0 laboratory angle is measured. The cross section saturates and is interpreted a strong absorption of the η in the nuclear medium. Another way is to note that the data are very close to a low in $A^{2/3}$ which means a surface effect. This aspect is more seriously obtained in intra-nuclear cascade models [37]. It should be noted that in this paper some doubt exist on the conclusion about the ηN cross section found twice smaller than in the predictions of Liu. This doubt comes from the fact that the $NN \rightarrow NN\eta$ cross section used at 1 GeV is two large by a factor around 4 compared to the PINOT data. So the negative consequence for the possible existence of η-nucleus bound states cannot be taken as final.

The Giessen folding model [42] [43] with proper primary cross sections and Hartree-Fock momentum distributions of the nucleons gives a rather correct cross section for a light nucleus (Carbon) above the elementary threshold. Below threshold (at 1 GeV for example as a function of A) it underestimates the production by a factor of 2 for all nuclei. It is a coherent feature seen also on the energy spectrum of η produced at 1 GeV from ^{10}B. This could be due to a lack of high momentum components in the Hartree-Fock density or to a bad account of the self energy of the η. This last mechanism was recently stressed [44] to give more cross section below the resonance.

Above the elementary threshold (1.3 GeV for example),the folding calculation is consistently too large especially for heavy nuclei. This can be cured by increasing the ηN absorption (taken originally from [5]) by a factor two (consistent with the scattering length of [10] and [28]. The PINOT experiment is still taking data this year. At the total, they will produce a large amount of results on the η production from protons on many nuclei (from Lithium to Gold) and for energies from 0.8 GeV to 1.5 GeV.

6 The η Production in the Physics of Heavy Ion Collisions

In the Physics of heavy ion collisions for 1 or 2 GeV per nucleon, one of the more interesting question is to obtain information of the high nuclear matter density (of the order of two times the normal one) produced during the collision. The relative production of η to π^0 due to there specific coupling to $N^*(1535)$ and $\Delta(1232)$ gives an information on the relative abundance of these resonances (recently measured around 1%) [3].

The η is also a potential candidate to obtain information on the high density produced during the collision. This has been studied with transport equation (BUU) and dynamical evolution of the elementary processes involving Δ and N^* [2]. In this calculation the η appears to be formed during the high density phase, and the outgoing η's have experienced rather few collisions (contrary to the π^o's). However, due to its strong absorption, the number of outgoing η is smoothed and rather few density dependent.

More hopes are based on the dileptons which, as electromagnetic particles, interact weakly after there formation and can be classified with respect to there invariant mass. The η is found to be the main contributor to this spectra (through its Dalitz decay) up to missing mass of 450 MeV. The published data on η production by heavy ions remains rather scarce at the moment but will certainly come soon from GSI.

7 Strangeness of the η

The strange component of the η is frequently mentioned, but it does not appear explicitly in the experiments we have mentioned up to now. The strange component is only hidden in the values of the various coupling constants.

An interesting suggestion was done [4] to use it explicitly. Considering the $s\bar{s}$ component in the η and the η', and in the limit of the OZI rule, it will be possible to have a sensitivity to the $s\bar{s}$ quark component of the proton. While the ratio $\sigma(\pi^- p \to \eta' n)/\sigma(\pi^- p \to \eta n)$ is sensitive to $s\bar{s}$ in a 0^{-+} state, the ratio $\sigma(\eta p \to \eta' p)/\sigma(\eta p \to \eta p)$ will be sensitive to a 0^{++} coupling of the $s\bar{s}$.

In spite of the very attractive idea, the measurement appears rather difficult. To be done, it needs a tagging of the η produced with an energy above the $\eta p \to \eta' p$ threshold. With this energy, the η will travel only a few Å before decaying. It will make the effective proton target very thin. This energy of the η means also that the primary tagging reaction should be well above its threshold with the consequence of a large emission angle for the tagging particle in the lab and probably a smaller cross section.

8 Conclusions

In summary, the production of the η in the energy range of a few GeV is governed by the S_{11} resonance very close to the ηN threshold. This can be corrected in addition by the P_{11} and the D_{13} resonances.

A strong attraction ηN is found with the possibility of bound or quasi bound states. The η^3He seems a good candidate while it was thought before that only heavier nuclei (at least ^{12}C) could possibly be bound with η. This should certainly be confirmed experimentally in other channels like the $\pi^{-3}\text{He} \to \eta^3\text{H}$ with measurements closer to its threshold. The precise study of the $dd \to \alpha\eta$ will also be a test for the coherence of the theoretical approach.

For the production from the NN interaction a coherent picture is obtained with a strong coupling of the ρ evaluated from vector meson dominance and with a constructive $\pi\rho$ interference for the np interaction and a destructive one for the pp. A precise, energy dependent experiment on the $np \to d\eta$ process will be certainly welcome for a better knowledge of the production from the NN interaction.

In the production on nuclei, the two step process with an intermediate π is favored. The η is strongly absorbed through the $\eta N \to \pi N, 2\pi N$ processes. Below the elementary threshold the experimental η production is frequently above the calculation, but the self energy of the η is to be investigated in more details. It could result in some information on the NN^* interaction.

The coupling constant $g_{\eta NN}$ poorly known at present could be determined with reactions like $^4\text{He}(p, p'\eta)^4\text{He}$ [44].

For the inclusive production in p-nucleus interaction, an important amount of data has and will be provided soon by the PINOT collaboration.

The strangeness in the η is still to be used in some cleaver experiment.

I apologize to have given more details on the Physics of SATURNE which I know the best, and this small review is certainly not exhaustive. Interesting results will also come soon from COSY (where η's have been recently produced at the internal cluster target facility) and from CELSIUS where Dalitz plots of the $pp \to pp\eta$ have been measured.

Acknowledgement. I have extensively used other review done by Y. LeBornec and C. Wilkin on the same subject [45].

References

1. P. Berthet et al: Nucl. Phys. **A443**, 589 (1985)
2. G.Y. Wolf et al: Prog. Part. Nucl. Phys. **30**, 273 (1993)
3. F.D. Berg at al: Phys. Rev. Lett. **72**, 977 (1994)
4. C.B. Dover and P.M. Fishbane: Phys. Rev. Lett. **64**, 3115 (1990)
5. R.S. Bhalerao and L.C. Liu: Phys.Rev¿ Lett. **54**, 865 (1985)
6. Q.Haider and L.C. Liu: Phys. Lett. **B172**, 257 (1986)
7. H.C.Chiang, E. Oset and L.C.Liu: Phys. Rev. **C44**, 738 (1991)
8. M. Kohno and H. Tanabe: Nucl. Phys. **A519**, 755 (1990)

9. M. Clajus and B.M.K. Nefkens: $\pi - N$ Newsletter N^0 **7**, 76 (1992)

10. Ch. Sauermann, B.L. Friman and W. Nörenberg: Phys. Lett. **B341**, 261 (1995)

11. B.J. Lieb et al: "Progress at LAMPF", Lampf report unpublished.

12. R.E. Chrien et al: Phys. Rev. Lett. **60**, 2595 (1988)

13. A.M. Bergdolt et al: "Nouvelles de SATURNE" **13**, 72 (1989), unpublished.

14. J.D. Johnson et al: Phys. Rev. **C47**, 2571 (1993)

15. A. Taleb: These de doctorat "Etude de la réaction $pp \to pp\eta$" CRN 94-61-N1925, C.R.N. Strasbourg, FRANCE.

16. A.M. Bergdolt et al: Phys. Rev. **D 48**, R2969 (1993)

17. E. Chiavassa et al: Phys. Lett. **B322**, 270 (1994)

18. V. Flaminio et al: Compilation of cross section, Report CERN-HERA 83-01, 248 (1983); E. Pickup et al: Phys. Rev. Lett. **8**, 329 (1962)

19. J.F. Germond and C. Wilkin: Nucl. Phys. **A518**, 308 (1990)

20. A. Moalem et al: Nucl. Phys. A, in press.

21. J.M. Laget, F. Wellers and J.F. Lecolley: Phys. Lett. **B257**, 254 (1991)

22. F. Plouin, P. Fleury and C. Wilkin:Phys. Rev. Lett. **65**, 690 (1990)

23. T. Vetter et al: Phys. Lett. **B263**, 153 (1991)

24. F. Plouin et al: Nucl. Phys. **A302**, 413 (1978)

25. E. Chiavassa et al: Phys. Lett. **B337**, 192 (1994)

26. J.M. Laget and J.F. Lecolley: Phys. Rev. Lett. **61**, 2069 (1988)

27. K. Kilian and H. Nann, AIP Conf. Proc. No 221, 185 (1990)

28. C. Wilkin Phys. Rev. **C47**, R938 (1993)

29. G. Fäldt and C. Wilkin: Preprint Uppsala, ISSN 0284-2769

30. B. Mayer et al: Preprint Saclay (1995).

31. L.A. Kondratyuk et al, S. Wycech et al, V.B. Belyaev: 15th European Conference on Few-Body Problems in Physics, Peniscola (1995)

32. J.C. Peng et al: Phys. Rev. Lett. **63**, 2353 (1989)

33. S.S. Kamalov, L. Tiator and C. Bennhold: Phys. Rev. **C47**, 941 (1993)

34. L.C. Liu: Phys. Lett. **B288**, 18 (1992)

35. R. Frascaria et al: Phys. Rev. **C50**, R537 (1994)

36. N. Willis et al: SPES 3 collaboration, private communication.

37. Ye S. Golubeva et al: Nucl. Phys. **A562**, 389 (1993)

38. E. Chiavassa et al: Z. Phys. **A342**, 107 (1992)

39. E. Chiavassa et al: Z. Phys. **A344**, 345 (1993)

40. E. Chiavassa et al: Il Nuovo Cimento **107A**,1195 (1994)

41. N. De Marco: Communication to Winter Meeting in Nuclear Physics, Bormio, (1995)

42. W. Cassing et al: Z. Phys. A-Hadrons and Nuclei **340**, 51 (1991)

43. E. Vercellin et al: Il Nuovo Cimento **106A**, 861 (1993)

44. B.L. Alvaredo and E. Oset: Phys. Lett. **B324**, 125 (1994)

45. Y. Le Bornec: "The interactions of heavy mesons with nuclei"; Y. Le Bornec and C. Wilkin: "heavy meson production at SATURNE: the role of baryon resonances", (1991)

Few-Body Systems Suppl. 8, 299–308 (1995)

Few-Body Systems

η–Physics with Electromagnetic Probes

G. Anton[1], B. Krusche[2]

[1] Physikalisches Institut, University of Bonn, Nussallee 12, D-53115 Bonn, Germany
[2] II. Physikalisches Institut, University of Giessen, Heinrich-Buff-Ring 16, D-35392 Giessen, Germany

Abstract. New eta photoproduction experiments have been performed at MAMI (Mainz) and ELSA (Bonn). The parameters of the resonance $S_{11}(1535)$ have been determined with high reliability. The contribution of other resonances is discussed. Eta photoproduction from the deuteron allows to deduce the isoscalar and isovector amplitude of the excitation of the $S_{11}(1535)$.

1 Introduction

The study of nucleon resonances has mainly been performed by pion production reactions. As an example, the total cross section of photoproduction of positive pions from protons is shown in Fig. 1. The contribution of the resonances to this process is indicated qualitatively by Breit-Wigner curves. The bumps in the cross section are obviously due to the excitation of dominating resonances while there is a large number of weakly excited states which overlap to a great amount and it is hard to disentangle them. One of this weakly excited resonances is the $S_{11}(1535)$.

Also shown in Fig. 1 is the total cross section $\gamma p \rightarrow \eta p$ with the same vertical and horizontal scale as in the upper plot. The $S_{11}(1535)$ is dominating in this reaction so that this channel is ideally suited for the investigation of this state. Accordingly, the $S_{11}(1535)$ plays a role in eta production which is analogous to the role of the $\Delta(1232)$ in pion production. The excitation of other resonances in the eta channel is not well known until recently because only very few experimental data existed. However, new experiments have been performed with significantly improved experimental techniques to measure eta photoproduction reactions over a large kinematic range with high precision. These are the TAPS experiment at MAMI (Mainz) and the PHOENICS experiment at ELSA (Bonn).

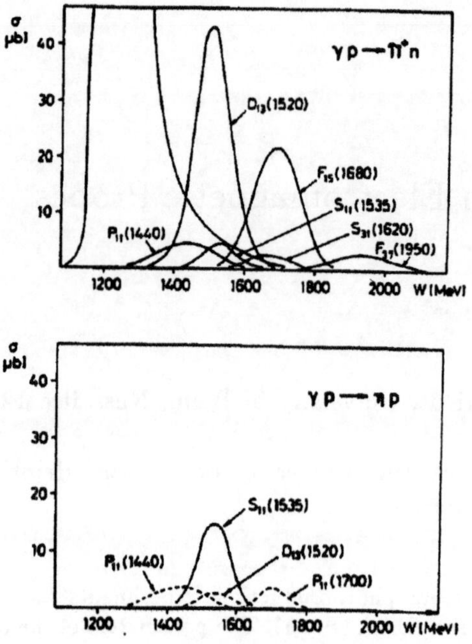

Figure 1. Qualitative picture of the excitation of baryon resonances in pion photoproduction and eta photoproduction. The strength of weakly excited resonances like the $P_{11}(1440)$,$D_{13}(1520)$ and $P_{11}(1700)$ is not well known for $\gamma p \rightarrow \eta p$.

2 Eta Experiments with Real Photons

The new generation of eta photoproduction experiments employs wide band energy tagged photon beams and large acceptance detectors.

The TAPS set-up at Mainz consisted of 5 detector blocks arranged at 55 cm distance around the target [1]. Each block was composed of 64 BaF_2- scintillators with 6 cm diameter hexagonal front face and 25 radiation length thickness. An individual plastic veto detector was positioned in front of each scintillator. Etas were detected via their decay modes $\eta \rightarrow 2\gamma$ and $\eta \rightarrow 3\pi^o$. The invariant mass spectrum reconstructed from 2 detected photons is shown in Fig. 2. Etas could be identified nearly free of background with an invariant mass resolution of typically 45 MeV. The experiment covered a photon energy range from threshold (i.e. $E_\gamma = 707$ MeV for a proton target) to the maximum energy of the Mainz tagger $E_\gamma = 790$ MeV. Differential and total cross section data were obtained with the targets H_2, D_2, C, Ca, Nb, Pb. Final results for the H_2 target have been published recently [2].

Another series of eta photoproduction experiments has been performed at ELSA. The set-up consisted of the PHOENICS tagging system [3], the pho-

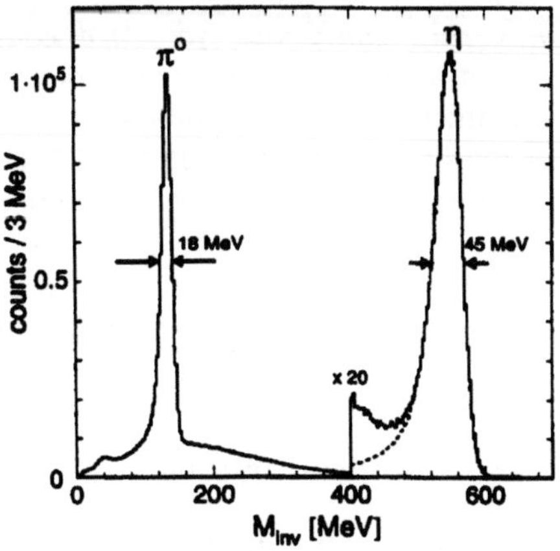

Figure 2. Invariant mass distribution of 2 photons seen by the TAPS detector

ton detector SPES0 from the Institut de Physique Nucleaire Orsay (France) [4], the neutron detector SENECA from the University of Goettingen [5] and the PHOENICS scintillation detectors from Bonn [6]. The SPES0 detector was composed of 24 lead glass bars which were arranged in two shells around the target covering 60 % of the total solid angle (see Fig. 3). The 64 liquid scintillators of the SENECA and the 32 plastic scintillator bars of PHOENICS served to reconstruct the recoil particles. The set-up enabled the detection of

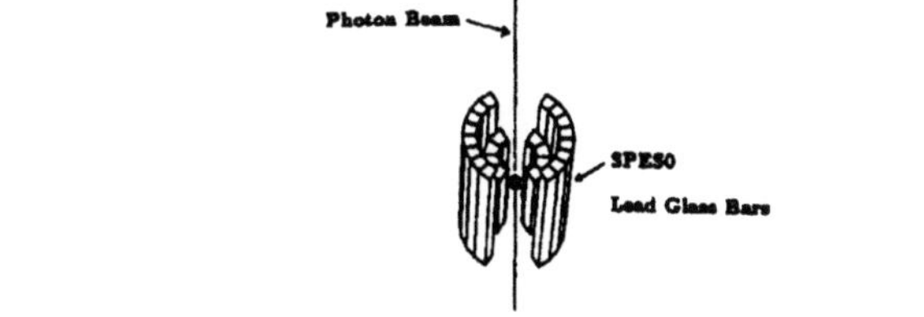

Figure 3. Top view on the experimental set-up at ELSA to measure eta photoproduction from proton, deuteron and nitrogen

the $\eta \to \gamma\gamma$ decay in coincidence with the recoil nucleon. Accordingly, reactions like $\gamma p \to \eta p$, $\gamma d \to \eta p X$, $\gamma d \to \eta n X$ and $\gamma d \to \eta d$ could be measured. Due to the intrinsic properties of lead glass as a photon detector the resolution of the eta is relatively poor. But the additional precise measurement of the recoil particle four momentum gives a very good eta missing mass resolution of about $\sigma_\eta \simeq 5 - 10 MeV/c^2$ depending on the photon beam energy. The experiment covered a photon beam energy range from threshold to 1150 MeV. The data were taken during the past 2 years and most of them are still under analysis. Accordingly, most of the results presented in the following sections are from the TAPS experiment.

3 The Physics of Eta Photoproduction

A number of different mechanisms contribute to the photoproduction of etas. They are related to the diagrams in Fig. 4 which are Born terms a) and b), resonance excitation in s- and u-channel c) and d), vector meson exchange e), the Primakoff effect f) and final state interaction g). Some aspects of these mechanisms will be discussed in the following sections. The properties of etas compared to pions are listed in Table 1 as a review.

Table 1. Properties of pions and etas

	mass (Mev/c^2)	I, I_3	flavour	decay fraction (%)
π^o	134.97	1 , 0	$\lvert\pi^o\rangle = \frac{1}{\sqrt{2}}(u\bar{u} - d\bar{d})$	$\gamma\ \gamma$ (98.8)
η	547.5	0 , 0	$\lvert\eta\rangle_8 = \frac{1}{\sqrt{6}}(u\bar{u} + d\bar{d} - 2s\bar{s})$	$\gamma\ \gamma$ (38.8) $3\ \pi^o$ (31.9) $\pi^+\pi^-\pi^o$ (23.6)

4 Born Terms

Important tests of the Born terms and related low energy theorems were performed for pion photoproduction near threshold. It is straight forward to look for the contribution of Born terms at another particle threshold as for instance the ηN -threshold. But the situation is very different in this case. The η is a relatively massive meson and chiral symmetry is broken to a large extent. Accordingly, not only pseudovector but also pseudoscaler ηNN - coupling may occur. Furthermore, not the Born terms but the $S_{11}(1535)$ resonance gives the largest contribution near threshold. Due to this, the information about $\gamma p \to \eta p$ Born terms can be extracted from experimental data with model assumptions only.

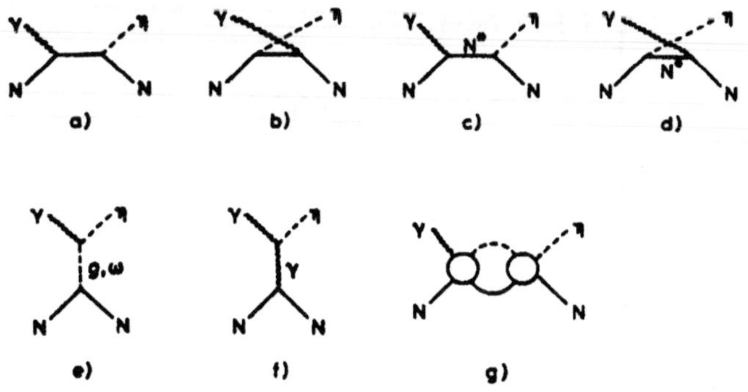

Figure 4. Diagrams contributing to eta photoproduction .

A calculation by L.Tiator et al. [7] takes into account nucleon resonances in a coupled channel formalism and adds Born terms and vector meson exchange terms as effective Langrangian diagrams. The forward backward asymmetry of the differential cross section $\gamma p \rightarrow \eta p$ shows significant sensitivity to a pseudoscaler resp. pseudovector ηNN- coupling leading to $g_\eta^2/4\pi \simeq 0.4$ and a pseudoscalar coupling. Not much sensitivity to Born terms is seen in an effective Lagrangian calculation by Benmerrouche el al. [8]. Here, the possible variation of resonance parameters and the fit to different data sets leaves open a large range for the ηNN- coupling. This situation should improve when only the new data are considered.

5 Resonance Excitation in $\gamma p \rightarrow \eta p$

Cross section data for the reaction $\gamma p \rightarrow \eta p$ measured with TAPS are shown in Fig. 5. The strong rise up to about 16 $\mu barn$ at 800 MeV is due to the excitation of the $S_{11}(1535)$ resonance. This is also seen in the angular distributions in Fig. 5, which are mainly isotropic reflecting s-wave interaction. A Breit-Wigner fit to the data respecting the energy dependence of the resonance width allowed to extract the helicity amplitude of the $\gamma p \rightarrow S_{11}(1535)$ transition. The resulting value of $A_{1/2}^P = 125 \pm 25\ (10^{-3}\ (GeV)^{-1/2})$ is considerably larger than the value extracted from pion photoproduction, which is quoted by the Particle Data Group to be 68 ± 10. The p-wave contribution to the differential cross section is very small, which means that obviously no significant excitation of the Roper resonance $P_{11}(1440)$ can be deduced. The d-wave gives a $cos^2(\Theta)$ dependence of the cross section. This contribution reaches about 10 % of the isotropic part at 800 MeV and is probably related to the excitation of the $D_{13}(1520)$.

The target asymmetry is a polarization observable which has enhanced sensitivity to the $D_{13}(1520)$ because it contains the interference between the d-wave with the strong s-wave. Therefore, a target asymmetry measurement

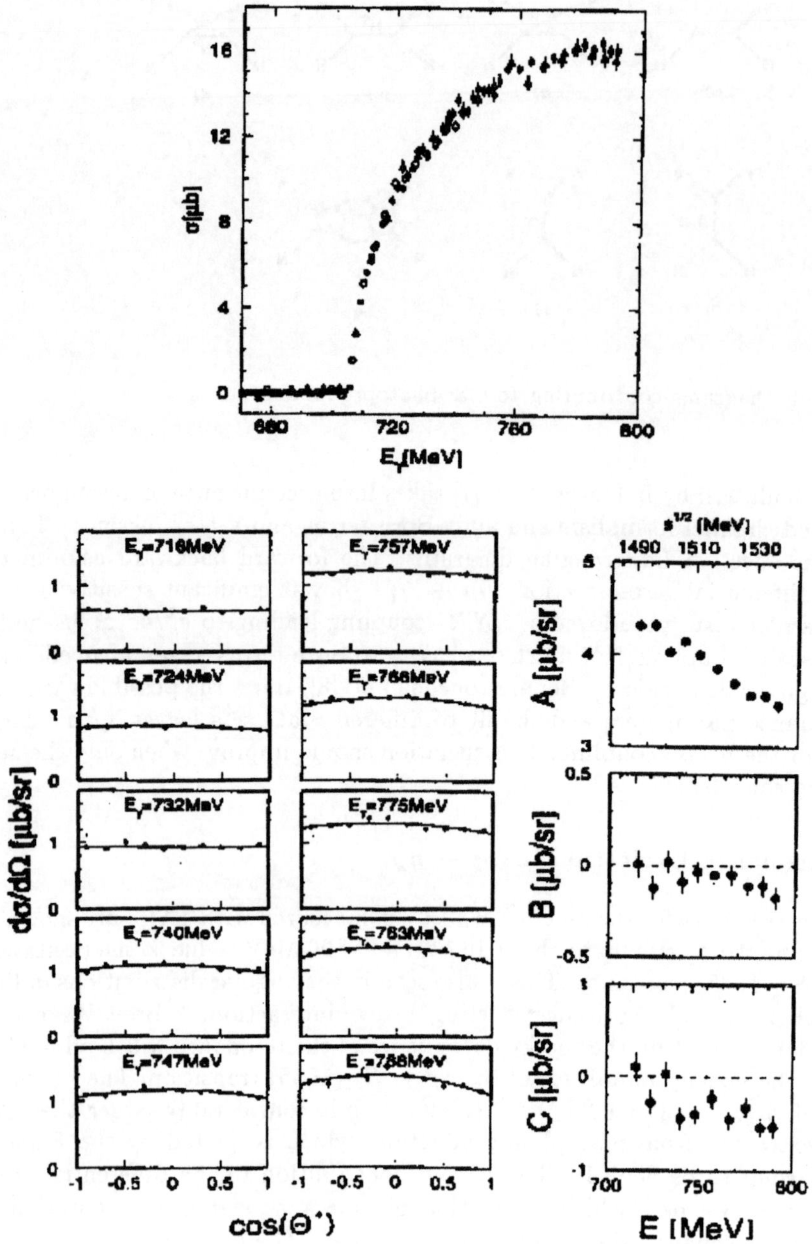

Figure 5. Cross section data for the reaction $\gamma p \to \eta p$ measured by TAPS [2]. Total cross section obtained by the analysis of $\eta \to \gamma\gamma$: full squares, $\eta \to 3\pi^\circ$ (4γ detected): open triangles, $\eta \to 3\pi^\circ$ (3γ detected): open circles. The angular distributions are fitted using the ansatz $d\sigma/d\Omega = q_\eta/k_\gamma(A + Bcos(\Theta) + Ccos^2(\Theta)$.

was performed with the above mentioned detectors SPES0 and PHOENICS in collaboration with the Bonn Polarized Target group. The target material (butanol) was implemented in a $^3He/^4He$ dilution cryostat oriented vertically inside and above the SPES0 detector. A pair of small superconducting coils inside the cryostat supplied the magnetic field to hold the spin of the polarized protons. The experiment took place at the beginning of 1995 and the data are still being analyzed.

6 Eta Production from the Deuteron

Photoproduction of etas can proceed via an isoscalar and isovector amplitude, T^0 resp. T^1. The determination of these amplitudes requires the amplitude from protons T^p and neutrons T^n: $T^0 = \frac{1}{2}(T^p + T^n)$ and $T^1 = \frac{1}{2}(T^p - T^n)$ The TAPS group measured the process $d(\gamma, \eta)$, see Fig. 6. The cross section of $\gamma p \rightarrow \eta p$ is used as an input in a calculation which considers the eta production from Fermi-moved nucleons inside the deuteron. The result is scaled by a factor $(1 + \sigma_n/\sigma_p)$ in order to fit the $d(\gamma, \eta)$ data, see curve in Fig. 6. Thus, a value of $\sigma_n/\sigma_p = 2/3$ is obtained.

The sensitivity to the process $\gamma n \rightarrow \eta n$ is enlarged in the PHOENICS experiment where the recoil proton respectively recoil neutron are explicitely detected. This will enable the determination of the cross sections $\gamma d \rightarrow \eta p X$ and $\gamma d \rightarrow \eta n X$ as a function of the photon energy and the recoil particle angle.

The coherent reaction $\gamma d \rightarrow \eta d$ was measured at $\Theta_\eta^{CM} = 90^o$ by Anderson et al. [10] with an untagged bremsstrahlung beam in 1969. They obtained

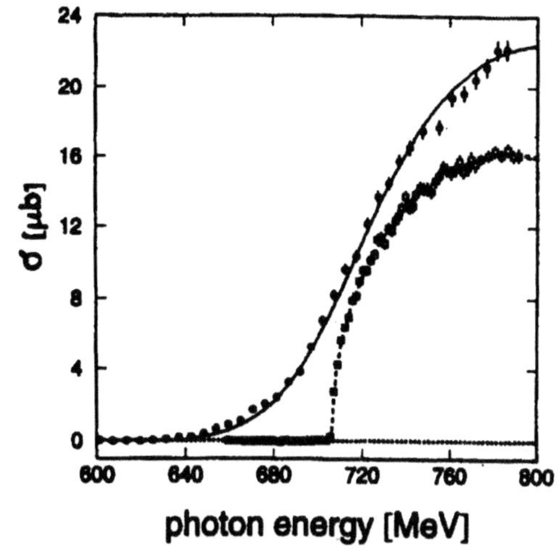

Figure 6. Total cross section data measured by TAPS; full circles: $d(\gamma, \eta)$, open squares: $\gamma p \rightarrow \eta p$. [9]

relatively large cross section values below $E_\gamma = 700$ MeV. This caused confusion because one needs a large isoscalar amplitude from the nucleon in order to describe the data. This result is not only seen in impulse approximation but also when rescattering terms are taken into account [11], see Fig. 7. The related value of $T^0/T^p \simeq 0.6$ is in strong contradiction to pion photoproduction data and quark model calculations which both quote $T^0/T^p \leq 0.2$ for the $S_{11}(1535)$ excitation. The TAPS group sees no signal for the $\gamma d \rightarrow \eta d$ reaction at $E_\gamma = 685 \pm 20$ MeV and $\Theta_\eta^{CM} = 90^o \pm 10^o$. An experiment at Bonn did also see no signal for $E_\gamma = 632$ MeV and $E_\gamma = 637$ MeV [12]. This findings indicate that the Anderson data are wrong and that T^o is really small.

The PHOENICS set-up was used to measure the process $\gamma d \rightarrow \eta d$ by detecting both particles in the final state, deuteron and eta. For the first time a signal for coherent eta photoproduction has been seen (see Fig. 8) from which the cross section will be determined. A new series of TAPS eta experiments with recoil particle detection has been started in spring 1995 at Mainz.

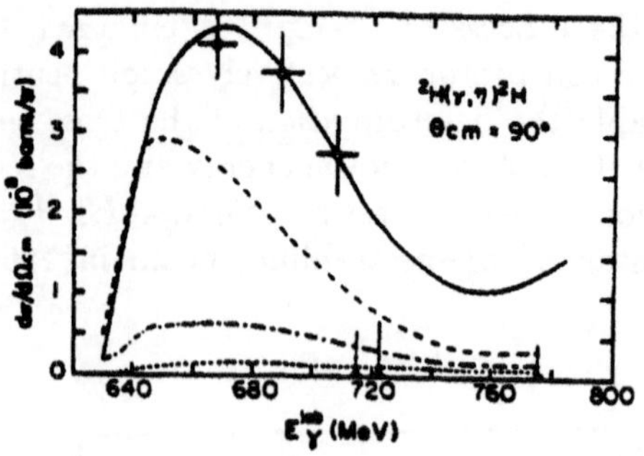

Figure 7. Differential cross section of the reaction $\gamma d \rightarrow \eta d$ [10] compared to the calculation of [11]. Dashed line: impulse term with $T^0/T^p = 0.6$, dotted line : pion rescattering, dot-dashed line: eta rescattering, solid line: all terms contributing.

7 Electroproduction of Eta Mesons

Only very few data exist for the eta production via electron scattering. A recent experiment at ELSA measured in the $S_{11}(1535)$ region at $Q^2 = 0.06\ (GeV/c)^2$ [13], see Fig. 9. The total cross section obtained at this low momentum tranfer is in very good agreement with the real photon cross section. The Q^2- dependence of the resonance excitation by virtual photons is of great interest for quark models. Further experiments for electroproduction of etas are planned at CEBAF.

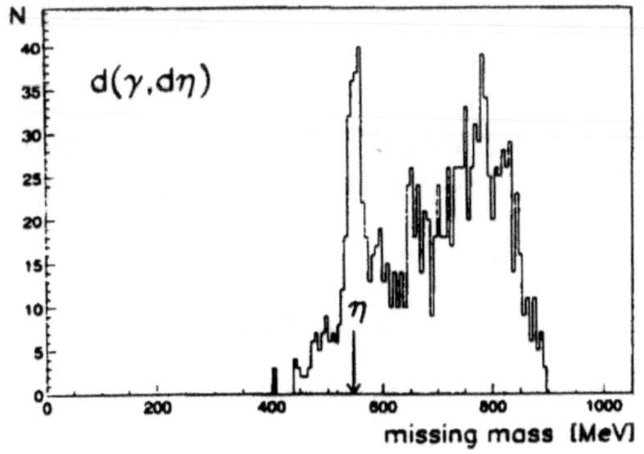

Figure 8. Missing mass spectrum from recoil deuterons measured at Bonn

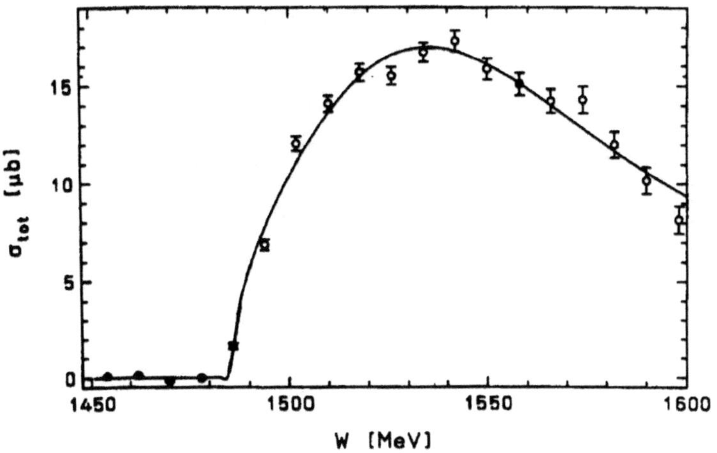

Figure 9. Total cross section of eta electroproduction from protons at $Q^2 = 0.06$ $(GeV/c)^2$.

8 Conclusions

The new eta photoproduction experiments allow an intense study of nucleon resonances. The TAPS group has determined the parameters of the $S_{11}(1535)$ resonance with high reliability. These values differ considerably from those extracted from pion photoproduction. Measurements from the deuteron give a small isoscalar and large isovector $S_{11}(1535)$ excitation. The PHOENICS (Bonn-Orsay- Goettingen) group is still analyzing the data; new results are expected for the higher mass resonances up to 1720 MeV.

References

1. R. Novotny, IEEE Trans. on Nucl. Science **38**, 379 (1991)
   A.R. Gabler et al.: Nucl. Instr. Meth. **A346**, 168 (1994)

2. B. Krusche et al.: Phys. Rev. Lett. **74**, 3736 (1995)

3. P. Detemple et al.: Nucl. Instr. Meth. **A321** 479 (1992)

4. G.S. Blanpied et al.: Few Body Systems Suppl **2**, 435 (1987)

5. G.v. Edel et al.: Nucl. Instr. Meth. **A332**, 224 (1993)

6. K. Buechler et al.: Nucl. Phys. **A570**, 580 (1994)

7. L. Tiator et al.: Nucl. Phys. **A580**, 455 (1994)
   G. Knoechlein et al.: Mainz Univ. preprint, MKPH-T-95-2, (1995)

8. M. Benmerrouche et al.: Phys. Rev. **D51**, 3237 (1995); Phys.Rev.Lett. **67**, 1070 (1991)

9. B. Krusche et al.: submitted to Phys. Lett. **B**

10. R.L. Anderson, R. Prepost: Phys. Rev. Lett. **23**, 46 (1969)

11. D. Halderson, A.S. Rosendahl: Nucl. Phys. **A 501**, 856 (1989)

12. W. Beulertz: PhD thesis, Bonn 1994, BN-IR-94-

13. M. Wilhelm: PhD thesis, Bonn, BN-IR-93-43, to be published

Few-Body Systems Suppl. 8, 309–316 (1995)

Few-Body Systems

Interaction of η-Meson with Light Nuclei

V.B. Belyaev [1], S.A. Rakityansky [2] *, S.A. Sofianos[2], M. Braun[2], W. Sandhas [3]

[1] Joint Institute for Nuclear Research, Dubna, 141980, Russia

[2] Physics Department, University of South Africa, P.O.Box 392, Pretoria 0001, South Africa

[3] Physikalisches Institut, Universität Bonn, D-53115 Bonn, Germany

Abstract. A microscopic treatment of η–nucleus scattering is presented. When applying the underlying exact integral equations, excitations of the target are neglected, and an input ηN amplitude is chosen which reproduces the $S_{11}(1535)$ resonance. It is shown that the η–nucleus scattering lengths are quite sensitive to the ηN parameters and the nuclear wave functions. For a special choice of these parameters an η^4He quasi–bound state occurs.

1 Introduction

There are at least three main reasons which motivate detailed investigations of the η–nucleus system.

The first one, being of fundamental character, is related to the possibility of studying the quark structure of the η meson and the nucleon $S_{11}(1535)$ resonance, especially the role played by the strange quarks in these systems. In recent experiments with $N\overline{N}$ annihilation into ϕ and ω channels [1], it was established that there is a strong polarization of the strange sea–quarks in nucleons. It is, therefore, interesting to investigate possible manifestations of this phenomenon also in η–nucleus systems.

The second reason concerns a pure nuclear problem, namely the possible formation of η-nucleus bound states. It was argued by Haider and Liu [2] that, due to the rather strong attraction in the ηN interaction at low energies, the η–meson, when immersed in the nucleus, might create a system which in some respect is analogous to a hypernucleus. Estimations in [2] indicated that such

* *Permanent address:* Joint Institute for Nuclear Research, Dubna, 141980, Russia

a possibiltity exists for nuclei with atomic number $A \geq 12$.

The third reason is that charge symmetry breaking (CSB) effects can be studied in processes involving η production. Indeed, CSB was observed at Saturne [3] in the reaction

$$d + d \longrightarrow {}^4\mathrm{He} + \pi^0 \tag{1}$$

near the η–production threshold. The pion rate found in this experiment indicates a much greater CSB than expected due to electromagnetic interaction. The experimental cross section is 0.97 pb/sr (at $E_d = 1100$ MeV) in contrast to 0.003 pb/sr [4] obtained from electromagnetic considerations at 600 MeV.

The CSB in the above reaction may be attributed to $\eta - \pi^0$ mixing. In fact, process (1) can proceed via the intermediate production of a pure η^0 state with isospin $I = 0$, i.e., via the allowed reaction

$$d + d \longrightarrow {}^4\mathrm{He} + \eta^0. \tag{2}$$

The state $|\eta^0\rangle$ is a linear combination of the physical states $|\eta\rangle$ and $|\pi^0\rangle$,

$$|\eta^0\rangle = \frac{1}{\sqrt{1+\lambda^2}}(\,|\eta\rangle + \lambda|\pi^0\rangle\,), \tag{3}$$

where the parameter λ characterises the $\eta - \pi^0$ mixing. From Eq. (3) it follows [5] that λ is the proportionality constant between the amplitude of reaction (1) and the amplitude of the corresponding η process,

$$f(dd \rightarrow \alpha\pi^0) \sim \lambda f(dd \rightarrow \alpha\eta). \tag{4}$$

Reaction (1), therefore, is no longer strongly forbidden, and a cross section much larger than the electromagnetic one is to be expected.
Based on these ideas, the authors of ref. [6] obtained 0.12 pb/sr for the cross–section of process (1). To remove the discrepancy between this value and the experimental one of 0.97 pb/sr quoted above, fully microscopic few–body calculations, which take into account the $\eta - \pi^0$ mixing, should be performed.

We would also like to mention the possibility of studying processes of the type

$$d + d \longrightarrow {}^4\mathrm{He}^*\,(\mathcal{J}^P, 0) + \pi^0, \tag{5}$$

where ${}^4\mathrm{He}^*\,(\mathcal{J}^P, 0)$ denotes excited states of ^{4}He with isospin $I = 0$ and different angular momenta and parities. These reactions might provide us with larger cross sections due to the extended size of the excited ^{4}He nucleus.

Having the above in mind, and in order to get some insight into the physics of η–nucleus systems, we undertook microscopic investigations concerning the low–energy behaviour of the η–nucleus elastic scattering amplitude and the positions of poles of the corresponding t-matrices in the complex k–plane. Another issue is up to which extent the attraction of the two–body ηN interaction has to be enhanced in order to produce quasi-bound states in the η–nucleus system.

2 The Method

Our method is based on the Finite–Rank–Approximation (FRA) of the Hamiltonian, proposed as an alternative to the multiple scattering theory of pion-nucleus scattering [7]. We, therefore, start by outlining this approach.

Consider the scattering of an η meson from a nucleus of atomic number A. The total Hamiltonian of this system is

$$H = H_0 + V + H_A, \tag{6}$$

where H_0 is the kinetic energy operator (free Hamiltonian) of the η–nucleus motion , $V = V_1 + V_2 + \cdots + V_A$ the sum of ηN potentials, and H_A the total Hamiltonian of the nucleus. The transition operator for η–nucleus collisions satisfies the Lippmann-Schwinger equation

$$T(z) = V + VG_A(z)T(z), \tag{7}$$

the Green function in its kernel being given by

$$G_A(z) = \frac{1}{z - H_0 - H_A}. \tag{8}$$

In order to bring Eq. (7) into a form appropriate for our purpose, we introduce the auxiliary transition operator $T^0(z)$ via

$$T^0(z) = V + VG_0(z)T^0(z), \tag{9}$$

where

$$G_0(z) = \frac{1}{z - H_0} \tag{10}$$

is the free Green function. From the definitions of $G_0(z)$ and V it is clear that T^0 describes the scattering of mesons by a nucleus in which the position of the nucleons is fixed. The operator T^0 differs however from the usual fixed center t-matrix in two respects: Equation (9) contains the kinetic energy operator H_0 describing the motion of the η meson with respect to the center of mass of the target, and the energy argument of T^0 is taken to be that of the total energy z of the system.

Using the resolvent equation

$$G_A(z) = G_0(z) + G_0(z)H_AG_A(z), \tag{11}$$

we easily infer from (7) and (9)

$$T(z) = T^0(z) + T^0(z)G_0(z)H_AG_A(z)T(z), \tag{12}$$

a form well suited for the low–energy approximation on which our approach will essentially be based. The spectral decomposition of the nuclear Hamiltonian H_A reads

$$H_A = \sum_n \mathcal{E}_n^A|\Psi_n^A\rangle\langle\Psi_n^A| + \int E|\Psi_E^A\rangle\langle\Psi_E^A|\,dE\,, \tag{13}$$

where $|\Psi_n^A\rangle$ are the bound–state eigenfuntions of H_A with $\mathcal{E}_n^A$ being the corresponding discrete energies. Since we are interested in processes at energies far below the excited states or break–up thresholds of the nuclei, we can neglect all contributions to (13) except the one for the ground state. That is, we use the approximation

$$H_A \approx \mathcal{E}_0^A |\Psi_0^A\rangle\langle\Psi_0^A|. \tag{14}$$

Inserting Eq. (14) into Eq. (12) and sandwiching with $|\Psi_0^A\rangle$, we obtain the Lippmann–Schwinger type equation

$$T(\boldsymbol{k}',\boldsymbol{k};z) = T^0(\boldsymbol{k}',\boldsymbol{k};z) + \mathcal{E}_0^A \int \frac{d^3q}{(2\pi)^3} \frac{T^0(\boldsymbol{k}',\boldsymbol{q};z)}{(z-\frac{q^2}{2\mu})(z-\mathcal{E}_0^A-\frac{q^2}{2\mu})} T(\boldsymbol{q},\boldsymbol{k};z), \tag{15}$$

where

$$T^0(\boldsymbol{k}',\boldsymbol{k};z) = \int d^{3(A-1)}r |\Psi_0^A(\boldsymbol{r})|^2 T^0(\boldsymbol{k}',\boldsymbol{k};\boldsymbol{r};z), \tag{16}$$

with $\boldsymbol{r}$ denoting all nuclear Jacobi variables.

From a practical point of view it is convenient to rewrite Eq. (9) by using the Faddeev-type decomposition

$$T^0(z) = \sum_{i=1}^{A} T_i^0(z).$$

Introducing the operators

$$t_i(z) = V_i + V_i \frac{1}{z - H_0} t_i(z),$$

we get for the Faddeev components T_i^0 the following system of integral equations

$$\begin{aligned} T_i^0(\boldsymbol{k}',\boldsymbol{k};\boldsymbol{r};z) &= t_i(\boldsymbol{k}',\boldsymbol{k};\boldsymbol{r};z) \\ &+ \int \frac{d^3k''}{(2\pi)^3} \frac{t_i(\boldsymbol{k}',\boldsymbol{k}'';\boldsymbol{r};z)}{z - \frac{k''^2}{2\mu}} \sum_{j\neq i} T_j^0(\boldsymbol{k}'',\boldsymbol{k};\boldsymbol{r};z). \end{aligned} \tag{17}$$

The amplitude t_i describes the scattering of the η meson off the i-th nucleon. It is expressed in terms of the corresponding two-body $t_{\eta N}$-matrix via

$$t_i(\boldsymbol{k}',\boldsymbol{k};\boldsymbol{r};z) = t_{\eta N}(\boldsymbol{k}',\boldsymbol{k};z)\exp\left[i(\boldsymbol{k}-\boldsymbol{k}')\cdot\boldsymbol{r}_i\right],$$

where $\boldsymbol{r}_i$ is the vector from the nuclear center of mass to the i-th nucleon. This vector can be expressed in terms of the Jacobi coordinates $\{\boldsymbol{r}\}$.

To get the nuclear wave function we use the Integrodifferential Equation Approach (IDEA)[8, 9]. In this approach the wave-function of a nucleus consisting of A nucleons is decomposed into Faddeev-type components

$$\Psi(\mathbf{r}) = \sum_{i<j\leq A} \psi_{ij}(\mathbf{r})$$

which obey

$$(h_0 - E)\psi_{ij}(\mathbf{r}) = -V(r_{ij}) \sum_{k<l\leq A} \psi_{kl}(\mathbf{r}),$$

with h_0 being the kinetic energy operator of the nucleus, and V_{ij} the potential between nucleons i and j. These components are written as products of harmonic polynomials $H_{[L_m]}$ and unknown functions P,

$$\psi_{ij}(\mathbf{r}) = H_{[L_m]}(\mathbf{r}) \frac{1}{\rho^{\mathcal{L}_0+1}} P(z,\rho),$$

where P depends only on the hyperradius $\rho = \left[2/A \sum r_{ij}^2\right]^{1/2}$ and on z via the pair separation $r_{ij} = \sqrt{\rho(1+z)/2}$. Eventually, one has to solve an integrodifferential equation for $P(z,\rho)$ which, for an A-boson system, reads

$$\begin{aligned} \left\{ \frac{\hbar^2}{m} \right. & \left[- \frac{\partial^2}{\partial \rho^2} + \frac{\mathcal{L}_0(\mathcal{L}_0+1)}{\rho^2} - \frac{4}{\rho^2} \frac{1}{W_0(z)} \frac{\partial}{\partial z}(1-z^2) W_0(z) \frac{\partial}{\partial z} \right] \\ & + \left. \frac{A(A-1)}{2} V_0(\rho) - E \right\} P(z,\rho) = \\ & - \left[V\left(\rho\sqrt{\frac{1+z}{2}}\right) - V_0(\rho) \right] \left\{ P(z,\rho) + \int_{-1}^{+1} f_0(z,z') P(z',\rho) dz' \right\}. \end{aligned}$$

The details of the method which takes into account the two-body correlations exactly, and the definition of the hypercentral potential $V_0(\rho)$ occurring in this equation can be found in refs. [8, 9]

3 Results

To solve Eq. (15) we need, as an input to Eq. (17), the two–body ηN amplitude (or the corrresponding t-matrix) and the ground state wave function Ψ_0^A of the nucleus. For $t_{\eta N}$ we employ the S-wave separable form

$$t_{\eta N}(k,k',z) = \frac{\lambda}{(k^2+\alpha^2)(z - E_0 + i\Gamma/2)(k'^2+\alpha^2)}, \tag{18}$$

where E_0 and Γ are position and width of the S_{11} resonance. Different sets of parameters α and λ are chosen such that the scattering lengths in [10] are reproduced. For this separable form Eq. (17) can be solved analytically. The

Table 1. The η-nucleus scattering lengths (in fm) for 9 combinations of the range parameter α and the $a_{\eta N}$ scattering lenght [10, 12, 13].

	α=2.357 (fm^{-1})	α =3.316 (fm^{-1})	α =7.617 (fm^{-1})	$a_{\eta N}$ (fm)
^{2}H	0.65+i0.86	0.62+i0.91	0.54+i0.97	
^{3}H	0.81+i1.90	0.66+i1.98	0.41+i2.00	0.27+i0.22
^{3}He	0.79+i1.90	0.64+i1.98	0.39+i2.00	
^{4}He	0.23+i3.54	-0.40+i3.43	-0.96+i2.95	
^{2}H	0.74+i0.77	0.73+i0.83	0.67+i0.91	
^{3}H	1.12+i1.82	0.99+i1.96	0.71+i2.07	0.28+i0.19
^{3}He	1.10+i1.83	0.97+i1.91	0.68+i2.07	
^{4}He	0.96+i3.99	0.06+i4.13	-0.91+i3.62	
^{2}H	1.38+i2.27	1.05+i2.45	0.57+i2.34	
^{3}H	-1.44+i4.60	-1.72+i3.76	-1.56+i3.00	0.55+i0.30
^{3}He	-1.42+i4.54	-1.68+i3.74	-1.53+i2.99	
^{4}He	-3.66+i1.78	-2.96+i1.39	-2.42+i1.25	

results obtained are given in Table 3.1.

The required nuclear wave functions were constructed by means of the IDEA, employing as the underlying NN interaction the Malfliet-Tjon (MT I+III) potential [11]. As can be seen there is a strong dependence of the resulting η-nucleus scattering length on the two-body input. A large ηN scattering length generates a strong attraction of the η meson by the three– and four–nucleon systems.

In Table 3.2 we present η–nucleus scattering lengths obtained by using two different sets of nuclear wave functions. One of them is constructed by mean of the IDEA with the MT potential. The other phenomenological one uses a Gaussian form, chosen to reproduce the root mean square radii of the nuclei obtained via the IDEA. We see that the considerable difference found for ^{2}H decreases when going over to heavier nuclei and practically vanishes for ^{4}He.

To find at which values of the ηN scattering length a quasi–bound state appears in the η nucleus system, we write this scattering length in terms of two parameters g_1 and g_2,

$$a_{\eta N} = (0.55 g_1 + i0.30 g_2)\ \text{fm}.$$

Note that $g_1 = g_2 = 1$ is just one of the choices in Table 3.1. These parameters are varied until the η-nucleus T matrix gets a pole at $Re\, E = 0$. Results of this search with Gaussian type wave functions are given in Table 3.3. It is seen that an η^4He quasi–bound state can exist for realistic values of $a_{\eta N}$. In

Table 2. The η-nucleus scattering lengths (in fm) for different nuclear wave functions. In all cases $a_{\eta N}$=0.55+i0.30 fm.

Model	IDEA	Gaussian
^{2}H	1.05+i2.45	0.57+i2.14
^{3}H	-1.72+i3.76	-1.42+i4.32
^{3}He	-1.68+i3.74	-1.39+i4.28
^{4}He	-2.96+i1.39	-3.07+i1.48

Table 3. Values of g_1 at which a quasi–bound state appears.

	$\alpha = 2.357$ (fm $^{-1}$)	$\alpha = 3.316$ (fm $^{-1}$)	$\alpha = 7.617$ (fm $^{-1}$)	g_2
d	1.610	1.613	1.547	0
	1.654	1.566	1.535	1
t	1.293	1.124	0.996	0
	1.361	1.310	1.260	1
^{3}He	1.302	1.204	1.130	0
	1.330	1.221	1.144	1
^{4}He	1.075	0.992	0.910	0
	0.955	0.911	0.899	1

all other cases the physical ηN input has to be modified in order to support a quasi–bound state.

Finally we mention that using the auxiliary matrix T^0 instead of the full solution of Eq. (15), completely different results are obtained.

4 Conclusions

We have investigated ηA systems with $A \leq 4$ in the framework of a microscopic few-body approach by using different two-body input parameters and nuclear bound state wave functions. The results can be summarized as follows:

- There is a remarkable increase of attraction with increasing atomic number.
- The same tendency appears for all nuclei when the ηN scattering length is increased.
- When calculating ηA observables it is important to use microscopically constructed wave functions.

- The fixed scatterer approximation, corresponding to the use of T^0 instead of T, is inadequate in describing the η–nucleus system.
- There is a strong indication that the $\eta\,^4$He system can form a quasi–bound state.

References

1. M.G. Sapozhnikov: Preprint JINR (Dubna) E15-94-501; J. Ellis et al.: CERN preprint, TH/94-7326, Geneva,1994
2. Q. Haider, L.C. Liu: Phys. Lett. **172 B**, 257 (1986)
3. R. Frascaria et al.: Phys. Rev. C **50**, R537 (1994)
4. C.Y. Cheung: Phys. Lett. **119 B**, 47 (1982)
5. C. Wilkin: Phys. Lett. **331 B**, 276 (1994)
6. S.A. Coon and B.M. Preedom: Phys. Rev. C **33**, 605 (1986)
7. V.B. Belyaev: in *Lectures on the Theory of Few-Body Systems*, Springer-Verlag, Heidelberg, 1990
8. M. Fabre de la Ripelle, H. Fiedeldey, and S.A. Sofianos: Phys. Rev. C **38**, 449 (1988)
9. W. Oehm, S.A. Sofianos, H. Fiedeldey, and M. Fabre de la Ripelle: Phys. Rev. C**44**, 81 (1991)
10. R.S. Bhalerao, L.C. Liu Phys: Rev. Lett. **54**, 865 (1985)
11. R.A. Malfliet and J.A. Tjon: Nucl. Phys. **A127**, 161 (1969); Ann. Phys. (N.Y.) **61**, 425 (1970)
12. C. Wilkin: Phys. Rev. C **47**, R938 (1993)
13. C. Bennhold, H. Tanabe: Nucl. Phys. **A 530**, 625 (1991)

Few-Body Systems Suppl. 8, 317–328 (1995)

Few-Body Systems

Nuclear Reactions of Astrophysical Interest Involving Light Nuclei

H. Oberhummer*, W. Balogh, V.D. Efros†, H. Herndl, R. Hofinger

Institute of Nuclear Physics, Technical University Vienna, Wiedner Hauptstr. 8–10, A–1040 Wien, Austria

Abstract. An introduction to nucleosynthesis, the creation of the elements in the big bang, in interstellar matter and in stars is given. The two–step process ^{4}He($2n$,γ)^{6}He and the reverse photodisintegration ^{6}He(γ,$2n$)^{4}He involving the halo nucleus ^{6}He could be of importance in the α–process in type–II supernovae. The reaction rates for the above processes are calculated using three–body methods and show an enhancement of more than three orders of magnitude compared to the previous adopted value. Direct–capture calculations give similar values for the above reaction rates. Therefore, this method was also used to calculate the reaction rates of the two–step processes ^{6}He($2n$,γ)^{8}He and ^{9}Li($2n$,γ)^{11}Li and the reverse photodisintegration of ^{8}He and ^{11}Li that could be also of importance in the α-process.

1 Introduction

One of the main driving forces for the evolution of our universe are nuclear processes. These nuclear processes are responsible for the energy production in stars as well as for the production of the elements in our universe. The process of the creation of the elements is called nucleosynthesis. We have three different scenarios, where nucleosynthesis occurs: in the big bang (primordial nucleosynthesis), in interstellar space (interstellar nucleosynthesis) and in stars (stellar nucleosynthesis). The result of nucleosynthesis is reflected today by the observed elemental abundances in different objects (sun and other stars, planets, meteorites, interstellar matter, ...).

The light elements from hydrogen to lithium with mass numbers $A = 1$ to $A = 7$ were mainly synthesized in primordial nucleosynthesis. In interstellar nucleosynthesis primarily elements from lithium to boron with mass numbers

* *E-mail address*: ohu@ds1.kph.tuwien.ac.at

† *Permanent address*: Kurchatov Institute, Institute for General and Nuclear Physics, SU–123182 Moscow, Russia

$A = 6$ to $A = 11$ were produced. Finally, all the heavier elements from carbon to uranium (in the astrophysical jargon these elements are called metals) with mass numbers $A = 12$ to $A = 238$ were created in stellar nucleosynthesis. Even though the heavier elements amount to about only 1% of the total observed abundance they are obviously indispensable for our environmemt. Without stellar nucleosynthesis our universe would be a boring place with the two gases hydrogen and helium (with a negligible amount of lithium, beryllium and boron) and without all the elements necessary for the existence of life.

One of the essential inputs for the investigation of different astrophysical scenarios for nucleosynthesis are cross sections, astrophysical S–factors and reaction rates of nuclear reactions as well as half lives and decay rates of nuclear decays. These quantities are determined using different experimental facilities (accelerators, reactors, ...). However, in many cases it is not possible to obtain the needed cross sections directly from experiments. This is due to the fact that astrophysically relevant reactions take place mainly at thermonuclear energies. These energies are mostly well below the Coulomb and/or centrifugal barrier, and in this region the cross sections are often too small for experimental determination. Furthermore, in nucleosynthesis often radioactive nuclei are involved, for which experimental information is only scarcely available. For nuclei far–off stability often experimental information is not attainable at all.

In the next section a short introduction to nucleosynthesis is given emphasizing those astrophysical scenarios, where light nuclei are involved and methods developed in the few–body field could be applied. In Sect. 3 we show as an example for an application of few–body calculations for astrophysically relevant nuclear reactions the two–step process $^4\text{He}(2n,\gamma)^6\text{He}$. This reaction involves also the halo nucleus ^{6}He. In the last section we calculate the reaction rate for this reaction as well as other two–step reactions involving halo nuclei, like $^6\text{He}(2n,\gamma)^8\text{He}$ and $^9\text{Li}(2n,\gamma)^{11}\text{Li}$, using the simpler direct–capture (DC) model.

2 Nucleosynthesis

In this section we only can give a short introduction to nucleosynthesis. There are many textbooks where this subject can be found in more detail (e.g. [1]).

Primordial nucleosynthesis took place as the cooling early universe reached a temperature of about 10^9 K about three minutes after the big bang. Before that time the temperature was high enough that formed deuterons were immediately again destroyed through photodisintegration. From primordial nucleosynthesis calaculations we can deduce abundance ratios of the light stable nuclides ^{1}H, ^{2}H, ^{3}He, ^{4}He, ^{6}Li and ^{7}Li. Even some primordial abundance ratios may be altered somewhat by interstellar and stellar nucleosynthesis, the agreement between the abundances obtained from primordial nucleosynthesis with the observed abundances is remarkable.

In interstellar nucleosynthesis the nuclides ^{6}Li, ^{7}Li, ^{9}Be, ^{10}B and ^{11}B are produced. In this scenario high–energy (larger than about 1 GeV) cosmic rays hit interstellar matter and create through spallation processes the above nuclei.

Table 1. Nuclear burning phases and corresponding temperatures

Nuclear burning phase	Temperature (T_9)
Hydrogen burning	0.01–0.04
Helium burning	0.1–0.2
Carbon burning	0.6–0.8
Neon burning	1.2–1.4
Oxygen burning	1.5–2.2
Silicon burning	3–4

Stellar nucleosynthesis begins when through gravitational contraction in a star the density and temperature in the core gets high enough so that nuclear processes are ignited. Stellar nucleosynthesis takes place in succesive burning phases. When the fuel of a preceding burning phase has been consumed, the ash of this burning phase is the fuel for the next burning phase. In the outer layers the previous burning phases still continue. The different burning phases are hydrogen burning (*pp*–chain, CNO–cycle, Ne–Na cycle, Mg–Al cycle), helium, carbon, neon, oxygen and silicon burning. In hydrogen burning helium is produced, in helium burning carbon and oxygen are synthesized, in the remainig advanced burning phases elements up to elements in the region of iron are produced. In successive burning phases the temperature must get larger, because of the increasing Coulomb barriers of the fusing heavier nuclei (Table 1).

The stars in which hydrogen burning takes place are named main–sequence stars, whereas the other burning phases occur in red giants. Not in all stars all the burning phases are ignited. For stars with less than approximately 8 solar masses, after helium burning, the outer part of the star that is blown off by strong stellar winds is called a planetary nebula, whereas the core remains as a white dwarf. For stars with more than about 8 solar masses all the above cited burning phases take place. Finally, after silicon burning the core of the star consists of elements with mass numbers in the region of iron. Then it is not possible to gain energy anymore through nuclear fusion and the star collapses. Through the rebounce effect after the implosion the stellar mantle is ejected in a gigantic explosion, whereas the core remains as a neutron star or black hole. This dramatic event is called a supernova of type II (a supernova of type I takes place in a binary system consisting of two stars, where triggered through mass accretion from the accompaning star the star is disrupted completely).

There are astrophysical scenarios in which free neutrons play a dominating role. Most elements heavier than iron have been created through neutron capture in two processes: in the s–process occuring in helium burning of red giants and in the r–proccess occuring in a supernova of type II. The neutron flux in the s–process of helium burning is comparetively low ($N_{\rm n} \simeq 10^8\,\rm cm^{-3}$), whereas in the r–process it is much higher ($N_{\rm n} \simeq 10^{20}\,\rm cm^{-3}$).

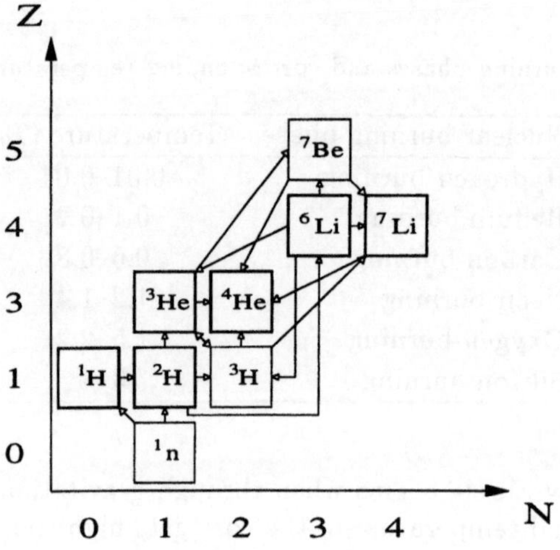

Figure 1. Primordial nucleosynthesis in the Standard Big Bang (SBB)

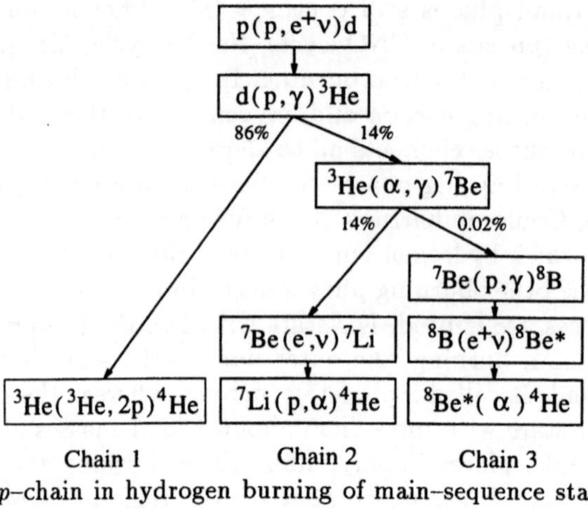

Figure 2. The *pp*–chain in hydrogen burning of main–sequence stars

In Fig. 1 the nuclear reaction network for primordial nucleosynthesis in the Standard Big Bang (SBB) is shown. Figure 2 shows the *pp*–chain dominating in low–mass main–sequence stars, like our sun, whereas the CNO–cycle takes over for more massive stars. There are two other scenarios additionally to the before discussed s– and r–process where neutron capture plays also a dominating role. The first one is an alternative to the standard Big Bang and is called the Inhomogenous Big Bang (IBB) with neutron densities up to about $N_n \simeq 10^{20}\,\mathrm{cm}^{-3}$. The second one is the so-called α–process occuring in type–II supernovae with neutron densities of about $N_n \simeq 10^{20-30}\,\mathrm{cm}^{-3}$. Figure 3 shows the net–work up to a mass number of $A = 12$ which should be taken into ac-

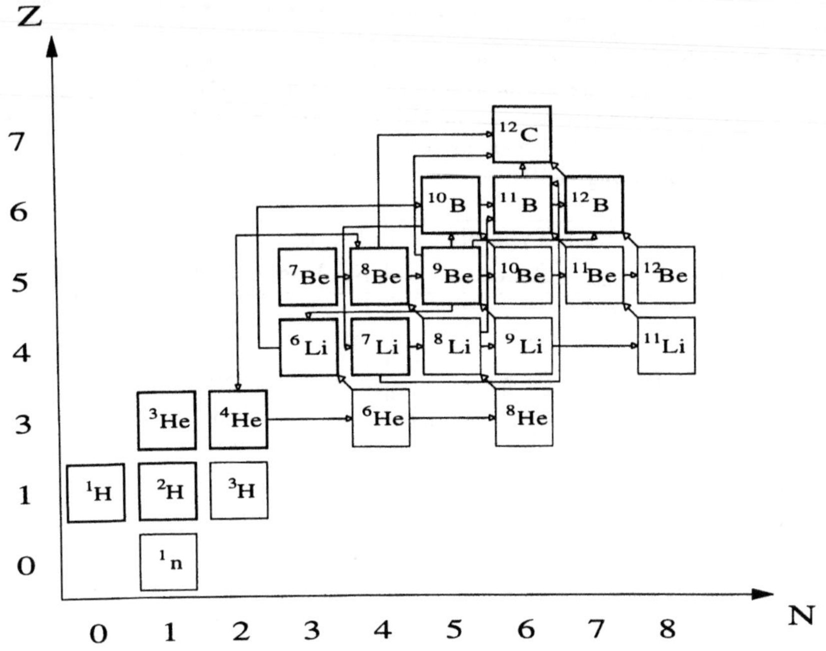

Figure 3. Additional reaction network in primordial nucleosynthesis in the α–process in type II supernovae or the Inhomogenous Big Bang (IBB)

count additionally to Fig. 2 for primordial nucleosynthesis in the IBB and the α–process. These are neutron–rich scenarios, where reactions involving light nuclei on the neutron–rich side of the region of stability have to be included. The main nuclear reactions up to a mass number of $A = 12$ taking place in the SBB, IBB, *pp*–chain, helium burning and the α–process are listed in Table 2.

We want to discuss the α–process a little more, because the reaction rates calculated in the next section occur in this scenario. The α–process takes place in the neutrino–heated hot bubble between the nascent neutron star and the overlying stellar mantle of a type–II supernova [2, 3, 4, 5, 6]. During the gravitational collapse in the supernova the temperature in the core of the star gets so high (up to 10^{10} K) that all the nuclei are disassembled through photodissociation into protons and neutrons. Afterwards, the material being originally in nuclear statistical equilibrium (NSE) at high temperature, is expanded and cooled so rapidly that not all the α–particles have time to reassemble. In this environment most nucleons are either in the form of free neutrons or bound in α–particles. The nucleosynthesis in the neutrino bubble takes place in subsequent steps: first, α–particles are formed from the free nucleons. Then, in the following α–process, nuclei up to about $A \approx 100$ are produced [2]. Finally, the neutrino bubble is also an ideal site for the r–process synthesizing the elements of about $A \geq 100$ [2, 7, 8].

Table 2. Nuclear reactions of astrophysical interest for light nuclei up to $A = 12$. The left column lists the major reactions for nucleosynthesis in the Standard Big Bang (SBB), additional reactions in the *pp*–chain and helium burning. The right column lists reactions which can be additionaly activated in the α–process or the Inhomogenous Big Bang (IBB).

Target Nucleus	Residual Nucleus	Type of Reaction	Target Nucleus	Residual Nucleus	Type of Reaction
	SBB			α–process, IBB	
^{1}n	^{1}H	β^-	^{4}He	^{6}He	$(2n,\gamma)$
^{1}H	^{2}H	(n,γ)	^{6}He	^{8}He	$(2n,\gamma)$
^{2}H	^{3}H	(n,γ), (d,p)	^{6}He	^{6}Li	β^-
^{2}H	^{3}He	(p,γ), (d,n)	^{6}Li	^{10}B	(α,γ)
^{2}H	^{6}Li	(α,γ)	^{7}Li	^{8}Li	(n,γ)
^{3}H	^{3}He	β^-	^{7}Li	^{11}B	(α,γ)
^{3}H	^{4}He	(d,n), (p,γ)	^{7}Be	^{8}Be	(n,γ)
^{3}H	^{7}Li	(α,γ)	^{8}He	^{8}Li	β^-
^{3}He	^{3}H	(n,p)	^{8}Li	^{9}Li	(n,γ)
^{3}He	^{4}He	(d,p), (n,γ)	^{8}Li	^{8}Be	β^-
^{3}He	^{7}Be	(α,γ)	^{8}Li	^{11}B	(α,p)
^{6}Li	^{3}H	(n,α)	^{8}Be	^{9}Be	(n,γ)
^{6}Li	^{3}He	(p,α)	^{9}Li	^{9}Be	β^-
^{6}Li	^{7}Li	(n,γ)	^{9}Li	^{11}Li	$(2n,\gamma)$
^{6}Li	^{7}Be	(p,γ)	^{9}Be	^{6}Li	(p,α)
^{7}Li	^{4}He	(p,α)	^{9}Be	^{10}Be	(n,γ)
^{7}Be	^{7}Li	(e^-,ν)	^{9}Be	^{10}B	(p,γ), (d,n)
^{7}Be	^{4}He	(n,α)	^{9}Be	^{12}B	(α,p)
^{7}Be	^{7}Li	(n,p)	^{9}Be	^{12}C	(α,n)
	pp–chain		^{10}Be	^{11}Be	(n,γ)
^{1}H	^{2}H	$(p,e^+\nu)$	^{10}Be	^{10}B	β^-
^{2}H	^{3}He	(p,γ)	^{10}B	^{7}Li	(n,α)
^{3}He	^{4}He	$(^3\text{He},2p)$	^{10}B	^{11}B	(n,γ)
^{3}He	^{7}Be	(α,γ)	^{11}Li	^{11}Be	β^-
^{7}Be	^{8}B	(p,γ)	^{11}Be	^{12}Be	(n,γ)
^{8}B	^{8}Be	$\beta+$	^{11}Be	^{11}B	β^-
^{8}Be	^{4}He*	decay	^{11}B	^{12}B	(n,γ)
	helium burning		^{11}B	^{12}C	(p,γ), (d,n)
^{4}He	^{12}C	$(2\alpha,\gamma)$	^{12}Be	^{12}B	β^-
^{8}Be	^{4}He	decay	^{12}B	^{12}C	β^-, (p,n)

3 Few–Body Calculations

Few–body methods have been used only partially for the calulation of astrophysical S–factors or reaction rates of astrophysically relevant nuclear reactions. An example is an exact three–nucleon calculation for radiative capture by deuterons starting from realistic NN–potentials. In this calculation the astrophysically relevant reaction rates for ^{2}H(p,γ)^{3}He at thermonuclear energies and ^{2}H(n,γ)^{3}H at thermal energies have been determined [9]. Another example is the calculation of the non–resonant part of the cross section for ^{4}He(d,γ)^{6}Li in the thermonuclear energy region. It was calculated in an α+p+n model by solving the three–body Faddeev equations [10] and using a three–body variational approach in the asymptotic region [11].

In the following we present the two–step processes ^{4}He($2n$,γ)^{6}He and the photodisintegration of ^{6}He as an example for an application of few–body calculations to astrophysically relevant nuclear reactions. A more extensive presentation of this calculation is given in ref. [12].

The reaction rate per particle triplet for ^{4}He($2n$,γ)^{6}He, in analogy to the triple–alpha process, is given by [13]

$$\begin{aligned}\left\langle 2n^4\mathrm{He}\right\rangle &= 2\int_0^\infty dE_1 \frac{\hbar}{\Gamma\left(^5\mathrm{He}, E_1\right)} \frac{d\left\langle n^4\mathrm{He}\right\rangle(E_1)}{dE_1} \\ &\quad \int_0^\infty dE_2 \frac{d\left\langle n^5\mathrm{He}\right\rangle(E_1,E_2)}{dE_2} \quad , \end{aligned} \tag{1}$$

where E_1 and E_2 are the relative energies in the (^{4}He+n)– and (^{5}He+n)–system, respectively. To determine the reaction rate in Eq. (1), an integration over both energies E_1 and E_2 has to be performed. The quantity $\Gamma\left(^5\mathrm{He}, E_1\right)$ is the energy–dependent width of ^{5}He, whereas $d\left\langle n^4\mathrm{He}\right\rangle(E_1)/dE_1$ and $d\left\langle n^5\mathrm{He}\right\rangle(E_1,E_2)/dE_2$ are the integrands of the reaction rates $\langle\sigma v\rangle$ [1] for the first and second step, respectively.

To calculate the second step reaction rate $\left\langle n^5\mathrm{He}\right\rangle$, the ^{5}He($n$,$\gamma$)^{6}He cross section is required. One may estimate that the electric dipole transition E1 from the incident s–wave is dominating the reaction. This is confirmed by both three–body and direct–capture calculations. The cross section is given by

$$\sigma_{\mathrm{E1}}\left(E_1, E_2\right) = \frac{2\pi}{81}\left(\frac{E_\gamma}{\hbar c}\right)^3 \frac{e^2}{\hbar v_2} \mid I \mid^2 \tag{2}$$

with

$$I = \frac{1}{k_2}\int_0^\infty dr_1 \int_0^\infty dr_2 u_{\mathrm{b}}(E_1,r_1) r_1 u_{\mathrm{sc}}(E_2,r_2) r_2^2 f_{\ell=1,j=3/2}(r_1,r_2) \quad , \tag{3}$$

where $E_\gamma = E_1 + E_2 + Q_{12}$ is the energy of the photon, with Q_{12} being the Q–value of the reaction ^{4}He($2n$,γ)^{6}He. In Eq. (3) the coordinates r_1 and r_2 are the distances of the valence neutrons from the α–core and ^{5}He, respectively. The quantities v_2 and k_2 are the relative velocity and wave number in the entrance

Table 3. Relevant input parameters for our calculations: Q_{12} is the Q–value for both steps, J^{Π}, $E_{\rm R}$ and $\Gamma_{\rm R}$ are the spin–parity assignments, resonance energies and widths of the indermediate nuclei ^{5}He, ^{7}He and ^{10}Li

reaction	Q_{12} [MeV]	J^{Π}	$E_{\rm R}$ [MeV]	$\Gamma_{\rm R}$ [MeV]
^{4}He(2n,γ)^{6}He	0.98[a])	$\frac{3}{2}^-$[b])	0.89[b])	0.76 ± 0.11[c])
^{6}He(2n,γ)^{8}He	2.14[a])	$\frac{3}{2}^-$[b])	0.44[b])	0.16 ± 0.03[b])
^{9}Li(2n,γ)^{11}Li	0.31[a])	1^+[d])	0.54[d])	0.22 ± 0.01[d])

[a]) Ref. [17]; [b]) Ref. [18]; [c]) Ref. [19]; [d]) Ref. [20]

channel (^{5}He+n) of the second step. The quantity $u_{\rm b}(E_1, r_1)/r_1$ represents the radial part of the quasi–bound wave function of the ^{5}He–resonance, and $u_{\rm sc}(E_2, r_2)/r_2$ is the radial part of the (^{5}He+n)–scattering wave function. The function $f_{\ell=1,j=3/2}(r_1, r_2)$ is the radial component in the expansion

$$\Psi\,(^6\mathrm{He}) = \sum_{\ell j} f_{\ell j}(r_1, r_2)\,[\chi_{\ell j}\,(\hat{\mathbf{r}}_1, \sigma_{1z})\,\chi_{\ell j}\,(\hat{\mathbf{r}}_2, \sigma_{2z})]_{J=0} \tag{4}$$

of the α+n+n ground–state wave function. Here $\chi_{\ell j}$ are the angular–momentum functions with orbital angular momentum ℓ and total angular momentum $j = \ell \pm 1/2$, and the brackets [...] denote vector coupling. The expansion includes only equal ℓ–values for both neutrons and only the ($\ell = 1, j = 3/2$)–component contributes to the cross section.

To calculate the first step reaction rate $\langle n^4\mathrm{He}\rangle$ we need the (^{4}He+n)–potential. It was obtained by fitting a folding–type potential [14, 15] to the experimental energy and width of the ^{5}He ground state. For the calculation of the energy dependent width $\Gamma\,(^5\mathrm{He}, E_1)$ of ^{5}He in Eq. (1) we used an equivalent square well potential with the same r.m.s. radius of the density distribution as for the folding potential.

In the second step the ground–state wave function $u_{\rm b}(E_1, r_1)$ of ^{5}He was determined by solving the single–particle Schrödinger equation, using the (^{4}He+n)–potential and a boundary condition for a decaying system. It was then cut at a reasonable cut–off radius. For the determination of the scattering wave function $u_{\rm sc}(E_2, r_2)$ we used again folding–type potentials. An accurate three–body wave function of ^{6}He (see [16]) was used in our calculations.

The relevant input parameters for the calculations of the reaction rate are shown in Table 3. The calculated cross section for ^{5}He(n,γ)^{6}He can be parametrized by $\sigma = \sigma_1/\left(E_2\,[\mathrm{MeV}]\right)^{1/2}$ [μb]. The ^{5}He(n,γ)^{6}He cross section exhibits the well–known $1/v$–behavior for incident s–waves.

The obtained reaction rate can be parametrized in the following way:

$$N_{\rm A}^2\,\langle 2n^4\mathrm{He}\rangle = aT_9^b \exp\left(\frac{c}{T_9}\right)\,10^{-8}\,\mathrm{cm}^6\,\mathrm{s}^{-1}\,\mathrm{mol}^{-2}\quad , \tag{5}$$

where the parameters a, b, c are shown in Table 4 for the two temperature regions $0.1 \le T_9 < 2$ and $2 \le T_9 \le 15$ (T_9: in units of 10^9 K). The inverse

Table 4. Parameters of the cross section for ^{5}He$(n,\gamma)^6$He and the reaction rates for ^{4}He$(2n,\gamma)^6$He and the reverse photodisintegration of ^{6}He calculated in our three–body model and in direct capture

	Three–body calculation		Direct–capture calculation	
σ_1 [MeV$^{1/2}\mu$b]	0.152		0.270	
	$0.1 \le T_9 \le 2$	$2 < T_9 \le 15$	$0.1 \le T_9 \le 2$	$2 < T_9 \le 15$
a	0.00265	0.293	0.00471	0.520
b	2.55	−0.351	2.55	−0.351
c	0.181	−5.24	0.181	−5.24
d	3.14	0.0286	3.14	0.0286
e	−1.05	1.85	−1.05	1.85
f	−21.9	−16.5	−21.9	−16.5

reaction rate λ_γ per nucleus per second can be calculated by using the RevRatio [21]

$$\lambda_\gamma = \text{RevRatio} \times N_A^2 \, \langle 2n^4\text{He} \rangle \ \text{s}^{-1} \quad , \tag{6}$$

where we assumed that the photodisintegration proceeds through the ^{5}He resonance. The RevRatio is parametrized in the following way

$$\text{RevRatio} = dT_9^e \exp\left(\frac{f}{T_9}\right) 10^{23}\,\text{cm}^{-6}\,\text{mol}^2 \quad . \tag{7}$$

The parameters d, e, f are also shown in Table 4.

The reaction rate for ^{4}He$(2n,\gamma)^6$He calculated with the help of our three–body model (second column of Table 4) is at least more than three orders of magnitude larger than the previously adopted value [21]. This is mainly due to the non–resonant E1–transition of the second step ^{5}He$(n,\gamma)^6$He, which dominates the cross section, and which was not taken into account previously.

4 Direct–Capture (DC) Calculations

We also calculated the reaction rates for ^{4}He$(2n,\gamma)^6$He and the reverse photodisintegration of ^{6}He by using the DC model for the second step of the reaction. In this case Eq. 3 reduces to

$$I = \frac{S_{\ell=1,j=3/2}}{k_2} \int_0^\infty dr u_{\text{sc}}(E_2, r) r^2 u_{\ell=1,j=3/2}(r) \quad , \tag{8}$$

where $r \equiv r_2$ and $S_{\ell=1,j=3/2}$ is the spectroscopic factor and $u_{\ell=1,j=3/2}(r)$ is the wave function for ^{6}He = ^{5}He+n.

The spectroscopic factor for ^{5}He$(n,\gamma)^6$He was calculated by shell–model calculations using the (6–16) 2BME interaction of Cohen and Kurath [22] and is given by $S_{\ell=1,j=3/2} = 1.3139$. All the other parameters (see Table 3) for

Table 5. Parameters for the cross section of the second step and reaction rates of ^{6}He$(2n,\gamma)^{8}$He and ^{9}Li$(2n,\gamma)^{11}$Li, and the reverse photodisintegration of ^{8}He and ^{11}Li calculated in direct capture

	^{6}He$(2n,\gamma)^{8}$He		^{9}Li$(2n,\gamma)^{11}$Li	
$\sigma_{1\,\text{MeV}}$ [μb]	1.828		0.2435	
	$0.1 \leq T_9 \leq 2$	$2 < T_9 \leq 15$	$0.1 \leq T_9 \leq 2$	$2 < T_9 \leq 15$
a	8.015	36.91	0.651	0.956
b	−0.209	−1.187	0.042	−0.412
c	−2.514	−4.262	−3.394	−3.432
d	0.013	0.003	0.004	0.003
e	1.709	2.687	1.458	1.912
f	−27.427	−25.679	−6.470	−6.432

the DC calculation of ^{4}He$(2n,\gamma)^{6}$He were assumed to be the same as in the three-body calculation. For the DC calculation we used the code TEDCA [23].

As can be seen from Table 4 the cross section for the second step and the total reaction rate of ^{4}He$(2n,\gamma)^{6}$He is enhanced by 78% enhanced compared to the three-body calculations. Therefore, we assumed that the DC model is also adequate within a factor of two to calculate the reaction rates for two other two-step processes involving halo nuclei like ^{6}He$(2n,\gamma)^{8}$He and ^{9}Li$(2n,\gamma)$, and the photodisintegration of ^{8}He and ^{11}Li. These reactions could also be of importance in the α-process.

We assume as for ^{4}He$(2n,\gamma)^{6}$He that again two p-neutrons are transferred in the two-step process and that in the second step the E1-transition is dominating. For ^{9}Li$(2n,\gamma)^{11}$Li we did not consider yet possible predicted low-energy intruder s-states in ^{10}Li [24, 20] and/or ^{11}Li [25, 26]. Such states could lead to s-wave resonances in ^{11}Li near the (^{10}Li$+n$)-threshold modifying our calculated non-resonant cross section for ^{10}Li$(n,\gamma)^{11}$Li. Furthermore, by considering such s-wave states the two-step process could also proceed over 1^{-}- or 2^{-}-states in ^{10}Li and not only over the 1^{+}-state at 0.54 MeV above threshold (see Table 3) that we considered in our DC-calculation.

The used input parameters are again shown in Table 3. The spectroscopic factors are given by $S_{\ell=1,j=3/2} = 3.2047$ for ^{7}He$(n,\gamma)^{8}$He, and $S_{\ell=1,j=3/2} = 0.647$ and $S_{\ell=1,j=1/2} = 0.024$ for ^{10}Li$(n,\gamma)^{11}$Li. For the potentials in the exit channels we used the folding method adjusted to the separation energy and the r.m.s. radii of the neutrons for the residual nuclei.

The results for the parameters of the cross sections for the second step and the reaction rates are also shown in Table 5. The reaction rate for ^{6}He$(2n,\gamma)^{8}$He is at $T_9 \approx 1$ about a factor 40 larger than for ^{4}He$(2n,\gamma)^{6}$He. Very similar results for the reaction rate of ^{4}He$(2n,\gamma)^{6}$He have also been obtained in another DC calculation by Görres et al. [27]. However, our reaction rate for ^{6}He$(2n,\gamma)^{8}$He is more than a factor 10 larger than in ref. [27], mainly because we used a larger spectroscopic factor. The reaction rate for ^{9}Li$(2n,\gamma)^{11}$Li (Table 5) is at

$T_9 \approx 1$ almost a factor two enhanced compared to ^{4}He$(2n,\gamma)^6$He (Table 4), even though the total Q–value for the Li–reaction is a factor three smaller (Table 3). This is mainly due to the more pronounced halo structure of the Li–nuclei as compared to the He–nuclei.

Acknowledgement. The authors are indebted to I.J. Thompson for help in their work and to J.S. Vaagen and M.V. Zhukov for valuable discussions on halo nuclei. This work was supported by the Fonds zur Förderung der wissenschaftlichen Forschung in Österreich (project P10361–PHY), by the Österreichische Nationalbank (project 5054) and by the International Science Foundation (grant J4M100). Partial financial support by the Russian–British–Northic Theory collaboration (RNBT) is acknowledged.

References

1. C.E. Rolfs, W. Rodney: *Cauldrons in the Cosmos.* Chicago: University of Chicago Press 1988
2. S.E. Woosley, R.D. Hofmann: Astrophys. J. **395**, 202 (1992)
3. B.S. Meyer et al.: Astrophys. J. **399**, 656 (1992)
4. W.M. Howard et al.: Astrophys. J. **417**, 713 (1993)
5. S.E. Woosely et al.: Astrophys. J. **433**, 229 (1994)
6. J. Witti, H.-Th. Janka and K. Takahashi: Astron. and Astrophys. **286**, 841 (1994)
7. K.L. Kratz et al.: Astrophys. J. **403**, 216 (1993)
8. K. Takahashi, J. Witti and H.-Th. Janka: Astron. and Astrophys. **286**, 857 (1994)
9. J.L. Friar et al.: Phys. Rev. Lett. **66**, 1827 (1991)
10. L.D. Blokhintsev et al.: Phys. Rev. C **48**, 2390 (1993)
11. V.I. Kukulin et al.: Nucl. Phys. A **417**, 128 (1984); **453**, 365 (1986)
12. V.D. Efros et al.: Phys. Lett. B (submitted)
13. K. Nomoto, F.-K. Thielemann, S. Hiyaji: Astron. and Astrophys. **149**, 239 (1985)
14. A.M. Kobos et al.: Nucl. Phys. A **425**, 205 (1984)
15. H. Oberhummer, G. Staudt: In: H. Oberhummer, (ed.): *Nuclei in the Cosmos*, p.29. Heidelberg: Springer Verlag 1991
16. M.V. Zhukov et al.: Physics Rep. **231**, 151 (1993)

17. G. Audi, A.H. Wapstra: Nucl. Phys. A **565**, 1 (1993)

18. F. Ajzenberg–Selove: Nucl. Phys. A **490**, 1 (1988)

19. V.D. Efros et al.: Phys. Rev. C (to be submitted)

20. B.M. Young et al.: Phys. Rev. C **49**, 279 (1994)

21. W.A. Fowler, G.R. Caughlan and B. Zimmerman: Ann. Rev. Astron. Astrophys. **13**, 69 (1975)

22. S. Cohen, D. Kurath: Nucl. Phys. **73**, 1 (1965)

23. H. Krauss: Computer code TEDCA. Technical University Vienna, 1992

24. R. Kryger et al.: Phys. Rev. C **47**, R4239 (1993)

25. I.J. Thompson, M.V. Zhukov: Phys. Rev. C **49**, 1904 (1994)

26. S. Shimumora et al.: Phys. Lett. B **348**, 29 (1995)

27. J. Görres et al.: Phys. Rev. C (submitted and private communication)

Few-Body Systems Suppl. 8, 329–334 (1995)

Few-Body Systems

Optical Theorem and Finite-Range Effect for Nuclear Reactions in Astrophysics

Yeong E. Kim, Alexander L. Zubarev

Department of Physics, Purdue University, West Lafayette, IN 47907, U.S.A.

Abstract. Optical theorem formulations of low-energy nuclear two-body and three-body reactions are presented. We obtain analytical formulae for the cross-section $\sigma(E)$ which exhibit explicitly the energy and charge dependences of $\sigma(E)$ and may provide a better physical understanding of anomalous low-energy enhancement of $\sigma(E)$ observed in nuclear fusion reactions relevant for the primodal nucleosynthesis and stellar evolution.

1 Introduction

There are many remaining uncertainties in nuclear astrophysics calculations due to unknown rates and cross-sections, $\sigma(E)$, for many nuclear reactions. Since $\sigma(E)$ at energies (a few keV) relevant to the primodal and stellar nucleosynthesis cannot be measured in the laboratory, they are extracted from the laboratory measurements of $\sigma(E)$ at higher energies by an extrapolation procedure based on nuclear theory. For non-resonance reactions, it is customary to extract the S-factor, $S(E)$, from the experimentally measured $\sigma(E)$ using the following formula

$$\sigma_G(E) = \frac{S(E)}{E} e^{-2\pi\eta(E)} \tag{1}$$

where $\eta(E) = Z_i Z_j e^2/\hbar v$, $e^{-2\pi\eta(E)}$ is the Gamow factor representing the probability of bringing two charged nuclei to zero separation distance, and $S(E)$ is expected to be a slowly varying function of E.

The energy dependence of the nuclear reaction cross-section $\sigma(E)$ cannot be obtained rigorously from first principles. Therefore, one must rely on physically reasonable nuclear reaction models, such as P-matrix method [1]. Recently, we have developed a new improved theoretical formulation of non-resonance nuclear reactions, which is based on the optical theorem and is nearly model-independent [2]. It can offer a better physical understanding of existing data exhibiting anomalous enhancement of the astrophysical S-factor at low energies which cannot be explained as the electron screening effect. The optical theorem formulation of three-body fusion reactions is also described.

2 Optical Theorem Formulation

The optical theorem for two-potential scattering problems involving both the strong and Coulomb interactions is given by $\mathrm{Im} f_\ell^{N(\mathrm{el})} = k(\sigma_\ell^{(r)} + \sigma_\ell^{N(\mathrm{el})})/\, 4\pi \approx k\sigma_\ell^{(r)}/4\pi$ at low energies, where $\sigma_\ell^{(r)}$ is the $\ell -$ th partial wave reaction cross-section and $\sigma_\ell^{N(\mathrm{el})} = |f_\ell^{N(\mathrm{el})}|^2$ is a part of the total elastic cross-section excluding the Rutherford scattering and interference term [2]. Using the optical theorem, we obtain the partial-wave cross-section $\sigma_\ell(E)(= \sigma_\ell^{(r)}(E))$ as

$$\sigma_\ell(E) \approx \frac{4\pi}{kE}\int_0^\infty \psi_\ell^C(r) U_\ell(r,r') \psi_\ell^C(r') dr dr' \tag{2}$$

where $E = \hbar^2 k^2/2\mu$ with the reduced mass μ, $U_\ell(r,r') = -\mathrm{Im}\langle r|T_\ell|r'\rangle$ with T_ℓ representing the $\ell -$ th partial wave contribution of the T-matrix operator, and ψ_ℓ^C is the $\ell -$ th partial wave Coulomb wave function. The total cross-section $\sigma(E)$ is given by $\sigma(E) = \sum(2\ell+1)\sigma_\ell(E)$. It can be shown [2] that $U_\ell(r,r')$ has a separable form for the case of two-channel (elastic and fusion). Our optical theorem formulation of nuclear reactions, Eq. (2), can be applied to both non-resonance and resonance reactions using the T-matrix or $U_\ell(r,r')$ given in Eq. (2). In the following, we consider only the non-resonance case.

To estimate the S-wave cross-section, $\sigma_0(E)$, we parametrize $U_0(r,r')$ by two parameters λ (strength/length) and β^{-1} (range) in a separable form as

$$U_0(r,r') = \lambda g(r) g(r') \ . \tag{3}$$

For $g(r) = e^{-\beta r}/r$, the integral in Eq. (2) can be carried out analytically using Eq. (3) and the exact form of $\psi_0^C(r)$. The result for the low-energy case is given by

$$\sigma_0(E) = \frac{4\pi\lambda}{kE}\left[\int_0^\infty dr \psi_0^c(r) e^{-\beta r}/r\right]^2 = \frac{4\lambda\pi^2}{E} R_B \frac{\left(e^{-2\phi\eta}-1\right)^2}{(e^{2\pi\eta}-1)} e^{4\phi\eta} \tag{4}$$

with

$$e^{4\phi\eta} = \exp\left[4\alpha\frac{\mu c^2}{\hbar c}(\frac{Z_i Z_j}{k})\tan^{-1}(\frac{k}{\beta})\right] \tag{5}$$

where $\phi = \tan^{-1}(k/\beta)$ and $R_B = \hbar^2/(2\mu Z_i Z_j e^2)$. The energy dependence of λ is expected to be weak. The use of a more general form for $g(r) = (e^{-\beta r}/r)\Sigma_{i=0}^N a_i r^i$ instead of $e^{-\beta r}/r$ in Eq. (3) also leads to the same enhancement factor $e^{4\phi\eta}$. Therefore, the enhancement factor $e^{4\phi\eta}$ is independent of shape of the separable function $g(r)$ used in Eq. (3). For $g(r) = (e^{-\beta r}/r)\sum_{i=0}^N a_i r^i$, we obtain for $\sigma_0(e)$ the following result

$$\sigma_0(E) = \frac{\tilde{S}_0(E)}{E} e^{4\phi\eta} e^{-2\pi\eta} \tag{6}$$

where

$$\tilde{S}_0(E) = \sum_{i,j} c_i c_j F_i^{\beta}(E) F_j^{\beta}(E) \tag{7}$$

with $F_0^{\beta}(E) = (e^{-2\phi\eta} - 1)$, $F_1^{\beta}(E) = (\beta^2 + k^2)^{-1}$, $F_2^{\beta}(E) = (\beta^2 + k^2)^{-2}$, etc. The enhancement factor $e^{4\phi\eta}$ is $e^{2/(R_B\beta)}$ at zero energy and decreases as E increases to $e^{\pi/kR_B} = 1$ for large E, and thus can account for the increase of experimentally extracted $S(E)$ toward lower energies. The enhancement factor $e^{4\phi\eta}$ can be applied both to light nuclei reactions (small Z_i and Z_j) and also to heavy ion reactions (larger Z_i and Z_j) such as sub-barrier heavy ion fusion where $e^{4\phi\eta}$ can be very large. Our derivation of Eq. (6) is the first derivation of the Gamow factor based on the optical theorem, and is much more rigorous than other previous derivations, most of which are based on the barrier transmission coefficient derived in the WKB approximation. The previous nuclear reaction models do not yield an explicit analytical form of the Gamow factor.

3 Application

As an application of our formula, Eq. (6), we consider the S-factor for $^6\text{Li}(d,\alpha)^4\text{He}$ reaction [3]. In the previous analysis [3] of this reaction, the S-factor, $S(E)$, in Eq. (1) is modified to $S_G(E) \approx f_U(E)S(E)$ where an enhancement factor $f_U(E) \approx \exp(\pi\eta U/E)$ is due to the electron screening effect, U is the electron screening energy, and $S(E)$ is parametrized as $S(E) = \sum_{i=0}^{n} b_i E^i$ (up to 4 terms, $n = 3$). Fits to the experimental data [3] yield unphysically large values (by a factor of ~ 2) of $U \approx 380$ eV than the expected values ($\approx$ 162 eV) from the Thomas-Fermi model. Using $U = 162$ eV and eqs. (5) and (6) we obtain a reasonable fit to the experimental data for $^6\text{Li}(d,\alpha)^4\text{He}$ ($\chi^2 \approx 0.35$ per data point for 10 keV < E < 500 keV with $\beta^{-1} = 5.0$ fm, $c_0 = 8.0$ (keV b)$^{1/2}$, and $c_1 = -1.5$ (keV b)$^{1/2}$/fm^2) yielding $S(0) = (24.1 \pm 2.1) \times 10^3$ keV b, which is substantially larger than the previous result of $S(0) = 17.4 \times 10^3$ keV b [3]. This may have important implications for the primodal nucleosynthesis and stellar evolution.

4 Three-Body Fusion Reactions

Recently, energetic protons ($\lesssim$ 17 MeV) and α particles ($\lesssim$ 6.5 MeV) have been observed in experiments in which deuterated Ti target is bombarded with 150 keV deuteron beam [4]. The observed proton and α-particle energy spectrum cannot be explained with existing two-body fusion reactions but is consistent with a three-body fusion reaction, $d+d+d \rightarrow p+n+^4\text{He}$ [4]. The observed rate $R^{(3)}$ is $\sim 10^{-6} R^{(2)}$ where $R^{(2)}$ is the observed fusion rate for D(d,p)T with 150 keV deuteron beam, i.e., $R^{(3)}/R^{(2)} \approx 10^{-6}$. However, the conventional theoretical estimate for $R^{(3)}/R^{(2)}$ is $\sim 10^{-17}$ [4] or much less, and hence the observed value of 10^{-6} for $R^{(3)}/R^{(2)}$ is anomalous. In this section, we describe the $3d$ fusion cross-section $\sigma^{(3)}(E)$ and rate $R^{(3)}(E)$ based on the optical theorem and a solution of three-body Schrödinger equation.

For the $3d$ fusion reaction, we consider two-channel case,

$$d+d+d \longrightarrow \begin{cases} d+d+d \\ p+n+{}^4\mathrm{He} \quad (Q=21.62\ \mathrm{MeV})\ . \end{cases} \tag{8}$$

Three-body Schrödinger equation in the elastic channel is (we suppress the spin and isospin degrees of freedom for simplicity)

$$-\frac{\hbar^2}{2m}(\nabla_x^2+\nabla_y^2)\Psi(\boldsymbol{x},\boldsymbol{y})+V\Psi(\boldsymbol{x},\boldsymbol{y})=E\Psi(\boldsymbol{x},\boldsymbol{y}) \tag{9}$$

where $\boldsymbol{x}=\sqrt{\frac{1}{2}}(\boldsymbol{r}_2-\boldsymbol{r}_3)$, $\boldsymbol{y}=\sqrt{\frac{2}{3}}[\frac{1}{2}(\boldsymbol{r}_2+\boldsymbol{r}_3)-\boldsymbol{r}_1]$, $V=V^C+V^S=(V_{12}^C+V_{23}^C+V_{13}^C)+(V_{12}^S+V_{23}^S+V_{13}^S)$ with $V^S=V_{\mathrm{Re}}^S-iV_{\mathrm{Im}}^S$. We introduce the hyperspherical harmonics expansion for $\Psi(\boldsymbol{x},\boldsymbol{y})$ as $\Psi(\boldsymbol{x},\boldsymbol{y})=\sum_{K\ell_x\ell_y}\Psi_K^{\ell_x\ell_yLM}(\rho)\Phi_K^{\ell_x\ell_yLM}(\Omega)$, where $\rho^2=x^2+y^2$ (hyperradius), and write $\Psi_K^{\ell_x\ell_yLM}(\rho)=U_K^{\ell_x\ell_yLM}(\rho)/\rho^{5/2}$.

For $K=0(\ell_x=0,\ell_y=0,L=0)$ case (known as "true" [5] or "democratic" state [6]), Eq. (9) reduces to

$$\frac{d^2U}{d\rho^2}+(k^2-\frac{15/4}{\rho^2})U=\frac{2m}{\hbar^2}\tilde{V}U\ , \tag{10}$$

where $\tilde{V}=\tilde{V}^C+\tilde{V}^S$ with $\tilde{V}=V_{00}^{00,00}$ with $V_{KK'}^{\ell_x\ell_y,\ell_x'\ell_y'}=\langle\Phi^{\ell_x\ell_yLM}|V|\Phi^{\ell_x'\ell_y'LM}\rangle$. For $\tilde{V}=\tilde{V}^C$(i.e. $\tilde{V}^S=0$), the solution $U^c(\rho)$ of Eq. (10) can be obtained in terms of the Whittaker function.

In terms of T-matrix, the $K=0$ elastic $3d$ nuclear scattering amplitude, $f_{3B}^{(\mathrm{el})}$, can be written as $f_{3B}^{(\mathrm{el})}=-2m\langle U^C|T|U^C\rangle/(\hbar^2k^{7/2})$. Using the optical theorem [7], $\sigma_r=\sqrt{2}\pi(\pi^2/k^{5/2})\mathrm{Im}(f_{3B}^{(\mathrm{el})})$, we obtain the following expression for the $3d$ fusion cross-section,

$$\sigma^{(3)}(E)=\sigma_r\approx\frac{2m}{\hbar^2k^6}\pi^2\sqrt{2\pi}\int_0^\infty U^C(\rho)W(\rho,\rho')U^C(\rho')d\rho d\rho' \tag{11}$$

where $W(\rho,\rho')=-\mathrm{Im}\langle U^C|T|U^C\rangle$. If we parametrize $W(\rho,\rho')=-\mathrm{Im}\langle U^C|T|U^C\rangle=\tilde{\lambda}g(\rho)g(\rho')$ with $g(\rho)=e^{-\beta\rho}\rho^{-3/2}$, we obtain

$$\sigma^{(3)}(E)\approx\tilde{\lambda}\frac{2m}{\hbar^2k_3}\pi^2\sqrt{2\pi}C^2\frac{(4!)^2}{(\beta^2+k_3^2)^5}e^{4\phi\eta_3} \tag{12}$$

where

$$|C|^2=\frac{1}{36}(\frac{9}{4}+\eta_3^2)(\frac{1}{4}+\eta_3^2)\frac{\pi e^{-2\pi\eta_3}}{1+e^{-2\pi\eta_3}}\ . \tag{13}$$

Other quantities in eqs. (12) and (13) are $\phi=\tan^{-1}(k_3/\beta)$, $k_3=(2mE/\hbar^2)^{1/2}$, $\eta_3=1/(2k_3R_B^{(3)})$, $R_B^{(3)}=\hbar^2/(2mZ_{eff}e^2)$, $Z_{eff}=16/(\pi\sqrt{2})\approx3.6$, and $\tilde{\lambda}=V_0/a^4$.

We note that $\sigma^{(3)}(E)$ in Eq. (12) has a dimensionality of the 5th power of length $[L^5]$ since $f_{3B}^{(\mathrm{el})}$ has a dimensionality $[L^{5/2}]$ expected from the solution,

$\Psi(\rho) = e^{ik\cdot\rho} + f_{3B}^{(el)} e^{ik\rho}/\rho^{5/2}$, while two-body fusion cross-section $\sigma^{(2)}(E)$ has a dimensionality of $[L^2]$.

To make an order of magnitude estimate for the ratio $R = (R^{(3)}/R^{(2)})$, we approximate the ratio as $R \approx f n_D \sigma^{(3)}(E_{cm}^{(3)})/\sigma^{(2)}(E_{cm}^{(2)})$ where $E_{cm}^{(3)} = (2/3)E_d$, $E_{cm}^{(2)} = E_d/2$, n_D is the target deuteron number density, and f is the relative strength of $K = 0$ component in $\Psi(\boldsymbol{x}, \boldsymbol{y})$. In Eq. (12), $\tilde{\lambda}$ (or V_0 and a for $\tilde{\lambda} = V_0/a^4$) and β are parameters to be determined from theory and/or experiment. In order to fit the experimental value of $R \approx 10^{-6}$ and the observed energy dependence of R [4], we need $V_0 \approx 10$ MeV, $a = 2$ fm, and $\beta = 0.065\ \text{fm}^{-1}(\beta^{-1} = 15\text{fm})$ with $S^{(2)}(E) \approx S_0 \approx 50$ keV b and $f = 1$ (this is the upper limit).

The quantum number K is closely related to the impact parameter. In terms of classical orbits, a large value of K implies that three particles cannot come close together. For $K = 0$, all three particles seem to converge to or diverge from a scattering center [8]. Therefore the $K = 0$ "true" (or "democratic") three-body state is important for the three-body fusion reaction.

5 Conclusion

Since the enhancement factor $e^{4\phi\eta}$ is a general feature arising from the strong and Coulomb interactions (i.e. not from the electron screening effect) for bare nuclei without electrons, it is applicable to all strong-interaction non-resonance nuclear reactions (excluding radiative captures) occurring in the primodal nucleosynthesis and stellar evolutions. Therefore, it may have far-reaching astrophysical importance.

If the experimental data of Kasagi et al. [4] is conclusively determined to be due to $d+d+d \rightarrow p+n+{}^4\text{He}$ fusion by future $p-{}^4\text{He}$ coincidence measurements by Kasagi et al. and independently by other groups, the experimental result and its theoretical understanding based on "true" (or "democratic") three-body state may have very important implications for nucleosynthesis and stellar evolutions, since there are possibilities that other three-body fusion reactions may be involved in astrophysical problems, such as pycnonuclear triple-alpha fusion [9].

References

1. Y.E. Kim, A.L. Zubarev: Int. J. Theor. Phys. **33**, 1889 (1994)

2. Y.E. Kim, A.L. Zubarev: PNTG-95-3 (1995) to be published

3. S. Engstler et al.: Z. Phys. **A342**, 471 (1992)

4. J. Kasagi et al.: J. Phys. Soc. Japan **64**, 777 (1995)

5. R.G. Newton: Ann. Phys. (N.Y.) **74**, 324 (1972); R.G. Newton, R. Shtochamer: Phys. Rev. **A14**, 642 (1976)

6. R.I. Jibuti, K.I. Sigua: Preprint IF AN GSSR 31-YaF (1977)

7. R.I. Jibuti, Ya. Keserashvili: Czech. J. Phys. **B30**, 1090 (1980)

8. E.W. Schmid, H. Ziegelmann: *The Quantum Mechanical Three-Body Problem*, Vieweg, Braunschweig, Germany, 1974

9. A.G.W. Cameron: Ap. J. **130**, 916 (1959); E.E. Salpeter, H.M. van Horn, Ap. J. **155**, 183 (1969)

Few-Body Systems Suppl. 8, 335–339 (1995)

Few-Body Systems

Peripheral Astrophysical Radiative Capture Processes

A.M. Mukhamedzhanov, R.E. Tribble*

Cyclotron Institute, Texas A&M University, College Station, Texas, 77843, USA

Abstract. We present a survey of the latest results for the astrophysical S-factors for the peripheral radiative capture reactions ^{7}Be$(p,\gamma)^8$B and $\alpha(d,\gamma)^6$Li.

1 Introduction

Many important astrophysical radiative capture reactions (RCR), such as $p + {^7}\mathrm{Be} \to {^8}\mathrm{B} + \gamma$, $\alpha + d \to {^6}\mathrm{Li} + \gamma$, $\alpha + t \to {^7}\mathrm{Li} + \gamma$, $\alpha + {^3}\mathrm{He} \to {^7}\mathrm{Be} + \gamma$, ${^{12}}\mathrm{C} + \alpha \to {^{16}}\mathrm{O}^* + \gamma$ and others are peripheral due to the low relative kinetic energy of the colliding particles compared to the Coulomb barrier and the small binding energy of the final nuclei. The overall normalization of the astrophysical S factors for such reactions is totally defined by only one quantity, the asymptotic normalization coefficient (ANC) of the overlap function of the bound state wave functions of the final and initial particles [1, 2, 3, 4]. The ANC's can be found from: 1) analysis of classical nuclear reactions such as elastic scattering (by extrapolation of the experimental scattering phase shifts to the bound state pole in the energy plane [5, 6]) or peripheral transfer reactions whose amplitudes contain the same overlap function as the amplitude of the corresponding astrophysical radiative capture process [3]; 2) theoretical calculations [7]. Recently using the concept of the ANC we have recalculated the astrophysical factors for two important RCR's ^{7}Be$(p,\gamma)^8$B and $\alpha(d,\gamma)^6$Li [2, 4]. Here we provide an updated survey of the latest developments for S-factors for these reactions.

2 The ^{7}Be$(p,\gamma)^8$B Reaction

The ^{7}Be$(p,\gamma)^8$B reaction, which occurs in the third $p-p$ chain of hydrogen burning of main-sequence stars, is of fundamental importance, both for defining the branching ratios between the different $p-p$ chains and for calculating the

* *E-mail address:* tribble@comp.tamu.edu

high-energy solar neutrino flux, which is of special interest for the solar neutrino problem. At astrophysical energies ($E_{cm} < 25\,keV$), the cross section for this reaction is so small that its measurement has not been possible to date. Among the several direct experiments, two were performed at the lowest energies [8, 9]. The energy dependence in the two sets agrees. However, one observes a 25% systematic difference in the overall normalization. The latest direct capture experiment at low energies, $E_{cm} > 117\,keV$, gives after extrapolation down to zero energy, the astrophysical factor $S_{17} = 21.7 \pm 2.5 eVb$ [9]. The overall uncertainty in the astrophysical factor, $S_{17}(0)$, derived in such a procedure has been estimated to be about 30% [10]. Very recently the first result from the Coulomb breakup of ^8B in the Coulomb field of ^{208}Pb has been reported [11]. The extracted S_{17}-factor appears to support the values that recently have been obtained of $S_{17}(0) = 17.6\,eVb$ [2] and $S_{17}(0) = 17 \pm 3\,eVb$ [13].

Theoretical calculations have also attempted to determine $S_{17}(0)$. Calculations have been made in the potential model and in the microscopic generator-coordinate method (GCM). Potential model calculations gave $S_{17}(0) = 24.7\,eVb$ [14] and $S_{17}(0) = 23.6 - 24.9\,eVb$ [15]. We note that in the potential approach the overall normalization of the cross section of the peripheral radiative capture, $^7Be(p,\gamma)^8B$, is defined by the product of two model quantities, the spectroscopic factor and the single-particle ANC of the $^7Be - p$ bound state wave function. This causes an uncertainty in the absolute normalization of the cross section, which has been explicitly demonstrated by Barker [12], who estimated the range of values for $S_{17}(0)$ to be between 13.5 and 22.3 eVb, based on uncertainties for the spectroscopic factor and geometrical parameters of the Woods-Saxon potential. The microscopic GCM is supposed to be a more accurate method for determining the reaction rate than the potential approach, but the overall normalization of the cross section is very sensitive to the adopted NN-potential [16]. Microscopic calculations gave $S_{17}(0) = 30\,eVb$ [17], $S_{17}(0) = 27 - 30\,eVb$ [18], $S_{17}(0) = 25.4\,eVb$ [19] and $S_{17}(0) = 22.4 \pm 2.1\,eVb$ [20]. We believe that the Gauss type potentials used in these calculations overestimate the S_{17}-factor. To compensate for the unrealistic fall-off of the Gauss type potentials, one should increase the normalization of the tail of the bound state ^8B wave function in the two-body channel $^7Be + p$, i. e. the S_{17}-factor.

In our work [1, 2], we pointed out that $S_{17}(0)$, due to the peripheral character of the reaction $^7Be(p,\gamma)^8B$, is defined solely by the ANC's for the virtual decay $^8B \to {}^7Be + p$, and calculated the S_{17}-factor using a simple relation between the ANC's for the virtual decay $^8B \to {}^7Be + p$ and $S_{17}(0)$. Using the microscopically calculated ANC's [7], we found in updated calculations that take into account s- and d-wave contributions in the initial channel, the value of $S_{17}(0) \approx 17.6\,eVb$ [2]. The calculation of S_{17} within the framework of the R-matrix approach, using the parameters derived from an analysis of the mirror reaction $^7Li(n,\gamma)^8Li$, gives $S_{17}(0) = 17 \pm 3\,eVb$ [13]. The information about the tail of the ^8B wave function extracted from ^8B breakup gives the interval $S_{17} = 12 - 20\,eVb$ [21]. Thus the latest results definitely favor the value of $S_{17}(0) \approx 17\,eVb$, which is smaller than the value currently adopted in astrophysical calculations, $S_{17}(0) = 22.5\,eVb$ [22, 23], by 25%. The value

$S_{17}(0) \approx 17\,\mathrm{eVb}$, if confirmed experimentally, would reduce the theoretical values of the solar neutrino flux detectable in the chlorine experiment calculated in the different models: 1) in the standard solar model (SSM) without diffusion from 7.2 SNU (for $S_{17}(0) = 22.5\,\mathrm{eVb}$) [22] to 5.85 SNU (for $S_{17}(0) = 17\,\mathrm{eVb}$); 2) in the model of Turck-Chieze et al. from 6.36 SNU (for $S_{17}(0) = 22.5\,\mathrm{eVb}$) [23] to 5.23 SNU (for $S_{17}(0) = 17\,\mathrm{eVb}$). The difference between the averaged chlorine experimental data, which gives $2.33 \pm 0.25\, SNU$, and theoretical predictions is still significant but not so severe as it was. As for the ratio of the experimental high energy neutrino flux detected in the Kamiokande III experiment [24] to the theoretical predictions, we find that the ratio of the experimental solar neutrino flux R_{KIII}, detected by the water detector at Kamiokande, to the theoretical prediction R_{SSM} calculated in the SSM with $S_{17}(0) = 17\,\mathrm{eVb}$ reaches $R_{KIII}/RSSM = 0.85$. If we use as the theoretical prediction the solar neutrino flux calculated in the model of Turck-Chieze et. al. [23] with $S_{17}(0) = 17\,\mathrm{eVb}$, then this ratio is of 0.95. Thus for the Kamiokande III experiment, there is no discrepancy between the experimental data and the theoretical predictions if the lower value of the S_{17}-factor is used. The intriguing situation surrounding the $^8\mathrm{B}$ S-factor calls for further experiments. The introduction of the ANC allows new possibilities, such as transfer reactions, to measure $S_{17}(0)$ indirectly with cross sections similar to those in Coulomb breakup reactions. These new possibilities measure the absolute value of S_{17} at zero energy directly and do not need to be extrapolated. Experiments to extract the ANC's for $^8\mathrm{B} \to {}^7\mathrm{Be} + p$ using different peripheral proton transfer reactions induced by $^7\mathrm{Be}$ radioactive beams have been proposed by us [3]. Experimental studies to directly measure the ANC's for $^8\mathrm{B} \to {}^7\mathrm{Be} + p$, and therefore the $S_{17}(0)$ factor, are currently under way using radioactive $^7\mathrm{Be}$ beams generated by the Texas A&M recoil spectrometer MARS and the K500 superconducting cyclotron.

3 The $\alpha(d,\gamma)^6\mathrm{Li}$ Reaction

The radiative capture reaction $\alpha(d,\gamma)^6\mathrm{Li}$ is the only process that produces $^6\mathrm{Li}$ in the Big Bang model. The astrophysically important α-d relative kinetic energies for the process are $E_{\mathrm{cm}} \leq 300\,\mathrm{keV}$ [25] with the most effective energy for the radiative capture being $E_{\mathrm{cm}} \approx 70\,\mathrm{keV}$. The available experimental data for this reaction cover the region $100\,\mathrm{keV} \leq E_{\mathrm{cm}} \leq 4\,\mathrm{MeV}$ [26, 27]. The currently accepted values of the cross section in the astrophysical region have been derived by the extrapolation of the experimental data for energies $E_{\mathrm{cm}} > 710\,\mathrm{keV}$ down to astrophysical energies [26]. The low energy experimental data have been obtained from measurements using the Coulomb break-up process $^6\mathrm{Li} + {}^{208}\mathrm{Pb} \to \alpha + d + {}^{208}\mathrm{Pb}$ with 30% uncertainty in the absolute normalization and about 15% statistical uncertainty [27]. The accuracy of that data was not enough to find a significant energy dependence of the S_{17}-factor at $E_{\mathrm{cm}} \leq 300\,\mathrm{keV}$, found in theoretical calculations [28, 4]. It is also important to note that in Coulomb breakup the $E1$ contribution is suppressed relative to $E2$ [27]. Thus in the analysis of the Coulomb break-up data in ref. [27], it was assumed that only the $E2$ transition contributes [27]. Hence, at energies E_{cm}, where $E1$ and $E2$ contributions are comparable in direct RCR, Coulomb dissociation will not provide an accurate determination of S_{24}. The theoretical calculations of the astrophysical factor, $S_{24}(E_{cm})$, have been done within the framework of the microscopic resonating group method (RGM) [29] and in the quasimicroscopic potential model [30, 28]. It was shown that, despite the fact that the $E1$ transition is forbidden in the long wave length approximation, the $E1$ multipole can dominate over the $E2$. Due to the small binding energy of ^{6}Li in the two-body channel $\alpha + d$, the RCR $\alpha(d,\gamma)^6$Li becomes almost totally peripheral at astrophysical energies. However, the work in refs. [29, 30, 28] did not take this fact into account. In these calculations the bound state wave function of ^{6}Li in the channel $\alpha + d$ had a Gauss type asymptotic behavior that caused the calculated S_{24} to be underestimated [4]. The very latest calculations of the S_{24}-factor have been done in [4], using the peripheral character of the reaction under consideration at astrophysical energies and the well established value of the ANC for the s-wave virtual decay $^6\text{Li} \rightarrow \alpha + d$ [6]. It was shown that one ANC for $^6\text{Li} \rightarrow \alpha + d$ defines the overall normalization of both $E1$ and $E2$ transitions. A significant energy dependence of S_{24} has been found with a 58% contribution from the $E1$-component at $E_{cm} = 70\,\text{keV}$. Our recommended value for the S_{24} at $E_{cm} = 70\,\text{keV}$ is 2.57 MeVnb, which is higher than $S_{24}(70\,\text{keV}) \approx 1.7\,\text{MeVnb}$ found in [28] and $S_{24}(70\,\text{keV}) \approx 0.7\,\text{MeVnb}$ calculated in [29]. The uncertainty of that S_{17}-factor calculated in [4] at astrophysical energies is about 12%. Our results confirm the conclusion of [26] that the production rate of ^{6}Li via the reaction $\alpha(d,\gamma)^6$Li in the Big Bang is a small contribution to the abundance of ^{6}Li in the universe, which is presumably produced mainly by galactic cosmic rays. It clearly would be useful to repeat the ^{6}Li break-up experiment to extract more reliable information about the astrophysical factor S_{24} at energies $E_{cm} \geq 300\,\text{keV}$, where the $E2$ transition dominates over $E1$. Any other experimental confirmation of the ANC value for $^6\text{Li} \rightarrow \alpha + d$ would be very important.

Acknowledgement. This work was supported in part by the U.S. Department of Energy under Grant No. DE-FG05-86ER40256 and the Robert A. Welch Foundation.

References

1. A. M. Mukhamedzhanov, N. K. Timofeyuk: JETP Lett. **51**, 282 (1990)
2. H. M. Xu et al.: Phys. Rev. Lett. **73**, 2027 (1994)
3. A. M. Mukhamedzhanov, R. E. Tribble, N. K. Timofeyuk: Phys. Rev. **C51**, 3472 (1995)
4. A. M. Mukhamedzhanov et al.: Submitted to Phys. Rev. **C**
5. Z. R. Irwinski, L. Rosenberg, L. Spruch: Phys. Rev. **C29**, 349 (1984)

6. L. D. Blokhintsev et al.: Phys. Rev. **C48**, 2390, (1993)

7. A. M. Mukhamedzhanov, N. K. Timofeyuk: Sov. J. Nucl. Phys. **51**, 431 (1990)

8. R. W. Kawanagh et al.: Bull. Amer. Phys. Soc. **14**, 1209 (1969)

9. B. W. Filippone et al.: Phys. Rev. **C28**, 2222 (1983)

10. F. C. Barker, R. H. Spear: Ap. J. **307**, 847 (1986)

11. T. Motobayashi et. al.: Phys. Rev. Lett. **73**, 2680 (1994).

12. F. G. Barker: Aust. J. Phys. **33**, 177 (1980); Phys. Rev. **C37**, 2920 (1988)

13. F. G. Barker: Nucl. Phys. (in print)

14. K. H. Kim, M. H. Park, B. T. Kim: Phys. Rev. **C35**, 363 (1987)

15. H. Krauss et al.: Ann. der Phys. **2**, 258 (1993)

16. T. Kajino: Nucl. Phys. **A460**, 559 (1986)

17. P. Descouvemont, D. Baye: Nucl. Phys. **A487**, 420 (1988)

18. P. Descouvemont, D. Baye: Nucl. Phys. **A567**, 341 (1994)

19. E. Kolbe, K. Langanke, H. J. Assenbaum: Phys. Lett. **B214**, 169 (1988)

20. C. W. Johnson et al.: Ap. J. **392**, 320 (1992)

21. K. Riisager, A. S. Hansen: Phys. Lett. **B301**, 6 (1993)

22. J. N. Bahcall, M. Pinsonneault: Rev. Mod. Phys. **64**, 885 (1992)

23. S. Turck-Chieze et al.: Phys. Rep. **230**, 57 (1993)

24. Y. Totsuka: Nucl. Phys. (Proc. Suppl.) **B19**, 69 (1991)

25. D. N. Schramm, R. V. Wagoner: Ann. Rev. Nucl. Sci. **27**, 37 (1977)

26. R. G. Robertson et al.: Phys. Rev. Lett. **47**, 1867 (1981)

27. J. Kiener et al.: Phys. Rev. **C44**, 2195 (1991)

28. M. A. Jusupov et al.: Izvestia Akademii Nauk Kaz. SSR. (in Russian) **2**, 30 (1991)

29. S. Typel et al.: Z. Phys. **A339**, 335 (1991)

30. N. A. Burkova et al.: Phys. Lett. **B248**, 15 (1990)

Few-Body Systems Suppl. 8, 340–344 (1995)

Few-Body Systems

Towards a Unified Description of the Baryon Spectrum and the Baryon-Baryon Interaction within a Potential Model Scheme

A. Valcarce[1], P. González[2], F. Fernández[1], V. Vento[2]

[1] Grupo de Física Nuclear, Universidad de Salamanca, E-37008 Salamanca, Spain

[2] Departamento de Física Teórica e IFIC, Universidad de Valencia-CSIC, E-46100 Burjassot, Valencia, Spain

Abstract. We study the low energy part of the nucleon and Δ spectra by solving the Schrödinger equation for the three-quark system in the hyperspherical harmonic approach. The quark-quark hamiltonian considered includes, besides the usual one-gluon exchange, pion and sigma exchanges generated by the chiral symmetry breaking. This quark-quark potential reproduces, in a Resonating Group Method calculation, the nucleon-nucleon scattering phase shifts and the deuteron properties. The baryonic spectrum obtained is quite reasonable and the resulting wave function is consistent with the ansatz used in the two baryon system.

1 Introduction

Models have been widely used to study the properties of the hadron spectrum due to the impossibility to solve Quantum Chromodynamics (QCD) at the current moment. In particular, the quark potential models incorporate the perturbative one-gluon exchange quark-quark (qq) potential (V_{OGE}) derived from QCD as well as a parametrization of some nonperturbative effects through a qq confining potential (V_{con}) [1]. Such an *effective theory* [the quark-gluon coupling constant is taken as an effective one and the constituent quark masses (m_q) are parameters fitted from the baryon magnetic moments] provides a reasonable understanding of the baryon spectrum and the hadron static properties [2].

The idea of an interquark potential has been also used to study baryon-baryon interactions. More explicitly, the repulsive core of the nucleon-nucleon (NN) force has been shown to arise from the color-spin structure of the V_{OGE} [3]. Nevertheless, the same scheme has been proven uncapable of describing the long range part of the NN interaction unless pion exchange between quarks is introduced. From the basic theory, the origin of the one-pion exchange potential ($V_{\rm OPE}$) is associated with the spontaneous chiral symmetry breaking of QCD. Moreover, the inclusion of the qq sigma potential ($V_{\rm OSE}$) consistently with its chiral partner (the pion) allows to reproduce the intermediate range NN interaction as well as the deuteron properties [4].

In this paper we examine the consistency of the two scenarios. We determine a qq interaction by studying two nucleon properties and charge exchange reactions and proceed to analyze the baryon spectrum for this interaction. Since we shall discuss the predictions for the baryon masses and the baryon wave functions, our procedure not only provides a stringent test of the potential but also a consistency test of the formalism used to generate the NN interaction [resonating group method (RGM)].

2 The Quark-Quark Potential

The starting point of our description is a quark-quark interaction of the following form:

$$V_{\rm qq}(\boldsymbol{r}_{ij}) = V_{\rm con}(\boldsymbol{r}_{ij}) + V_{\rm OGE}(\boldsymbol{r}_{ij}) + V_{\rm OPE}(\boldsymbol{r}_{ij}) + V_{\rm OSE}(\boldsymbol{r}_{ij}) \,, \tag{1}$$

where $\boldsymbol{r}_{ij}$ is the interquark distance.

The confinement is chosen to be linear as suggested by the meson spectrum and lattice calculations:

$$V_{\rm con}(\boldsymbol{r}_{ij}) = -a_c\,(\boldsymbol{\lambda}_i \cdot \boldsymbol{\lambda}_j)\, r_{ij} \,, \tag{2}$$

where $\lambda's$ are the SU(3) color matrices.

$V_{\rm OGE}$, $V_{\rm OPE}$ and $V_{\rm OSE}$ have been derived in detail elsewhere [1, 4] and we shall limit here to write the final expressions:

$$V_{\rm OGE}(\boldsymbol{r}_{ij}) = \frac{1}{4}\,\alpha_s\,\boldsymbol{\lambda}_i \cdot \boldsymbol{\lambda}_j \left\{ \frac{1}{r_{ij}} - \frac{1}{4\,m_q^2}\left[1 + \frac{2}{3}\boldsymbol{\sigma}_i \cdot \boldsymbol{\sigma}_j\right] \frac{e^{-r_{ij}/r_0}}{r_0^2\, r_{ij}} - \frac{1}{4m_q^2\, r_{ij}^3}\, S_{ij} \right\}, \tag{3}$$

where α_s is the effective quark-quark-gluon coupling constant, r_0 is the range of a smeared δ function in order to avoid and unbound spectrum [5], the $\sigma's$ stand for the spin Pauli matrices and S_{ij} is the quark tensor operator $S_{ij} = 3(\boldsymbol{\sigma}_i \,.\, \hat{r}_{ij})(\boldsymbol{\sigma}_j \,.\, \hat{r}_{ij}) - \boldsymbol{\sigma}_i \,.\boldsymbol{\sigma}_j$.

$$\begin{aligned} V_{\rm OPE}(\boldsymbol{r}_{ij}) \;=\; & \frac{1}{3}\,\alpha_{\rm ch} \frac{\Lambda^2}{\Lambda^2 - m_\pi^2}\, m_\pi \left\{ \left[Y(m_\pi\, r_{ij}) - \frac{\Lambda^3}{m_\pi^3}\, Y(\Lambda\, r_{ij}) \right] \boldsymbol{\sigma}_i \cdot \boldsymbol{\sigma}_j \,+ \right. \\ & \left. \left[H(m_\pi\, r_{ij}) - \frac{\Lambda^3}{m_\pi^3}\, H(\Lambda\, r_{ij}) \right] S_{ij} \right\} \boldsymbol{\tau}_i \cdot \boldsymbol{\tau}_j \,, \end{aligned} \tag{4}$$

$$V_{\rm OSE}(\boldsymbol{r}_{ij}) = -\alpha_{\rm ch} \frac{4\,m_q^2}{m_\pi^2} \frac{\Lambda^2}{\Lambda^2 - m_\sigma^2}\, m_\sigma \left[Y(m_\sigma\, r_{ij}) - \frac{\Lambda}{m_\sigma}\, Y(\Lambda\, r_{ij}) \right] , \tag{5}$$

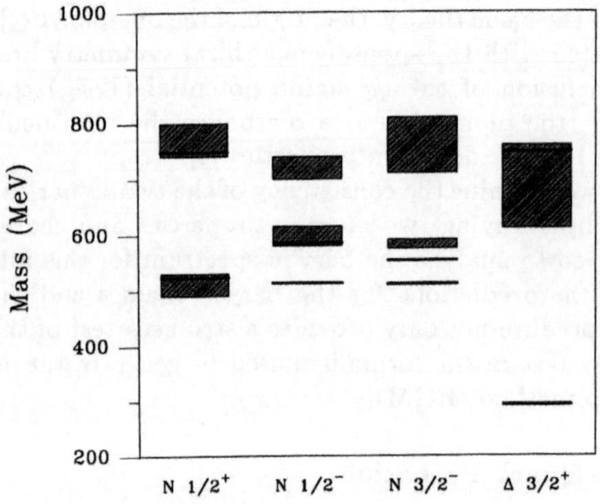

Figure 1. Relative energy N and Δ spectrum up to 0.7 GeV excitation energy. The solid line corresponds to the results of our model. The boxes represent the experimental data with the corresponding uncertainties.

where m_π (m_σ) is the pion (sigma) mass, $\alpha_{\rm ch}$ is the chiral coupling constant (related to the πNN coupling constant), Λ is a cutoff parameter and $Y(x)$, $H(x)$ are the Yukawa functions defined as:

$$Y(x) = \frac{e^{-x}}{x} \quad , \quad H(x) = \left(1 + \frac{3}{x} + \frac{3}{x^2}\right) Y(x) \, , \tag{6}$$

With this interaction, the two baryon system has been studied in the RGM framework assuming for the spatial part of the wave function of the quarks a harmonic oscillator ground state,

$$\eta_{\rm os}(\boldsymbol{r}_i - \boldsymbol{R}) = \left(\frac{1}{\pi b^2}\right)^{3/4} e^{-(\boldsymbol{r}_i - \boldsymbol{R})^2/2b^2} \, , \tag{7}$$

where the parameter $\boldsymbol{R}$ determines the position of the baryon and b is the harmonic oscillator constant.

Using the same set of parameters given in Table 1, the NN scattering phase shifts, the static and electromagnetic properties of the deuteron [4] and reactions that take place with the excitation of the Δ resonance [6] are reasonably reproduced.

3 Results

To study the baryonic spectrum with the potential just described (parameters as in Table 1) we solve the Schrödinger equation in the hyperspherical harmonic

Table 1. Value of the potential parameters

m_q	b	α_s	α_{ch}	a_c	m_π	m_σ	Λ
MeV	fm			$\mathrm{MeV}\cdot\mathrm{fm}^{-1}$	fm^{-1}	fm^{-1}	fm^{-1}
313	0.518	0.485	0.027	91.488	0.7	3.42	4.2

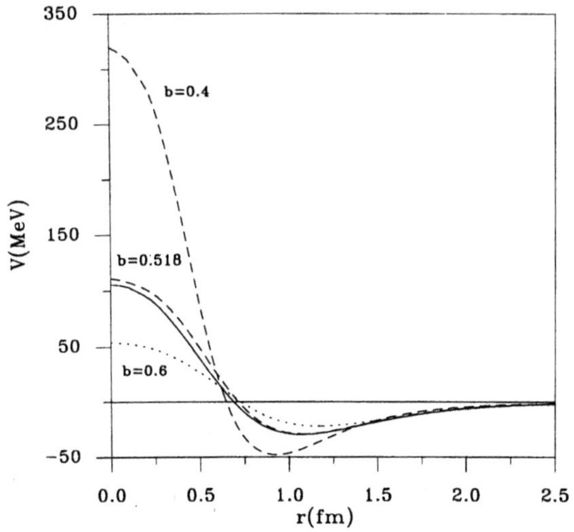

Figure 2. Adiabatic NN potential for different nucleon wave functions for the channel (S,T)=(1,0). The solid line corresponds to the solution of the Schrödinger equation. The others to a gaussian ansatz with the values of b given in the figure.

approach [7]. The low energy N and Δ spectrum obtained, a part of which is shown in Fig. 1, is quite reasonable though some small discrepancy remains concerning the relative energy position of the Roper resonance [$N^*(1440)$] and the first negative parity state excitation [$N^-(1535)$], as it is common in two body potential models.

Regarding the wave function and in order to check the internal RGM consistency, it makes sense to compare the NN potential one obtains from the wave function solution of the Schrödinger equation, with the potential obtained from the ansatz RGM wave function for different values of the parameter b. For the sake of technical simplicity we do this in the Born-Oppeheimer approximation. The results appear in Fig. 2, where we can see that it is precisely the value of b that gives the best overall fit to the NN data with RGM, the one which provides a better approximation. This confers to b a self-consistent character and solves the controversy about its possible values.

Certainly, more work and further refinements are needed along this direction. Meanwhile, the proposed model points out the plausability of a unified description of the baryon structure and the baryon-baryon interaction.

Acknowledgement. This work has been partially funded by Dirección General de Investigación Científica y Técnica (DGICYT) under Contract Nb. PB91-0119

References

1. A. de Rújula, H. Georgi, S.L. Glashow: Phys. Rev. **D12**, 147 (1975)
2. A. Le Yaouanc, Ll. Oliver, O. Pène, J.C. Raynal: *Hadron Transitions in the Quark Model.* London: Gordon and Breach 1988, and references therein
3. A. Faessler, F. Fernández, G. Lübeck, K. Shimizu: Phys. Lett. **B112**, 201 (1982)
4. F. Fernández, A. Valcarce, U. Straub, A. Faessler: Jour. of Phys. **G19**, 2013 (1993)
   A. Valcarce, A. Buchmann, F. Fernández, A. Faessler: Phys. Rev. **C50**, 2246 (1994)
5. R.K. Bhaduri, L.E. Cohler, Y. Nogami: Phys. Rev. Lett. **44**, 1369 (1980)
6. F. Fernández, A. Valcarce, P. González, V. Vento: Phys. Lett. **B287**, 35 (1992); Nucl. Phys. **A567**, 741 (1994)
7. J. L. Ballot, M. Fabre de la Ripelle: Ann. Phys. (NY) **127**, 62 (1980)

Few-Body Systems Suppl. 8, 345–349 (1995)

Few-Body Systems

Nucleon and Pion Electromagnetic Structure and Constituent Quark Form Factors

F. Cardarelli [1], E. Pace [1,2], G. Salmè [3], S. Simula [3]

[1] Istituto Nazionale di Fisica Nucleare, Sezione Tor Vergata, Via della Ricerca Scientifica 1, I-00133 Roma, Italy

[2] Dipartimento di Fisica, Università di Roma "Tor Vergata", Via della Ricerca Scientifica 1, I-00133 Roma, Italy

[3] Istituto Nazionale di Fisica Nucleare, Sezione Sanità, Viale Regina Elena 299, I-00161 Roma, Italy

Abstract. We have evaluated nucleon and pion electromagnetic form factors using for the first time eigenfunctions of a light-front mass operator, reproducing a large set of energy levels, and constituent quark form factors in the one-body current. The comparison with the experimental data yields valuable information on the electromagnetic structure of light constituent quarks and on the different role played by their Dirac and Pauli form factors.

The investigation of the electromagnetic (e.m.) form factors of hadrons, within relativistic constituent quark (CQ) models, has been widely [1, 2, 3, 4] pursued for extending the range of applicability of the CQ picture of hadrons. Our approach [3, 4] is based on the CQ model embedded in the light-front (LF) formalism, which has many interesting features [5]. Among the properties relevant for what follows, there are: i) the spacelike e.m. form factors, evaluated in a frame where the plus component of the momentum transfer $q_+ = q_0 + q_z = 0$, are not affected by the contribution of the pair creation from the vacuum [6], and ii) the centre of mass motion in the wave function of the interacting system can be exactly separated. As to the latter, the LF hamiltonian of the interacting system, $P_- = \frac{\mathcal{M}^2 + P_\perp^2}{P_+}$, allows a straightforward separation of the coordinates of the centre of mass, ($P_\mp = P_0 \mp P_z$, $P_\perp \equiv [P_x, P_y]$), from the intrinsic ones contained in the mass operator defined by, e.g., $\mathcal{M} = M_0 + \mathcal{V}$, with M_0 the free mass and $\mathcal{V}$ a Poincaré invariant interaction.

The main features of our approach are: i) the hadron state is eigenvector of a LF mass operator, constructed from the effective $q\bar{q}$ and qq-interaction of [7], that reproduces a lot of energy levels; ii) the configuration mixing, due to the one-gluon-exchange (OGE) part of the effective interaction, leads to high momentum components and SU(6) breaking terms in the hadron wave function; iii) Dirac and Pauli form factors for the CQ's, as suggested by their quasi-particle nature (cf. [8]), are introduced in the one-body e.m. current.

The LF hadron wave functions are eigenvectors of the mass operator and of the non-interacting angular momentum operators j^2 and j_z. Disregarding the color degree of freedom, the LF wave function is (see [1, 4])

$$\langle\{\xi_i \mathbf{q}_{i\perp}; \nu_i \tau_i\}|\Psi_{\mathcal{H}}\rangle = \mathcal{N}_{\mathcal{H}} \sum_{\{\nu_i'\}} \langle\{\nu_i\}|\mathcal{R}^\dagger|\{\nu_i'\}\rangle\langle\{\mathbf{q}_i; \nu_i' \tau_i\}|\psi_{\mathcal{H}}\rangle \tag{1}$$

where the curly braces { } indicates $i = 1, n_q$ with $n_q = 2(3)$ for the mesons (baryons); ξ_i, $\mathbf{q}_{i\perp}$ and q_{iz} are intrinsic LF variables; $\mathcal{N}_{\mathcal{H}}$ is the proper normalization factor; $\nu_i'(\tau_i)$ is the third component of the quark spin (isospin); $\mathcal{R}^\dagger = \left[\prod_{j=1}^{n_q} R_M^\dagger(\mathbf{q}_{j\perp}, \xi_j, m_j)\right]$ is a unitary operator, with $R_M(\mathbf{q}_{j\perp}, \xi_j, m_j)$ being the generalized Melosh rotation. The equal-time state $|\psi_{\mathcal{H}}\rangle$ is an eigenvector of the transformed mass operator $M = \mathcal{R}\mathcal{M}\mathcal{R}^\dagger = \mathcal{R}M_0\mathcal{R}^\dagger + \mathcal{R}\mathcal{V}\mathcal{R}^\dagger$, that becomes $M = M_0 + V$. The interaction $V = \mathcal{R}\mathcal{V}\mathcal{R}^\dagger$ must fulfil the translational and rotational invariances, and be independent of the total momentum (cf. [5]). Since it turns out that $M_0 = \sum_{i=1}^{n_q}\sqrt{m_i^2 + q_i^2}$, we have chosen [3, 4] M equal to the effective hamiltonian proposed for mesons and baryons in [7]. Thus, our equal-time hadron state $|\psi_{\mathcal{H}}\rangle$ is eigenvector of the following mass operator

$$H_{\mathcal{H}}\,|\psi_{\mathcal{H}}\rangle \equiv \left[\sum_{i=1}^{n_q}\sqrt{m_i^2 + q_i^2} + \sum_{i\neq j=1}^{n_q} V_{ij}\right]|\psi_{\mathcal{H}}\rangle = M_{\mathcal{H}}\,|\psi_{\mathcal{H}}\rangle \tag{2}$$

where $M_{\mathcal{H}}$ is the mass of the hadron, and V_{ij} the effective $q\bar{q}$ or qq potential for mesons or baryons, respectively. The interaction is composed by a OGE term (dominant at short separations) and a linear confining term (dominant at large separations). As in [7], the values $m_u = m_d = 0.220$ GeV have been adopted. The wave equation (2) has been solved by expanding the wave function onto a (truncated) harmonic oscillator basis and then applying the Rayleigh-Ritz variational principle (see [4]). It should be pointed out that the relativistic mass operator (2) reproduces a large set of hadron energy levels [7] and generates a huge amount of configuration mixing, due to the presence of the OGE part of the interaction, that also determines the hyperfine mass-splitting.

In [3, 4] we have evaluated the nucleon and pion form factors using the eigenstates of Eq.(2), whereas in the current literature only the effects of the confinement scale are considered, through gaussian or power-law wave functions (cf. for the nucleon [1, 2]), without taking care of the mass eigenvalues. In particular we have shown that, assuming pointlike CQ's, the calculated nucleon and pion form factors sharply differ from the experimental data, because of configuration mixing effects. Therefore the possibility of a non trivial e.m. structure for extended CQ's, summarizing the underlying degrees of freedom, has been considered, adopting the following one-body component of the hadron

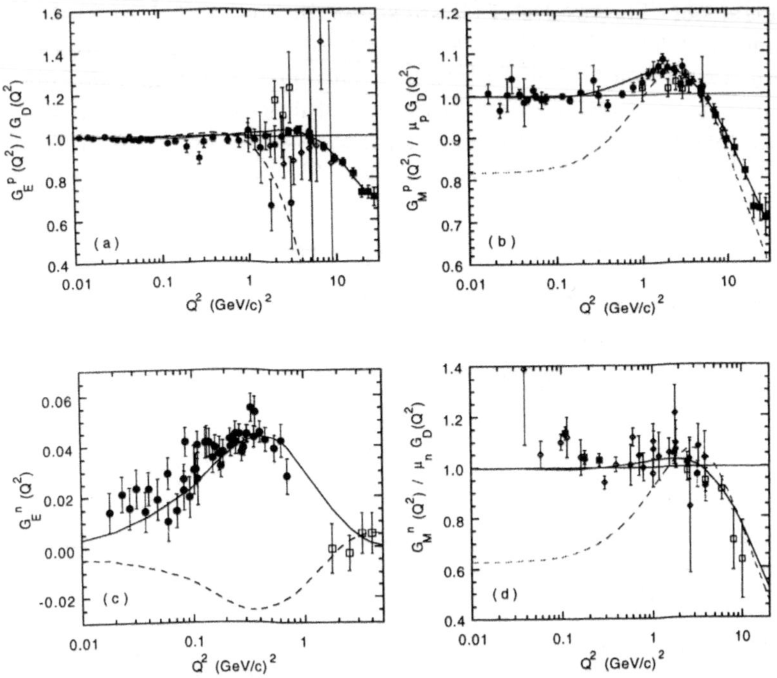

Figure 1. - (a) The proton form factor G_E^p/G_D vs. Q^2. Solid line: G_E^p/G_D obtained in [4] by using the nucleon wave function corresponding to the interaction of [7] and the nucleon e.m. current with CQ form factors. Dashed line: the same as the solid line, but with $\kappa_{u(d)} = 0$. Experimental data are quoted in detail in [4]. - (b) The same as in Fig. 1a, but for $G_M^p/(\mu_p\ G_D)$.- (c) The same as in Fig. 1a, but for G_E^n. - (d) The same as in Fig. 1a, but for $G_M^n/(\mu_n\ G_D)$.

e.m. current

$$\mathcal{I}_{\mathcal{H}}^{\nu} = \sum_{j=1}^{n_q} I_j^{\nu} = \sum_{j=1}^{n_q} \left(e_j \gamma^{\nu} f_1^j(Q^2) + i\kappa_j \frac{\sigma^{\nu}{}_{\mu} q^{\mu}}{2m_j} f_2^j(Q^2) \right) \tag{3}$$

where e_j is the charge of the j-th quark, κ_j the corresponding anomalous magnetic moment, $f_{1(2)}^j$ its Dirac (Pauli) form factor, and $Q^2 \equiv -q \cdot q$ the squared four-momentum transfer. The values of κ_u and κ_d can be fixed by the request of reproducing the experimental nucleon magnetic moments, obtaining $\kappa_{u(d)} = 0.085\ (-0.153)$ [4]. In spite of their smallness such values affect the nucleon anomalous magnetic moments by $\sim 40\%$. The CQ Dirac and Pauli form factors have been parametrized in a simple way (the Dirac form factor = monopole + dipole, and the Pauli one = dipole + octupole); the values of the parameters have been estimated through a standard minimization procedure using the experimental nucleon and pion form factors in a wide range of momentum transfer. The results of [4] are shown in Figs. 1 and 2 (solid lines). It can be seen that our calculations, though obtained in the framework of the one-body approximation for the hadron e.m. current (i.e. by disregarding the two-body currents necessary for fulfilling both the gauge and rotational invariances, see [5]), are in nice agreement with the data. The CQ form factors, appearing in the effective one-body current Eq. (3), are presented in Fig. 3.

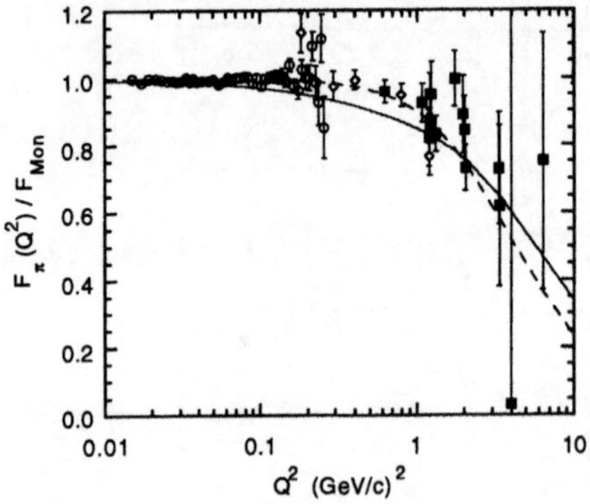

Figure 2. The charge form factor of the pion $F_\pi(Q^2)/F_{Mon}(Q^2)$ vs. Q^2, with $F_{Mon}(Q^2) = 1/(1+Q^2/0.54)$. The solid line has been obtained in [4] by using the pion wave function corresponding to the interaction of [7], and CQ form factors. Dashed line : the same as the solid line, but with $\kappa_u = \kappa_d = 0$. Experimental data are quoted in detail in [4].

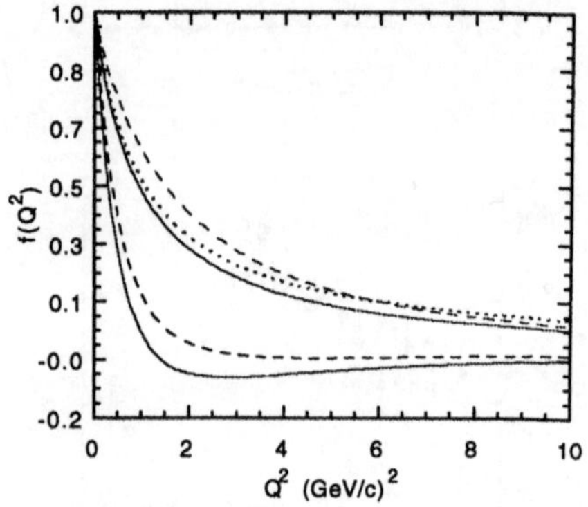

Figure 3. The constituent quark form factors, extracted from the analysis of the nucleon and pion form factors (see Figs. 1 and 2), vs. Q^2. Solid lines represent $f_1^u(Q^2)$ (upper) and $f_2^u(Q^2)$ (lower), respectively; dashed lines represent $f_1^d(Q^2)$ (upper) and $f_2^d(Q^2)$ (lower), respectively. For comparison $f_1^u(Q^2) = f_1^d(Q^2)$ obtained in [3] is also shown (dotted line). (After [4])

As previously mentioned, the underlying degrees of freedom frozen in the extended CQ influence their properties, such as the form factors and the anomalous magnetic moments. In order to undertake a more detailed investigation, in this contribution we would focus on the different role played by the Dirac and Pauli form factors of the CQ's. To this end the comparison with the experimental data is carried out in Figs. 1 and 2, considering both the full calculation and the one performed by retaining the CQ Dirac form factor only, i.e. assuming $\kappa_{u(d)} = 0$ (dashed lines). The following comments are in order: i) G_E^p is strongly affected by f_2^q for $Q^2 > 1(\mathrm{GeV}/c)^2$; ii) G_E^n changes its sign when f_2^q is switched off; iii) the nucleon magnetic form factors are very sensitive at low Q^2 to the CQ Pauli form factor and dominated by the Dirac one at high Q^2, iv) the pion form factor is only slightly influenced by f_2^q, according to [3]. It is worth noting that the CQ form factors, shown in Fig. 3, yield a quark radius, defined as

$< r_1^{u(d)} >^2 = -6\, df_1^{u(d)}(Q^2)/dQ^2$ at $Q^2 = 0$, equal to $< r_1^{u(d)} > = 0.51\ (0.42)$fm; such values are similar to the ones obtained in refs. [8, 9] and in our exploratory analysis of the pion data [3] (cf. also the dotted line in Fig. 3).

To sum up, within the LF formalism, we have given a simultaneous description of the pion and nucleon form factors adopting for the first time the eigenfunctions of a LF mass operator, with correct mass eigenvalues, and a one-body approximation for the e.m. current, containing CQ form factors. The same model has been applied in a parameter-free calculation of the magnetic form factor of the $N - \Delta$ transition, which has been recently completed [10] achieving a very satisfactory result. The improvement of our approach including the contribution of two-body e.m. currents is in progress.

References

1. P.L. Chung, F. Coester: Phys. Rev. **D44**, 229 (1991); S. Capstick, B. Keister: Phys. Rev. **D51**, 3598 (1995)
2. Z. Dziembowski : Phys. Rev. **D37**, 778 (1988); I.G. Aznaurian: Phys. Lett. **B316**, 391 (1993); H. J. Weber: Phys. Rev. **D49**, 3160 (1994); F. Schlumpf: J. Phys. **G20**, 237 (1994)
3. F. Cardarelli, I.L. Grach, I.M. Narodetskii, E. Pace, G. Salmè, S. Simula: Phys. Lett. **B332**, 1 (1994), and submitted to Phys. Rev. **D**, brief report; F. Cardarelli, I.L. Grach, I.M. Narodetskii, G. Salmè, S. Simula: Phys. Lett. **B349**, 393 (1995)
4. F. Cardarelli, E. Pace, G. Salmè, S. Simula: Phys. Lett. **B357**, 267 (1995); F. Cardarelli, S. Simula: to be published
5. For a review, see B.D. Keister, W.N. Polyzou: Adv. Nucl. Phys. **20**, 225 (1991); F. Coester: Progress Part. Nucl. Phys. **29**, 1 (1992)
6. See for instance M. Sawicki: Phys. Rev. **D46**, 474 (1992)
7. a) S. Godfrey, N. Isgur: Phys. Rev. **D32**, 185 (1985); b) S. Capstick, N. Isgur: Phys. Rev. **D34**, 2809 (1986)
8. U. Vogl, M. Lutz, S. Klimt, W. Weise: Nucl. Phys. **A516**, 469 (1990)
9. B. Povh, J. Hufner: Phys. Lett. **B245** (1990) 653
10. F. Cardarelli, E. Pace, G. Salmè, S. Simula: Preprint INFN-ISS 95/12, submitted to Phys. Lett. **B**

Few-Body Systems Suppl. 8, 350–353 (1995)

Few-Body Systems

On the Dynamical Retardation Effect for the Mass Spectrum of Heavy Quarkonia

T. Kopaleishvili, A. Rusetsky

High Energy Physics Institute, Tbilisi State University, 380 086 Tbilisi, Georgia

The Bethe–Salpeter equation provides the most natural basis for the investigation of the properties of the bound $q\bar{q}$ systems within the constituent quark model. For the three–dimensional regularization of this equation in the kernel the instantaneous (static) approximation is usually used, i.e. the dependence of the kernel on the energy variables is suppressed. To our knowledge, the first attempt to take into account this dependence for the confining kernels (the dynamical retardation effect) was made in refs. [1, 2]. The results turned out to be somewhat unnatural from the physical point of view. In our previous paper [3] a simple "covariantization" procedure for the confining static potential is used in order to construct the first–order quasipotential. In a result physically acceptable results for the meson spectrum at least in the light–quark sector are obtained. Now we suggest a new prescription for the construction of the relativistically invariant $q\bar{q}$ interaction kernels, corresponding to the confining static potentials. Namely, we assume that the kernel of the Bethe–Salpeter equation, being the covariant 4–dimensional generalization of the power–law potential $V_\alpha(r) = r^\alpha$ has the form:

$$\bar{K}_\alpha(x) = \frac{\Gamma(1+\frac{\alpha}{2})}{\sqrt{\pi}\Gamma(\frac{1}{2}+\frac{\alpha}{2})}\theta(-x^2)(-x^2)^{(\alpha-1)/2}; \quad \alpha \neq -1, -2\cdots \tag{1}$$

In the 2–dimensional space–time expression (1) can be interpreted, as the *average* of the static power–law potential over all possible directions of the quantization axe. Note that for the case $\alpha = 1$ (linear confinement) the Fourier–transform of $\bar{K}_1(x)$ reads:

$$\bar{K}_1(q) = -4\pi\left(\frac{1}{(q_0^2-(|\mathbf{q}|+i0)^2)^2} + \frac{1}{(q_0^2-(|\mathbf{q}|-i0)^2)^2}\right) \tag{2}$$

i.e., in other words, in order to obtain the kernel (2) in the ladder approximation, the principal–value prescription must be used in the gluon propagator instead of the familiar causal one. One can find here a close analogy with the 2-dimensional QCD [4]. It is easy to ensure that due to the principal-value prescription in (2) the most severe infrared singularities in the equations for the Green's functions are avoided.

The retardation corrections to the heavy quarkonia mass spectrum, corresponding to the kernels of the type (1) are studied with the use of the Logunov–Tavkhelidze quasipotential approach, described in the ref. [3]. The first-order quasipotential equation for the ${}^{2S+1}L_J$ meson radial wave function $\tilde{R}^{(+)}_{LSJ}(p)$ for the equal-mass case $m_1 = m_2 = m$ reads as:

$$[M_B - 2w(p)]\tilde{R}^{(+)}_{LSJ}(p) = \frac{4}{3}\int_0^\infty q^2dq\left(\tilde{V}^{(+)}_L(p,q)\frac{1}{2}\left(1+\frac{m^2}{w(p)w(q)}\right)+\right. \tag{3}$$

$$\left.+(L-J)\frac{4J(J+1)}{(2J+1)^2}(\tilde{V}^{(+)}_{J-1}(p,q)-\tilde{V}^{(+)}_{J+1}(p,q))\frac{(w(p)-m)(w(q)-m)}{4w(p)w(q)}\right)\tilde{R}^{(+)}_{LSJ}(q)$$

where M_B is the mass of the bound $q\bar{q}$ state, $w(p) = \sqrt{m^2+\mathbf{p}^2}$, the spin structure of the confining kernel is taken to be the equal-weight mixture of the scalar and the fourth component of vector $\hat{O}_c = \frac{1}{2}(I_1 \otimes I_2 + \gamma_1^0 \otimes \gamma_2^0)$ and, for the case of local quasipotential

$$\tilde{V}^{(+)}_L(p,q) = \int r^2 dr j_L(pr)\tilde{V}_c(r;M_B)j_L(qr) \tag{4}$$

j_L being the spherical Bessel function.

The positive-energy component of the first-order quasipotential, corresponding to the interaction kernel (1) can be written in the following form:

$$\underline{\tilde{V}}^{(+)}(M_B;\mathbf{p},\mathbf{q}) = \int d^3\mathbf{x}e^{-i(\mathbf{p}-\mathbf{q})\mathbf{x}}\frac{\Gamma\left(1+\frac{\alpha}{2}\right)}{\sqrt{\pi}\Gamma\left(\frac{1}{2}+\frac{\alpha}{2}\right)}r^\alpha\times \tag{5}$$

$$\times\int_{-1}^{1} d\tau(1-\tau^2)^{(\alpha-1)/2}(\theta(\tau)e^{i(M_B-w(p)-w(q))r\tau}+\theta(-\tau)e^{-i(M_B-w(p)-w(q))r\tau})$$

Neglecting relativistic corrections in the exponentials $M_B - w(p) - w(q) = M_B - 2m + O(\frac{1}{m}) = -\epsilon_B$, as well as the imaginary part of this expression, we arrive at the local first-order quasipotential:

$$\underline{\tilde{V}}^{(+)}(r;\epsilon_B) = \frac{\Gamma\left(1+\frac{\alpha}{2}\right)}{\sqrt{\pi}\Gamma\left(\frac{1}{2}+\frac{\alpha}{2}\right)}r^\alpha\int_{-1}^{1} d\tau(1-\tau^2)^{(\alpha-1)/2}\cos\epsilon_B r\tau =$$

$$= \left|\frac{2r}{\epsilon_B}\right|^{\alpha/2}\Gamma\left(1+\frac{\alpha}{2}\right)J_{\alpha/2}(|\epsilon_B r|) \tag{6}$$

$J_\nu(x)$ being the Bessel function.

Using this expression for the relativistic generalization of the static potential $V_c(r) = kr + c$, we obtain the first-order local quasipotential

$$\tilde{V}_c(r,\epsilon_B) = k\frac{\sin\epsilon_B r}{\epsilon_B} + cJ_0(\epsilon_B r) \tag{7}$$

Table 1. The dynamical retardation corrections to the heavy quarkonia mass spectrum. 1) The meson mass in the static approximation, $M_B^{(st)}$ (GeV), 2) The meson mass with an account of the retardation effect, $M_B^{(ret)}$ (GeV), 3) The size of the retardation correction, $M_B^{(ret)} - M_B^{(st)}$ (GeV) and 4) $|M_B^{(ret)} - M_B^{(st)}|/M_B^{(st)}$.

Mesons	J^{PC}	$N^{2S+1}L_J$	1	2	3	4
$c\bar{c}$						
$\eta_c(2.980)$	0^{-+}	1^1S_0	3.095	3.204	0.109	$3.5\cdot10^{-2}$
$\eta_c'(3.590)$	0^{-+}	2^1S_0	3.690	3.739	0.049	$1.3\cdot10^{-2}$
$J/\psi(3.097)$	1^{--}	1^3S_1	3.096	3.204	0.108	$3.5\cdot10^{-2}$
$\psi'(3.686)$	1^{--}	2^3S_1	3.691	3.742	0.051	$1.3\cdot10^{-2}$
$h_{c1}(3.526)$	1^{+-}	1^1P_1	3.445	3.460	0.014	$4.2\cdot10^{-3}$
$\chi_{c0}(3.414)$	0^{++}	1^3P_0	3.445	3.460	0.014	$4.2\cdot10^{-3}$
$\chi_{c1}(3.511)$	1^{++}	1^3P_1	3.445	3.460	0.014	$4.2\cdot10^{-3}$
$\chi_{c2}(3.556)$	2^{++}	1^3P_2	3.445	3.461	0.014	$4.2\cdot10^{-3}$
$b\bar{b}$						
η_b	0^{-+}	1^1S_0	9.463	9.619	0.156	$1.7\cdot10^{-2}$
η_b'	0^{-+}	2^1S_0	9.899	9.966	0.067	$6.8\cdot10^{-3}$
$\Upsilon(9.460)$	1^{--}	1^3S_1	9.463	9.619	0.156	$1.7\cdot10^{-2}$
$\Upsilon'(10.023)$	1^{--}	2^3S_1	9.899	9.966	0.067	$6.8\cdot10^{-3}$
Υ''	1^{--}	1^3D_1	9.938	9.993	0.055	$5.5\cdot10^{-3}$
$\Upsilon'''(10.355)$	1^{--}	3^3S_1	10.250	10.255	0.005	$5.1\cdot10^{-4}$
Υ^{IV}	1^{--}	2^3D_1	10.277	10.291	0.014	$1.4\cdot10^{-3}$
h_{b1}	1^{+-}	1^1P_1	9.720	9.835	0.115	$1.2\cdot10^{-2}$
$\chi_{b0}(9.860)$	0^{++}	1^3P_0	9.720	9.835	0.115	$1.2\cdot10^{-2}$
$\chi_{b1}(9.892)$	1^{++}	1^3P_1	9.720	9.835	0.115	$1.2\cdot10^{-2}$
$\chi_{b2}(9.913)$	2^{++}	1^3P_2	9.720	9.835	0.115	$1.2\cdot10^{-2}$
h_{b1}'	1^{+-}	2^1P_1	10.097	10.109	0.013	$1.3\cdot10^{-3}$
$\chi_{b0}'(10.232)$	0^{++}	2^3P_0	10.097	10.109	0.013	$1.3\cdot10^{-3}$
$\chi_{b1}'(10.255)$	1^{++}	2^3P_1	10.097	10.109	0.013	$1.3\cdot10^{-3}$
$\chi_{b2}'(10.268)$	2^{++}	2^3P_2	10.097	10.109	0.013	$1.3\cdot10^{-3}$

which accounts for the retardation effect and reduces to $V_c(r)$ in the limit $\epsilon_B \to 0$.

As it can be seen from Eq. (7), the account for the dynamical retardation effect in the case of pure linear potential effectively leads to the colour screening at an intermediate distances. At a larger distances the deviation of the retarded potential from the static one becomes significant. Note, however, that for the case of heavy quarkonia the latter difficulty causes no trouble since the wave function of the $q\bar{q}$ bound system in this case rapidly vanishes with the increase of r and, therefore, does not "feel" the oscillating "tail" of the potential at a large distances.

It is not obvious from the beginning whether the potential (7) leads to the discrete energy levels due to its oscillating behaviour at $r \to \infty$ We attempted to solve the Eq. (3) numerically, expanding the unknown radial wave function $\tilde{R}^{(+)}_{LSJ}(p)$ in the orthonormalized basis of the nonrelativistic oscillator wave functions $R_{nL}(p)$ [3]

$$\tilde{R}^{(+)}_{LSJ}(p) = p_0^{-3/2} \sum_{n=0}^{\infty} c^{(+)}_{nLSJ} R_{nL}(p/p_0) \tag{8}$$

where p_0 is an arbitrary scale parameter. Substituting (8) into the equation (3), we arrive at the (infinite) system of linear homogeneous algebraic equations for the coefficients $c^{(+)}_{nLSJ}$, which is solved by the standard technique. The calculations show that the final results do not depend on the scale parameter p_0, but the appropriate choice of this parameter leads to the faster convergence of the series (8).

Since the potential (7) depends on the unknown binding energy, $\epsilon_B = 2m - M_B$, of the $q\bar{q}$ system, the equation (3) is solved with the use of the iteration method. In Table 1 the results of calculations of the dynamical retardation corrections to the heavy quarkonia mass spectrum are presented. In these calculations the parameters k and c were taken to be $k = 0.21$ GeV^2, $c = -1.0$ GeV. The constituent quark masses were chosen to be $m_c = 1.72$ GeV and $m_b = 5.10$ GeV in order to fit J/ψ and Υ meson masses. As we see from Table 1, this set of parameters gives the reasonable description of heavy meson mass spectrum in the static approximation. As expected, the dynamical retardation corrections turn out to be small (typically a few percent) for all low–lying quarkonia states given Table 1.

Note that, unlike the cases considered in refs. [1-3], an additional mass scale parameter does not appear in the "relativized" counterpart of the static potential. Moreover, this approach possesses a smooth static limit and yields small corrections due to the retardation in the heavy quarkonia mass spectrum. This, unlike the results obtained in refs. [1-3], does not manifestly contradict with our expectation at least in the heavy–quark sector.

References

1. T. Kopaleishvili, A. Rusetsky: Talk presented at the international Symposium on Clusters in Hadrons and Nuclei, Tübingen, Germany, July 15-17, 1991; J. Nucl. Phys. (Russia) **56**, 168 (1993)

2. N. Brambilla, G.M. Prosperi: Phys. Rev. **D46**, 1096 (1992); **D48**, 2360 (1993)

3. T. Kopaleishvili, A. Rusetsky: Nucl. Phys. **A** (1995) (in print)

4. G. t'Hooft: Nucl. Phys. **B75**, 468 (1974); C.G. Callan, N. Coote, D.J. Gross: Phys. Rev. **D13**, 1649 (1976); M.B. Einhorn: Phys. Rev. **D14**, 3451 (1976); R.C. Brower, W.L. Spence, J.H. Weis: Phys. Rev. **D18**, 199 (1978)

Few-Body Systems Suppl. 8, 354–358 (1995)

Few-Body Systems

Relativistic Quark Model Baryon Analysis

Stephen R. Cotanch, Adam Szczepaniak, Chueng-Ryong Ji

Department of Physics, North Carolina State University, Raleigh, NC 27695-8202, USA

Abstract. Using a relativistic quark model approach that successfully described flavored meson systems, electromagnetic form factors and structure functions are calculated for the low mass baryons. The observed proton, neutron and lambda moments, rms radii and form factors are all reproduced and predictions are made for other baryon observables which will be measured at CEBAF. Most significantly, by properly including the quark potential energy the long standing neutron to proton structure function ratio anomaly for $x \sim 1$ has been resolved.

1 Introduction

In several recent papers [1] we have completed a comparative study of alternative, relativistic quark models and analyzed observables for the low mass, flavored, pseudo-scalar meson and baryon spectrum. Specifically we have described the nucleon as a few-body bound state that is an element of the unitary representation of the Poincaré group with the group generators being interaction-dependent. Our approach is based on the Bakamjian-Thomas (BT) construction of the Poincaré group generators for a fixed number of constituents ($N = 2$ for mesons and $N = 3$ for baryons) with direct interactions [2]. The representation space and the wave function are constructed for a mass operator which is a relativistic generalization of the Isgur-Karl constituent quark model Hamiltonian [3, 4]. We also have generalized the BT construction by introducing a variable quantization surface. This permits a comparative study of the conventional (instant) and light front schemes and also documents quantization surface (frame dependence) sensitivity.

The following section contains details of the nucleon wave function. In Sect. 3 the results of the nucleon structure calculations are presented along with an analysis of the deep inelastic structure functions. The validity of the single particle current approximation in both instant and light front quantization is also discussed. Finally the main findings are summarized in Sect. 4

2 Relativistic Nucleon Wave Function

The formulation of relativistic dynamics for a fixed number of particles originated with the pioneering work of Dirac [5] and since has been extensively studied on both classical and quantum levels. To obtain a representation of the Poincaré group, a quantization surface is first specified and the generators are then arranged into an interaction-independent subgroup and an interaction-dependent set that involve the evolution of the system in time defined as the direction perpendicular to the quantization surface. For a sensible phenomenology a key requirement is the existence of a set of relative and center of mass variables such that the generators of the symmetry group can be written in a macroscopic, one-body form representing the hadron center of mass motion while the bound state mass is governed by an operator that depends only on the relative variables. For example if the energy operator is expressed in terms of the individual quark mass m_a, momenta, $\mathbf{p}_a$, position, ρ_a and spin, $\mathbf{s}_a$ variables as $P^0 = \sum_a \sqrt{m_a^2 + \mathbf{p}_a} + V(\mathbf{p}_a, \rho_a, \mathbf{s}_a)$ with V representing the interquark potential, the effective one-body, macroscopic Hamiltonian should then have the form $P^0 = \sqrt{\mathbf{P}^2 + M^2(\mathbf{k}_a, \mathbf{x}_a, \mathbf{s}'_a)}$. Here $\mathbf{P}$ is the total bound state momentum and M is the mass operator depending only on the relative variables, momenta, $\mathbf{k}_a$, conjugated positions, $\mathbf{x}_a$, and internal spins $\mathbf{s}'_a$. As discussed elsewhere [6] for given V there is no unique way in which the relative variables can be defined in terms of the individual variables so that the energy operator and all remaining generators of the Poincaré group take this macroscopic form. Furthermore the most general relation may depend on the interaction itself. A particular set of relative and center of mass variables that render the generators in macroscopic form has been introduced by Bakamjian and Thomas [7]. In the BT construction the relative variables are obtained from the individual, constituent variables through an interaction independent Lorentz transformation to the rest frame of the N particle system. Interactions are incorporated into the generators through the replacement of the free mass $\mathcal{M} = \sum_a \sqrt{m_a^2 + \mathbf{k}_a^2}$, by an interaction dependent operator $\mathcal{M}(\mathbf{k}_a) \to M = \mathcal{M}(\mathbf{k}_a) + V$. The replacement of $\mathcal{M}$ with M in the single particle representation of the Poincaré algebra preserves the interaction-free relation between the relative and individual particle variables. This is the essence of the BT construction.

The simple version of the Isgur-Karl Hamiltonian adopted here for the u and d quarks ($m_a = m$) in the nucleon is given by [3, 4]

$$M = \sum_a \omega(\mathbf{k}_a) + \sum_{a<b} \frac{\beta^4}{6m}(\mathbf{x}_a - \mathbf{x}_b)^2 + \frac{2}{3}(2\pi)^{3/2}\frac{\delta}{\beta^3}\sum_{a<b} \mathbf{s}_a \cdot \mathbf{s}_b \delta^3(\mathbf{x}_a - \mathbf{x}_b). \quad (1)$$

Diagonalization of this mass operator gives the nucleon wave function in the relative variable basis, $\psi(\mathbf{P}, \mathbf{k}_a, \sigma_a)$ where σ_a are the eigenvalues of s'_{za}. In order to calculate current matrix elements it is necessary to express the nucleon wave function in terms of the individual particle basis, $\psi(\mathbf{p}_a, \lambda_a)$ with λ_a being the eigenvalues of s_{za}, the constituent spin variables. The resulting wave function is given by

$$\psi(\mathbf{p}_a, \lambda_a) = \sum_{\sigma_a} \prod_a D(\mathbf{p}_a, \mathbf{k}_a)_{\lambda_a \sigma_a} \psi(\mathbf{P}, \mathbf{k}_a(\mathbf{p}_a), \sigma_a) \tag{2}$$

where D are the Wigner rotations and the flavor and color indices have been suppressed.

3 Electromagnetic Structure of the Nucleon

Under a Poincaré transformation, hadronic states should transform as elements of a unitary representation while a matrix element is expected to transform covariantly (i.e. consistent with the tensoral rank of the matrix element operator). These conditions are not satisfied by the free field, one-body current operator frequently used to calculate matrix elements and form factors because the generators, $M_{\mu\nu}$, contain interactions. Thus two particle and perhaps even more complex currents must be included to restore covariance. Consequently, hadronic matrix elements expressed in terms of 4-vectors representing the particle's momenta and polarizations will violate covariance. This in turn leads to an incorrect dependence of matrix elements on the vector n^μ which defines the orientation of the quantization surface. Violations of covariance can most efficiently be studied by allowing the matrix element of the electromagnetic current to explicitly depend upon n^μ having most general form $\langle \mathbf{P}'\lambda' | J^\mu_{em} | \mathbf{P}\lambda \rangle = \bar{u}(\mathbf{P}'\lambda') \Gamma^\mu u(\mathbf{P}, \lambda)$ with

$$\Gamma^\mu \equiv \gamma^\mu F_1 + \frac{i\sigma^{\mu\nu}}{2M} q_\nu F_2 + \frac{n^\mu |\Sigma|}{\Sigma \cdot n} F_3 + \frac{n^\mu \Sigma^2 n \cdot \gamma}{(\Sigma \cdot n)^2} F_4 + \frac{\Sigma^\mu n \cdot \gamma}{\Sigma \cdot n} F_5, \tag{3}$$

$\Sigma^\mu \equiv P'^\mu + P^\mu$, $|\Sigma| = 2\sqrt{M^2 + Q^2/4}$, $Q^2 = -q^2 = -(P' - P)^2$ and $M = M_N$ being the nucleon mass. The form factors F_i, $i = 1 \cdots 5$ will in general depend not only on Q^2 but also on the scalar products of the final and initial nucleon momenta with n. Here the analysis is restricted to spacelike momentum transfer with $q \cdot n = 0$. Expanding the nucleon wave function in the $SU(6)$ basis and keeping the **56** and **70** representations corresponding to the relevant $2\hbar\omega$ states, we have varied the quark mass m and the oscillator parameter β to obtain the best fit to the proton and neutron electric and magnetic form factors. Not including the orbital symmetric $2\hbar\omega$, **56'**, contribution is reasonable for this application since it does not change significantly the behaviour of form factors at momentum transfer $Q^2 \sim 1\text{GeV}^2$. The addition of the mixed symmetric $2\hbar\omega$ excitations of the **70** representation is crucial for the proper description of the neutron electric form factor G_E^n at low Q^2 and also for the right behavior of the neutron to proton structure function ratio $F^n(x)/F^p(x)$ in the valence region as $x \to 1$. For $n^2 \to 0$ corresponding to the infinite momentum frame limit or light cone quantization and for quark mass 200MeV, oscillator parameter 400MeV and mixing angle $\phi = -14^0$ between the **56** and **70** representations, all of the nucleon form factors are well reproduced. It is interesting to note however, that in the instant quantization limit, corresponding to $n^2 = 1$, and for typical quark mass and oscillator parameters, the predicted magnetic form factors are 30% smaller than observed. Conversely, Lorentz symmetry breaking effects as measured by the size of the unphysical form factors F_i, $i = 3 - 5$, are more pronounced in the light cone quantization, particularly at larger momentum transfer.

In the scaling limit the leading twist nucleon structure functions $F^{p,n}(x)$ are given in terms of quark probability distributions determined from

$$q_N(x) = \langle N, P_N|\bar{\psi}_q \delta(x - \hat{x})\psi_q|N, P_N\rangle, \; q = u, d, \tag{4}$$

where x is the Bjorken variable and $\hat{x} = (E_a + k_a^z)/M_N$, with $E_a = \sqrt{m_a^2 + \mathbf{k}_a^2}$ and M_N being the nucleon mass. The operator $\hat{x}$ can be replaced by its expectation value in the infinite momentum frame. The interaction independent relation in the BT construction between the individual and relative particle momenta is equivalent to an interaction-independent boost. Thus the light cone wave function is given by a free Lorentz transformation of the rest frame wave function to the infinite momentum frame. The quark distributions in the scaling variable x, in the BT construction, correspond to distributions in the rest frame wave function for the joint variable $\omega_a(\mathbf{k}_a) + k_a^z$. The negative charge radius of the neutron implies that in the neutron there is an excess of the d quarks over the u at larger distances. This in turn implies that the neutron wave function generates a u_n quark momentum distribution peaking at larger momentum than the d_n quark distribution. Thus the ratio

$$R^{np} = \frac{F^n(x)}{F^p(x)} = \frac{1 + 2u_n(x)/d_n(x)}{4 + u_n(x)/2d_n(x)} \tag{5}$$

will not approach 1/4 as $x \to 1$

This example shows the limitation of the BT construction which includes only the free quark kinetic energy in the boost transformation relating wave functions in different frames or quantization schemes. There is a physical reason why a simple boost approximation cannot properly describe R^{np}. The spin-spin interaction in Eq. (1) responsible for the $N - \Delta$ mass splitting and the spatially asymmetric quark wave function gives an asymmetry in the quark spin potential energy which is defined as an average over the nucleon wave function of the spin-spin interaction of a quark with the spectators. These give bound single u and d quark total energies of $E_{s,u/N} = -\frac{1}{3}(M_\Delta - M_N)$, $E_{s,d/N} = -\frac{1}{12}(M_\Delta - M_N)$, respectively. If the potential energy contribution to the quark energy is now properly included i.e. $E_a \to E_a + E_{s,a/N}$ in the definition of $\hat{x}$ in Eq.(4), $d_n(x)$ peaks at larger x and is shifted to the right compared to $u_n(x)$ to produce a physically correct structure function ratio as shown by the solid line in Fig.1. Accordingly, a long standing anomaly has been resolved as the present quark model analysis simultaneously describes both the neutron charge radius and neutron to proton structure function ratio for x near 1.

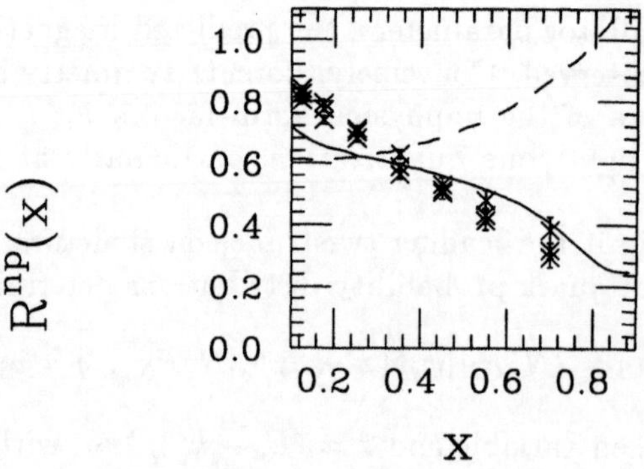

Figure 1. $R^{np}(x)$ ratio calculated with the full bound state quark energy (solid line) and with kinetic energy only (dashed line).

4 Conclusion

In summary, we have satisfactory calculated many baryon observables using a relativistic quark model with parameters effectively determined from our previous meson analysis. Further, by correctly including the quark potential energy, we have resolved a long standing anomaly by simultaneously describing the neutron charge radius and neutron to proton structure function ratio for $x \to 1$. In anticipation of new CEBAF measurements future work will focus on more sophisticated confinement schemes including specific color interactions and extend applications to more exotic systems such as hybrids and glueballs.

References

1. C.-R. Ji, S.R. Cotanch: Phys. Rev. **D41**, 2319 (1990); C.-R. Ji, P.L. Chung, S.R. Cotanch: Phys. Rev. **D45**, 4214 (1992); A. Szczepaniak, C.-R. Ji, S.R. Cotanch: Phys. Rev. **D49**, 3466 (1994)

2. A. Szczepaniak, C.-R. Ji, S.R. Cotanch: Phys. Rev. **C** (to appear)

3. N. Isgur, G. Karl, R. Koniuk: Phys. Rev. Lett. **41**, 126 (1978)

4. N. Isgur, G. Karl: Phys. Lett. **B72**, 109 (1977); **B74**, 353 (1978); Phys. Rev. **D18**, 4187 (1978); **D19**, 2653 (1979); **D20**, 1119 (1979)

5. P.A.M. Dirac: Rev. Mod. Phys. **21**, 392 (1949)

6. R.A. Krajcik, L.L. Foldy: Phys. Rev. **D10**, 1777 (1974)

7. B. Bakamjian, L.H. Thomas: Phys. Rev. **92**, 1300 (1953)

Few-Body Systems Suppl. 8, 359–363 (1995)

Few-Body Systems

Polarized Structure Functions and the Nucleon Spin

M. Traini [1] *, M. Ropele [2], V. Vento[3] †

[1] ECT★, European Centre for Theoretical Studies in Nuclear Physics and Related Areas, Villa Tambosi, I-38050 Villazzano (Trento), Italy and I.N.F.N., g.c. di Trento

[2] Dipartimento di Fisica, Università degli Studi di Trento, I - 38050 Povo (Trento), Italy

[3] Departament de Física Teòrica, Universitat de València and Institut de Física Corpuscular, Centre Mixt Universitat de València, Consejo Superior de Investigaciones Científicas, E-46100 Burjassot (València), Spain

Abstract. The link between the light-cone momentum distributions of constituent quark models and their corresponding parton distributions is used to study the polarized structure functions g_1 of proton and neutron. The formalism allows to analyze the effects of the structure of the quark model wave function and thus we discuss the $SU(6)$ breaking mechanism associated with the *One Gluon Exchange* and large D-state admixtures originated in pionic models. We analyze the contribution to these structure functions of the gluon polarization arising from the anomaly.

1 Introduction

We investigate the interpretation of the polarization experiments in terms of the constituent degrees of freedom. In our approach one considers the nucleon to be entirely consistent of valence quarks at some static point $Q^2 = \mu_0^2$ and generates the gluon and sea content at $Q^2 > \mu_0^2$ dynamically via bremsstrahlung of gluons off the valence system [1]. At the low scale we calculate the quark content from a constituent quark model. The leading order evolution procedure of the renormalization group equations, is used to generate the gluon and the sea content at Q^2 from the scale μ_0^2. In the present calculation we use non relativistic models which contain $SU(6)$ breaking terms leading to a non vanishing polarized structure function $g_1^{\rm n}(x)$ for the neutron.

* *Permanent address:* Dipartimento di Fisica, Università degli Studi di Trento, I - 38050 Povo (Trento), Italy

† Supported in part by DGICyT under contract PB94-0080 and CICyT under contract AEN93-0234

The non relativistic quark model assumption for the nonperturbative part of our calculation is mainly motivated by the success of the model and it has no fundamental justifications. In fact it has been demonstrated [2] that polarization observables can be strongly influenced by relativistic effects. However the largest sensitivity is shown for the transverse polarization (and consequently for the definition and properties of $g_2(x)$ response) rather than for $g_1(x)$ which is the main topic of the present work.

A delicate point which has been much discussed, and which we shall also address, is the role of the gluons. It has been suggested that gluons might contribute significantly to the polarization data due to the anomaly [3]. Even if there are no gluons at the non perturbative scale, quarks generate them at higher Q^2 by evolution modifying therefore the spin dependent distributions of the quarks. We therefore address the question of the compatibilty of the gluon and sea content at large Q^2 with the constituent quark picture at the low energy scale.

2 Polarized Distributions and Constituent Quarks

The polarized parton distributions at the hadronic scale μ_0^2 are calculated using a recent approach [4, 5] which relates the parton model to the light-cone momentum fraction of the valence quarks as evaluated within standard quark models. Relevant information come from the momentum distribution of the constituent quarks [6] which can be connected to the valence parton distribution at μ_0^2 scale

$$x\, q_V^{\uparrow(\downarrow)}(x,\mu_0^2) = \frac{m_q}{M} \int n_q^{\uparrow(\downarrow)}(\mathbf{p})\, \delta(x - p_+/M)\, d\mathbf{p} \qquad (1)$$

where $q^{\uparrow(\downarrow)}$ denotes the q-th valence quark whose spin is aligned (anti-aligned) to the nucleon's spin while $n_q^{\uparrow(\downarrow)}(\mathbf{p})$ is the momentum density distribution for the valence quark of momentum $\mathbf{p}$ and equivalent spin-projection. The q-th valence (constituent) quark is assumed to have a mass $m_q = M/3$ (where M is the nucleon mass), $p_+ = p_0 - p_z$ is the light-cone fraction of the initial quark, and $x = Q^2/2M\nu$ is the Bjorken variable.

The deep inelastic formalism here presented does not entirely fulfill the requirement that the structure functions are defined in the region $0 \le x \le 1$. We have discussed in detail the relevance of such a spuriosity [5]. In addition, in a recent investigation [7] it has been demonstrated that the modification induced by the correct support can be incorporated by using the prescription $q_V(x) \to 1/(1-x)^2 \times q_V\,(x/(1-x))$. As a consequence the relevant effects are limited to the high x region ($x \ge 0.6$) and our conclusions are not invalidated.

The polarized proton and neutron structure functions, at a given Q^2, can be expressed in terms of the polarized distributions of quarks $\Delta q(x,Q^2) = q^{\uparrow}(x,Q^2) - q^{\downarrow}(x,Q^2)$ and gluons $\Delta G(x,Q^2)$, namely

$$g_1(x,Q^2) = \frac{1}{2} \sum_{q=u,d,s} e_q^2 \left[\Delta q(x,Q^2) + \Delta\bar{q}(x,Q^2) - \frac{\alpha_s(Q^2)}{2\pi} \Delta G(x,Q^2) \right] \qquad (2)$$

where we have assumed $SU(3)$ flavor symmetric polarized sea ($\Delta\bar{q} \equiv \Delta\bar{u} = \Delta\bar{d} = \Delta\bar{s} = \Delta s$). In addition we assume, that the sea and gluon contributions vanish at $Q^2 = \mu_0^2$ [8, 9], i.e.

$$\Delta\bar{q}(x, \mu_0^2) = \Delta G(x, \mu_0^2) = 0\,. \tag{3}$$

We note that $g_1^{\mathrm{n}}(x, \mu_0^2)$ vanishes in the exact $SU(6)$ limit of the wave function. On the contrary the main contribution to $g_1^{\mathrm{p}}(x, \mu_0^2)$ comes from $SU(6)$-symmetric components. In the following we will assume that the nucleon wave function can be approximated by a superposition of few $SU(6)$ configurations

$$|N\rangle = a_S|N,\,56, 0^+\rangle_S + a_{S'}|N,\,56', 0^+\rangle_S + a_M|N,\,70, 0^+\rangle_M + a_D|N,\,70, 2^+\rangle_M \tag{4}$$

where $N \equiv \mathrm{p}, \mathrm{n}$; S, M indicate the space part symmetry of the $SU(6)$ representation (the antisymmetric color wave function has not been explicitly written). Eq.(4) includes, in particular, the Isgur-Karl model ($a_S = 0.931, a_{S'} = -0.274, a_M = -0.233, a_D = -0.067$) [12].

According to the nucleon wave function (4) the contribution of the quark to the total spin of the nucleon results

$$\langle\, S_z \,\rangle = \langle\, \frac{1}{2} \sum_i \sigma_i^z \,\rangle = \frac{1}{2}\, \left[1 - 2\, a_D^2\right]\,. \tag{5}$$

Eq.(5) is a particular case of the more general expression $\langle\, J_z \,\rangle = \frac{1}{2} = \langle\, S_z \,\rangle + \langle\, L_z \,\rangle + \langle\, \Delta G \,\rangle$, connecting the total spin (1/2) of the nucleon to the total spin of quarks and antiquarks $\langle\, S_z \,\rangle$, the gluon polarization contribution $\langle\, \Delta G \,\rangle$ and the angular momentum component $\langle\, L_z \,\rangle$. At the static point μ_0^2 (where $\langle\, \Delta G \,\rangle = 0$) the presence of the D-state component introduces an angular momentum contribution $\langle\, L_z \,\rangle = a_D^2$. The relevance of this D-state admixture has been discussed in the past [13] and the value $a_D^2 \approx 0.20$ in a simple model (called hereafter D-model) with a_S and a_D components only ($a_S^2 + a_D^2 = 1$, $a_{S'} = a_M = 0$) would reproduce the experimental value of the axial-vector coupling constant g_a.

We have performed our analysis in the leading order evolution formalism. Of course higher order effects might be non negligible. Certainly the static point and the high momentum scale are quite separated and moreover the confinement scale is low in absolute value. Next-to-Leading-Order formalism is not yet available for polarized scattering structure functions, but we can check the relevance of higher order corrections in the unpolarized case. As a result the LO and NLO evolutions differ in the peak region by 15% showing the size of corrections beyond LO within the radiatively generated parton distribution approach. Even if the static point is low, the change in going from the LO to the NLO scheme produces effects of higher order corrections that are less than 15%.

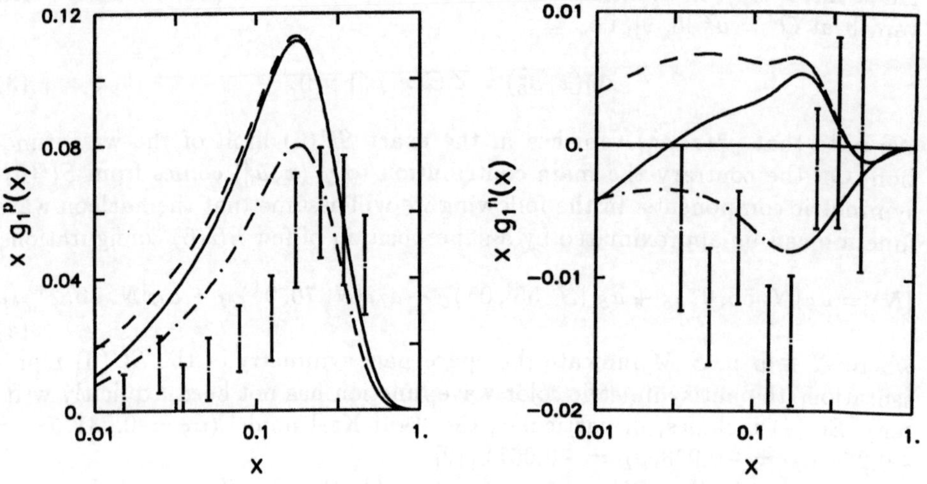

Figure 1. The proton and neutron structure functions predicted by the IK and D models are compared with the experimental data from EMC and SLAC, respectively. For the IK model we show results obtained including valence and sea components only (dashed curves) in addition to the complete calculation which includes the anomalous contribution also (full curves). The D-model results (dot-dashed curves) include anomaly. $Q^2 = 10.7\,GeV^2$ for the proton, and $Q^2 = 2\,GeV^2$ for the neutron is assumed as mean value of the momentum transfer.

3 Comparison with Experimental Data and Discussion

In Fig.1 we show the results of the IK and D-models for the neutron and proton structure functions. The predictions are compared with the available experimental data.

The inclusion of the gluon polarization lowers the values of $g_1^{\mathrm{N}}(x)$ at low x, as expected, because the radiative $\Delta G(x)$ is positive, but the contribution is not enough to reproduce the data. We could speculate on the presence of gluon polarization at the static point of the evolution procedure, but we prefere to remain within a more conservative point of view: the breaking of $SU(6)$ symmetry in the IK model seems to induce an excessive valence contribution, in particular in the u channel. In order to study the $SU(6)$ breaking effects we can resort to the D-model and the analysis is also shown in Fig.1. The improvement in the comparison with the experimental data is evident for both the proton and neutron structure functions indicating that the symmetry breaking terms needed to reproduce the data are more complex than those predicted by the naive OGE mechanism.

Summarizing we have found that if we modify this breaking in a more naive model (the D-model) we are able to come closer to the data. This does not mean that the puzzling behaviour of $g_1^{p,n}(x)$ can be explained by the simple D-state admixture into the nucleon wave function, but rather it implies that the breaking term in the wave function is more complex than that predicted by the naive one-gluon-exchange mechanism. In the same spirit certain recent calculations with unpolarized structure functions have pointed out that these naive models may lack high momentum components and/or sea and gluons ([10, 7].We may therefore conclude that once a link between low energy descriptions of QCD and high energy data is established, the latter impose on the former severe restrictions which may be extremely useful to find better models of hadron structure. In particular, low energy data seem to be blind to high momentum components and sea or gluon contributions at the static scale, features which are coming up over and over again in the type of analysis we have here performed.

References

1. G. Parisi, R. Petronzio: Phys. Lett. **B62**, 331 (1976)
2. J.D. Jackson, G.G. Ross, R.G. Roberts: Phys. Lett. **B226**, 159 (1989)
3. G. Altarelli, G.G. Ross: Phys. Lett. **B212**, 391 (1988); R.D. Carlitz, J.C. Collins, A.H. Mueller: Phys. Lett. **B214**, 229 (1988); A.V. Efremov, O.V. Teryaev: Dubna Report No.E2-88-287, 1988 (unpublished)
4. M. Traini, L. Conci, U. Moschella: Nucl. Phys. **A544**, 731 (1992)
5. M. Ropele, M. Traini, V. Vento: Nucl. Phys. **A584**, 634 (1995)
6. L. Conci, M. Traini: Few-Body Syst. **8**, 123 (1990)
7. M. Traini, V. Vento, A. Zambarda: Mod. Phys. Lett. (in press)
8. M. Glück, E. Reya, W. Vogelsang: Nucl. Phys. **B329**, 347 (1990)
9. Recent work has relaxed this assumption, introducing gluons and sea contributions at the static scale and shown that some features of the unpolarized distributions are better explained [10, 11]
10. M. Glück, E. Reya, A. Vogt: Z. Phys. **C53**, 127 (1992)
11. A. Zambarda: Thesis Università di Trento 1994, unpublished
12. N. Isgur, G. Karl: Phys Rev. **D18**, 4187 (1978); Phys Rev. **D19**, 2653 (1979); Erratum Phys. Rev. **D23**, 817 (1981)
13. S. Glashow: Physica **A96**, 27 (1979); V. Vento, G. Baym, A.D. Jackson: Phys. Lett. **B102**, 97 (1981); A. Abbas: Europhys. Lett. **5**, 287 (1988); J. Phys. G **15**, L195 (1989)

Few-Body Systems Suppl. 8, 364–368 (1995)

Few-Body Systems

Quark-Gluon Point of View on the Spin Structure of the Deuteron from Experiments with Relativistic Deuteron Beams

A.P. Kobushkin*, A.I. Syamtomov

Bogolyubov Institute for Theoretical Physics, National Academy of Sciences of Ukraine, 252143 Kiev, Ukraine

Abstract. We summarize the present status of experimental and theoretical studies of the deuteron spin structure by a hadronic probe. They include differential cross sections, tensor analyzing power, T_{20}, deuteron–proton polarization transfer, κ_0, in deuteron breakup (d,p) and dp backward elastic scattering, respectively. We draw the attention to the existence of the three kinematical regions, where different points-of-view on the deuteron spin structure and/or deuteron interactions with hadronic probe should be distinguished: the standard meson-nucleon, constituent quark model and asymptotic QCD.

1 Introduction

In recent experimental and theoretical investigations (see [1-4] and references therein) of the inclusive breakup (BU) (d,p) and the deuteron backward elastic scattering (BES) $dp \to pd$ it has been demonstrated that their cross sections, as well as polarization observables (for example, the tensor analyzing power, T_{20}, and the coefficient of polarization transfer, κ_0), are deeply connected to the deuteron structure. The main conclusions of this study are as follows:
(i). The empirical proton momentum distributions in the deuteron, extracted from BU cross section in the IA, coincides with that, extracted from BES cross section in the one-neutron exchange (ONE) [5], in the light-cone dynamics formalism.
(ii). Although the qualitative behavior of T_{20} and κ_0 are rather similar for both reactions, some differences in their detailed structure were established.
(iii). If the deuteron wave function has a standard $(S+D)$ components, IA/ONE describes the data only for $k \leq 0.1$ GeV/c [3]. [1]

* *E-mail address:* kobushkin@gluk.apc.org

[1] k stands for the light-cone internal momentum in the deuteron.

One can think that additional contributions beyond IA/ONE, as well as relativistic effects, generating P-wave components of the deuteron wave function [6], should be crucial to account for the discrepancy between theory and experiment for $k > 0.1$ GeV/c [7]. Of course, effects connected to color degrees of freedom in the deuteron [8-11] cannot be excluded *a priori* in this region.

2 Quark Effects in Intermediate Region

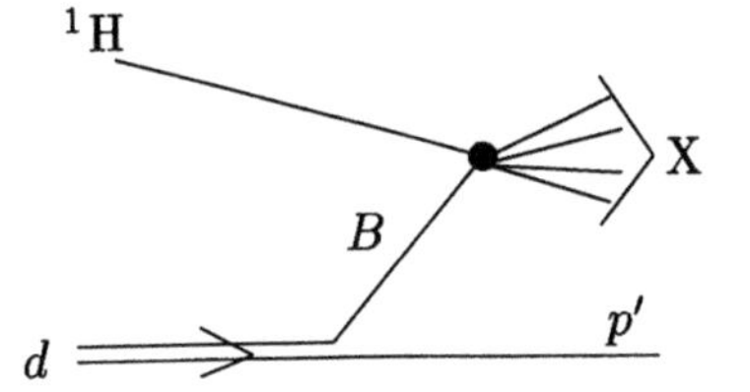

Figure 1. Deuteron inclusive breakup in spectator approximation

Now let's discuss the deuteron breakup at the region (k between 0.1 and 1 GeV/c) where the nucleons in the deuteron are assumed to loose gradually their individuality, but the pure perturbative QCD does not yet work.

The deuteron wave function, being considered as a six-quark ($6q$) object, was shown [12] to be qualitatively equivalent to the wave function of Resonating Group Method (RGM)

$$\psi^d(1,2,\ldots,6) = \hat{A}\left\{\varphi_N(1,2,3)\varphi_N(4,5,6)\chi(\mathbf{r})\right\}, \tag{1}$$

where $\hat{A} = \frac{1}{\sqrt{10}}(1 - 9\hat{P}_{36})$ is a quark antisymmetrizer and $\varphi_N(1,2,3)$ and $\varphi_N(4,5,6)$ are the wave functions of the nucleon three quark ($3q$) clusters; $\chi(\mathbf{r})$ is the RGM distribution function. In the absence of the quark antisymmetrizer the RGM distribution function would coincide, up to the renormalization effects at short range, with the conventional NN deuteron wave function, $\chi_{NN}(\mathbf{r})$.

Due to the antisymmetrizer the deuteron wave function (1), being decomposed into $3q\times 3q$ clusters, includes, apart from the standard pn component, $NN^\star$, $N^\star N$ and $N^\star N^\star$ components which respond to all possible nucleon resonance states, $N^\star$ (see, *e.g.*, [13]). Because the set of resonances $N^\star$ includes states with negative parity the antisymmetrizer $\hat{A}$ in (1) generates effective P-wave components of the deuteron wave function.

As a first step to choose $\chi(\mathbf{r})$ we have used [10] the function $\chi_{NN}(\mathbf{r})$, modified by the RGM renormalization condition [2]. The further calculations were done in the spectator approximation (Fig. 1).

Figure 2 displays the results of the model [10]. One concludes that:

(i) When the internal momentum in the deuteron is of a few hundreds MeV/c, the effects of the quark exchange between three-quark clusters in the deuteron are as important as the ones originating from the relative motion of the nucleons (estimated by the standard two-nucleon potentials).

(ii) The effective P-waves generated by the antisymmetrization of the $6q$ deuteron wave function notably change the momentum distribution of protons in the deuteron and polarization observables of the BU.

[2] Authors [11] fix $\chi(\mathbf{r})$ from microscopic calculations in the framework of constituent quark model, without reference to phenomenological NN deuteron wave function.

(iii) The results of this approach depend weekly on the choice of an input NN potential, but very sensitive to the quark core radius of the nucleon, b.

The antisymmetrization (1) destroys the simple relations between observables calculated in IA/ONE: $n^{\rm BES}_{\rm ONE} = n^{\rm BU}_{\rm IA}$, $T^{\rm BES}_{20}({\rm ONE}) = T^{\rm BU}_{20}({\rm IA})$, $\kappa^{\rm BES}_0({\rm ONE}) = \kappa^{\rm BU}_0({\rm IA})$, and $K^{y,0}_{y,0}({\rm ONE}) = \frac{4}{9}\left(\kappa^{\rm BU}_0({\rm IA})\right)^2$, where $K^{y,0}_{y,0}$ is deuteron–deuteron polarization transfer of BES and $n^{\rm BES}(k)$ and $n^{\rm BU}(k)$ are effective momentum distributions of the protons in the deuteron for BES and BU, respectively. In this case one gets the appropriate relations expressed in terms of three independent functions $\varepsilon_{(1,0)}(k)$, $\varepsilon_{(-1,0)}(k)$ and $\varepsilon_{(1,-1)}(k)$ (relative 'discrepancies') [15]:

$$\left(\frac{n^{\rm BU}}{n^{\rm BES}}\right)^2 - 1 = 2(\varepsilon_{(1,-1)} + \varepsilon_{(-1,0)} + \varepsilon_{(1,0)}) \tag{2}$$

$$T^{\rm BES}_{20} - \left(\frac{n^{\rm BU}}{n^{\rm BES}}\right)^2 T^{\rm BU}_{20} = \sqrt{2}(2\varepsilon_{(1,-1)} - \varepsilon_{(-1,0)} - \varepsilon_{(1,0)}) \tag{3}$$

$$\kappa^{\rm BES}_0 - \left(\frac{n^{\rm BU}}{n^{\rm BES}}\right)^2 \kappa^{\rm BU}_0 = \frac{3}{2}(\varepsilon_{(1,0)} - \varepsilon_{(-1,0)}) \tag{4}$$

$$K^{y,0}_{y,0} - \frac{4}{9}\left(\frac{n^{\rm BU}}{n^{\rm BES}}\right)^2 (\kappa^{\rm BU}_0)^2 = \frac{1}{3}\varepsilon_{(-1,1)}. \tag{5}$$

If $T^{N^\star}_{Mm'}$ is the virtual disintegration amplitude of the deuteron in the $p + N^\star$ channel (where M and m' are spin projections of the deuteron and the proton, respectively) the relative discrepancies are defined as

$$\begin{aligned}\varepsilon_{(MM')} \;=\;& \left(n^{\rm BES}\right)^{-2}\left[\left(\sum_{N^\star} T^{(N^\star)}_{M\frac{1}{2}} T^{(N^\star)^*}_{M\frac{1}{2}}\right)\left(\sum_{N^\star} T^{(N^\star)}_{M'\frac{1}{2}} T^{(N^\star)^*}_{M'\frac{1}{2}}\right)\right.\\ &\left. - \left|\sum_{N^\star} T^{(N^\star)}_{M\frac{1}{2}} T^{(N^\star)^*}_{M'\frac{1}{2}}\right|^2\right].\end{aligned} \tag{6}$$

If the sum (6) contains only one term (corresponding to the ONE-approximation), all relative discrepancies turn to zero. In this case one comes to the results of IA(ONE).

3 Asymptotic Region

One sees that at $k \geq 500$ MeV/c the constituent quark model does not describe, even *qualitatively*, polarization observables of the BU. It was argued [9] that this is a region of a smooth transition to perturbative QCD: estimations made in a very simplified model of reduced nuclear amplitudes in QCD give rise to $T_{20} \to -0.3$ and $\kappa_0 \to 1.06$ at $k \to \infty$. If the latest data (Aono et al. [1]) are in good agreement with this predictions for T_{20}, the asymptotic region for κ_0 seems not to be reached by experiment.

4 Conclusions

- We have estimated the contribution of the quark exchange effects in deuteron breakup and (d,p) elastic backward scattering. At the region of the internal momentum in the deuteron $0.1 \leq k \leq 0.5$ GeV/c they are comparable with the contribution of the IA/ONE estimated using the standard NN potential.

- At $k > 0.5$ GeV/c the behavior of polarization observables of these reactions cannot be explained by quark exchange corrections to the deuteron wave function. There are some arguments that in this region one has smooth transition to the picture defined by asymptotic QCD.

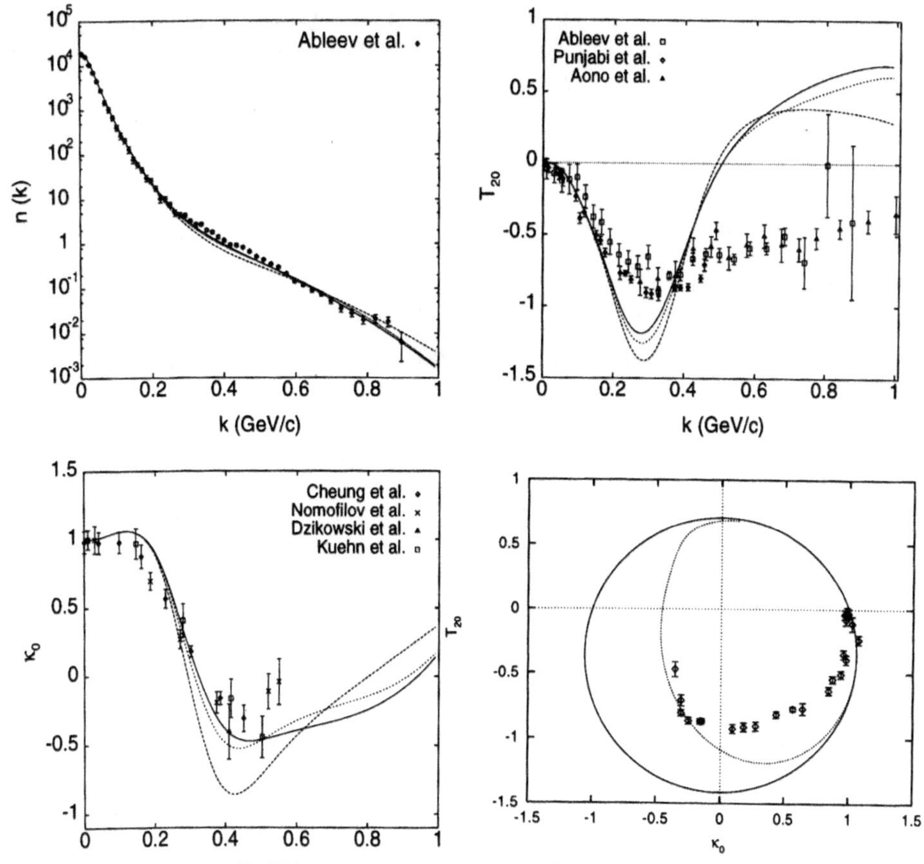

Figure 2. The effective momentum distribution of protons in the deuteron, $n(k)$, T_{20}, κ_0, and $T_{20} - \kappa_0$ plot, calculated in the approach of [10]. The solid and dotted lines correspond to Paris- and RSC- input deuteron wave functions, respectively. Dashed line denotes the IA with Paris potential.

Acknowledgement. We are grateful to E.A. Strokovsky for useful discussions. One of the authors (A.P.K) thanks also the organizers of this Conference for financial support during his stay in Peniscola. This work was supported in part by grants RFU000 from the International Science Foundation and RFU300 from the International Science Foundation and Russian Government.

References

1. V. Punjabi et al.: In:*14-th Int. FB Conf.* Ed. by F.Gross, Williamsburg, 1994, p.161; L.S. Azhgirei et al.: *ibid.* p.18; V.A. Ableev et al.: JINR Rapid Comm. **1[52]-92**, 10 (1992); Ableev et al.: Pis'ma v ZhETF **47**, 558 (1988); B. Kuehn et al.:Phys. Lett. **B334**, 298 (1994); A.A. Nomofilov et al.: Phys. Lett. **B325**, 327; N.E. Cheung et al.: Phys. Lett. **B284**, 210 (1992); T. Aono et al.: Phys. Rev. Lett. **74**, 4995 (1995); T. Dzikovsky et al.: In *Proc. of Int. Workshop DEUTERON-93.* Dubna, 1994, p.181
2. A.P. Kobushkin: In *14-th Int. FB Conf.* Ed. by F.Gross, Williamsburg, 1994, p.99; In *Proc. of Int. Workshop DEUTERON-93.* Dubna, 1994, p.71
3. B. Kuehn, E.A. Strokovsy, C.F. Perdrisat: *ibid.*, p.31
4. C.F. Perdrisat, V. Punjabi: In *Proc. of the XI Int. Seminar on High Energy Physics Problems.* Dubna, 1994, p.491
5. A.P. Kobushkin: J. Phys. G: Nucl.Phys. **12**, 485 (1986)
6. W. Buck, F. Gross: Phys. Lett. **63B**, 286 (1976)
7. B.D. Keister, J.A. Tjon: Phys. Rev. **C26**, 578 (1982); G.I. Lykasov, M.G. Dolidze: Z. Phys. **A336**, 339 (1990)
8. A.P. Kobushkin, L. Vizireva: J. Phys. G: Nucl. Phys. **8**, 893 (1983)
9. A.P. Kobushkin: J. Phys. G: Nucl. Part. Phys. **19**, 1993 (1993)
10. A.P. Kobushkin, A.I. Syamtomov, L.I. Glozman: Preprint ITP-95-9E, Bogolyubov Institute for Theoretical Physics, 1995; submitted to Yad. Fiz.
11. I.T. Obukhovsky, V.S. Gorovoy: talk at: Int. Seminar on High Energy Physics Problems, Dubna, 1994 (in print) and Private Communication
12. L.Yu. Glozman, V.G. Neudatchin and I.T. Obukhovsky: Phys. Rev. **C48**, 389 (1993)
13. L.Yu. Glozman, E.I. Kuchina: Phys. Rev. **C49**, 1149 (1994)
14. L.Ya. Glozman et al.: Phys. Lett. **B200**, 983 (1988); I.T.Obukhovsky, A.M. Kusainov: Yad. Fiz. **47**, 494 (1988) and Phys. Lett. **B238**, 142 (1990)
15. A.P.Kobushkin, A.I. Syamtomov: in preparation

Few-Body Systems Suppl. 8, 369–373 (1995)

Searching for ^{6}Li D-state Effects

A.M. Eiró[1], F.D. Santos[1], I.J. Thompson[2,3]

[1] Departamento de Fisica and Centro de Fisica Nuclear da Universidade de Lisboa, 1700 Lisboa, Portugal

[2] Department of Physics, University of Notre Dame, IN 46556, USA

[3] Department of Physics, University of Surrey, Guildford GU2 5XH, UK

Abstract. It is shown that transfer reactions induced by polarized ^{6}Li can be used as a tool to determine the sign of the D/S state asymptotic ratio in the $\alpha + d$ configuration in ^{6}Li. T_{20} is found to be the observable with larger sensitivity both in (^{6}Li, d) and (^{6}Li,α) reactions.

The characterization of the D-state components in the light nuclei provides direct information on the tensor part of the NN interaction and has been widely studied both theoretically and experimentally [1]. Although the presence of D-states in the A=2, 3 and 4 nuclei is well established, experiments are underway to improve the accuracy of the determinations of D-state parameters in those nuclei [2]. The situation in the ^{6}Li nucleus is at a very different stage. Apart from the measurement of the quadrupole moment, which is small and negative, there is very little experimental evidence of the D-state components in this nucleus. The effects of the tensor force on the structure of nuclei with A>4 are qualitatively quite different from those in the 1s shell. The mixing of S- and D-state in a given cluster configuration is determined to a large extent by the shell structure and not by the tensor force. In the case of ^{6}Li, D-state components exist in principle for two cluster configurations: $\alpha - d$ and ^{3}H-^{3}He. So far experimental and theoretical efforts have focused solely on the $\alpha - d$ configuration, where the relative motion between the α particle and the deuteron is described by the overlap

$$\langle \phi_\alpha \phi^{\sigma_d} | ^6\mathrm{Li}; 1M \rangle = \sum_{L=0,2} (L\lambda 1 \sigma_d | 1M) u_L(r) Y_L^\lambda(r)$$

with σ_d and M being the deuteron and ^{6}Li spin projection. Asymptotically for large r the radial wave functions $u_L(r)$ behave as

$$u_L(r) = N_L u_{NL}(r) \longrightarrow \frac{N_L}{\beta r} \mathrm{W}_{\eta_c, L+1/2}(2\beta r)$$

Here $W_{\eta_c,L+1/2}$ is a Whittaker function with a Coulomb parameter η_c and a wave number β calculated at the ^{6}Li$\rightarrow \alpha + d$ vertex. The asymptotic D/S ratio is defined as η=N_2/N_0 and $D_2 = \lim_{k\rightarrow 0}[(u_2(k)/k^2 u_0(k)]$ where $u_L(k)$ are the radial wave functions in momentum space.

^{6}Li has been frequently described as an $\alpha - d$ cluster because of the tight binding energy of the ^{4}He and the small separation energy between ^{4}He and ^{2}H in ^{6}Li. This $\alpha - d$ cluster model has been very successful in describing the elastic and inelastic scattering of polarized ^{6}Li from heavy targets [3]. In this approach the amplitude of the D-state component is adjusted to reproduce the ^{6}Li quadrupole moment, and the calculation predicts $\eta = -0.014$. Three body (αnp) models have been developed [4], providing reasonable description of many ^{6}Li properties, of the $d - \alpha$ scattering and of $(e, e'd)$ data on ^{6}Li. However in spite of the improvements in the treatment of the $\alpha - N$ effective interaction and of the use of realistic NN potentials, the experimental value of the quadrupole moment is not reproduced. These calculations [5] predict η=0.0194 or η=0.0169 and D_2=0.226 fm^2 or D_2=0.218 fm^2 depending on the type of the $\alpha - N$ potential used. A very recent [6] six nucleon calculation performed with Green's function Monte Carlo method using Argonne v_{18} potential, reproduces the binding energy of the ^{6}Li, but neither the quadrupole moment nor the overlap $\langle ^6\text{Li}|\alpha d\rangle$ are calculated.

The experimental information is very limited, and leads to contradictory results. From forward dispersion relation analysis of elastic $\alpha + d$ scattering, Bornand et al. [7] extracted η =0.005±0.017. The analysis of tensor analyzing power data of ^{6}Li$(d,\alpha)^4$He reaction by Santos et al [8] gives a negative value for D_2 and predicts for the asymptotic ratio η a value between -0.015 and -0.010. Finally, from the analysis of T_{20} in the inclusive breakup of polarized ^{6}Li at 4.5 GeV, Punjabi et al. [9] conclude for a positive value of D_2.

In the present paper we address the question of determining η from the analysis of the tensor analyzing powers (TAP) of (^{6}Li,d) and (^{6}Li,α) reactions. The underlying hypothesis is that the probability for a cluster transfer is relatively high. Although other components may be present, we assume an α particle cluster transfer in the case of the (^{6}Li,d) reactions and a deuteron cluster transfer in the case of (^{6}Li,α) reactions. Under these assumptions, the transition amplitudes for the (^{6}Li,d) and (^{6}Li,α) reactions involve explicitly in the form factor the $\langle \alpha d|V_{d\alpha}|^6\text{Li}\rangle$ matrix element which is directly related with the $\langle \alpha\ d|^6\text{Li}\rangle$ overlap. It is important to emphasize that we are not assuming a $\alpha + d$ cluster model for the ^{6}Li. However we are probing the $\alpha + d$ cluster configuration in ^{6}Li by studying transitions which involves the $\langle \alpha\ d|^6\text{Li}\rangle$ overlap.

As in other transfer reactions [1], the tensor analysing powers of these reactions are expected to be sensitive to the presence of the D-state component. In fact, using the peripheral model of Santos et al. [10], for the bound states, one can prove that the TAP for A(^{6}Li,d)B reactions can be expressed as

$$T_{2q} = -\sqrt{\frac{8\pi}{5}} A Y_2^q(Q)$$

with $\boldsymbol{Q} = \boldsymbol{k}_6 - \frac{m_A}{m_B}\boldsymbol{k}_d$ being the momentum transfer between the momenta in the entrance channel $\boldsymbol{k}_6$ and the outgoing channel $\boldsymbol{k}_d$, and

$$A = -\sqrt{2}\,\frac{\gamma^2}{\beta^2}\left(\eta\frac{I_2}{I_0} - \frac{1}{2\sqrt{2}}\frac{\gamma^2}{\beta^2}\eta^2\frac{I_2^2}{I_0^2}\right) \Big/ \left(1 + \frac{\gamma^4}{\beta^4}\eta^2\frac{I_2^2}{I_0^2}\right).$$

In the above expression γ is the wave number corresponding to the binding energy of the transferred cluster in the target and $I_L(Q)$ are defined in Eq. (50) of ref. [10]. In the absence of spin dependent distortions and D-state effects T_{2q} are therefore zero, and they should be proportional to η in first order. It is therefore likely to expect that the measurement of such observables shows sensitivity to the sign and magnitude of η.

To test this conclusion we analyze (^{6}Li,d) reactions on heavy targets. These reactions have been used extensively in the past [11] to probe the α-cluster structure in nuclei. The cross sections have been analyzed assuming a one step DWBA direct process at energies from 30 to 50 MeV in several targets of the sd and the fp shells. Here we consider a 0^+ target leading to a 0^+ residual nucleus, which correspond to a J=0 and L=0 transfer. Specifically we analyze the reactions ^{58}Ni(^{6}Li,d)^{62}Zn and ^{28}Si(^{6}Li,d)^{32}S induced by a 50 MeV polarized beam. Theoretical results of T_{20} and T_{21} are shown in Fig.1. The calculations are DWBA analysis using the code FRESCO [12]. The dotted curves corresponds to the description of the $\langle\alpha d|^6\mathrm{Li}\rangle$ wave function with no D-state, the dashed and full lines account for the inclusion of the D-state with $\eta = -0.014$ (dashed) and η =+0.014 (full). The optical potentials used are the global potentials from Cook [13] in the case of the entrance channel and from Daehnick et al. [14] in the case of the outgoing channel. The bound states are described with a Wood-Saxon model using in the case of the ^{6}Li the geometry

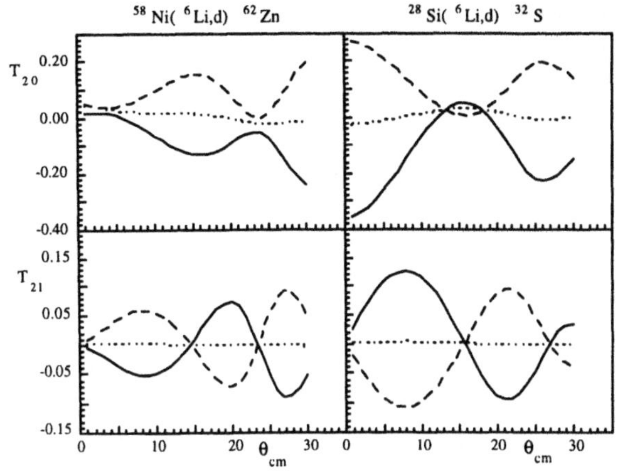

Figure 1. DWBA predictions for T_{20} and T_{21} in (^{6}Li,d) reactions at 50 MeV. The dotted curves include no D-state in the $\langle\alpha d|^6\mathrm{Li}\rangle$ overlap. The full and dashed curves include D-state with η=+0.014 and η=−0.014 respectively.

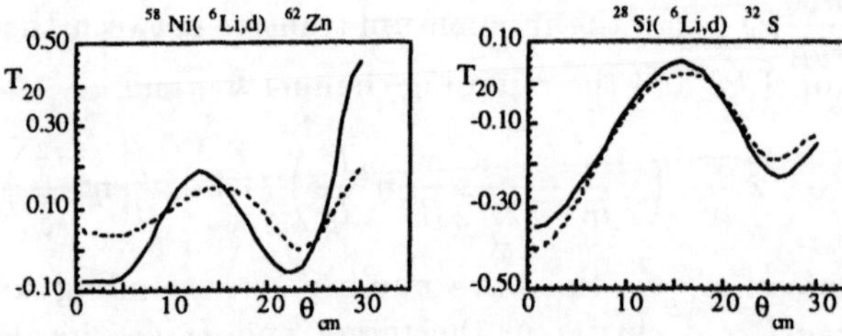

Figure 2. Comparison between one-step (dashed curves) and two-step (full curves) calculations as described in the text. All the curves include D-state.

suggested in ref. [3] (r=1.90, a=0.65), which is consistent with the calculations from ref. [4]. We can observe a distinct sensitivity to the sign of η, particularly in the case of T_{20} which is predicted to be larger in magnitude than the other TAP.

Due to the importance of the coupling to the excited states of the ^{6}Li, in ^{6}Li elastic scattering, no conclusions should be drawn from the analysis of these tensor observables in transfer reactions without testing the importance of the coupling between the ground state (1^+) and the low lying excited states of the ^{6}Li. We consider the coupling with the 3^+ excited state (E_x=2.18 MeV) as caused by the interaction with the target, within the weak binding energy approach, allowing no reorientation of the projectile, and including an overall normalization factor of 0.45. The comparison between the one-step and the two-step processes is shown in Fig.2 for the T_{20} in the case of the two reactions studied, showing some sensitivity, that nevertheless does not destroy the possibility to determining the sign of η.

In the case of (^{6}Li,α) reactions there is a spin one transfer and, for a given J transfer, the L transfer may not be unique. There are natural parity transitions with $L = J$ and unnatural parity transitions with $L = J \pm 1$. Also, even in the absence of spin dependent distortions and D-state, the TAP can be non zero. A recent analysis by Mendez et al. [15] of ^{12}C(^{6}Li,α)^{14}N at 33 MeV, that assumed a DWBA direct mechanism in several unnatural parity transitions, leads to the conclusion that the TAP show strong sensitivity to the L mixing in both the projectile and the target. To get rid of the dependence on the L mixing in the target, we analyse a non-mixed transition, namely a natural parity transition. As a general rule, in these natural parity transitions D-state effects are expected to be less important than in the unnatural parity[10], but the effect should increase for reactions with a large Q-value. It is therefore crucial to measure a transition where the binding energy of the transferred cluster in the target is large when compared with the separation energy $B_{^6Li}-B_\alpha-B_d$=1.474 MeV. This is the case of ^{28}Si(^{6}Li,α)^{30}P (2^+, E_x=1.454 MeV). In Fig.3(a) we show a DWBA calculation of the T_{20} for that reaction where we see, as expected, a very strong sensitivity to the sign of η. The inclusion of the coupling with the 3^+ state, leads to changes in the curves as shown in Fig.3(b), but does not

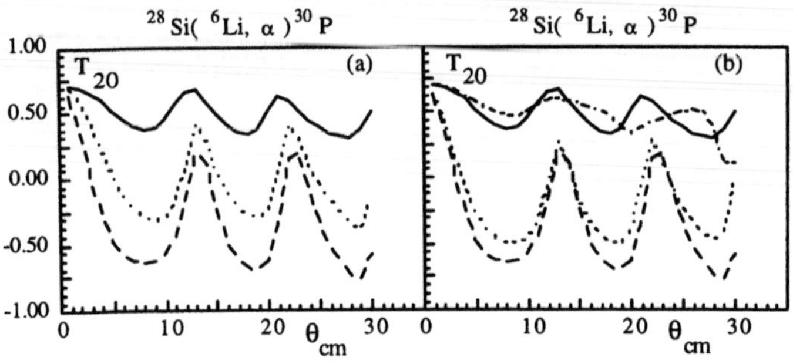

Figure 3. DWBA predictions for T_{20} in (^{6}Li,α) reaction at 50 MeV. The curves in (a) have the same meaning as in Fig.1. In (b) it is shown a comparison between one-step (full and dashed lines) and two-step (dash-dotted and dotted lines) calculations. All the curves include D-state. In the upper curves η=+0.014 and in the lower curves η=–0.014.

destroy the large sensitivity in T_{20} to the sign of η. It is therefore expected that measurements of the TAP in well chosen (^{6}Li,d) and (^{6}Li,α) reactions will give information on η. Measurements have been started at FSU by E.J. Ludwig and collaborators [16].

References

1. A.M. Eiró, F.D. Santos: J.Phys. G: Nucl. Part. Phys. **16**, 1139 (1990)
2. B. Kowslowka et al.: Phys. Rev **C50**, 2695 (1994)
3. H. Nishioka et al.: Phys. Lett. **124B**, 17 (1983)
4. D.R. Lehman et al.: Phys. Rev. **C31**, 1920 (1985); Phys. Rev. **C37**, E2266 (1988); N.W. Schellingerhout et al.: Phys. Rev. **C48**, 2714 (1993)
5. D.R. Lehman: Colloque de Physique Suppl. n 22, **52**, C6-47 (1990)
6. B.S. Pudliner et al.: Phys. Rev. Lett. **74**, 4396 (1995)
7. M.P. Bornand et al.: Nucl. Phys. **A294**, 492 (1978)
8. F.D. Santos et al.: Colloque de Physique Suppl. n 22, **52**, C6-443 (1990)
9. V. Punjabi et al.: Phys. Rev. **C46**, 984 (1992)
10. F.D. Santos, A.M. Eiró: Port. Phys. **15**, 65 (1984)
11. T. Yamaya et al.: Phys. Rev. **C42**, 1935 (1990)
12. I.J.Thompson: Comput. Phys. Report. **7**, 167 (1988)
13. J. Cook: Nucl. Phys. **A388**, 153 (1982)
14. W.W. Daehnick et al.: Phys. Rev. **21**, 2253 (1980)
15. A.J. Mendez et al.: Phys. Rev. **C51**, 651 (1995)
16. E.J. Ludwig: Private Communication

Few-Body Systems Suppl. 8, 374–378 (1995)

Few-Body Systems

Comparison between Three-Body and Many-Body Descriptions of ^{11}Li

J.M.G. Gómez[1], M. Casas[2]

[1] Departamento de Física Atómica, Molecular y Nuclear, Universidad Complutense de Madrid, 28040 Madrid, Spain

[2] Departament de Física, Universitat de les Illes Balears, 07071 Palma de Mallorca, Spain

Abstract. Most of the theoretical work on the halo nucleus ^{11}Li has used three-body models based on the $^{9}\text{Li} + n + n$ configuration. We show that a self-consistent mean-field description of the neutron halo of ^{11}Li in fair agreement with the experimental mass density is possible using density-dependent interactions with a symmetry energy $a_{\text{sym}} \sim 26$ MeV at saturation density. A substantial improvement is also obtained for the binding energy of asymmetric nuclei through the nuclear chart. The three-body and the many-body approaches to ^{11}Li are complementary.

1 Introduction

The development of new experimental techniques in the last few years, especially the study of nuclear reactions with radioactive nuclear beams pioneered by Tanihata *et al.* [1], has revealed very unusual properties for some nuclei close to the neutron drip line. The study of these exotic nuclei has become a very active field of theoretical and experimental research. One of the most interesting novel phenomena is the existence of a *neutron halo* stretching far outside the core of some neutron drip line nuclei, of which ^{11}Li is an archetype. In ^{11}Li the single-neutron separation energy S_n is about 730 keV, but $S_{2n} = 310 \pm 40$ keV and ^{10}Li is thus unbound [2]. The properties of halo nuclei are extensively discussed in refs. [3-4].

Several theoretical approaches have been applied to describe the ground state of ^{11}Li, mainly three-body models of type $^{9}\text{Li} + n + n$ [5-7], and mean field models. But Hartree-Fock (HF) theory with density-dependent forces has been unsuccessful up to date to reproduce the neutron halo. The inclusion of correlations in shell model + self-consistent HF calculations has improved the agreement for binding energies of neutron-rich Li and Be isotopes, but has also failed to reproduce the ^{11}Li halo [8].

In Sect. 2 we outline the three-body approach to ^{11}Li. In Sect. 3 we outline the mean-field approach to the structure of this nucleus and discuss how to improve the results of self-consistent calculations for nuclei far from stability. A comparison of the three-body and many-body descriptions of ^{11}Li is discussed in Sect. 4.

2 Three-Body Models

The ^{11}Li nucleus can be considered a three-body system consisting of ^{9}Li+n+n. The three-body procedures used for ^{11}Li have been recently reviewed by Zhukov *et al.* [5]. The basic assumption in all these models for ^{11}Li is that its core is identical to ^{9}Li, e.i. its structure is not modified by the presence of the two additional neutrons. This is not in agreement with HF predictions, and the existence of an excited state of ^{9}Li at 2.6 MeV also shows that it is a relatively soft core. On the other hand, the assumption is based on the weak binding of the halo neutrons to the core, and on the use of an effective core-neutron potential that would include the core polarization effects. It has also been argued that the mean-field approach can be questioned close to the nucleon drip lines where anomalous nuclear properties like neutron halos appear.

The essential limitation of the ^{9}Li+n+n models is the lack of unambiguous experimental data on the core-neutron interaction. The results of three-body calculations depend very sensitively on the potentials used and thus can not be considered conclusive.

As an example, Fig. 1 shows the ^{11}Li mass density obtained in a coordinate space Faddeev calculation for the ^{9}Li-n potential L5D of ref. [5], compared to the empirical density which fits reaction cross sections [9]. The halo is well reproduced, with a mass radius r_m=3.04 fm, a halo radius r_{halo}=5.16 fm and S_{2n}=198 keV, in good agreement with experiment.

A similar good fit of these data can be obtained with other potentials. However, the ^{11}Li spatial correlations are strongly dependent on the potentials assumed and the halo structure is thus ambiguous.

3 Mean-Field Models

In the simplest shell model, ^{11}Li is just a closed-shell nucleus plus a single $0p_{3/2}$ proton, and the HF wave function should be a good approximation to the ground state. This picture is supported by experimental electric quadrupole and magnetic dipole moments in fair agreement with the $0p_{3/2}$ single-proton values [10].

Table 1 shows a comparison of experimental energies and radii of ^{11}Li with the HF predictions for several forces of Skyrme type. The forces SIII, Ska and Skm* are some of density-dependent Skyrme interactions frequently used in the literature. It is seen that the results are poor compared with the good standards obtained in stable nuclei. We have found that the results are also

Table 1. Comparison between experimental properties of ^{11}Li and Hartree-Fock calculations with several Skyrme interactions. B is the total binding energy, S_n the neutron separation energy, r_p, r_n, r_m and $r_{\rm halo}$ the proton, neutron, mass and halo radii, respectively, and Q the quadrupole moment.

	SIII	Ska	Skm*	Skm1	Expt.	Ref.
B(MeV)	49.3	52.9	54.6	46.9	45.5	[2]
S_n(MeV)	2.0	4.1	2.9	0.3	0.73 ± 0.07	[2]
r_p(fm)	2.19	2.26	2.26	2.26		
r_n(fm)	2.85	2.84	2.90	3.43		
r_m(fm)	2.69	2.69	2.74	3.15	3.12 ± 0.30	[9]
$r_{\rm halo}$(fm)	3.60	3.36	3.55	5.17	4.8 ± 0.8	[9]
					5.03 ± 0.16	[12]
					5.1 ± 0.9	[13]
Q(mb)	-25.4	-28.4	-27.5	-27.0	-31.2 ± 4.5	[10]

poor for other highly asymmetric, non drip-line nuclei. This suggests that the *symmetry energy* of these interactions could be improved.

For the forces used above, SIII, Ska and Skm*, the symmetry energy coefficient in asymmetric nuclear matter is given by

$$a_{\rm sym} = \frac{\hbar^2 k_F^2}{6m} - \frac{1}{4} t_0 (x_0 + \frac{1}{2}) \rho - \frac{1}{24} t_3 (x_3 + \frac{1}{2}) \rho^{1+\alpha} + \frac{1}{6} t_2 \rho k_F^2 \ , \tag{1}$$

where $k_F = (\frac{3}{2}\pi^2\rho)^{1/3}$. For fixed values of the parameters α and t_i, variation of x_0 and x_3 changes asymmetric nuclear matter properties without changing the symmetric limit. In finite nuclei the symmetry energy has both surface and volume contributions, and the combined variation of x_0 and x_3 thus offers the possibility of changing simultaneously the force both near the saturation and at low densities.

Thus, starting e.g. from the force Skm*, we have obtained a new force Skm1 with parameters identical to Skm* except $x_0 = 1.3$ and $x_3 = 1.734$. For N=Z nuclei, the forces Skm1 and Skm* are completely equivalent. At saturation density, $a_{\rm sym} = 25.2$ MeV for Skm1 and 30.0 MeV for Skm*. At *low density*, however, the symmetry energy is *larger* in Skm1 than in Skm*. In ^{208}Pb the volume and surface contributions compensate to give the same total binding energy with both forces, but the lower density effect is dominant for light nuclei.

The ^{11}Li properties obtained with Skm1 are given in Table 1. The results are clearly better than with the other forces.

Fig. 1 shows the matter density distribution of ^{11}Li calculated with the Skm1 and Skm* interactions. The Skm* density becomes too small at distances beyond 6 fm, but Skm1 reproduces the empirical density distribution determined by Tanihata [9] in the whole region.

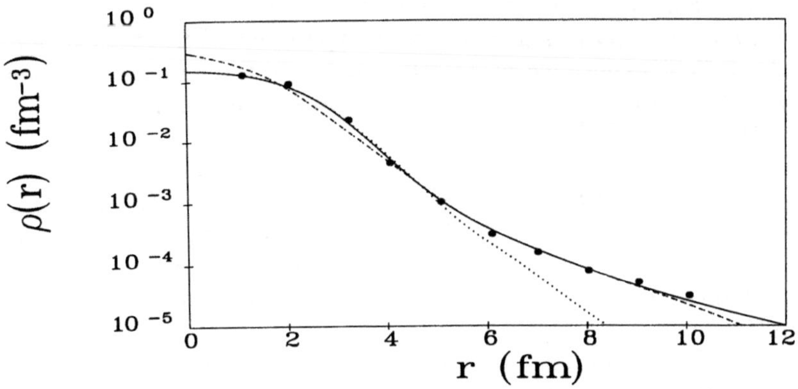

Figure 1. Empirical and calculated mass density distributions of ^{11}Li. The solid and the dotted curves give the HF results obtained with the Skm1 and Skm* interactions, respectively, and the dashed curve is a Faddeev calculation [5]. The limits of the empirical density [9] are given by the error bars.

Table 2. Experimental and Hartree-Fock binding energies (MeV) obtained with the interactions Skm* (a_{sym} = 30.0 MeV) and Skm1 (a_{sym} =25.2 MeV).

	Skm*	Skm1	Expt.		Skm*	Skm1	Expt.
^{4}He	26.7	26.7	28.3	^{32}Mg	253.1	247.9	250.6
^{8}He	37.9	31.0	31.4	^{40}Ca	341.1	341.1	342.0
^{8}C	31.0	24.2	24.8	^{48}Ca	420.0	417.5	416.0
^{12}C	93.3	93.3	92.2	^{56}Ni	485.4	485.4	484.0
^{20}C	127.6	118.1	119.7	^{80}Zn	680.7	674.7	674.1
^{16}O	127.7	127.7	127.6	^{90}Zr	783.0	782.2	783.9
^{24}O	178.2	171.9	168.7	^{132}Sn	1110.7	1106.4	1102.9
^{28}Si	236.5	236.5	235.5	^{208}Pb	1636.4	1636.5	1636.5

Table 2 shows the total binding energies for several asymmetric nuclei. The overall agreement of Skm1 with experiment is good, for light and heavy nuclei, and for symmetric and asymmetric nuclei as well, in contrast to the previous situation with Skm* and other Skyrme forces.

The symmetry energy of the force has implications for many other nuclear properties of the asymmetric nuclei. But for Skyrme forces with a_{sym} in the 25-26 MeV range, charge radii and other bulk properties of stable nuclei are in good agreement with experimental data [11].

4 Discussion

The existence of a neutron halo in ^{11}Li is well described by three-body models, with a radial mass density in agreement with experiment. The weakest point is the lack of sufficient experimental data to determine the core-neutron potential, because the results, e.g. the spatial and momentum correlations, are very sensitive to the potential choice.

The mean-field approach to ^{11}Li has been questioned by some authors. It is a very light nucleus at the neutron drip line, and the earlier HF calculations with conventional forces gave a poor description of the ground state. We have shown that the results for ^{11}Li are much better for the modified force Skm1, which further more also improves the results for other asymmetric nuclei far from stability.

The three-body calculations take into account correlations between the halo neutrons and treat well the center of mass motion, but contain approximations for the Pauli principle between the core and valence nucleons. The HF calculations treat exactly the Pauli principle, but make approximations for the center of mass motion. Thus the three-body and the many-body methods are complementary descriptions of ^{11}Li.

Acknowledgement.

This work has been supported in part by DGICYT (Spain), grants PB92-0021-C02-02 and PB93-0263; and by CICYT (Spain), grant AEN93-0776.

References

1. I. Tanihata et al.: Phys. Rev. Lett. **55**, 2676 (1985)
2. G. Audi and A.H. Wapstra: Nucl. Phys. **A565**, 1 (1993)
3. P.G. Hansen: Nucl. Phys. **A553**, 89c (1993)
4. K. Riisager: Rev. Mod. Phys. **66**, 1105 (1994)
5. M.V. Zhukov et al.: Phys. Rep. **231**, 151 (1993)
6. G.F. Bertsch, H. Esbensen, Ann. Phys. (NY) **209**, 327 (1991)
7. L. Johannsen, A.S. Jensen, P.G. Hansen: Phys. Lett. **B244**, 357 (1991)
8. J.M.G. Gómez, C. Prieto, A. Poves: Phys. Lett. **B295**, 1 (1992)
9. I. Tanihata et al.: Phys. Lett. **B287**, 307 (1992)
10. E. Arnold et al.: Phys. Lett. **B281**, 16 (1992); Phys. Lett. **B197**, 311 (1987)
11. J.M.G. Gómez, J. Martorell: Nucl. Phys. **A410** , 475 (1983)
12. A.K. Chaudhury: Phys. Rev. **C49**, 1603 (1994)
13. W.R. Gibbs, A.C. Hayes: Phys. Rev. Lett. **67**, 1395 (1991)

Few-Body Systems Suppl. 8, 379–383 (1995)

Few-Body Systems

Light Ion Induced Reactions at Intermediate Energies

S. Hirenzaki* †

Departamento de Física Teórica and IFIC, Centro Mixto, Universidad de Valencia - CSIC, 46100 Burjassot (Valencia) Spain

Abstract. We study the light ion induced reactions at intermediate energies theoretically. We show that $(^3\mathrm{He}, t)$ reactions on unstable nuclei are useful tool to investigate the surface of the unstable nuclei and isospin dependence of the Δ excitation mechanism in nuclei. We also show that (α, α') reactions on proton target are interesting since we can observe Roper resonance excited by σ-meson exchange. In both cases we can get interesting results by making use of properties of the light ions.

1 Introduction

We study two kinds of light ion induced reactions, namely, $(^3\mathrm{He}, t)$ reactions on unstable nuclei and (α, α') reactions on proton, at intermediate energies. In both cases we show that we can get interesting information by making use of properties of the light ions.

First, the charge-exchange $(^3\mathrm{He}, t)$ reactions at intermediate energies are suitable to study the surface of the unstable nuclei, which are known to have interesting structure like neutron halo and skin [1], since (1) the reaction takes place in the nuclear surface due to the large distortion effects of ^{3}He and t, (2) the charge-exchange reaction depends strongly on the numbers of neutrons and/or protons which participate in the reaction, (3) the reaction mechanism is relatively simple at intermediate energies and (4) we can perform the experiments using the inverse kinematics with unstable-nuclear beams [2]. These conditions are also satisfied for other charge-exchange reactions. We consider here the $(^3\mathrm{He}, t)$ reactions since the reaction has been studied extensively for stable nuclei [3], and we can calculate the reaction cross section reliably using the model of ref. [4].

In addition to the possible sensitivity to the nuclear surface of the unstable nuclei, the Δ-excitation by the $(^3\mathrm{He}, t)$ reaction is an interesting subject since

*JSPS research fellow

†*E-mail address:* zaki@evalvx.ific.uv.es & Hirenzaki@rikvax.riken.go.jp

the Δ-peak shifts were found in stable nuclei [5]. We think that to observe the Δ-resonances in unstable nuclei is a method to see the nucleon isospin dependence of the reaction mechanism since the unstable nuclei are expected to have a thick neutron/proton skin. In other words we can observe the Δ-excitation on a neutron target by $(^3\mathrm{He}, t)$ reactions on neutron-rich nuclei because of the neutron skin. We can expect to obtain new information on the Δ-excitation mechanism by $(^3\mathrm{He}, t)$ reactions on unstable nuclei.

Second, we investigate theoretically the (α, α') reaction on a proton target at intermediate energies in order to obtain new information on the reaction mechanism and the properties of hadron resonances. The fact that the α particle has isospin $T = 0$ is particularly useful, since, due to isospin conservation, it reduces the number of reaction mechanism which contributes to the reaction and allows an easier interpretation of the results.

The experimental study of the (α, α') reaction on the proton target was done in ref. [6]. Two clear peaks were observed there; a large one, which was associated in ref. [6] with Δ-excitation in the projectile (DEP), and a small one, at higher excitation energies, which was attributed in ref. [6] to the Roper excitation in the target. Because of the isospin conservation, the Δ-excitation in the target does not contribute to the reaction. A theoretical study was done in ref. [7] and the large peak was nicely reproduced by the DEP mechanism. In the present paper we study the different mechanisms that can lead to the Roper excitation in the (α, α') reaction on the proton to explain the second peak found in the experiment.

2 $(^3\mathrm{He}, t)$ Reactions on Unstable Nuclei

To investigate the $(^3\mathrm{He}, t)$ reaction, we use three kinds of target nuclear density distributions which are calculated using the standard model, (1) the Woods-Saxon (WS) density distribution with the radius parameter $R = 1.18A^{1/3} - 0.48$ fm and the diffuseness $a = 0.5$ fm. This distribution was established by the study of the stable nuclei and does not provide the neutron skin, (2) the calculated density distribution by the relativistic mean-field model (RMF) [8]. All the details of the RMF are described in ref. [9] including the definitions of parameter sets used in the model. Here we used the parameter set NL1. It provides the thick neutron skin of neutron-rich nuclei, and (3) the calculated density distributions by Hatree-Fock theory with the standard SIII Skyrme interaction (SHF) [10]. All the details of the calculation are given in ref. [11]. This model also provides the thick neutron skin.

In order to show the sensitivity of the $(^3\mathrm{He}, t)$ reactions to the nuclear surface, we calculate the effective numbers of protons and neutrons which participate in the reaction [12]. We have shown that that the existence of the neutron skin changes the effective numbers considerably. Then we have studied the reaction using a realistic cascade model which was developed to study the inclusive $(^3\mathrm{He}, t)$ reaction on nuclei in ref. [4]. The results show that (1) the isospin dependence of the reaction mechanism, especially in the Δ-excitation region, will be seen in the shape of the energy spectrum of the reaction, and (2) the energy-integrated cross section will be a good observable to check the model of the nuclear structure [2]. For the study of the reaction mechanism, the present work is complementary to the work of ref. [13]

Experiments of this (^{3}He, t) reaction can be performed for unstable nuclei using the inverse-kinematics method in which the target ^{3}He is bombarded by a secondary unstable nuclear beam. The reaction, therefore, can be a powerful tool for a systematic study of the neutron skin of unstable nuclei. Finally we mention that heavier projectiles are more sensitive to the nuclear surface than ^{3}He and t because of larger distortion effects. Hence, it is also interesting to use heavier particles in experiments. In these cases, however, the Δ-resonance peak may not be seen well because of the projectile-ejectile form-factor effect.

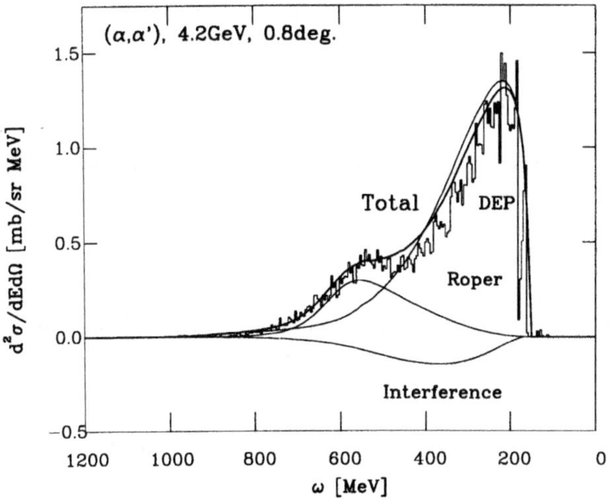

Figure 1. Calculated cross sections of the target Roper process [14] and the projectile Δ process [7]. The variable ω is the energy transfer in laboratory system. The thick line indicates the sum of all contributions. Experimental data are taken from ref. [15]

3 (α, α') Reactions on Proton

To calculate the (α, α') reaction cross sections, we consider all processes which may be relevant in the Roper excitation region, namely, Roper excitation in the target proton, Roper excitation in the projectile, and double Δ excitation processes. The theoretical investigation shows that the Roper excitation in the proton target mediated by σ exchange is the dominant mechanism in the process [14]. The experimental cross sections are well reproduced as shown in Fig. 1 with Roper mass $M^* = 1430$ MeV, Roper width $\Gamma^*(s = M^{*2}) =$ 300 MeV, and σNN^* coupling constant $g^2_{\sigma NN^*}/4\pi = 1.33$. This value is very close to the value $g_{\sigma NN^*}^2/4\pi = 1.26$ which is evaluated from the σNN coupling constant by a scale factor $(f'/f)^2$, where f' is the coupling constant of the πNN^*-vertex and f is that of the πNN-vertex. This observation would give support to the interpretation of a σ-meson as a correlated two pion exchange.

We would like to stress here that we can extract the σNN^* coupling constant from the experimental data since the dominant contribution only comes from the target Roper excitation by σ exchange. Because of this simplifying property, the (α, α') reaction is particularly suited to study the σ-meson and the Roper resonance [14].

4 Summary

We have studied the $(^3\text{He}, t)$ reactions on unstable nuclei and the (α, α') reactions on proton at intermediate energies. We find that the $(^3\text{He}, t)$ reaction is suitable to study the surface of the unstable nuclei and isospin dependence of the Δ-excitation mechanism [2]. We also find that (α, α') reaction is a good tool for the study of Roper excitation and σ meson [14].

Acknowledgement. We acknowledge fruitful collaborations with P. Fernández de Córdoba, N. Fukunishi, D. Hirata, H. Kitagawa, E. Oset, and M. J. Vicente-Vacas.

Note added at proof
The "σ" exchange potential which we have obtained in the study of (α, α') reaction should be interpreted as an effective interaction which accounts for all the ingredients in the $T = 0$ exchange channel of $NN \rightarrow NN^*$ transition potential, see ref. [14] for details.

References

1. I. Tanihata, D. Hirata, T. Kobayashi, S. Shimoura, K. Sugimoto, H. Toki: Phys. Lett. **B289**, 261 (1992)
2. S. Hirenzaki, P. Fernández de Córdoba, N. Fukunishi, D. Hirata, H. Kitagawa, M.J. Vicente-Vacas: Nucl. Phys. **A587**, 787 (1995)
3. C. Gaarde: Nucl. Phys. **A478**, 475c (1988)
4. P. Fernández de Córdoba: PhD thesis, University of Valencia (1992); P. Fernández de Córdoba, E. Oset, M.J. Vicente-Vacas, submitted to Nucl. Phys. **A**; E. Oset, P. Fernández de Córdoba, J. Nieves, M.J. Vicente-Vacas: Phys. Scripta **48**, 101 (1993)
5. D. Contardo et al.: Phys. Lett. **B168**, 331 (1986)
6. H.P. Morsch et al.: Phys. Rev. Lett. **69**, 1336 (1992)
7. P. Fernández de Córdoba, E. Oset, M.J. Vicente-Vacas, Yu. Ratis, J. Nieves, B. López-Alvaredo, F. Gareev: Nucl. Phys. **A586**, 586 (1995)

8. B.D. Serot, J.D. Walecka: Adv. Nucl. Phys. **16**, 1 (1986)

9. D. Hirata, H. Toki, T. Watabe, I. Tanihata, B.V. Carlson: Phys. Rev. **C44**, 1467 (1991)

10. M. Beiner, H. Flocard, N. Van Giai, P. Quentin: Nucl. Phys. **A238**, 29 (1975)

11. N. Fukunishi, T. Otsuka, I. Tanihata: Phys. Rev. **C48**, 1648 (1993)

12. G.F. Bertsch, O. Scholten: Phys. Rev. **C25**, 804 (1982)

13. P. Fernández de Córdoba, E. Oset: Nucl. Phys. **A544**, 793 (1992)

14. S. Hirenzaki, P. Fernández de Córdoba, E. Oset: submitted to Phys. Rev. **C**

15. H.P. Morsch: Private Communication

Few-Body Systems Suppl. 8, 384–388 (1995)

Few-Body Systems

An Analysis of the Use of the Cluster Separability in Scattering Theory

M. Damjanović[1] *, Z. Marić[2] †

[1] Faculty of Pharmacy, Dept. of Physics, Belgrade University, P.O.B. 146, Vojvode Stepe 450, Yu-11000 Belgrade, Yugoslavia

[2] Institute of Physics, P.O.B. 57, Pregrevica 118, Yu-11001 Belgrade, Yugoslavia

Abstract. In the framework of the Time Dependent Scattering Theory we discuss three forms of Cluster Separability as well as the conditions for the representation of the scattering system dynamics implied by their respective use.

For a broad class of theories applied to the studies of the scattering processes of particles and their bound states, and in particular for the Quantum Field Theory (QFT), one needs to include the procedure which regulates a segmentation and clusterisation of the scattering process. Implications of the implementation of such a procedure as well as its relationship to the general formalism deserve a careful analysis. In this paper we would like to point out the relationship existing between Cluster Separability (CS), Adiabatic Cluster Separability (ACS) and Disturbative Adiabatic Cluster Separability (DACS). The last one being introduced by us in the previous publication [1]. These three competing procedures are stated in the form of a principle for scattering theory as conditions for the representation of the scattering system dynamics.

The common feature of all of them is that they imply distinction between central interaction channel of the system and two channels corresponding to the initial and final subsystems respectively. Channels could be described by corresponding Hamiltonians and states in their corresponding Hilbert spaces.

What makes a difference between CS, ACS and DACS is the degree and a way of describing the "phase transitions" between the corresponding segments of the scattering process.

We will consider the scattering process $\alpha \rightarrow \beta$ in the framework of Time Dependent Propagator Theory. Initial and final channels of the process are

* *E-mail address:* edamjano@ubbg.etf.bg.ac.yu

† *E-mail address:* emaricz@ubbg.etf.bg.ac.yu

defined with the sets of the process constituents - particles and their bound states: $\{I_1, ..., I_k\} = \alpha$; $\{F_1, ..., F_l\} = \beta$. A corresponding Hilbert space state is denoted by $\mid \alpha\rangle$. The Bethe-Salpeter amplitudes χ_α, $\bar{\chi}_\alpha$ for channel α and process propagator $G_{\alpha\beta}$ are:

$$\chi_\alpha(n) = \langle 0 \mid T(\psi(x_1)...\psi(x_n)) \mid \alpha\rangle; \; \bar{\chi}_\alpha(n) = \langle \alpha \mid T(\psi(x_1)...\psi(x_n)) \mid 0\rangle, \quad (1)$$

$$G_{\alpha\beta}(n;m) = \langle 0 \mid T\Big(T(\psi(x_1)...\psi(x_n)) \; T(\psi(x_1')...\psi(x_m'))\Big) \mid 0\rangle. \quad (2)$$

In order to reduce general forms (1), (2) to the working expressions, one needs to perform as well the following steps:
1. Functional derivation of the $G_{\alpha\beta}(n;m)$. It allows an expression for the process propagator $G_{\alpha\beta}(n;m)$ through the cluster propagators G_i, G_f and establishes the initial frame for the cluster decomposition of the process.
2. Corresponding Gell-Mann–Low limiting procedure [2]. It assures the QFT consistent contact with states of the system for remote times.
3. Application of the one of the above mentioned procedures (CS, ACS, or DACS). It results in the segmentation of the scattering process and also defines the final form of the cluster decomposition.

The exposed steps which are defined for the general propagator imply the decomposition of the corresponding S-matrix as well as the corresponding Feyman developement.

We will focus our attention on the third procedure which equips the system with the corresponding form of the cluster separability: unspecified, adiabatic or disturbative adiabatic.

Cluster Separability is the characteristic that a system of particles when broken into spatially remote subsystems should be such that the dynamics of each subsystem are independent of each other [3]-[4]. (This property is also denoted as Separability of the Interaction or Cluster Property.)

The outcome of the CS is a flexible frame for the representation of the systems dynamics. In addition it does not introduces any demands for the "contact processes". A freedom for the interpretation of that part of a scattering formalism still remains.

The definite demands for the segmentation of the scattering process are established with the implementation of ACS.

Adiabatic Cluster Separability: "No error is introduced into the treatment of the physically realizable scattering process by a formulation of the theory in which the interactions among the particles of interest are 'turned off' at remote times-provided, of course, that the turning-off procedure is sufficiently gradual that it does not, of itself, create disturbances" [5]. (The standard name is Adiabatic Hypothesis.)

The configurational content of ACS is a regionalization of the process in a space-time. The initial (or incoming) region of propagation $\mathcal{R}_i$, the interaction region $\mathcal{R}_k$ and the final (or outgoing) region of propagation $\mathcal{R}_f$ are distinguished. The consequent factorisation of the process representatives (propagator and S-matrix) is expressed in relations:

$$G_{\alpha\beta} = G_\beta R_{\alpha\beta} G_\alpha = \prod_{f=1}^{l} G_f R_{\alpha\beta} \prod_{i=1}^{k} G_i; \tag{3}$$

$$S_{\beta\alpha} = \bar{\chi}_\beta^{out} R_{\alpha\beta} \chi_\alpha^{in} = \prod_{f=1}^{l} \bar{\chi}_{\beta f} R_{\alpha\beta} \prod_{i=1}^{k} \chi_{\alpha i}. \tag{4}$$

In Eqs. (3)-(4) G_α, G_β are channel propagators, χ_α^{in}, $\bar{\chi}_\beta^{out}$ are the remote times amplitudes, and $R_{\alpha\beta}$ is a truncated propagator. G_i, G_f, $(\chi_{\alpha i}, \bar{\chi}_{\beta f})$ are propagators, (amplitudes) of the scattering process constituents. A segmentation of the scattering process is expressed on the left-hand side of the above expressions. A cluster decomposition is seen on their right-hand side.

A strict application of the ACS implies the sharp distinction of three propagation regions. Since the propagation is adiabatic, for the whole regions $\mathcal{R}_i$ and $\mathcal{R}_f$ the observable characteristics of the system are the same as they were or wili be at the remote times $\mp\infty$.

We argue that such a feature introduces serious limitations for the consistent representation of the scattering processes involving bound states. Particularly critical would be the description of the intermediate objects: transition operators or segments of the convolutional expresions for the form factors, for instance. Additional problems are related to the interpretation of the contact processes which are reduced to a cut-off.

In this way one can recognise a limitations for the incorporation of the ACS into the formalisms which include deconfinement and hadronization. In particular, it concerns the formalisms which contain on the basic level different or hybrid levels of compositeness.

We extract three additional groups of motives for the reformulation of ACS with a less rigid principle.

I) Motives of conceptual nature. The intermediate phases have been already tacitly introduced (by using, for example, off mass shell contributions or even intermediate clusterisations of the channels) without explication whether it is consistent or not with the general procedure.

II) In applications of the standard formulation one often finds the incompatibility of the interaction effects originating from scattering amplitudes and from truncated propagators.

III) The idea was to find consistent and in the same time flexible formulation which could be applied to many concrete scattering situations.

The rigidity of the ACS and its limitations in encompassing the intermediate segments led us to the conclusion that one should formulate a different principle for the segmentation of the scattering processes.

We retained the supposition that interaction between constituents of scattering processes could be switched off at the remote times, or precisely for the initial and final space-time regions of propagation. But in addition we included a new one by which the mechanism of inclusion/exclusion is such that it creates the disturbances for initial/final channels in the vicinity of the interaction region. We supposed also that disturbation effects could be represented by segments obtained in a factorization of the scattering picture i.e. that the transition from initial/final regions of propagation to the region of "pure" interaction, generally, is not direct but it goes over the intermediate disturbed phases.

Disturbative Adiabatic Cluster Separability: In the framework of the Quantum Field Theory, propagation process can be, generally, represented by a few-step mechanism. Finite intermediate regions of disturbance which can be represented by a superposition of the physically realizable configurations for the corresponding channels are compatibly connected with the central, interaction region. Interaction among the constituents of the scattering process is 'turned-off' for the initial and final propagation regions, so that initial and final configurations are corresponding to them [1]. (The other name is Disturbative Adiabatic Hypothesis.)

By incorporation of the DACS in the scattering theory space-time content becomes correspondingly reacher. Two additional domains of scattering process appear. Disturbed channels of the initial and final configurations correspond to the initial and final disturbation regions $\mathcal{R}_{id}$ i.e. $\mathcal{R}_{fd}$. So one can visualize the sequence of the five regions: $\mathcal{R}_i, \mathcal{R}_{id}, \mathcal{R}_k, \mathcal{R}_{fd}$ and $\mathcal{R}_f$ separated by the four space-like surfaces. DACS causes the corresponding factorization of the representatives of the scattering theory.

In our formulation the propagator for the process $\alpha \to \beta$ is written

$$G_{\alpha\beta} = G_\beta \bar{K}^d_\beta R_{\alpha\beta} K^d_\alpha G_\alpha = G_\beta M_{\alpha\beta} G_\alpha . \tag{5}$$

Here $\bar{K}^d_\beta$ and K^d_α describe the effect of disturbation introduced. On the right hand side of Eq. (5) the expanded interaction term, $M_{\alpha\beta}$, $M_{\alpha\beta} = \bar{K}^d_\beta R_{\alpha\beta} K^d_\alpha$, is defined.

The S-matrix gets the form

$$S_{\beta\alpha} = \bar{\chi}^{out}_\beta \bar{K}^d_\beta R_{\alpha\beta} K^d_\alpha \chi^{in}_\alpha = \bar{\chi}^{out}_\beta M_{\alpha\beta} \chi^{in}_\alpha . \tag{6}$$

The new objects, "disturbation amplitudes" read: $\chi^d_\alpha \equiv K^d_\alpha \chi^{in}_\alpha$; $\bar{\chi}^d_\beta \equiv \bar{\chi}^{out}_\beta \bar{K}^d_\beta$.

We derive mathematical basis for the disturbative adibatic (DA) formalism on the level of the QM S-matrix theory by using the method of the wave (or Møller) operators.

The contact with experimental situation is model dependent. The interaction is given with the form of the expanded interaction term $M_{\alpha\beta}$. In addition it is necessary to define the disturbation effects. The remote time amplitudes χ^{in}_α and $\bar{\chi}^{out}_\beta$ are determined phenomenologically.

The general scheme of the DACS could be projected on the various scattering formalisms. The disturbation input is determined with appropriate superposition of the physically realizable configurations for the corresponding channels. The concrete examples of possible applications are the off mass shell effects in nuclear reactions, the formation of the new particles states, the transition amplitudes at the quark level and the appropriate intervention in the Fock expansion. The interaction content could be described by the corresponding Lagrangian.

The general formalism contains also two procedures which are independent and mutually compatible.

One of them consist in trying to find "natural" representation of disturbation effects in the form: $K^d = G^d V; \quad \bar{K}^d = \bar{V} G^d$.

Propagators G^d are related to the finite domains. They have the same physical meaning as propagational segments of a truncated propagator $R_{\alpha\beta}$ which appear in the course of the functional derivation. V, $\bar{V}$ are the vertex type operators defined by the algoritm. Expanded interaction term $M_{\alpha\beta}$ becomes:

$$M_{\alpha\beta} = \bar{V} G^d R_{\alpha\beta} G^d V. \tag{7}$$

One can recognize that the form of the expanded interaction term (7) corresponds to the convolution structure of the form factor. As an example of this structure one can use the nucleon form factor described in elastic electron-nucleon scattering within the QCD hard scattering scheme.

The second procedure consists in the series expansion of $M_{\alpha\beta}$ and reads

$$M_{\alpha\beta} = (\bar{K}_\beta^{d(0)} + \bar{K}_\beta^{d(1)} + ...)(R_{\alpha\beta}^{(0)} + R_{\alpha\beta}^{(1)} + ...)(K_\alpha^{d(0)} + K_\alpha^{d(1)} + ...). \tag{8}$$

This expansion is not perturbative series but it reflects the content of Bethe-Salpeter equation with the precise meaning of spatio-temporal composition of interaction effects.

One needs an additional analysis of terms in (8) in order to achieve a compatibility of the interaction contributions independently of their origin.

The CS, ACS and DACS could be considered as three different formulations of the general Principle of the segmentation of scattering process. This principle one could include among the first principles of the Quantum Field Theory.

The configurational features of this analysis are the part of more general approach which also includes the studies of covariant propagation and its correlation to the forms of dynamics. Some segments of that study have been already refered [6], [1].

References

1. M. Damjanović, Z. Marić: In: *Advances in Fundamental Physics*, M. Barone, F. Selleri (eds.), p. 349. Palm Harbor: Hadronic Press 1995
2. M. Gell-Mann, F.E. Low: Phys. Rev. **84**, 350 (1951)
3. L.L. Foldy: Phys. Rev. **122**, 275 (1961)
4. S.N. Sokolov: Teor. Mat. Phys. **23**, 355 (1975)
5. A. Klein, C. Zemach: Phys. Rev. **108**, 126 (1957)
6. M. Damjanović, Z. Marić: XIth European Conference on Few-Body Physics, Fontevraud, France, Abs. of contr. papers 20, 21 (1987)

Few-Body Systems Suppl. 8, 389–393 (1995)

Few-Body Systems

Method of Integral Transforms for Calculating (e, e') and $(e, e'N)$ Reactions

V.D. Efros[1], W. Leidemann[2], G. Orlandini[2,3]

[1] Russian Research Centre, Kurchatov Institute, Kurchatov Square 1, 123182 Moscow, Russia

[2] Dipartimento di Fisica, Università di Trento, I-38050 Povo (Trento), Italy

[3] Istituto Nazionale di Fisica Nucleare, Gruppo collegato di Trento

Abstract. The method of obtaining inclusive electromagnetic response functions of nuclei from their integral transform is described. The success of the method for two- and three-body nuclei is illustrated. It is shown how the approach can be applied to the response of ^{4}He, which is under investigation at present. Furthermore, the applicability for exclusive $(e, e'N)$ reactions is discussed.

The response of a system to an external probe is an excellent tool to study details of the electromagnetic interaction of nuclei and some underlying dynamics. Unfortunately, in case of an energy transfer ω calculations can become quite complicated, since the final state wave function is requested for a direct calculation of the response function $R(\omega)$. To avoid this problem Efros proposed in 1985 to perform an indirect calculation [1]. The idea is to calculate an integral transform of the response function at a constant momentum transfer q

$$\Phi(\sigma) = \int d\omega R(\omega) f(\omega, \sigma) \tag{1}$$

in a first step and to extract $R(\omega)$ in a second step by inversion of $\Phi(\sigma)$. Of course, such an approach is only helpful if the transform can be calculated without explicit knowledge of the final state spectrum. In fact Efros could show that this is possible for the Stieltjes transform ($f(\omega, \sigma) = (\omega + \sigma)^{-1}$, $\sigma > 0$). However, it took quite some time before such an approach was used for a realistic calculation. In 1992 the Euclidean response (Laplace transform, $f(\omega, \sigma) = \exp(-\omega\sigma)$) was considered for the longitudinal (e, e') response of ^{4}He [2]. This transform can be evaluated with the Green Function Monte Carlo method.

In order to explore the method of integral transforms better we initiated a rigorous study with the aim to clarify whether $R(\omega)$ can be reliably obtained within this approach [3, 4]. We chose the deuteron as test nucleus, since its final NN scattering state is well under control, thus $R(\omega)$ can be also calculated directly. In ref. [3] we first evaluated the Stieltjes transform without using $R(\omega)$ and then we performed the inversion. Concerning the inversion one has to realize that the Stieltjes transform - as the Laplace transform - in practice cannot be retransformed with a simple mathematical operation like, e.g., the Fourier transform. For the methods of inversion we refer to [1, 3]. The results of our study were as follows. Though $\Phi(\sigma)$ was calculated with a rather high precision (error less than 0.5%), the inversion of the Stieltjes transform gives only a very good estimate of $R(\omega)$, but does not uniquely lead to the exact result. Even the quasielastic peak height has an uncertainty of about 10%. Here it is very important to notice that this is not a problem of the inversion method itself, but rather of the kernel, i.e. that quite different $R(\omega)$ can lead to almost identical $\Phi(\sigma)$. We also found that the same problem exists for the Laplace transform. The reason is that both transforms have almost no resolution, e.g., the quasielastic peak is smeared out in a way that no structure is left in the transform.

At this point we started to search for a transform, which on the one hand has a good energy resolution and which on the other hand can be calculated without explicit knowledge of the final state spectrum. As a matter of fact we found a very good case [4]. It is a transform with a Lorentz kernel ($f(\omega,\sigma) = (\sigma_{\mathrm{I}}^2 + (\omega - \sigma_{\mathrm{R}})^2)^{-1}$, σ_{I}, $\sigma_{\mathrm{R}} > 0$), which, for simplicity we will call Lorentz transform in the following. Fig. 1a shows for σ_{I} equal to 5 and 10 MeV the calculated transform of the deuteron longitudinal response function $R_{\mathrm{L}}(E_{np}, q_{\mathrm{c.m.}}^2)$, where E_{np} is the relative kinetic energy of the outgoing np pair in the c.m.-system. Since our intention was to check if even in the presence of errors the transform can be inverted in a reliable way we did not calculate $\Phi(\sigma_{\mathrm{R}}, \sigma_{\mathrm{I}})$ with a very high precision, but allowed for errors (for details see [4]). In the σ_{R} range up to 100 MeV the resulting error is of the size of 1% for $\sigma_{\mathrm{I}} = 10$ MeV, while for $\sigma_{\mathrm{I}} = 5$ MeV one already has errors up to 5%. Nevertheless, even in the case of $\sigma_{\mathrm{I}} = 5$ MeV the inversion of $\Phi(\sigma_{\mathrm{R}}, \sigma_{\mathrm{I}})$ leads to an extremely good result for $R(\omega)$ as Fig. 1b shows.

In the meantime the method has also been applied to the three-body system [5]. In this case it is already much more difficult to calculate $R(\omega)$ directly. There exists only a single calculation [6] for an effective central NN potential and a recent calculation by the Bochum group [7] for realistic NN interactions, where with a large numerical effort just a few points have been obtained for one momentum transfer q. As illustrated in Fig. 1c [8] also in this case the inversion of the Lorentz transform leads to a very good agreement with the direct calculation, while the inversion of the Stieltjes transform suffers from the same problems as for the deuteron. More results for the longitudinal response of the three-body system from the inversion of the Lorentz transform can be found in [5, 8].

Microscopic studies of the (e, e') response functions of heavier nuclei would be of importance. At present we investigate the ^{4}He response with the help of the method discussed. The Lorentz transform $\Phi(\sigma_{\mathrm{R}}, \sigma_{\mathrm{I}})$ is calculated (see

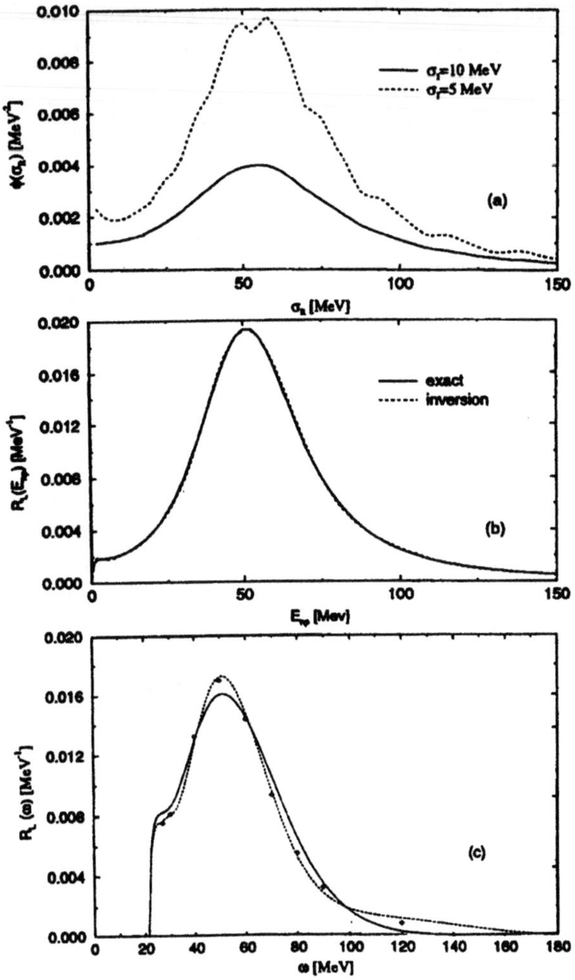

Figure 1. (a) Lorentz transform of the deuteron $R_{\rm L}$ at $q^2_{\rm c.m.} = 5\ {\rm fm}^{-2}$ with $\sigma_{\rm I} = 5$, 10 MeV. (b) $R(E_{np})$ from inversion of the Lorentz transform above with $\sigma_{\rm I} = 5$ MeV (dotted curve) and $R(E_{np})$ from direct calculation (full curve). (c) $R_L(\omega)$ from inversions of Stieltjes (full curve) and Lorentz (dotted curve) transforms for ^{3}He at $q = 300$ MeV. The dots are the results of a direct calculation [7]

[4]) as $\langle\tilde{\Psi} \mid \tilde{\Psi}\rangle$ where $\tilde{\Psi}$ is a *localized* solution to the four-body inhomogenous equation

$$(H - E_0 - \sigma_{\rm R} + i\sigma_{\rm I})|\tilde{\Psi}\rangle = \hat{O}|\Psi_0\rangle\,. \qquad (2)$$

Here $\hat{O}$ is a transition operator and Ψ_0 is an α-particle wave function, while

H is the hamiltonian of the system and E_0 its ground state energy. We are seeking for the solution $\tilde{\Psi}$ in the form of an expansion over hyperspherical harmonics and hyperradial functions. The expansion coefficients are found from the requirement

$$\langle F \mid F \rangle = \min, \qquad |F\rangle = (H - E_0 - \sigma_R + i\sigma_I)|\tilde{\Psi}\rangle - \hat{O}|\Psi_0\rangle \,. \tag{3}$$

This leads to a system of linear equations and the matrix elements entering this system are calculated with the Monte Carlo method.

In the present study we use central NN potentials as a dynamical input. Together with the low-energy data the potentials reproduce the S-wave scattering phases up to pion threshold. One can anticipate, and the previous 3-body studies [6, 7] testify to this, that the response functions are rather insensitive to the replacement of a non-central NN force by an effective central one. For our purposes super-soft core type smooth potentials are convenient. We have constructed such potentials starting from the super-soft core interaction of ref. [9]. They were obtained by modification of the short-range behaviour of the central components of the interaction in order to obtain smoother potentials with lower amplitudes of the cores and by introducing additional terms of the form $-V_T^2(r)/E_{\rm eff}$. Here V_T is the tensor component of the interaction of ref. [9] and $E_{\rm eff}$ are fitting parameters. In the singlet case such a term simply compensates the performed modification of the short-range behaviour of the potential while in the triplet case it mainly describes the effect of the tensor force on the $^3S_1$ component of the wave function in the $^3S_1$-$^3D_1$ NN state. Indeed, excluding the $^3D_1$ component of the wave function from the coupled-channel equations one gets an effective interaction of the form $-V_T G(E) V_T \simeq -V_T^2/E_{\rm eff}$ in the $^3S_1$ channel. The NN potentials obtained lead to reasonable values of the binding energy and the matter radius of the α-particle, i.e. $E_b = 31.3$ MeV and $< r^2 >^{\frac{1}{2}} = 1.40$ fm.

The system of our four-body equations is split with respect to the orbital momentum L and the isospin $T = 0, 1$ values, L being a component in the multipole expansion of the transition operator. We get a reasonable behaviour of the monopole contribution to the response but the convergence of the calculation is reached at a rather large number of basis functions retained. In order to speed-up the convergence we are planning to insert the Jastrow-type correlation factor into our basis functions.

After having discussed the inclusive (e, e') reaction, one may ask if it is also possible to consider exclusive $(e, e'N)$ reactions with the method of integral transforms. Already in 1985 it was shown that this is possible for the Stieltjes transform [1]. This can also be done for the Lorentz transform. In the present work we explain how it can be done in a simpler way in the framework of the so called spectral function approximation. In this case the exclusive cross section is given in terms of the spectral function

$$S(p, E) = \langle \Psi_0^A |\, a_p^\dagger \, \delta(E - H + E_0^A)\, a_p \,|\Psi_0^A \rangle \,, \tag{4}$$

where $a_p^\dagger$ and a_p are creation and annihilation operators of nucleons with momentum p. Within the present method $S(p, E)$ can be calculated with full treatment of the final state of the (A-1) system. Such a calculation has not yet been done for A > 3. To this end one first evaluates the Lorentz transform of $S(p, E)$

$$\Sigma(\sigma_{\rm R}, \sigma_{\rm I}) = \int dE\, S(p, E) \frac{1}{(E - \sigma_{\rm R})^2 + \sigma_{\rm I}^2} \tag{5}$$

at fixed $\sigma_{\rm I}$. This quantity can be represented as

$$\Sigma(\sigma_{\rm R}, \sigma_{\rm I}) = \langle \Psi_0^{\rm A} |\, a_{\rm p}^\dagger \frac{1}{H - E_0^{\rm A} - \sigma_{\rm R} - i\sigma_{\rm I}} \frac{1}{H - E_0^{\rm A} - \sigma_{\rm R} + i\sigma_{\rm I}} a_{\rm p} \,| \Psi_0^{\rm A} \rangle \tag{6}$$

which may be rewritten as $\langle \tilde{\Phi} | \tilde{\Phi} \rangle$ where $\tilde{\Phi}$ is the localized solution to

$$(H - E_0 - \sigma_{\rm R} + i\sigma_{\rm I})\, |\tilde{\Phi}\rangle = a_{\rm p}\, |\Psi_0^{\rm A}\rangle\,. \tag{7}$$

Thus we are able to obtain the left-hand side of Eq. (5) without solving the final-state continuum spectrum problem. One can then invert Eq. (5) along the same lines as for the inclusive case.

In conclusion we may say that the integral transform approach is very powerful. Here we did only consider the electromagnetic interaction with nuclei, but the method can also be fruitfully applied in other fields.

Acknowledgement. This work was supported by INFN and by the International Science Foundation (grant J4M100).

References

1. V.D. Efros: Sov. J. Nucl. Phys. **41**, 949 (1985)
2. J. Carlson, R. Schiavilla: Phys. Rev. Lett. **68**, 3682 (1992)
3. V.D. Efros, W. Leidemann, G. Orlandini: Few-Body Syst. **14**, 151 (1993)
4. V.D. Efros, W. Leidemann, G. Orlandini: Phys. Lett. B **338**, 130 (1994)
5. S. Martinelli, H. Kamada, G. Orlandini, W. Glöckle: Phys. Rev. **C**, in press
6. E. van Meijgaard, J.A. Tjon: Phys. Rev. **C45**, 1463 (1992)
7. J. Golak et al.: Phys. Rev. **C51**, 1638 (1995)
8. S. Martinelli: thesis, University of Trento (1995)
9. R. de Tourreil, B. Rouben, D.W.L. Sprung: Nucl. Phys. **A242**, 445 (1975)

Few-Body Systems Suppl. 8, 394–398 (1995)

Few-Body Systems

Effect of the Three-Nucleon Force Based on Δ-Isobar Excitation in Neutron–Deuteron Scattering

K. Chmielewski, J. Haidenbauer*, U. Meyer, S. Nemoto, P.U. Sauer, N.W. Schellingerhout

Institute for Theoretical Physics, University of Hannover, Appelstraße 2, D-30167 Hannover, Germany

Abstract. Elastic neutron–deuteron scattering is described. The description is based on a two-baryon coupled-channel potential which allows for Δ-isobar excitation; an effective three-nucleon force results. The prediction for the analyzing power at 10 MeV neutron lab energy does not remove the long-standing and non-understood disagreement with experimental data. The calculation employs a separable expansion of the coupled-channel potential; the quality of the expansion is studied in detail.

1 Introduction

A two-baryon interaction allowing for Δ-isobar excitation yields an effective three-nucleon force in many-nucleon systems. A Hamiltonian with such an additional Δ-isobar degree of freedom was extensively used [1, 2, 3] in describing the properties of the three-nucleon bound state. This Hamiltonian is now applied to neutron–deuteron scattering.

The AGS formulation [4] of three-particle scattering is employed. The two-baryon coupled-channel potential is separably expanded according to the EST method [5]. The quality of the separable expansion is tested in two-nucleon scattering and in the trinucleon bound state. Spin-averaged cross sections and spin observables of neutron–deuteron scattering are calculated and compared with data. The few observables [6] for which the traditional description in terms of purely nucleonic potentials is not satisfactory are given particular attention. The role of the virtual excitation of a Δ-isobar in low-energy three-nucleon scattering is explored.

**Present address:* KFA Jülich GmbH, IKP, D-52425 Jülich, Germany

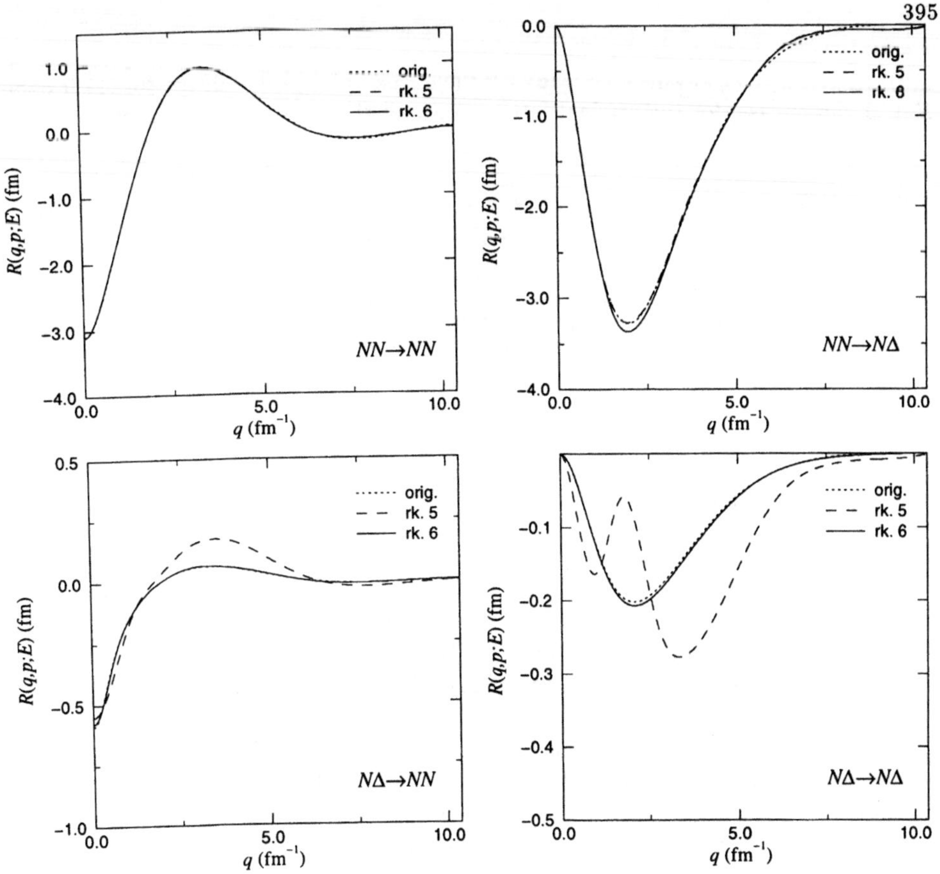

Figure 1. Plots of the R matrix of the ${}^1S_0$–${}^5D_0$ partial wave for $E = -10$ MeV and $p = 0.3472\,\mathrm{fm}^{-1}$, for ranks (rk.) five and six.

2 Test of the Separable Expansion

We have tested the separable expansion at the two- and three-nucleon level, by comparing two-nucleon scattering and three-nucleon bound-state and scattering results for the original potential and its separable expansion.

2.1 Nucleon–Nucleon Scattering

In Fig. 1 the two-body R matrices are shown for the four parts of the potential in the coupled ${}^1S_0$–${}^5D_0$ partial wave. For the purely nucleonic and the transition parts, five ranks suffice, but for the $N\Delta \to N\Delta$ part of the potential, an additional rank is required. (Note that four ranks were designed to fit the $NN \to NN$ and $NN \to N\Delta$ potential, so that the rank-five and rank-six expansions have just one or two ranks dedicated to the $N\Delta \to NN$ and $N\Delta \to N\Delta$ parts of the potential.) The overall agreement for the rank-six expansion is excellent. We did not yetcoplete similar convergence studies for the P waves.

Table 1. Energy eigenvalues (in MeV) for five cases and probabilities (in %) for the $N+\Delta$, $j \le 2$ case, for the original interaction and the separable expansion.

Interaction		E_{orig}	E_{sep}
N only	$j \le 2$	−7.388	−7.360
$N+\Delta$	dispersive $j \le 2$	−6.800	−6.805
$N+\Delta$	${}^3S_1$–${}^3D_1$ + ${}^1S_0$–${}^5D_0$	−6.825	−6.829
	$j \le 1$	−7.101	−7.114
	$j \le 2$	−7.696	−7.688

	P_{orig}	P_{sep}
P_S	87.68	87.72
$P_{S'}$	1.28	1.32
P_P	0.09	0.09
P_D	8.64	8.59
P_Δ	2.31	2.28

2.2 Three-Nucleon Bound State

We have calculated the energy E of the triton using both the original potential of ref. [1] incorporating virtual Δ-isobar excitation of the nucleon, and the separable expansion thereof. The results are compared in the columns E_{orig} and E_{sep} of Table 1. The five cases are: (*i*) the purely nucleonic Paris potential, (*ii*) taking only the dispersive effect of Δ-isobar excitations into account, (*i.e.*, without the effective three-nucleon force), and (*iii–v*) including all Δ-isobar effects. (As in ref. [1] the $N\Delta$ channels with two-body orbital angular momentum greater than two were neglected. In contrast to ref. [1], the spectator nucleon was not allowed to be excited. This is the reason for the small difference between E_{orig} given here and the value from ref. [1].) The channels for $j \le 2$ are enumerated as the first 32 channels in Table 2 of ref. [1]. The error in the energy eigenvalue due to the separable expansion is less than 15 keV for all cases, *i.e.*, less than 0.2%.

To get a better measure for the quality of the three-body bound-state wave function, we also calculated the traditional wave function probabilities. These are shown in Table 1 for the original (P_{orig}) potential and its separable expansion (P_{sep}). The agreement is excellent, even for the Δ-isobar probability, which is sensitive to the quality of the model.

Finally, we calculated

$$\varepsilon_\alpha = \left(\int \mathrm{d}p\, p^2 \mathrm{d}q\, q^2 |\langle p\, q\, \alpha|\psi_{i,\text{orig}}\rangle - \langle p\, q\, \alpha|\psi_{i,\text{sep}}\rangle|^2\right)^{1/2},$$

and

$$\hat{\varepsilon}_\alpha = \varepsilon_\alpha \Big/ \left(\int \mathrm{d}p\, p^2 \mathrm{d}q\, q^2 |\langle p\, q\, \alpha|\psi_{i,\text{orig}}\rangle|^2\right)^{1/2},$$

where $|\psi_i\rangle$ is the normalized Faddeev amplitude, for various channels α. For the most important channels (${}^3S_1$–${}^3D_1$ and ${}^1S_0$–${}^5D_0$ with spectator in s waves) $\hat{\varepsilon}_\alpha$ varies from 0.2% to 0.3%. For the other channels, this number is larger, but ε_α is less than 0.007%. The sum of ε_α over all channels is 0.3%.

2.3 Neutron–Deuteron Scattering

As a final test, we compare neutron–deuteron scattering calculations using the nucleonic Paris potential and its separable expansion. The agreement is excellent, as can be seen from Fig. 2. The results for the original nucleonic potential are taken from ref. [6].

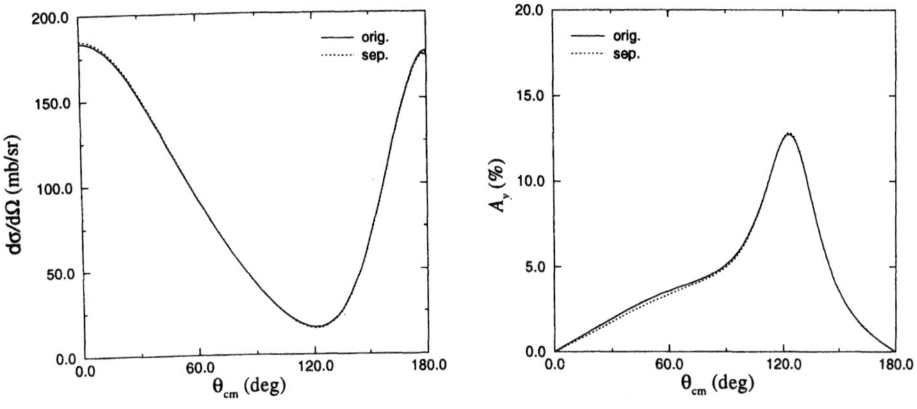

Figure 2. Comparison of the differential cross section and analyzing power for n–d scattering at 10.5 MeV using the nucleonic potential and its separable expansion.

3 Predictions for Neutron–Deuteron Scattering

In Fig. 3 we show our predictions for the model with virtual Δ-isobar excitation, as well as the purely nucleonic results, and some experimental data. The effect of the Δ-isobar excitation on the differential cross section is negligible. The experimental data, which were taken from ref. [7], appear to be unreliable at extreme backward angles. Other measurements at similar energies are significantly lower, and coincide with the theoretical curves [8]. The effect of the Δ-isobar excitation on A_y is small (compared to the discrepancy with experiment), and has the wrong sign. The experimental data was taken from ref. [9].

4 Conclusion

The effect of including virtual Δ-isobar excitation for the analyzing power A_y is smaller than that of the Tucson–Melbourne (TM) three-nucleon force. This appears to be consistent with the smaller effect of the Δ-isobar on the binding energy in the bound state compared with the TM force. Note that our results are preliminary, since we have not yet fully completed the convergence study.

The present calculation is to be extended to break-up. The wave function will be used to incorporate final-state effects in inelastic electron scattering.

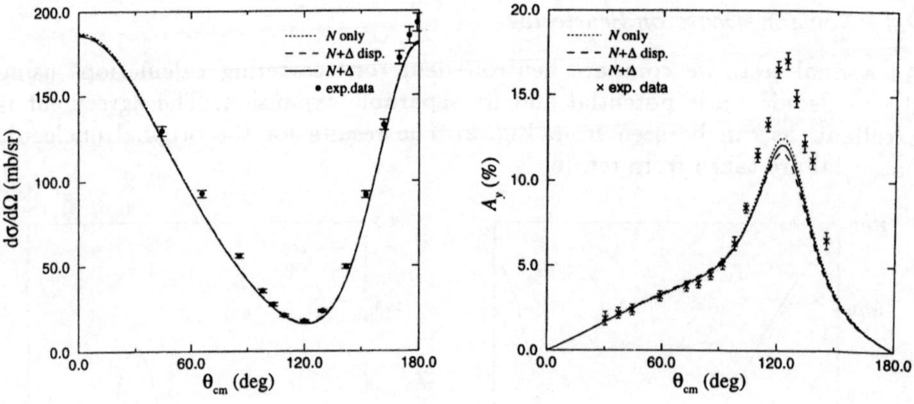

Figure 3. Comparison of the differential cross section (at 10.25 MeV) and analyzing power (at 10 MeV) for n–d scattering for the purely nucleonic case, the model with virtual Δ-isobar excitations, and experimental data. See text for details.

Acknowledgement. We thank A. Fonseca, Y. Koike and S. Oryu for sharing their expertise on the separable expansion of the two-baryon coupled-channel potential and its use in three-nucleon scattering with us. We are grateful to W. Glöckle and his collaborators for supplying us with their results for $n-d$ scattering, as displayed in Figs. 2 and 3. The work is supported by a graduate-student grant of the German state Lower Saxony (KC), by a grant of the German Bundesministerium für Forschung und Technologie Contract No. 06 OH 741 (SN) and by a grant of the Deutsche Forschungsgemeinschaft Contract No. Sa 247/7-4 (NWS).

References

1. Ch. Hajduk, P.U. Sauer, W. Strueve: Nucl. Phys. **A405**, 581 (1983)
2. H. Henning, P.U. Sauer, W. Theis: Nucl. Phys. **A357**, 367 (1992)
3. J. Adam, et al.: Nucl. Phys. **A531**, 623 (1991)
4. E.O. Alt, P. Grassberger, W. Sandhas: Nucl. Phys. **B2**, 167 (1967)
5. D.J. Ernst, C.M. Shakin, R.M. Thaler: Phys. Rev. **C8**, 46 (1973)
6. W. Glöckle, H. Witała, Th. Cornelius: Nucl. Phys. **A508**, 115c (1990)
7. P. Schwarz et al.: Nucl. Phys. **A398**, 1 (1983)
8. G. Janson et al.: *Few-Body Problems in Physics* (B. Zeitnitz ed.), North Holland 1984, p. 529
9. W. Tornow et al.: Phys. Rev. Lett. **49**, 312 (1982)

Few-Body Systems Suppl. 8, 399–403 (1995)

Few-Body Systems

Electromagnetic Associated Strangeness Production and Hadrons Form Factors

Ch. Charles[1], J.C. David[1], C. Fayard[2], M. Guidal[1], G.H. Lamot[2], B. Saghai[1]

[1] Service de Physique Nucléaire, CEA/DSM/DAPNIA, Centre d'Études de Saclay F-91191 Gif-sur-Yvette, France

[2] Institut de Physique Nucléaire de Lyon, IN2P3/CNRS, Université Claude Bernard, F-69622 Villeurbanne, France

Abstract. A formalism, based on an isobaric approach, via Feynman diagrammatic technique, including the nucleonic (spin $\leq 5/2$), hyperonic (spin 1/2) and kaonic resonances is developed. Using this formalism, a thorough investigation of *all* the electromagnetic strangeness processes, for which experimental results are available, is performed and satisfactory agreement with the data is obtained. The reaction mechanism is found to include a reasonable number of baryonic resonances, with the extracted main kaon-hyperon-nucleon coupling constants in agreement with the SU(3)-symmetry predictions. The sensitivity of this model to the form factors of the relevant hadrons is emphasized.

1 Introduction

Using a phenomenological approach based on an isobaric model including s, u, and t-channels, we have performed a thorough study of the associated strangeness electromagnetic production [1, 2], namely the following reactions:

$\gamma\, p \;\rightarrow\; K^+ \Lambda,\; K^+ \Sigma^o,\; K^o \Sigma^+;\; E_\gamma^{\rm lab} \leq 2.1$ GeV,
$e\, p \rightarrow e'\, K^+ \Lambda,\; e'\, K^+ \Sigma^o,$
$K^-\, p \;\rightarrow\; \gamma\, \Lambda,\; \gamma\, \Sigma^o$, branching ratios with stopped kaons.

Here, we focus on the sensitivity to the hadronic form factors of the reaction $ep \;\rightarrow\; e'K^+\Lambda$, in the few GeV region. The virtual photon has a longitudinal part besides the transverse component, and offers the possibility to vary independently the energy and momentum transfers. In this respect, the electrons are a finer probe for the strangeness realm. At the present time, the data in this field are scarce [3], but the foreseen experiments [4] at CEBAF are expected to produce much copious and accurate data in a near future.

2 Model

Our formalism is based on an isobaric approach using Feynman diagrammatic technique. To determine the reaction mechanism, two interconnected questions need to be addressed: which resonances are exchanged and what are the corresponding coupling constants? These latter are free parameters and their values are extracted through a least-squares fitting procedure. In our investigation, we have used all the roughly 500 available data points for the seven reactions mentioned in the Introduction, and examined the possible contributions of some 30 resonances. A model is obtained containing, besides extended Born terms, the following exchanged particles:

s-channel: $P_{11}(1440), P_{13}(1720), D_{15}(1675)$,
u-channel: $\Lambda(1405), \Lambda(1670), \Lambda(1810), \Sigma(1660)$,
t-channel: $K^*(892), K1(1270)$.

In addition, the values of the two main coupling constants $g_{K\Lambda N}$ and $g_{K\Sigma N}$, are in agreement with the broken SU(3)-symmetry constraints.

3 Form Factors

In the electroproduction reaction, the virtual photons probe the electromagnetic structure of the involved hadrons. This is taken into account in phenomenological models by incorporating appropriate form factors for each hadron. In the present work, we have selected some parametrizations proposed in the literature, without trying to re-adjust the parameters. We now describe briefly the form factors introduced at the relevant vertices.

3.1 Baryonic Form Factors

At the γpp vertex, we have to consider the two form factors of the proton, namely F_1^p and F_2^p, which are related to the electric (G_E^p) and magnetic (G_M^p) form factors:

$$F_1^p = \frac{1}{1-\tau}(G_E^p - \tau G_M^p) \quad , \quad F_2^p = \frac{1}{\kappa_p(1-\tau)}(G_M^p - G_E^p) \tag{1}$$

with $\tau = \frac{Q^2}{4M^2}$, and κ_p the anomalous magnetic moment of the proton.
These form factors have been extensively studied in the literature. We have retained the model developed by Gari and Krümpelmann [GK] in ref. [5] in its most recent version [6] which gives a good agreement with the available data. The basis of this approach is the extended vector meson dominance (EVMD) model, combining the vector meson dominance (VMD) hypothesis (appropriate at low Q^2) with the perturbative QCD approach describing the spacelike region. At the $YY\gamma$ vertices ($Y \equiv \Lambda,\ \Sigma^0$), we have used the two form factors of the neutron ($F_1^Y = F_1^n,\ F_2^Y = F_2^n$) as parametrized by GK [5, 6].
For the nucleonic (hyperonic) resonances, we choose the second Dirac form factor of the proton (neutron).

3.2 Kaonic Form Factors

The form factor of the K^+ has not been studied as extensively as the nucleon one. Moreover, the available experimental data are restricted to the low energy domain ($-Q^2 < 0.15$ GeV^2), which makes difficult to choose a model valid in the whole energy range considered in the present work. We have adopted the VMD model as proposed by Williams [7]: $F_{K^+} = (F_{is} + F_{iv})/2$, where the isoscalar and isovector parts are expressed in terms of the appropriate mesons contributions:

$$F_{is}(Q^2) = \sum_{v=\Phi,\omega} F_v(Q^2)\left(\frac{g_v}{f_v}\right)\frac{M_v^2}{M_v^2+Q^2-iM_v\Gamma_v} \tag{2}$$

$$F_{iv}(Q^2) = F_\rho(Q^2)\left(\frac{g_\rho}{f_\rho}\right)\frac{M_\rho^2}{M_\rho^2+Q^2-iM_\rho\Gamma_\rho} \tag{3}$$

The free parameters are determined by the known $\Phi \to K^+K^-$ decay width, plus the two normalization constraints $F_{K^+}(Q^2=0)=1$, $F_{K^0}(Q^2=0)=0$. The values obtained by Williams lead to a K^+ charge radius $< r^2_{K^+} >= 0.335$ fm^2, in good agreement with the experimental value [8] : 0.34 ± 0.05 fm^2.
For the K^* and $K1$ kaonic resonances, we have chosen the simplified VMD model used by Adelseck [9]. A monopole form factor was assumed: $F(Q^2) = 1/(1+Q^2/\Lambda^2)$, and Λ_{K^*} and Λ_{K1} were adjusted to give the best fit to the electroproduction data.

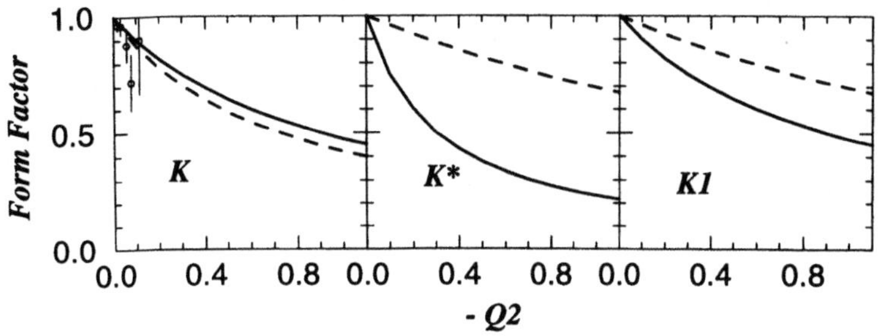

Figure 1. Form factors for kaon, K^* and $K1$. The solid and dashed curves come from the parametrizations of refs. [9] and [10], respectively

4 Results and discussion

In this contribution, using the reaction mechanism [2] outlined in Section 2, we report on the results corresponding to the baryonic form factors by Gari and Krümpelmann published in 1985 [ref. 5] and 1992 [ref. 6], hereafter referred to as GK85 and GK92, respectively. For the t-channel exchanged particles, namely K, K^*, $K1$, the expressions by Adelseck [9], and Williams [7] are used. Figure 1 shows these kaonic form factors.

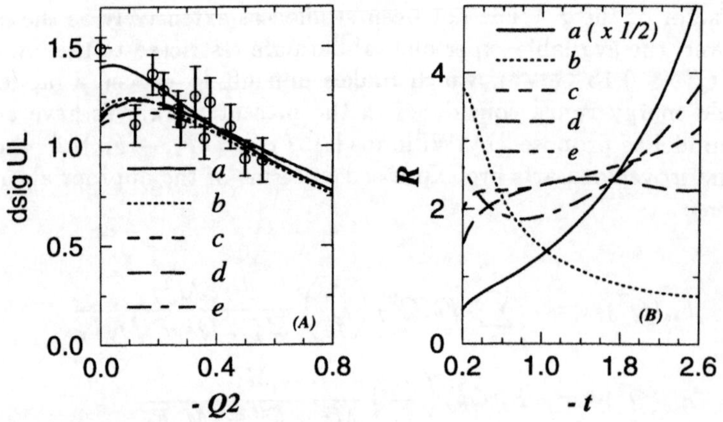

Figure 2. Cross section $d\sigma_{UL} \equiv d\sigma/d\Omega_K = \sigma_U + \varepsilon_L \sigma_L$ (in μb/GeV^2) as a function of momentum transfer at $t = -0.1$ (GeV/c)^2, and the ratio $R = \sigma_L/\sigma_U$ as function of Mandelstam variable t at $Q^2 = -1.$ (GeV/c)^2, curves are explained in the text. Curve *(a)* is divided by a factor of 2.

The existing data [8] concern only the kaon form factor and the 25 data points are in the range $-Q^2 \simeq 0.02 \div 0.1$ (GeV/c)^2. For the sake of clarity, we show in Fig.1 only a few of them including the extreme Q^2 values. Figure 2 shows our results for the cross section $d\sigma_{UL} = \sigma_U + \varepsilon_L \sigma_L$, and the ratio $R = \sigma_L/\sigma_U$, with σ_U and σ_L the differential cross sections for an unpolarized incident photon beam, and a longitudinally polarized virtual photon, respectively; ε_L is the longitudinal polarization parameter. The curves, at $W = 5.02$ (GeV/c)^2 and $\varepsilon = 0.7$, come from reaction mechanisms with the same exchanged particles [2], and differ in the expressions for the form factors. For each form factor combination, the free parameters (coupling constants) have been refitted to all the relevant data points. In Fig. 2, the curve *(a)* corresponds to our model as mentioned in Section 2, for which the reaction mechanism (set of exchanged particles and values of couplings) has been optimized. The form factors for baryons are from GK92, for kaon from Williams (W), and for K^* and $K1$ from Adelseck (A). This model called SL (for Saclay-Lyon) is identified as the following combination for form factors: GK92-WAA. The curve *(b)* is the same as SL, except that it does not contain the K1-exchange. The curves *(c)* to *(e)* incorporate the same resonances as the SL model but differ in the form factors, namely *(c)*: GK92-WAW, *(d)*: GK92-WWW, *(e)*: GK85-WWW.
We emphasize that all these curves have been obtained by fitting the complete set of data (photoproduction, electroproduction, and radiative capture) producing reasonable reduced χ^2's. For the $K\Lambda$ channels the values are: 1.73 for *(a)*, 2.47 for *(b)*, and 2.22 for the three others c to *e)*. Nevertheless, removing $K1$ from the model leads to large coupling constants for the $\Lambda(1810)$, and $\Sigma(1660)$ resonances as expected from the duality hypothesis [11], showing the need for

higher spin baryonic resonances [2] if the t-channel terms are discarded.
For the cross section $d\sigma_{UL}$, our optimized model SL (curve a) gives the best agreement with the data. The other curves (b–e) compare equally to the data and differ from the SL model at low $-Q^2$. At this level, different choices of the form factors do not lead to significant effects on the cross section. However, the ratio between the two components of this observable offers very interesting features (Fig. 2B), showing drastic differences according to the included form factors sets. The SL gives a large and monotonously increasing ratio. Switching off the K1 exchange (curve (b)) presents the opposite trend. Putting back the K1 and using for this resonance the expression obtained by Williams [7] rather than by Adelseck [9] produces the curve (c) which is much flatter than in the case of the SL model. Doing in addition the same substitution for the K^* (curve (d)) leaves the magnitude almost unchanged but reverses the curvature. Keeping these kaonic form factors and using GK85 instead of GK92 for the hadrons form factors (curve e) leads to a minimum around $-t \simeq 1.$ (Gev/c)^2. This latter shape is comparable to the one produced by Williams-Ji-Cotanch model [10, 4].
In summary, the longitudinal and transverse electroproduction cross sections show sizable sensitivities to the choice of hadrons form factors. The upcoming results from CEBAF [4] will therefore play a crucial role in our knowledge of some fundamental properties of the hadrons.

References

1. J-C David et al.: in SPIN '94 International Symposium , 1994, Bloomington, Indiana, Ed. S. Vigdor et al.: AIP (to appear)
2. J.C. David et al.: in preparation
3. P. Brauel et al.: Z. Phys. **C3**, 101 (1979) and references therein
4. O. Baker et al.: CEBAF proposal 93-018; K. Hicks, M Mestayer et al.: CEBAF proposal 93-030
5. M. Gari, W. Krümpelmann: Z. Phys. **A322**, 689 (1985)
6. M. Gari, W. Krümpelmann: Phys. Rev. **D45**, 1817 (1992)
7. R. Williams: Ph.D. Thesis, University of North Carolina, 1993
8. E.B. Dally et al.: Phys. Rev. Lett. **45**, 232 (1980); S. R. Amendolia et al.: Phys. Lett. **B178**, 435 (1986)
9. R. Adelseck: Ph.D. Thesis, Ohio University (1988)
10. R. Williams, C. Ji, S. Cotanch: Phys. Rev. **C46**, 1617 (1992)
11. B. Saghai, F. Tabakin: in SPIN '94 International Symposium , 1994, Bloomington, Indiana, Ed. S. Vigdor et al.: AIP (to appear); submitted to Phys. Rev. **C**

Few-Body Systems Suppl. 8, 404–408 (1995)

Few-Body Systems

η-Helium Scattering

J.A. Niskanen[1]*, S. Wycech[2], A.M. Green[3]

[1] Department of Physics, P.O. Box 9, FIN-00014 University of Helsinki, Finland

[2] Soltan Institute for Nuclear Studies, Warsaw, Poland

[3] Research Institute for Theoretical Physics, P.O. Box 9, FIN-00014 University of Helsinki, Finland

Abstract. The scattering lengths and energy dependence of the scattering of η mesons from helium nuclei are studied in various models.

1 Scattering Lengths

Due to the short lifetime of the η mesons there cannot be actual η beams. Still the scattering of these mesons can be studied as final state interaction effects in production reactions on nucleons or nuclei. In this way information on the ηN interaction has been extracted from η photo- and electroproduction. Although the variation e.g. between the extracted scattering lengths is quite large, it is clear that the s-wave interaction is much stronger than in the πN system but still not strong enough to bind the ηN system. However, it is plausible that the coherent interaction with several nucleons could be strong enough for binding. Therefore, it is important to look for experimental signatures of such a strong attraction in the final state interaction in nuclear production processes as well as to develop theoretical tools to study threshold scattering on few nucleons.

This work is an attempt in a simple model to evaluate zero-energy scattering of η mesons from helium nuclei to see whether the total scattering length can be considered to support the existence of a quasibound state or not [1]. Our method has the multiple scattering series

$$T(\eta A) = \sum_{i=1}^{A} T_i + \sum_{i=1}^{A} \sum_{j\neq i} T_j D T_i + \sum_{i=1}^{A} \sum_{j\neq i} \sum_{k\neq j} T_k D T_j D T_i + \ldots \qquad (1)$$

in coordinate space as the starting point in order to facilitate some systematic corrections. Here individual scatterings on different nucleons are iterated (with

*Supported by the Academy of Finland

the ηHe propagator D at zero energy), so that T involves multiple integrals with the dimension extending to infinity. This infinite operator series we sum approximately as a geometric series of expectation values of the lowest order ηHe scattering matrix as

$$\langle T(\eta A)\rangle = \langle T(\mathrm{IA})\rangle[1 + P + Q + R + ...]^{-1}, \tag{2}$$

Here $\langle T(\mathrm{IA})\rangle$ is simply the impulse approximation result, i.e. the expectation value of the first term in Eq. (1) and P, Q, R, are the multiple scattering terms

$$P = -\frac{\langle TDT\rangle}{\langle T\rangle} \quad , \quad Q = \frac{\langle TDT\rangle^2}{\langle T\rangle^2} - \frac{\langle TDTDT\rangle}{\langle T\rangle}$$

$$R = -\frac{\langle TDT\rangle^3}{\langle T\rangle^3} + 2\frac{\langle TDTDT\rangle\langle TDT\rangle}{\langle T\rangle^2} - \frac{\langle TDTDTDT\rangle}{\langle T\rangle}. \tag{3}$$

Here the unphysical "disconnected" product terms are included to avoid them from appearing after evaluating (2) into a power series. To the order of R the expansion (2) is now an exact reproduction of the relevant iterations. The ηHe scattering length is then obtained as $a(\eta A) = -\frac{m}{2\pi}\langle T(\eta A)\rangle$ with m the ηHe reduced mass and $A = 3$ or 4.

At first T may be taken as the impulse approximation scattering operator $T = -\frac{2\pi}{\mu} a_{\eta N} A\rho(r)$ acting as an optical potential with T_i expressed in terms of the ηN scattering length. Here μ is the ηN reduced mass and $A\rho$ the nuclear density taken to be a simple Gaussian. With this choice of T the multiple scattering series should be equivalent to the optical model and, indeed, a comparison of Table 1.1 with the optical potential result of Wilkin [2] shows the fast convergence of the series once terms up to the third order (R) are included

Table 1. Comparison for η-He scattering lengths $a(\eta\mathrm{He})$ (in fm) from various stages of the series expansion in Eq. (2) with the direct optical potential calculation giving $a(\eta^3\mathrm{He})$ = -0.88+i1.80 fm and $a(\eta^4\mathrm{He})$ = -1.42+i1.31 fm [2]. The results are illustrated with the ηN input $a(\eta N) = 0.291 + i\,0.360$ fm.

Optical	$a(\mathrm{IA})$	$Q = R = 0$	$R = 0$	$a(\eta A)$
^{3}He	1.16+i1.43	-0.93+i1.92	-0.90+i1.81	-0.89+i1.80
^{4}He	1.61+i1.99	-1.62+i1.38	-1.47+i1.28	-1.42+i1.30
$A \to A-1$	$a(\mathrm{IA})$	$Q = R = 0$	$R = 0$	$a(\eta A)$
^{3}He	1.16+i1.43	-0.13+i2.36	-0.16+i2.31	-0.16+i2.30
^{4}He	1.61+i1.99	-1.53+i2.25	-1.45+i2.10	-1.41+i2.09
Off-shell	$a(\eta N, \mathrm{off})$	—	$R' = 0$	$a_{A-1}(\eta A, \mathrm{off})$
^{3}He	0.29+i0.26	—	0.59+i2.17	0.58+i2.17
^{4}He	0.29+i0.23	—	-0.78+i2.95	-0.78+i2.93

in the denominator. Also this confirms Wilkin's result of a large negative real part of the η^3He scattering length indicating a quasibound state. Some other values of $a_{\eta N}$ in the literature give two times larger real parts for $a(\eta A)$.

The above optical model simulation simply iterates the whole sum $T(\mathrm{IA}) = \sum_i T_i$, whereas according to (1) successive scatterings are not allowed from the same nucleon. At zero energy the spatial dependence of the scatterings is to a large extent lost and this effect can be simulated simply by setting $A \to A - 1$ in the numerator terms P, Q, R. Then it can be seen from Table 1.1 that the real part has become much smaller making the indication of a bound state more vague in ^{3}He.

Some further correction should be due to the ηN system being off-shell. This we take approximately into account by considering ηN scattering at an imaginary momentum yielding the replacement

$$\frac{1}{a(0)} \to \frac{1}{a(\text{off})} = \frac{1}{a(0)} + \kappa_\eta, \tag{4}$$

with $\kappa_\eta = \sqrt{2\mu(E_{\text{sep}} + E_{\text{rec}})}$ and $E_{\text{sep,rec}}$ the separation and spectator recoil energies. This procedure is, however, model sensitive. For example, if the effective range is large, there could be more corrections in (4). Anyway, this simple replacement reduces the elementary input scattering length drastically and in Table 1.1 all signature of a quasibound state in ^{3}He has vanished, whereas for ^{4}He such a state is still possible. It should be further added that inelasticity could be essential in such considerations. We only include the inelasticity inherent in the open πN channels. Another important inelasticity could be due to absorption $\eta NN \to NN$ [3]

2 Energy Dependence of the Cross Section

The energy dependence due to the final state interaction in the production cross section is normally given by the simple expression

$$|f|^2 = \left| \frac{a(\eta A)}{1 - ipa(\eta A)} \right|^2, \tag{5}$$

which is valid for very large a as in e.g. NN scattering. However, this condition is not well satisfied here, since $a(\eta A)$ is not large, especially if compared with the size of the He nuclei. Then probably a more realistic energy dependence would require the use of the full wave function, which for a separable Yamaguchi form would be

$$\psi^-(r) = \frac{\sin(pr)}{pr} + f^* \, \frac{[\exp(-ipr) - \exp(-\beta r)]}{r}. \tag{6}$$

The first two terms give the asymptotic behaviour, while the last term is due to the potential form factor. In production reactions this wave function is to be used in the integral $\int (\psi^-(r))^* \, H(r) \, d^3r$, where $H(r)$ is the elementary production amplitude inside the nucleus. Lacking reliable models for $H(r)$ we take

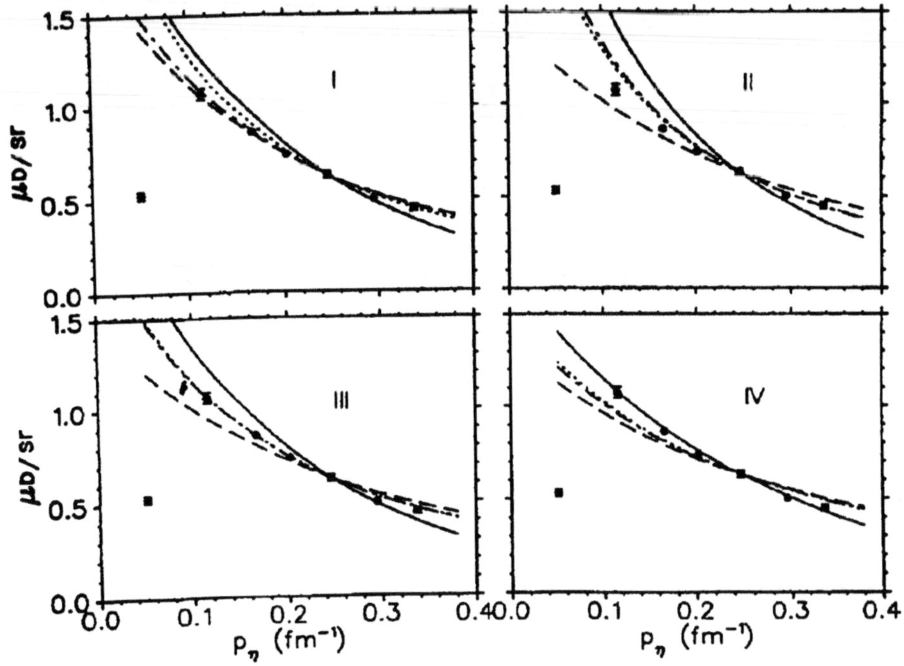

Figure 1. The squared amplitude in $pd \to \eta^3$He against the η momentum in the c.m. system for the four elementary ηN amplitudes I–IV given in the text and in Table 1.1. Dashed curve: optical model; Dotted: optical model corrected by $A \to A-1$ but described by (5); Dash-dot: off-shell effect also included in (5); Solid: the full model with corrections to the optical model and with the background term in the wave function (6).

this simply as proportional to the nuclear density normalizing the results to fit the data at one energy.

The results for the $pd \to \eta^3$He cross section using various models are shown in Fig. 1. (The trivial kinematic current factors are removed.) The models using the scattering lengths of Table 1.1 are given as IV. The others use as the input $a_{\eta N} = 0.476 + i\,0.279$ fm (I), $0.579 + i\,0.399$ fm (II), and $0.430 + i\,0.394$fm (III). The first one has been used by Wilkin in Eq. (5) giving a good fit with the optical model results. However, the corrections discussed above change the results considerably making the final slope too steep for this $a_{\eta N}$. The $a_{\eta N}$(IV) gives the best result of these for our full model. The theoretical results for ^{4}He have a similar energy dependence, but the presently available scarce cross section data do not show any significant slope.

A different way of looking at this for arbitrary scattering lengths is to display contours of equal $\sqrt{\chi^2/\mathrm{dp}}$ for the cross section in the complex $a_{\eta N}$ plane for various models in Fig. 2. This shows within each model long and narrow valleys of good fits. Each correction moves the best fit in the same direction away from

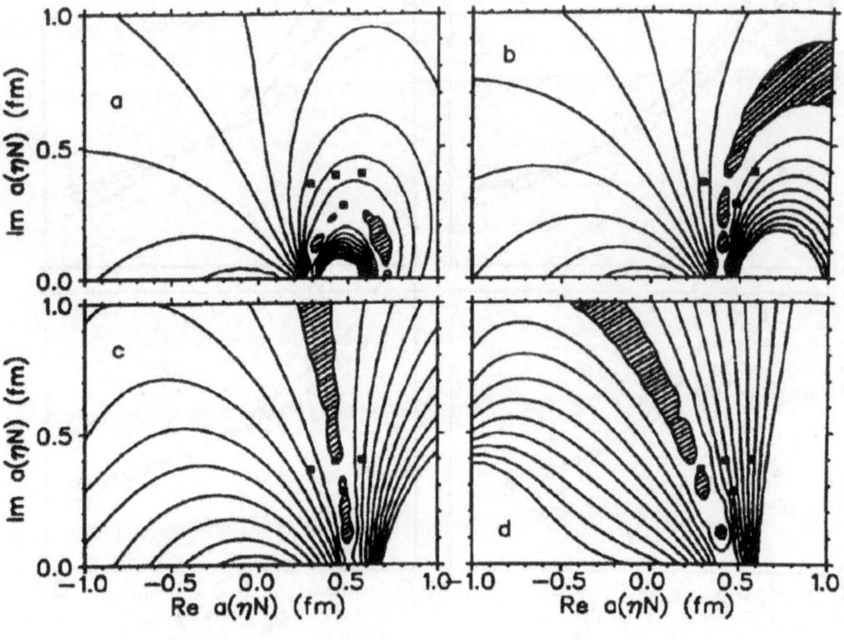

Figure 2. Contour diagrams of $\sqrt{\chi^2/dp} = 1, 2, 3, 4...$ for different models: a) optical model, b) optical model corrected by $A \to A - 1$, c) off-shell effect also included, d) the full model with corrections to the optical model and with the background term in the wave function (6). The hatched areas have this quantity less than one.

the optical model results and it is easy to understand the plots of Fig. 1 in this light.

The slope of the energy dependence in the $pd \to \eta^3$He cross section led Wilkin to conjecture a quasibound state was necessary [4]. After these corrections the slope is obtained also without a bound η^3He state. Further, given a model, it seems that data on η production on light nuclei can be used as a sensitive but not unique constraint also on the elementary ηN scattering length.

References

1. S. Wycech, A.M. Green, J.A. Niskanen: preprint HU-TFT-95-8
2. C. Wilkin: private communication
3. H.C. Chiang, E. Oset, L.C. Liu: Phys. Rev. **C44**, 738 (1991)
4. C. Wilkin: Phys. Rev. **C47**, R938 (1993)

Few-Body Systems Suppl. 8, 409–414 (1995)

Few-Body Systems

$\bar{p}N$ Scattering with Annihilation Channel in an Extended Hilbert Space Model

Yu.A. Kuperin, S.B. Levin, Yu.B. Melnikov, E.A. Yarevsky

Laboratory of Complex Systems Theory, St.Petersburg State University, St.Petersburg, 198904, Russia

Abstract. The expressions for scattering length and partial total cross-section for $\bar{p}n$ and $\bar{p}p$ systems are obtained in the frame of an extended Hilbert space model with annihilation channel. A numerical algorithm for scattering data calculation is suggested. The numerical calculations for $\bar{p}n$ and $\bar{p}p$ scattering data at orbital momentum $L = 0, 1, 2, 3$ are performed on this base. The interaction parameters are fitted via two-body scattering data. A satisfactory agreement between experimental and theoretical data is obtained.

1 Introduction

Few-body systems with antiproton are now the subject of intensive investigations [1, 2, 3]. Such systems include more information about particle structure and interparticle interaction, and are more complicated than the systems without antiparticles. The reason is not the annihilation description from QCD point of view, because in the annihilation region the description coincides with the nonperturbative QCD one. Because of the complication of the processes which include antiparticles, a lot of physical models for the description of such processes have been developed.

The current annihilation models are different variants of optical potential models [4, 5], coupling channel models [6], quark models [7] and others. There is a satisfactory description of the basic experimental scattering data for all these models. The total and differential elastic and annihilation cross sections, as well as scattering lengths, are reproduced with acceptable accuracy. But only certain approaches allow to calculate exactly more delicate characteristics such as, for instance, differential rearrangement cross-sections of the scattering process $\bar{p}p \to \bar{n}n$ [8].

Let us now consider few-body systems with $\bar{p}$. An important problem is the study of $\bar{p}d$-scattering. The precise analysis of this process allows to extract unknown data about $\bar{n}n$ process [2]. The use of standard approaches

[4, 5, 6, 7] for the $\bar{p}d$ system, however, has the difficulty of the construction of a correct three-body equation. Another interesting problem is the study of weakly-coupled bound states for $\bar{p}dd$, $\bar{p}pd$-systems and others [9, 10]. The annihilation effects may be rather large in such systems [10].

In the present work we suggest a model with annihilation interaction based on the operator extension theory [11]. We construct the annihilation interaction in the frame of this theory for $\bar{p}p$ and $\bar{p}n$ systems. This method has several advantages in comparison with some other methods. Firstly, it allows to consider both regular and singular interactions, as well as local and non-local interaction. Secondly, powerful and well-developed operator theory methods give the possibility of simplifying the calculations and, in some cases, of obtaining exact analytic formulae. Finally, a mathematically rigorous and a physically natural generalization of the scheme for three-body systems is constructed [12].

2 Description of the Model

Let us briefly describe the construction of the model Hamiltonian (for details see [11, 12]). First consider the *external* block of the Hamiltonian h in the Hilbert space $L^2(\mathbf{R}^3)$. For $\bar{p}n$ system we take h as the Laplace operator, $h = -\Delta$, and for $\bar{p}p$ system we add the Coulomb attraction, $h = -\Delta + ar^{-1}$, where $a = M_p e_p e_{\bar{p}}$ is the Coulomb constant. Next we introduce an abstract Hilbert space $\mathcal{H}^{\text{ad}}$ where an additional selfadjoint operator h^{ad} acts. This operator describes the annihilation channels. In the general case its spectrum may also include some eigenvalues interpreted as binding energies of quark compound bag [11]. Next, both selfadjoint operators h and h^{ad} are restricted to symmetric ones, h_0 and h_0^{ad} respectively. For the *external* channel $L^2(\mathbf{R}^3)$, the operator h_0 is determined in terms of the $W_2^2(\mathbf{R}^3)$ functions, vanishing in the vicinity of the sphere S_ρ of radius ρ. This sphere is called as the interaction surface, and ρ is the annihilation radius. The common extension of the symmetric operator $h_0 \oplus h_0^{\text{ad}}$ to some selfadjoint operator H gives the total Hamiltonian of the system. The corresponding theory as well as some calculations may be found in detail in refs. [11, 12]. In the present paper we use the results of [11] which allow to restrict the initial Schrödinger equation to the effective boundary problem in $\mathbf{R}^3$:

$$\begin{cases} h\Psi = z\Psi & (1.1)\\ \Psi^+(r,s;z)|_{r=\rho} = \Psi^-(r,s;z)|_{r=\rho}, & (1.2)\\ [\Psi'(r,s;z)]|_{r=\rho} = -Q(z)\int_{S_\rho} \Psi(\rho,s';z)\bar{\Phi}(s')ds'\Phi(s), & (1.3) \end{cases}$$

where $(r,s) = x \in \mathbf{R}^3$ are spherical coordinates, $\Psi^\pm(r,s;z) = \Psi(r\pm 0,s;z)$, $\Psi' = \partial\Psi/\partial r$, $[\Psi]|_{r=\rho} = \Psi^+|_{r=\rho} - \Psi^-|_{r=\rho}$. Function $\Phi(s) \in L^2(S_\rho)$ is the channel coupling function (parameter of the model), and its squared L^2-norm $||\Phi||^2 = \int_{S_\rho} |\Phi(s)|^2 ds$ is the coupling constant. The energy-dependent function $Q(z)$ is given by the Schwartz integral [11, 12]

$$Q(z) = \int_{-M}^{\infty} \frac{\eta^2 + z\zeta}{\zeta - z} d\mu(\zeta),$$

where $-M$ is the annihilation threshold, $M = m_{\bar{p}} + m_{n(p)} - \sum_i m_i$, where $\sum_i m_i$ is the sum of annihilation products masses. The constant $\eta^2 \in \mathbf{R}$ is the model parameter, and $d\mu(\zeta)$ is the absolute continuous measure. Its derivative $\mu'(\zeta) = d\mu(\zeta)/d\zeta$ has the meaning of the velocity of annihilation probability alteration, depending on energy. Effective energy-dependent selfadjoint boundary conditions (1.2), (1.3) simulate the short-range interaction including the annihilation channel with threshold $-M$.

After partial wave analysis we obtain the following boundary problem in the l-th partial wave ψ_l:

$$\begin{cases} (-\partial_r^2 + l(l+1)r^{-2})\psi_l(r,z) = z\psi_l(r,z) & \text{for } \bar{p}n \text{ system,} \\ (-\partial_r^2 + l(l+1)r^{-2} + ar^{-1})\psi_l(r,z) = z\psi_l(r,z) & \text{for } \bar{p}p \text{ system,} \\ \psi_l^+(r,z)|_{r=\rho} = \psi_l^-(r,z)|_{r=\rho}, & \\ [\psi_l'(r,z)]|_{r=\rho} = -(2l+1)|\phi_l|^2 Q(z)\psi_l(r,z)|_{r=\rho}, & \end{cases}$$

where ϕ_l are the components of the coupling function $\Phi(s)$. Let us denote by $h_l^{\pm}(x) = f_l^r(x) \pm i f_l^s(x)$, where $f_l^r(x)$ and $f_l^s(x)$ are regular and singular partial-wave solutions of Eq. (1.1). With this notation we obtain for partial-wave S-matrix element $S_l = e^{2i\delta_l}$ a representation:

$$S_l(k) = \frac{\left[|\phi_l|^2 Q(k^2) - \dfrac{f_l^{r\prime}(k\rho)}{f_l^r(k\rho)}\right] h_l^-(k\rho) + h_l^{-\prime}(k\rho)}{\left[-|\phi_l|^2 Q(k^2) + \dfrac{f_l^{r\prime}(k\rho)}{f_l^r(k\rho)}\right] h_l^+(k\rho) - h_l^{+\prime}(k\rho)}, \tag{2}$$

where primes designate derivatives. In the frame of our model one can calculate the scattering length for the $\bar{p}n$ system

$$a_0 = \frac{\rho^2 Q(0)}{\rho Q(0) - 1}, \tag{3}$$

and for the $\bar{p}p$ system

$$a_0^c \approx \frac{-1 + \alpha\rho}{\frac{1}{\rho^2 Q(0)} - \frac{1}{\rho} + 2\alpha[\ln(2\alpha\rho) + 2\gamma - 1] + \frac{\alpha}{Q(0)}\left(\frac{2}{\rho} - \frac{1}{\rho(1-\alpha\rho)}\right)} \tag{4}$$

at the first order of perturbation theory with respect to parameter $\alpha \equiv a/2$ ($\alpha = 0.035\,\text{fm}^{-1}$ for $\bar{p}p$ system).

3 Numerical Analysis

The model described in previous section includes a set of parameters. These are the interaction radius ρ, coupling constants $|\phi_l|^2$, extension parameter η^2 and functional measure parameter $\mu(\zeta)$. We choose the measure in the simple form:

$$\mu(\zeta) = \frac{\beta - M}{\beta} \frac{\zeta}{\zeta + \beta}.$$

Numerical experiments show that the description of the experimental data depends weakly on the choice of the form of $\mu(\zeta)$. The mentioned choice is quite convenient.

All used model parameters may be fitted via experimental data, such as complex scattering length and total scattering cross-section. Let us introduce the functional

$$\chi = \int_{k_{min}}^{k_{max}} \frac{|\sigma_{tot}(k) - \sigma_{tot}^{exp}(k)|}{\sigma_{tot}(k) + \sigma_{tot}^{exp}(k)} dk$$

where $\sigma_{tot}(k) = \sum_l (2l+1)\sigma_l^t$, and the quantities σ_l^t are given by

$$\sigma_l^t = 4\pi |f_l|^2 = \frac{\pi}{k^2}|1 - S_l|^2 + \frac{\pi}{k^2}(1 - |S_l|^2). \tag{5}$$

σ_{tot}^{exp} is the approximated experimental total cross-section [13]. We minimize the value χ as a function of $|\phi_l|^2$, (l=0,1,2,3), ρ, β, η^2 in the momentum region $k_{min} = 50$Mev/c, $k_{max} = 300$Mev/c.

We fit the parameters of our model β and μ^2 by complex $\bar{p}p$ scattering length $a_0 = (-0.93 + i0.95)$fm [2] and fit the parameters ρ and $|\phi_l|^2$, $l = 0, 1, 2, 3$ by approximation of experimental $\bar{p}N$ total cross-section [13, 14].

The results of the calculation of the partial waves for $\bar{p}n$ and $\bar{p}p$ systems are shown in the Table 1 and in Fig. 1, and in Table 2 and Fig. 2 respectively.

Table 1. The parametrization results for $\bar{p}n$ interaction

waves	χ	ρ, fm	η^2, fm^{-2}	β, fm^{-1}	$\lvert\phi_0\rvert^2$	$\lvert\phi_1\rvert^2$	$\lvert\phi_2\rvert^2$	$\lvert\phi_3\rvert^2$
s	12.5	0.70	12000	$1.2\cdot10^6$	8.8	–	–	–
s,p	3.78	0.90	1000	$2.0\cdot10^5$	14.0	5.0	–	–
s,p,d	0.48	1.25	24	$3.8\cdot10^4$	81.0	50.0	40.0	–
s,p,d,f	0.09	1.20	60	$4.2\cdot10^4$	38.3	20.0	24.0	1.0

Table 2. The parametrization results for $\bar{p}p$ interaction

waves	χ	ρ, fm	η^2, fm^{-2}	β, fm^{-1}	$\lvert\phi_0\rvert^2$	$\lvert\phi_1\rvert^2$	$\lvert\phi_2\rvert^2$	$\lvert\phi_3\rvert^2$
s	12.1	0.65	30000	$2.4\cdot10^6$	7.0	–	–	–
s,p	3.80	1.05	150	$8.5\cdot10^4$	35.0	8.1	–	–
s,p,d	0.90	1.35	12	$3.2\cdot10^4$	122.	20.0	2.0	–
s,p,d,f	0.11	1.59	470	$2.0\cdot10^4$	1.6	9.0	1.1	1.2

The results presented in the Figs. 1 and 2 show that the role of F-wave both in $\bar{p}p$ and $\bar{p}n$ cases is already small. This fact allows to limit ourselves to S, P, D and F waves.

The total cross-section have the resonance behavior near threshold. This fact confirms the prediction of authors of the ref.[6] about the existence of nearthreshold resonances (barionium).

The existence of the annihilation channel in the system produces the great difference when comparing with experimental data of NN scattering. It shows up, for example, in the behavior of higher partial waves. As was noted in [2], the P-wave contributes 40-60% to the total cross-section. The size of this contribution is related to the behavior of the ratio of real to imaginary part of the elastic scattering forward amplitude ρ as a function of $\bar{p}$-momentum transfer (see Fig. 3). This ratio rapidly increases from -1 at the threshold up to 0 at 200Mev/c. The reason of that is connected with the compensation of the real parts of S- and P-wave amplitudes, which have different signs [6]. (See the Argand diagrams of Fig. 4 for S- and P-waves.)

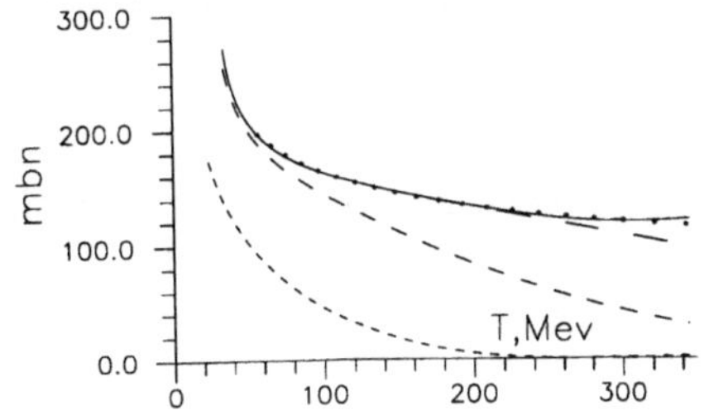

Figure 1. The total $\bar{p}n$ cross-section. Solid line is for the model with S, P, D and F waves, long-dashed line is for the model with S, P and D waves, dashed line is for the model with S and P waves, and short-dashed line is for the model with S wave only. The experimental values are from ref. [2]

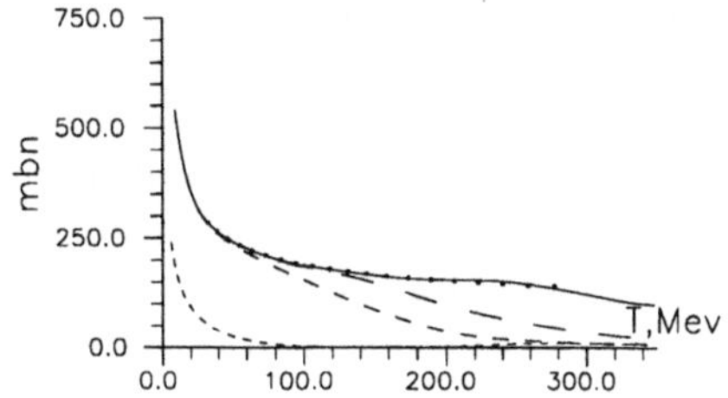

Figure 2. The total $\bar{p}p$ cross-section. The notations are the same as for Fig. 1

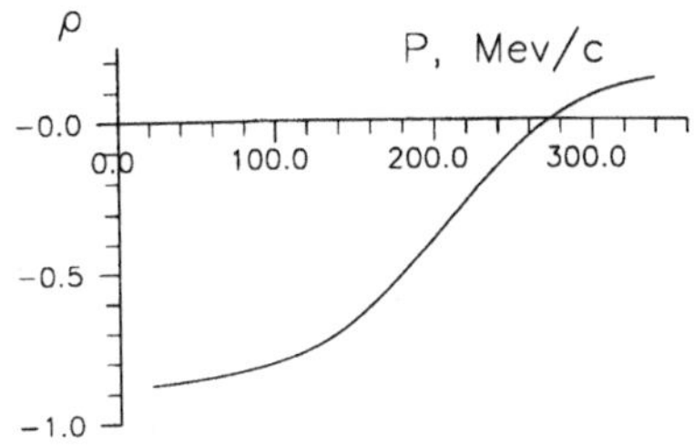

Figure 3. The ratio of real to imaginary part of the elastic scattering forward amplitude ρ as a function of $\bar{p}$-momentum transfer

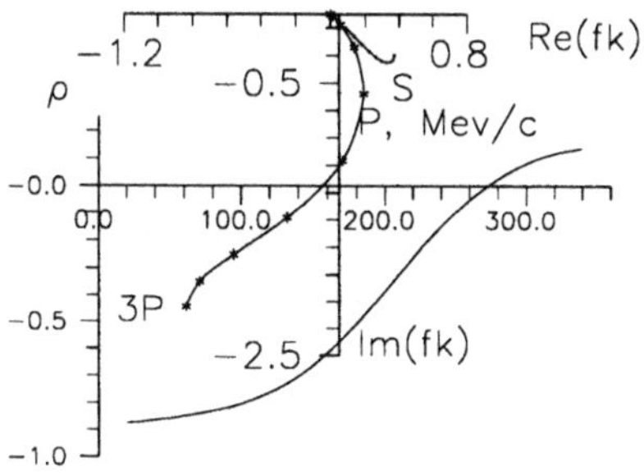

Figure 4. Argand diagrams for S- and P-waves

4 Conclusions

In the frame of our model we describe the scattering length and total cross-section both in $\bar{p}p$ and $\bar{p}n$ systems. The fitted results show that the contribution of high partial waves to the total cross-section are essential. This fact is in a good agreement with the experimental results [15]. The suggested model describes the qualitative as well as the quantitative picture of $\bar{p}N$ scattering processes.

The proprosed parametrization of the energy-dependent pair interactions is an adequate basis for the numerical investigation of three-body scattering problems with annihilation channels in the frame of Faddeev scheme in the operator extension theory approach [12]. This parametrization may be used for the description of bound states and scattering processes in the $\bar{p}d$, $\bar{p}t$ and $\bar{p}^3$He systems [10] as well as for the description of the annihilation processes in the $\mathrm{He}^{++}e^-\bar{p}$ atomcule [3].

Acknowledgement. We are very grateful to Prof. O.D. Dalkarov and Prof. A.E. Kudrjavtsev for fruitful discussions and interesting notes. We thank the Commission of the European Communities for support in the frame of EC-Russia Collaboration Contract ESPRIT P 9282 ACTCS. The research described in this publication was made possible in part by Grant # NX 1000 from the International Science Foundation.

References

1. C.J. Batty: Rep. Prog. Phys. **52**, 1165 (1989)
2. B.O. Kerbikov, L.A. Kondratyuk, M.G. Sapozhnikov: Usp. Fiz. Nauk. **159**, 1 (1989)
3. M. Iwasaki et al.: Phys. Rev. Lett. **67**, 1246 (1991) 1246; T. Yamazaki et al.: Nature. **361**, 238 (1992)
4. C.B. Dover, J.M. Richard: Phys. Rev. **C21**, 1466 (1980)
5. T.-A. Shibata: Phys. Lett. **B189**, 232 (1987)
6. I.S. Shapiro: Nucl. Phys. **A478**, 665 (1988); O.D. Dalkarov, J. Carbonell, K.V. Protasov: Yad. Fiz. **52**, 1670 (1990)
7. H.R. Rubinstein, H. Stern: Phys. Lett. **21**, 447 (1966); A.M. Green, J.A. Niskanen: Nucl. Phys. **A412**, 448 (1984)
8. W. Brückner et al.: Phys. Lett. **B169**, 302 (1986)
9. A.M. Frolov, A.J. Thakkar: Phys. Rev. **A46**, 4418 (1992)
10. V.I. Korobov, Yu.A. Kuperin, S.I. Vinitsky: Phys. Lett. **B315**, 215 (1993)
11. Yu.A. Kuperin, Yu.B. Melnikov: J. Math. Phys. **33(8)**, 2795 (1992)
12. Yu.A. Kuperin, S.P. Merkuriev: Amer. Math. Soc. Transl. **150**, 141 (1992)
13. D.V. Bugg et al.: Phys. Lett. **B194**, 563 (1987)
14. L.A. Kondratjuk, M.G. Sapozhnokov: Nucl. Phys. **A46**, 89 (1987)
15. C. Amsler, F. Myhrer: Annu. Rev. Nucl. Part. Sci. **41**, 219 (1991)

Few-Body Systems Suppl. 8, 415–418 (1995)

Few-Body Systems

$\Delta I = 1/2$ Rule: Interplay of Quark and Hadronic Degrees of Freedom

K. Maltman[1*] and M. Shmatikov[2†]

[1] York University, Toronto, Ontario M3J 1P3, Canada

[2] Russian Research Center *Kurchatov Institute*, 123182 Moscow, Russia

Abstract. The weak conversion process $\Lambda N \longrightarrow NN$ is investigated using the quark model and the QCD-modified effective weak Hamiltonian evaluated at a scale $\simeq 1$ GeV. The results show large violations of the $\Delta I = 1/2$ rule.

Strong violation of the $\Delta I = 1/2$ rule occurs also in the induced Λ-hyperon decay mediated by vector-meson exchanges. The modified factorization contributions to the $\Lambda \rightarrow N\rho$ and $\Sigma \rightarrow N\rho$ couplings are calculated to demostrate that these contributions naturally include $\Delta I = 3/2$ terms which are comparable in magnitude to the corresponding $\Delta I = 1/2$ counterparts.

The contribution of the $\Delta I = 3/2$ terms is sufficiently large so that the resulting violation should be easily detectable in light hypernuclear decay experiments.

The celebrated $\Delta I = 1/2$ rule seems to be one of the most prominent features of the observed $\Delta S = 1$ non-leptonic weak decays [1]. In absence of other evidences it became conventional to assume the validity of the $\Delta I = 1/2$ rule for all strangeness-changing process. In particular, this assumption was used as one of the cornerstones in the meson-exchange treatment of the $\Lambda N \rightarrow NN$ process.

More close examination of the nature of the $\Delta I = 1/2$ rule reveals, however, that mechanisms responsible for the dominance of the $\Delta I = 1/2$ amplitudes over their $\Delta I = 3/2$ counterparts in the K-meson and hyperon non-leptonic decays are quite different. Mechanism of the $\Delta I = 1/2$ amplitude enhancement which is common to both of these decays is known to be the QCD-induced corrections in the effective weak lagrangian $\mathcal{L}_{\text{eff}}$. The resulting enhancement ratio is only about $\simeq 4$, far below the experimentally observed factor of $\simeq 20$. Mismatch of the observed and measured enhancement factors signals the importance of other (nonperturbative) mechanisms which are operative in the nonleptonic decays of K-mesons and hyperons.

* *E-mail address*: FS300175@Sol.YorkU.ca

† *E-mail address*: MSH@ofpnp.kiae.su

In the latter decays the $\Delta I = 3/2$ piece of the effective weak-interaction lagrangian $\mathcal{L}_{\text{eff}}$ being symmetric under the interchange of the color labels of the quark fields vanishes in the so-called "internal" diagrams in the limit in which each baryon consists of three quarks in a color-antisymmetric state. Besides, the $\mathcal{L}_{\text{eff}}$ operator having the left-right structure which arises only from the purely $\Delta I = 1/2$ penguin diagrams gains large enhancement in the remaining "external" with respect to the left-left operators.

At the same time the required additional enhancement in the $K \to \pi\pi$ decay stems from the final-state interaction. Here the $\Delta I = 1/2$ transition produces the isoscalar $\pi\pi$ final state whereas the $\Delta I = 3/2$ transition ends in the $I = 2$ final state. This difference in the isospin of the $\pi\pi$ system proves to be crucially important: interaction between pions is attractive in the former case and repulsive in the latter.

Apparent difference in the origin of hadronic (as opposed to QCD-induced) enhancement in $K \to \pi\pi$ and hyperon decays casts doubts on the universal character of the $\Delta I = 1/2$ rule. With this in mind we have investigated the next simplest $\Delta S = 1$ process, induced Λ-hyperon decay $\Lambda N \to NN$.

First, we investigate the induced Λ-hyperon in the context of the quark model [2]. The basic ingredients of the calculations are as follows: 1) the effective weak Lagrangean, $\mathcal{L}_{\text{eff}}$; 2) a model for the structure of involved baryons with explicit quark degrees of freedom and 3) relative baryon-baryon initial- and final-state wave function.

We employ the effective weak Lagrangean, $\mathcal{L}_{\text{eff}}$, of the standard model as modified by the perturbative QCD up to a $\simeq 1$ GeV scale. The explicit form of the $\mathcal{L}_{\text{eff}}$ is then:

$$\mathcal{L}_{\text{eff}} = \frac{G}{\sqrt{2}} \sum_{r=1, r\neq 4}^{6} (C_r^c \xi_c + C_r^t \xi_t) Q_r \,, \tag{1}$$

where $\xi_\alpha = V_{\alpha d} V^*_{\alpha s}$ (V_{ij} being the elements of the CKM matrix), and

$$\begin{aligned} Q_1 &= (\bar{s}_\alpha d_\alpha)_L (\bar{u}_\beta u_\beta)_L \\ Q_2 &= (\bar{s}_\alpha d_\beta)_L (\bar{u}_\beta u_\alpha)_L \\ Q_3 &= (\bar{s}_\alpha d_\alpha)_L \textstyle\sum_{q=u,d,s} (\bar{q}_\beta q_\beta)_L \\ Q_5 &= (\bar{s}_\alpha d_\alpha)_L \textstyle\sum_{q=u,d,s} (\bar{q}_\beta q_\beta)_R \\ Q_6 &= (\bar{s}_\alpha d_\beta)_L \textstyle\sum_{q=u,d,s} (\bar{q}_\beta q_\alpha)_R , \end{aligned} \tag{2}$$

with α, β color labels. The values of the (scale-dependent) coefficients, $C_r^{c,t}$, at a $\simeq 1$ GeV scale are taken from the updated analysis of ref. [3].

The initial- and final-state baryonic systems are to be considered as six-quark ones. Corresponding wave functions have then the form

$$\begin{aligned} |\Lambda N(\alpha)\rangle &= \hat{A}[\Lambda(123) N(456) \Psi^{(\alpha)}_{N\Lambda}(\boldsymbol{R})] \\ |NN(\beta)\rangle &= \hat{A}[N(123) N(456) \Psi^{(\beta)}_{NN}(\boldsymbol{R})] \end{aligned} \tag{3}$$

where $\hat{A}$ is the antisymmetrizer for the identical light quarks, $\boldsymbol{R}$ is the separation of the centers of mass of the two three-quark clusters (123) and (456),

$\Psi^{(\alpha)}_{N\Lambda}$, $\Psi^{(\beta)}_{NN}$ are ΛN and NN wave functions obtained by solving the relevant Schrödinger equation with the phenomenologial ΛN and NN potentials, and (α), (β) label specific ΛN and NN partial waves, respectively. For the Λ, N wave functions we employ the constituent quark model. The color-spin-isospin part of the former functions is fixed unambigously by the quantum numbers of the corresponding baryon. The spatial wave functions are taken to be of Gaussian form corresponding to the oscillator-type confinement potential.

Given $\mathcal{L}_{\text{eff}}$ of Eq.(1) and wave functions of Eq.(3) one can, in principle, calculate the transition amplitudes

$$A_{\alpha\beta} = \langle NN(\beta)|\mathcal{L}_{\text{eff}}|\Lambda N(\alpha)\rangle \tag{4}$$

Our use of the constituent quark model wave functions requires rewriting $\mathcal{L}_{\text{eff}}$ in terms of constituent, rather than current, quark fields. This matching can be performed following the recipe of ref.[4] by replacing the current quark axial current, $A_\mu^{(q)a}$, by $g_A^{(c)} A_\mu^{(c)a}$ where q, c label current and constituent quarks, respectively, and $g_A^{(c)} \simeq 3/4$ is the constituent quark axial coupling.

Using standard techniques associated with the resonating group method the matrix elements of Eq.(4) can be cast into the form of matrix elements of nonlocal transition operators associated with the various quark pair labels and permutation labels. Results of calculations of the quark-mechanism contribution to the induced Λ-hyperon decays, $\Lambda N \to NN$, are summarized in Table 1. The are confronted to the combined contribution of the one-pion (OPE) and one-kaon (OKE) exchanges. Inspection of the Table 1 shows that the $\Delta I = 1/2$ rule is badly violated in transitions from the ${}^1S_0$ initial state in $\Lambda N \to NN$. Moreover, quark amplitudes in these channels noticeably exceed their meson-exchange counterparts, making the effect readily detectable in hypernuclear decay experiments which are sensitive to the n- to p-stimulated ratio in the ${}^1S_0$ conversion. Existing data already favor large violation of the $\Delta I = 1/2$ rule in these channels [5].

Table 1. Representative values of the $\Lambda N \to NN$ amplitudes (arbitrary units)

Transition	Quark	OPE + OKE
${}^1S_0 \to {}^1S_0$ (p)	-1.20	0.54
${}^1S_0 \to {}^1S_0$ (n)	1.09	0.76
${}^3S_1 \to {}^3S_1$ (p)	0.03+0.25i	0.17-0.16i
${}^3S_1 \to {}^3D_1$ (p)	-0.11-0.15i	-0.87-1.56i
${}^1S_0 \to {}^3P_0$ (p)	-2.15	-0.03
${}^1S_0 \to {}^3P_0$ (n)	-0.85	-0.05
${}^3S_1 \to {}^3P_1$ (p)	0.41	0.01
${}^3S_1 \to {}^3P_1$ (n)	0.61	0.02
${}^3S_1 \to {}^1P_1$ (p)	-0.94	-1.12

The $\Delta I = 1/2$ rule is known empirically, from hyperon decays, to be a valid assumption, but no similar experimental support exists for the assumption that vector meson coupling satisfy this rule. We investigated factorization contribution to the latter couplings to find out that the structure of QCD modifications to the weak interactions are such as to naturally distinguish the pseudoscalar and vector cases [6]. Factorization contributions to the weak $YN\rho$ vertices ($Y = \Lambda, \Sigma$) are considered (corresponding coupling constants involving ω-meson are small and, even for the $\Sigma N\omega$, satisfy the $\Delta I = 1/2$ rule). We employ the effective Lagrangean (1) and (2). In the modified factorization approximation the weak ρ-type couplings of the operators $Q_{5,6}$ vanish. This fact follows from the observation that these operators contain color-singlet flavor-octet non-strange currents only in Fierz-rearranged form, in which form, owing to the original left-right chiral structure, only scalar and pseudoscalar currents are involved, the vacuum-to-ρ matrix elements of which automatically vanish. Then the relative strength of the $\Delta I = 1/2$ to $\Delta I = 3/2$ contributions to the weak coupling is determined, in all cases, by the factor $-\frac{2}{3}C_1 + \frac{4}{3}C_2 + \frac{4}{3}C_3 + \frac{4}{3}C_4$ for the ρ^- couplings, and $-\frac{2}{3}C_1 + \frac{4}{3}C_2 + \frac{4}{3}C_3 - \frac{8}{3}C_4$ for the ρ^o couplings. Using the C_i values of [3] one comes to the conclusions that the weak $\Lambda N\rho$ and $\Sigma N\rho$ couplings badly violate the $\Delta I = 1/2$ rule: relative contribution of the $\Delta I = 3/2$ terms makes $\simeq 0.4$ to -0.8. The basic reason for this is the complete absence of the penguin contributions, which had large enhancements in the π coupling case. Strong violation of the $\Delta I = 1/2$ rule will, however, be reflected in the full couplings only if the factorization contributions represent a moderate to sizable fraction of the full couplings. Model predictions for this fraction differ considerably, most significantly for the parity-violating axial-vector coupling. The factorization contributions are $\simeq 1/3$ of the full predictions for the (parity-conserving) vector and tensor couplings and between $\simeq 1/4$ and 1 times the full prediction for the axial-vector coupling constant. Since the size of these contributions is not small on the scale of values to be expected for the full couplings, it appears very unlikely that it is safe to assume the validity of the $\Delta I = 1/2$ rule as an input when determining constants of weak coupling of the ρ-meson to a hyperon.

To summarize, the $\Delta I = 1/2$ rule is shown to be badly violated in (some channels of) the induced Λ-hyperon decay, $\Lambda N \to NN$. The ratio of the $\Delta I = 3/2$ and $\Delta I = 1/2$ contributions may be large and detectable in dedicated experiments.

References

1. L.B.Okun': *Leptons and quarks.* Amsterdam: North-Holland, 1982
2. K.Maltman, M.Shmatikov: Phys. Lett. **B331**, 1 (1994)
3. E.A.Paschos, T.Schneider, Y.L.Wu: Nucl. Phys. **B332**, 285 (1990)
4. A.Manohar, H.Georgi: Nucl. Phys. **B234**, 189 (1984)
5. R.Schumacher: Nucl. Phys. **A547**, 143c (1992)
6. K.Maltman, M.Shmatikov: Phys. Rev. **C51**, 1576 (1995)

Few-Body Systems Suppl. 8, 419–422 (1995)

Few-Body Systems

Asymptotic Behavior of the Deuteron Extra-Components in the Covariant Light Front Dynamics

B. Desplanques[1], V.A. Karmanov[2], J.F. Mathiot[3]

[1] Institut des Sciences Nucléaires, 53 avenue des Martyrs, F-38026 Grenoble Cedex, France

[2] Lebedev Physical Institute, Leninsky Prospect 53, 117924 Moscow, Russia

[3] Division de Physique Théorique, Institut de Physique Nucléaire, F-91406 Orsay Cedex, France

1 Introduction

In the covariant formulation of the light-front dynamics, the wave function of two nucleon systems such as the deuteron or any NN scattering state depends on two variables, the relative momentum, $\boldsymbol{k}$, and a unit vector, $\boldsymbol{n}$ [1]. This latter is related to the four-vector that allows one to separate in an invariant way two space-time events. Its appearance leads to the existence of extra-components in the wave function as compared to the usual non-relativistic wave functions. At first-sight, the arbitrariness in the orientation of this vector $\boldsymbol{n}$ introduces some uncertainty in the predictions. Practical ways to get rid of it in calculating observables have been proposed in the literature [2, 3, 4]. In this contribution we investigate how this $\boldsymbol{n}$ dependence is automatically removed in the observables related to the asymptotic behavior of the wave function.

It is known that the residues of the wave function in momentum space at $k = i\alpha$ for the deuteron ($\alpha = \sqrt{M\epsilon_d}$) or $k = \sqrt{ME_{\rm cm}}$ for a scattering state are observable quantities which can be related for instance to the deuteron asymptotic normalizations or to phase shifts. To prevent the extra-components to give rise to unphysical results, the corresponding residues have to vanish. We here sketch how this may be achieved for the so- called f_5 component, which is the largest of the deuteron extra-components when a pseudo-scalar πNN coupling is used [5]. The interest of this coupling here is to emphasize some of the questions to be answered such as: how the pion pair term is reproduced in the light-front dynamics or what about the asymptotic behavior of the extra-components?. All the developments are done at the lowest $1/m$ order.

2 The f_5 Component and the Reaction $ed \to e'np$

The expression of $f_5(k)$ (denoted f_5 for simplicity) is given at the lowest order in $g^2_{\pi NN}$ by:

$$f^0_5[\boldsymbol{k} \times \boldsymbol{n}]/k = -\frac{3g^2_{\pi NN}}{4\sqrt{3}m^2}\frac{m}{k^2+\alpha^2}\int \frac{d^3\boldsymbol{k}'}{(2\pi)^3}\frac{(\boldsymbol{k}-\boldsymbol{k}')\times\boldsymbol{n}}{\mu^2+(\boldsymbol{k}-\boldsymbol{k}')^2}\frac{(\boldsymbol{k}^2-\boldsymbol{k}'^2)}{m}u_S(\boldsymbol{k}')\,, \tag{1}$$

where $u_S(k')$ represents the non-relativistic deuteron S-wave. The D-wave contribution is ignored to simplify the discussion on the one hand and consistently with our requirement to retain lowest $1/m$ order terms on the other hand. Examination of Eq. 1 indicates that $f_5 \to \infty$ for $k \to i\alpha$, which motivated the present work.

It has been shown in [6] that the component f^o_5 (Eq. 1), in the limit where $(k^2-k'^2)/(k^2+\alpha^2) \to 1$, was providing one half of the usual pair term contribution to the deuteron electro-disintegration near threshold. This pair term is essential to account for experimental results in the range of Q^2 around 10 - 15 fm^{-2}. It was also shown that there exists a contact term in the electromagnetic current and that this one was providing another one half of the pair term contribution, allowing one to reconcile the description of the deuteron electro-disintegration near threshold in the light-front dynamics [7, 8] with the non-relativistic one completed by the pion pair term. Furthermore, it was found that the $\boldsymbol{n}$ dependent part of this contact term was cancelling a similar contribution due to the f^o_5 component [6]. This raises the question about the role that similar contact terms may play in the NN interaction when going beyond one meson exchange contributions usually retained.

3 Further Contributions to the f_5 Component

Examining the problem, we did find contributions that are of interest here and involve the one pion exchange on the one hand and the σ- and ω-exchanges on the other hand. These ones are chosen because they are relevant to the description of the dominant features of the NN strong interaction. The π-exchange does not appear a second time because it corresponds to a higher $1/m$ order. Their expression in momentum space is given by:

$$\delta V^{\pi,\sigma+\omega}(\boldsymbol{k},\boldsymbol{k}'') = \frac{g^2_{\pi NN}}{8m^2}\boldsymbol{\tau}_1.\boldsymbol{\tau}_2\,\boldsymbol{\sigma}_1\times\boldsymbol{\sigma}_2\int\frac{d\boldsymbol{k}'}{(2\pi)^3}$$

$$\left[V^{\sigma+\omega}(\boldsymbol{k}-\boldsymbol{k}')\frac{(\boldsymbol{k}'-\boldsymbol{k}'')\times\boldsymbol{n}}{\mu^2+(\boldsymbol{k}'-\boldsymbol{k}'')^2}-\frac{(\boldsymbol{k}-\boldsymbol{k}')\times\boldsymbol{n}}{\mu^2+(\boldsymbol{k}-\boldsymbol{k}')^2}V^{\sigma+\omega}(\boldsymbol{k}'-\boldsymbol{k}'')\right]\,. \tag{2}$$

The two terms in the last bracket, which transform into each other in a hermitic conjugation, can contribute to f_5 but somewhat differently. The contribution of the 2nd term (operating on the right) can be combined with f^0_5 Eq. (1) to give

$$f_5(\boldsymbol{k}\times\boldsymbol{n})/k = -\frac{3g^2_{\pi NN}}{4\sqrt{3}m^2}\frac{m}{k^2+\alpha^2}\int\frac{d^3\boldsymbol{k}'}{(2\pi)^3}\frac{(\boldsymbol{k}-\boldsymbol{k}')\times\boldsymbol{n}}{\mu^2+(\boldsymbol{k}-\boldsymbol{k}')^2}$$

$$\left[\frac{\boldsymbol{k}^2}{m}u_S(\boldsymbol{k}') - \int d^3\boldsymbol{k}''\left(\frac{\boldsymbol{k}'^2}{m}\delta(\boldsymbol{k}'-\boldsymbol{k}'') + \frac{1}{(2\pi)^3}V(\boldsymbol{k}'-\boldsymbol{k}'')\right)u_S(\boldsymbol{k}'')\right]. \quad (3)$$

where the second term in the last bracket is nothing but the Hamiltonian operating on the wave function, $u_s(k")$. The result, $\frac{\alpha^2}{m}u_s(k')$, can be combined with the first term in the bracket, $\frac{k^2}{m}u_s(k')$, to provide a factor $\frac{k^2+\alpha^2}{m}$, which cancels the front factor, $\frac{m}{k^2+\alpha^2}$, arising from the Green function. This leads to the following expression of f_5:

$$f_5(\boldsymbol{k}\times\boldsymbol{n})/k = -\frac{3g^2_{\pi NN}}{4\sqrt{3}m^2}\int\frac{d^3\boldsymbol{k}'}{(2\pi)^3}\frac{(\boldsymbol{k}-\boldsymbol{k}')\times\boldsymbol{n}}{\mu^2+(\boldsymbol{k}-\boldsymbol{k}')^2}u_S(\boldsymbol{k}'). \quad (4)$$

where there is no more pole at $k = i\alpha$, thus removing an undesirable feature. Furthermore, Eq. 4 for f_5 ensures the full equivalence of the calculation of the deuteron electro-disintegration near threshold in the light- front dynamics with the non-relativistic one including the pair term. It is remembered that in ref. [6] it had to be assumed that the factor $(k^2-k'^2)/(k^2+\alpha^2)$ appearing in f_5^0 (Eq.1) was equal to 1 (large k limit).

The first term in the bracket in (2) (operating on the right) has not a role as important as the second one. It also provides a contribution to f_5, but this one just cancels the first order contribution to f_5 (Eq. 4) arising from the rescattering of the nucleons due to the $\sigma+\omega$ -exchange:

$$\delta f_5(\boldsymbol{k}\times\boldsymbol{n})/k = \frac{-m}{\boldsymbol{k}^2+\alpha^2}\int\frac{d\boldsymbol{k}'}{(2\pi)^3}V^{\sigma+\omega}(\boldsymbol{k}-\boldsymbol{k}')f_5(\boldsymbol{k}'\times\boldsymbol{n})/k'. \quad (5)$$

We believe that this result is not fortuitous and contributes to the consistency with other relativistic approaches.

In [6], it was mentioned that the f_5 component could be calculated from a term in the interaction given by

$$\delta V^{\pi}(\boldsymbol{k},\boldsymbol{k}') = \frac{(\boldsymbol{k}^2-\boldsymbol{k}'^2)}{m}\frac{g^2_{\pi NN}}{8m^2}(\boldsymbol{\tau}_1.\boldsymbol{\tau}_2)\frac{\boldsymbol{\sigma}_1\times\boldsymbol{\sigma}_2.[(\boldsymbol{k}-\boldsymbol{k}')\times\boldsymbol{n}]}{\mu^2+(\boldsymbol{k}-\boldsymbol{k}')^2}. \quad (6)$$

It is interesting to notice that $\delta V^{\pi,\sigma+w}$ adds to δV^{π} so that to give the commutator of the total Hamiltonian, H, with the operator -i U:

$$\delta V^{\pi} + \delta V^{\pi,\sigma+\omega} = [H,-iU], \quad (7)$$

where

$$<\boldsymbol{k}|H|\boldsymbol{k}'> = \frac{k^2}{m}\delta(\boldsymbol{k}-\boldsymbol{k}') + \frac{1}{(2\pi)^3}V^{\sigma+\omega}(\boldsymbol{k}-\boldsymbol{k}'), \quad (8)$$

$$< \boldsymbol{k}'| - iU|\boldsymbol{k}'' >= \frac{g^2_{\pi NN}}{8m^2}\boldsymbol{\tau}_1.\boldsymbol{\tau}_2\, \boldsymbol{\sigma}_1 \times \boldsymbol{\sigma}_2.\frac{(\boldsymbol{k}' - \boldsymbol{k}'') \times \boldsymbol{n}}{\mu^2 + (\boldsymbol{k}' - \boldsymbol{k}'')^2}. \tag{9}$$

Considering (7) as a perturbation immediately allows one to recover the expression (4) for f_5. On the other hand, but not independently, Eq. 7 allows one to extend beyond the Born approximation the disappearance of the $\boldsymbol{n}$ dependence for an on - energy shell amplitude.

4 Conclusion

It is current practice to only retain one meson exchange contributions to the NN interaction. In this work, we considered selected contributions due to two meson exchange. Limiting ourselves to the lowest $1/m$ order, we found they were combining with one meson exchange contributions to get rid of undesirable features, otherwise evidenced by the asymptotic behavior of the deuteron extra components. They also make the light-front dynamics approach to the deuteron electro-desintegration near threshold more consistent with the non-relativistic one completed by the pion pair term. Higher $1/m$ order terms may be considered, but their role is likely to be more difficult to analyze. Some, such as the deuteron D-wave or the π exchange force in the Hamiltonian (8) should ensure the above results at higher $1/m$ orders, while other ones should correspond to corrections to f_5^o, implying to consider higher order meson exchanges and so on.

References

1. V.A. Karmanov Fiz. Elem. Chastits At. Yadra, **19**, 525 (1988) (transl.: Sov. J. Part Nucl. **19**, 228 (1988))
2. V.A. Karmanov, A.V. Smirnov, Nucl. Phys. **A546**, 691 (1992) and **A575**, 520 (1994)
3. M.G. Fuda: Phys. Rev. **D44**, 1881 (1991)
4. C.R. Ji, Y. Surya: Phys. Rev. **D46**, 3565 (1992)
5. V.A. Karmanov: Nucl. Phys. **A362**, 331 (1981); J. Carbonell and V.A. Karmanov: Nucl. Phys. A581, 625 (1995)
6. B. Desplanques, V.A. Karmanov and J.F. Mathiot, Nucl. Phys. **A**
7. B.D. Keister: Phys. Rev. **C37**, 1765 (1988)
8. A. Amghar et al.: Nucl. Phys. **A567**, 919 (1994)

Few-Body Systems Suppl. 8, 423–427 (1995)

Few-Body Systems

Nucleon-Nucleon Correlations and Inclusive Electron Scattering off Few-Nucleon Systems at $x > 1$

Silvano Simula *

Istituto Nazionale di Fisica Nucleare, Sezione Sanitá, Viale Regina Elena 299, I-00161 Roma, Italy

Abstract. Inclusive electron scattering off few-nucleon systems is investigated at $x > 1$ and high momentum transfer, including the contributions from quasi-elastic and deep inelastic scattering processes. It is shown that at $x > 1$ the inclusive cross section is dominated by the process of virtual photon absorption on a pair of correlated nucleons both in case of quasi-elastic and deep inelastic scattering events. The sensitivity of the nuclear response to the effects arising from the possible presence of multiquark cluster configurations at short inter-nucleon separations as well as from possible medium-dependent modifications of the nucleon structure function, is illustrated.

The aim of this contribution is to show that the investigation of inclusive electron scattering off few-nucleon systems at high values of Q^2 and $x > 1$ (where Q^2 is the squared four-momentum transfer and $x = Q^2/2M\nu$ the Bjorken scaling variable) could provide relevant information on the short-range properties of hadronic matter. According to the value of the invariant mass W produced by virtual photon absorption on a nucleon in the nucleus, the inclusive process $A(e,e')X$ is governed by the following two mechanisms: i) the quasi-elastic (QE) process, for which $W = M$, and ii) the deep inelastic scattering (DIS) corresponding to $W > M$, where M is the nucleon mass. The basic quantities that could be extracted from the analysis of QE and DIS processes at $x > 1$ are the nucleon momentum distribution in a nucleus and the valence quark distributions in a bound nucleon, respectively. However, present experimental data do not reach sufficiently high values of Q^2, so that at large values of x (> 1.3) they are mainly due to QE scattering and not to DIS [1]. Nevertheless, the analysis of existing data (limited to $Q^2 \sim$ few $(\mathrm{GeV/c})^2$) shows that the kinematical regions corresponding to $x > 1.3$ are strongly affected by high momentum and high removal energy components of the nuclear wave function

* *E-mail address:* simula@hpteo2.iss.infn.it

arising from nucleon-nucleon (NN) short-range and tensor correlations, provided that a consistent treatment of NN correlations both in initial and final nuclear states is considered. Such a result has been obtained in [2], where it has been shown that in case of light as well as complex nuclei the inclusive cross section at $1.3 < x < 2$ is dominated by virtual photon absorption on a pair of correlated nucleons and by their elastic rescattering in the continuum.

As Q^2 increases, the effects due to the final state interactions in the QE process are expected to decrease, whereas the DIS contribution increases. In what follows, the inclusive process $A(e,e')X$ will be investigated at $x > 1$ and for values of Q^2 in the range $10 \div 20$ $(\mathrm{GeV/c})^2$, adopting the impulse approximation. The cross section for the inclusive process $A(e,e')X$ can be written as: $d^2\sigma/dE_{e'}d\Omega_{e'} = \sigma_{\mathrm{Mott}}\,[W_2^A(x,Q^2)+2W_1^A(x,Q^2)\tan^2(\theta_e/2)]$, where $W_i^A(x,Q^2)$ $(i = 1,2)$ is the inclusive nuclear response function, given within the impulse approximation by the following convolution formula

$$W_i^A(x,Q^2) = \sum_{N=1}^{A} \int dW d\boldsymbol{k} dE\ \delta\left[\nu + k^0 - \sqrt{W^2+(\boldsymbol{k}+\boldsymbol{q})^2}\right]$$
$$P^N(k,E)\,\frac{M}{E_k}\,\frac{W}{\sqrt{W^2+(\boldsymbol{k}+\boldsymbol{q})^2}}\sum_{j=1,2} a_{ij}\ W_j^N(W,Q^2,...) \tag{1}$$

where $k \equiv |\boldsymbol{k}|$, $k^0 \equiv M_A - \sqrt{(M_A+E-M)^2+\boldsymbol{k}^2}$ is the initial nucleon energy in the lab system (with E being its removal energy) and the kinematical coefficients a_{ij} are: $a_{11} = 1$, $a_{12} = k_\perp^2/2M^2$, $a_{21} = (Q^2/|\boldsymbol{q}|^2)\ [1-(Q^2/\bar{Q}^2)]$, $a_{22} = (E_k - k_{\|}\bar{\nu}/|\boldsymbol{q}|)^2)/M^2\ +(Q^2/|\boldsymbol{q}|^2)\ k_\perp^2/2M^2$, with $\bar{Q}^2 \equiv |\boldsymbol{q}|^2 - \bar{\nu}^2$, $\bar{\nu} \equiv \nu + k^0 - E_k$, $E_k \equiv \sqrt{M^2+k^2}$, $k_{\|} \equiv \boldsymbol{k}\cdot\boldsymbol{q}/|\boldsymbol{q}|$ and $k_\perp^2 \equiv k^2 - k_{\|}^2$. In Eq. (1) $W_i^N(W,Q^2,...)$ is the inclusive response function of a bound nucleon, which can be written as the sum of an elastic term and an inelastic one, viz.

$$W_i^N = W_i^{N(\mathrm{el})} + W_i^{N(\mathrm{inel})} \tag{2}$$

Following the $cc1$ prescription of ref. [4], the elastic part reads as follows

$$W_1^{N(\mathrm{el})} = \frac{\bar{Q}^2}{4M^2}\,\{F_1^N(Q^2)+F_2^N(Q^2)\}^2\ \delta(W-M)$$
$$W_2^{N(\mathrm{el})} = \{[F_1^N(Q^2)]^2 + \frac{\bar{Q}^2}{4M^2}[F_2^N(Q^2)]^2\}\ \delta(W-M) \tag{3}$$

where $F_{1(2)}^N(Q^2)$ are the Dirac and Pauli (free) nucleon form factors. As far as nucleon inelastic channels are concerned, the free nucleon structure function $W_i^{N(\mathrm{inel})}(W,Q^2)$, parametrized as in ref. [3], will be adopted.

The relevant nuclear quantity appearing in Eq. (1) is the nucleon spectral function $P^N(k,E)$, which represents the joint probability to find in a nucleus a nucleon with momentum k and removal energy E. In presence of ground-state NN correlations, the nucleon spectral function can be written as $P^N(k,E) = P_0^N(k,E) + P_1^N(k,E)$, where P_0^N includes the ground and one-hole states of the (A-1)-nucleon system and P_1^N more complex configurations (mainly 1p-2h states) which arise from 2p-2h excitations generated in the target ground state by NN correlations. For $A = 3$ and $A = 4$ one gets

$$P_0^N(k, E) = n_0(k)\ \delta(E - E_{min}) \tag{4}$$

where $n_0(k)$ is the nucleon momentum distribution corresponding to the ground-to-ground transition and $E_{min} \simeq 5.5,\ 20$ MeV, respectively. As for the correlated part P_1^N, since high values of k ($k > 1.5\ fm^{-1}$) and E ($E \sim (A-2)k^2/2(A-1)M$) are considered in this contribution, the extended two-nucleon correlation ($2NC$) model of ref. [5] is adopted, viz.

$$P_1^{N_1}(k, E) = \sum_{N_2=n,p} \int d\boldsymbol{k}_{CM}\ n_{rel}^{N_1N_2}(\boldsymbol{k} - \frac{\boldsymbol{k}_{CM}}{2})\ n_{CM}^{N_1N_2}(\boldsymbol{k}_{CM})$$
$$\delta[E - E_{thr}^{(2)} - \frac{A-2}{2M(A-1)}(\boldsymbol{k} - \frac{A-1}{A-2}\ \boldsymbol{k}_{CM})^2] \tag{5}$$

where $n_{rel}^{N_1N_2}$ ($n_{CM}^{N_1N_2}$) is the momentum distribution of the relative (center-of-mass (CM)) motion of the two nucleons in a correlated pair, and $E_{thr}^{(2)} = M_{A-2} + 2M - M_A$ is the two-nucleon break-up threshold. Eq. (5) quantitatively reproduces the high momentum and high removal energy parts of the many-body spectral function calculated for ^3He, ^4He and nuclear matter, as well as the momentum sum rule of complex nuclei (see for details ref. [5]). In case of the deuteron one has $P_1^N = 0$ and $P_0^N(k, E) = n(k)\ \delta(E - E_{min})$, where $n(k)$ is the nucleon momentum distribution in the deuteron and $E_{min} = 2.226$ MeV is the deuteron binding energy. It should be pointed out that a numerical investigation of the integration limits over k and E in Eq. (1) shows that the use of a non-relativistic spectral function is grounded for $x < 1.7$ (see also [5]).

Within the $2NC$ model nuclear effects in the inclusive process $^4He(e, e')X$ have been investigated, using the relative and CM momentum distributions of ref. [6]. The results obtained using in Eq. (1) the ground-to-ground spectral function P_0^N only (Eq. (4)) and the full spectral function $P_0^N + P_1^N$, which includes the effects of NN correlations (Eqs. (4-5)), are shown in Fig. 1. In order to illustrate the role played by binding effects, the results of the calculations performed by applying the closure approximation to the final nuclear states (i.e., by disregarding in Eq. (1) the E-dependence in the energy-conserving delta function), are reported in the same figure. It can be seen that at $x > 1$: i) the inclusive nuclear response is dominated by the process of virtual photon absorption on a pair of correlated nucleons both in case of QE and DIS processes; ii) binding corrections sharply affect the inclusive cross section, i.e. the knowledge of the momentum and removal energy behaviours of the nucleon spectral function is needed.

Let us now illustrate the sensitivity of the inclusive nuclear response to possible modifications of the short-range structure of the NN cluster absorbing the virtual photon. To this end, let us consider the possibility that the nucleons in a correlated pair can loose their identity at short separations, so that the

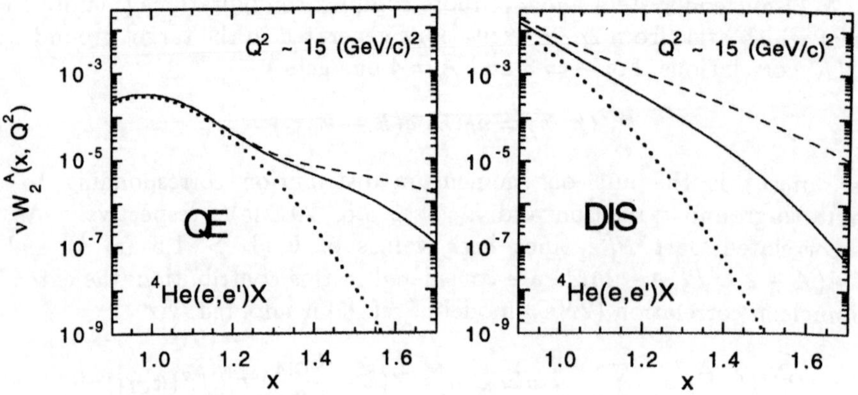

Figure 1. Nuclear structure function $\nu W_2^A(x, Q^2)$ for the inclusive process $^4\text{He}(e, e')X$ as a function of the Bjorken variable x at $Q^2 \sim 15$ (GeV/c)^2. The dotted and solid lines correspond to the calculations of the QE and DIS contributions performed using in Eq. (1) the ground-to-ground spectral function P_0^N only (Eq. (4)) and the full spectral function $P_0^N + P_1^N$ (Eqs. (4-5)), respectively. The dashed lines are the results obtained within the closure approximation for the final nuclear states.

incoming photon can interact with a six-quark ($6q$) bag structure. This is the mechanism proposed and applied to the investigation of inclusive DIS processes off the deuteron in refs. [7]. For $A > 2$ the effects from the Fermi motion of the CM of the $6q$ bag can be estimated by adopting the extended $2NC$ model, i.e. by using for the CM momentum distribution of a $6q$ bag the one of a correlated NN pair; this implies that the introduction of a $6q$ bag is assumed to modify only the intrinsic structure of a NN cluster at short separations (see for details [8]). Furthermore, a medium-dependent modification of the structure function of a nucleon bound in a nucleus has been proposed in [9], based on the observation that small size components in the nucleon wave function interact weakly with the surrounding medium. Thus, the valence quark distribution should be suppressed in a bound nucleon with respect to the free nucleon case by a factor, which is expected to depend on the nucleon momentum k [9]. The results of the calculations for the inclusive processes $^2\text{H}(e, e'X)$ and $^4\text{He}(e, e'X)$ are reported in Fig. 2 at $Q^2 \sim 15$ (GeV/c)^2. It can be seen that: i) at $x < 1.5$ the nuclear response could be sensitive to the medium-dependent modifications of the nucleon structure function, due to the point-like-configuration suppression of ref. [9]); ii) the kinematical regions corresponding to $x > 1.5$ appear to be appropriate for investigating the effects of the possible presence of multiquark cluster configurations at short NN separations; iii) the QE contamination could be relevant at large values of x.

In conclusion, inclusive electron scattering off few-nucleon systems has been investigated at $x > 1$ and high momentum transfer, showing that in such kinematical conditions the inclusive nuclear response is sensitive to the short-

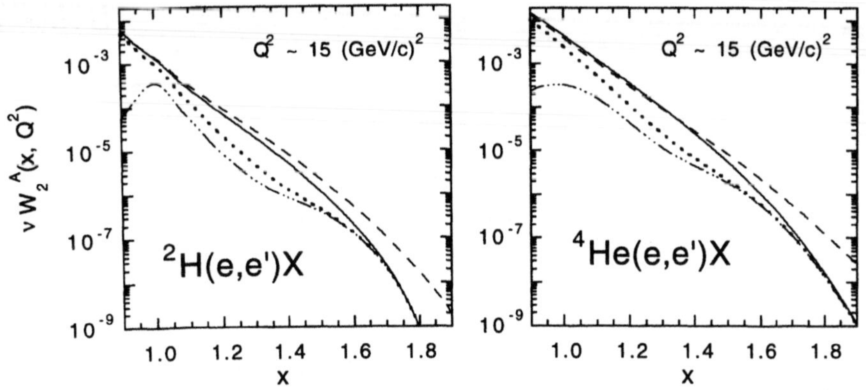

Figure 2. Nuclear structure function $\nu W_2^A(x, Q^2)$ for the inclusive processes ^{2}H$(e, e')X$ and ^{4}He$(e, e')X$ versus x at $Q^2 \sim 15$ (GeV/c)^2. The dot-dashed lines are the QE contribution, whereas the solid lines include the DIS contribution calculated using in Eq. (2) the free nucleon structure function of ref. [3]. The dotted lines correspond to the results obtained including in Eq. (2) the free nucleon structure function multiplied by the point-like-configuration suppression factor of ref. [9]. Finally, the dashed lines are the results obtained assuming a 5% (10%) probability of a $6q$ cluster admixture in ^{2}H (^{4}He) ground-state.

range structure of the two-nucleon cluster absorbing the virtual photon. It is therefore desirable that the investigation of inclusive $A(e, e')X$ processes at $x > 1$ will become part of the experimental program at existing and planned facilities (like, e.g., $CEBAF$, $HERA$ and $ELFE$).

References

1. D.B. Day et al.: Phys. Rev. **C48**, 1849 (1993)
2. C. Ciofi degli Atti, S. Simula: Phys. Lett. **325B**, 276 (1994)
3. L.W. Whitlow et al.: Phys. Lett. **282B**, 475 (1992)
4. T. de Forest: Nucl. Phys. **A392**, 232 (1983)
5. C. Ciofi degli Atti et al.: Phys. Rev. **C44**, R7 (1991); C. Ciofi degli Atti, S. Simula: Preprint INFN-ISS 95/4, 1995
6. H. Morita, Y. Akaishi, H. Tanaka: Prog. Theor. Phys. **79**, 863 (1988)
7. G. Yen et al.: Phys. Rev. **C42**, 1665 (1990)
8. S. Simula: to appear in Few Body Systems Suppl. **9**, (1995)
9. L.L. Frankfurt, M.I. Strikman: Phys. Rep. **160**, 235 (1988)

Few-Body Systems Suppl. 8, 428–432 (1995)

Few-Body Systems

Antiproton Helium Polarization States

J. Carbonell, F. Ciesielski, C. Gignoux

Institut des Sciences Nucléaires, 53, Av. des Martyrs, 38026 Grenoble, France

Abstract. A new family of long living antiprotonic states in He is presented. The origin of these states is the long range $\bar{p}He$ polarization potential which generates two-body $\bar{p}$–He and three-body $\bar{p}$ – He – He charged clusters

1 Introduction

One of the more interesting phenomena recently discovered is the so called antiproton delayed annihilation, i.e. the experimental observation that about 3% of $\bar{p}$ stopped in He have a lifetime of few microseconds. This phenomena, predicted by Condo [1, 2], has been first observed at KEK [3] and soon after confirmed at LEAR [4]. Up to now the only reliable explanation is the formation of a metastable neutral $(\alpha e^- \bar{p})$ state. This model is strongly supported by the observation of the isotopic effect and the laser induced transitions, both in close agreement with the predictions. It should be however mentioned that some questions remain open concerning the formation mechanism, the collisional stability and the Auger cascade [5, 6].

The observation of metastable antiprotonic states having an enhancement factor of $\sim 10^6$ with respect to their usual lifetime in matter raised the interest on exotic atoms. We have considered the different possibilities for an $\bar{p}$ to form a long living state in matter [7]. It turned out that the long range $\bar{p}$–atom polarization potential together with the heavy antiproton mass generates a series of two and three body states living in the tail of the potential. In the He case, these sates are of big size compared to the electronic density and we have calculated radiative decay times of the order of the μs. The main properties of these states will be presented in what follows.

2 The $\bar{p}$-He Interaction

The $\bar{p}$–He interaction has been approximated by means of a local potential $V_{\bar{p}\mathrm{He}} = V_p + V_s$ properly taking into account the short and long range part of the interaction. The long range part is due to the polarization forces V_p which create a potential having the asymptotic form

$$V_p(x) = -\frac{1}{2}\frac{\alpha_d}{x^4} \tag{1}$$

α_d being the atomic dipole polarizability ($\alpha_d = 1.38$ a.u. for He). This potential is regularized below some cutoff radius x_c by a $C^{(1)}$ polynomial matching that gives negligible contribution for $x << x_c$.

The short range part is described by the static approximation with the He ground state given by the Hylleraas variational wavefunction. It results into the following Coulomb screened potential (a.u.)

$$V_s(x) = -2\left(Z_e + \frac{1}{x}\right)e^{-2Z_e x} \tag{2}$$

with $Z_e = \dfrac{27}{16}$.

The range of the polarization forces is much greater than the range of other components of the interaction, including the inelastic channels (~ 1 a.u.). This fact is essential for the validity of our treatment which implicitely assumes process which are not sensitive to the details of the short range forces.

3 Two Body $\bar{p}$–He States

Let us first consider the $\bar{p}$-He spectrum generated by the polarization potential alone. The results, with $x_c = 2$ Å, are shown in Fig. 1. One can see a family of states with binding energies smaller than 0.2 eV, level spacing $\Delta E \approx 0.05$ eV and angular momentum up to $L_{\max} = 11$. These states are extended in comparison to the He size, wiht typical r.m.s radius of 2Å (for L=2 e.g. it varies from 1.26 Å up to 4.05 Å). Although their exact position is quite sensitive to the cutoff function, their existence is independent of the short range details of the interaction. When the static potential V_s is switched on, the spectrum is strongly perturbed and great number of deeply bound states appears. However a family of states, similar to those displayed in Fig. 1 remains. These states, called hereafter polarization sates, are generated by the long range part of the potential and their main properties are independent of V_s. In Table 1 are given the energies of the $\bar{p}$-He states (L=2) calculated respectively with V_s, V_p and $V_s + V_p$ potentials. The distinction between both types of states is clear. The polarization states in column V_p are still present with only small changes in their binding energies when V_s is added. The states generated by V_s alone have different energy and size scales and cannot be confidently described by the simplified form of the interaction. The main characteristics of the polarization states, small binding energies and great sizes, suggest to look for their possible metastability. On one hand their peripheral character gives small overlaping with the electronic density and can lead to a small Auger width. On the other hand the Stark mixing is expected to be weak due to the absence of L-degeneracy in a $1/x^4$ potential as can be seen in Fig. 1. The electromagnetic

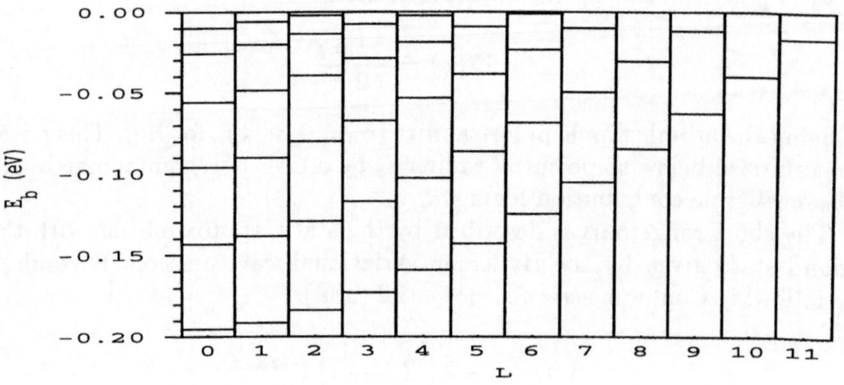

Figure 1. Spectrum of L=2 $\bar{p}$-He states with the polarization potential V_p ($x_c = 2\AA$)

Table 1. Energies (eV) for L=2 $\bar{p}$-He states calculated with the polarization V_p, static V_s and $V_p + V_s$ potentials; n is the corresponding radial quantum number

n	V_p	n	V_s	n	$V_p + V_s$
6	-0.0023			42	-0.0013
5	-0.0137			41	-0.0111
4	-0.0392			40	-0.0391
3	-0.0798	37	-0.0710	39	-0.0933
2	-0.1294			38	-0.1656
1	-0.1835			37	-0.2683
	-	36	-0.4020	36	-0.5621
		35	-1.0187	35	-1.1485
		34	-1.9388	34	-2.0458
		33	-3.1965	33	-3.2873
		...	...	...	...

dipole radiative transitions have been calculated [7] and we have found that several polarization states have a lifetime of the order of μs.

We would finally remark that the antiprotonic polarization states should also exist for other systems. On this regard He is only the less polarizable atom and was first considered due to the KEK results. It is interesting to consider the other extrem case, the Li atom, with a polarizability $\alpha_d = 165$ two orders of magnitude greater. The results are displayed in Table 2. The same conclusion holds but with different energy and size scales. The big sizes obtainied (up to 20 $\AA$) and the smallness of the binding energies ~ 0.001eV suggest an e.m. and

Table 2. Binding energies (eV) and r.m.s. radius (Å) for the $\bar{p}$-Li polarization states

L=0		L=5		L=11	
B	$< x >$	B	$< x >$	B	$< x >$
0.0002	22.2	0.0002	21.5	0.0007	15.1
0.0008	18.7	0.0009	17.3	0.0024	12.3
0.0018	15.1	0.0021	14.1	0.0054	10.4
0.0035	12.7	0.0042	11.9	0.0102	9.1
0.0062	11.0	0.0076	10.3	0.0172	8.1
0.0103	9.7	0.0126	9.1	0.0266	7.3
...		...		...	

Auger stability even greater than for He. They indicate also a great collisional unstability which will make difficult their experimental observation.

4 Three Body $\bar{p}$–He–He States

A similar situation occurs with the $\bar{p}$–^{4}He–^{4}He system. It is known that ^{4}He could form a loosely bound trimer with binding energy B≈ 0.1^{0}K. This structure is an extended (~ 10Å) three-body-like system, in which the He atom stays frozen in its ground state. The binding is ensured by the He-He Van der Waals forces, i.e. long range $1/x^6$ attraction due to dynamical polarization. When one of the He is replaced by an antiproton, the $\bar{p}$–^{4}He–^{4}He system is expected to form several bound and resonant states. Some of them could be metastable for the same reasons that in the two-body case above discussed. The first results concerning L=0 states are given in Table 3. They have been obtained by solving the Faddeev equations with the He-He potential [8]. The $\bar{p}$–He interaction is limited to the polarization potential V_p with cutoff radius varying from 1.75 to 3.0 Å. Column E_2 contains the corresponding $\bar{p}$–He binding energy, colmnn E_3 and E_3^* respectively the ground and first excited three body binding energies and $< x >$ is the mean value of the $\bar{p}$–He distance in the 3-body state. The units are eV and Å. Only the states which are below the corresponding two body threshold have been calculated.

5 Conclusion

We have presented a family of two-body like $\bar{p}$–He bound states created by the long range polarization forces. These states have angular momenta up to $L \sim 10$ and binding energies lying on a narrow band of 0.2 eV width. They have a radiative lifetime of the order of 1 μs and their spatial extension (4-8 a.u.) advocates a weak Auger transition.

This kind of states are also formed with other elements. The case of Li,

Table 3. Energies (eV) and r.m.s. radius (Å) for the L=0 ground (E_3) and first excited (E_3^*) He−$\bar{p}$−He states for several x_c values. E_2 are the $\bar{p}$−He binding energies

x_c	E_2	E_3	$< x >$	E_3^*	$< x >$
1.75	-0.3386	-0.3548	1.41	-0.3390	1.86
2.00	-0.1952	-0.2155	1.54	-0.1979	1.78
2.50	-0.0773	-0.0936	1.80	-0.0819	1.98
3.00	-0.0360	-0.0471	2.09	-0.0395	2.29

for instance, could exhibit a greater Auger and electromagnetic stability. Its polarizability, two orders of magnitude greater than for He, generates states which are consequently more extended, thus minimizing the overlaping with the electronic wavefunction.

The $\bar{p}$−He polarization states could play a role in the recently discovered delayed annihilation phenomena. On one hand by modifying the whole picture of low energy $\bar{p}$ interacting with media. On the other hand by forming two- or three-body metastable clusters.

To disentangle the possible mechanisms leading to metastable clusters we suggested [7] to look for delayed annihilation components using very low energy antiprotons. If the antiprotons are slowed down enough, the formation of states with high angular momentum, a key point in the Condo-Russell explanation, is strongly inhibited and the phenomena should not be observed.

References

1. G. T. Condo: Phys. Lett. **9**, 65 (1964)
2. J. E. Russell: Phys. Rev. Lett. **23**, 63 (1969); Phys. Rev. **188**, 187 (1969); Phys. Rev. **A1**, 721 (1970); Phys. Rev. **A1**, 742 (1970)
3. M. Iwasaki et al.: Phys. Rev. Lett. **67**, 1246 (1991)
4. A. Adamo et al.: Nucl. Phys. **A558**, 665 (1993)
   E. Widmann et al.: Nucl. Phys. **A558**, 679 (1993)
   J. Eades et al.: Hyperfine Interactions **81**, 227 (1993)
   T. Yamazaki: contribution to LEAP'94, September 12-17, Bled (1994)
5. See contibution of O.I. Kartavtsev to this conference
6. See contibution of O.D. Dalkarov to this conference
7. J. Carbonell, F. Ciesielski, C. Gignoux, A.Yu. Voronin: I.S.N. Prep. 95.23
8. R.A. Aziz and M.J. Slaman: J. Chem. Phys. **94**, 8047 (1991)

Few-Body Systems Suppl. 8, 433–437 (1995)

Few-Body Systems

Antiproton Delayed Annihilation: the Problem of the Description in Isolated Atom Model and Other Possible Scenarios

O.D. Dalkarov, A.Yu. Voronin

P.N. Lebedev Physics Institute, 53 Leninsky pr., 117924 Moscow, Russia

Abstract. The internal Auger de-excitation mechanism in the cluster $(e^- \bar{p}He^{++})$ is examined in terms of the coupled channels approach. An important feature of $\bar{p}He^+$ atom is that there are two types of three-body clusters with drastically different Auger widths. The significant importance of virtual cascade Auger mechanism is shown, especially in case of multielectron atoms. The lifetime of fraction of antiprotons, stopped in Li, is estimated to be no longer than 10^{-7} s.

1 Introduction

In recent experiments [1, 2] it was discovered that the small fraction of antiprotons, stopped in He (about 3%) has the lifetime of order of 10^{-6} s, which is million times longer than the usually observed lifetimes of antiprotons stopped in other gas or liquid media. The delayed fraction was observed only in He, while in the multielectron atoms like Ar, Ne no such effect has been detected.

Our theoretical examination of the phenomenon, presented here, is based on the Condo's concept [3] of longliving metastable states of the cluster $e^- \bar{p}$ He^{++}, which is formed in the reaction:

$$\mathrm{He} + \bar{p} \rightarrow (e^- \bar{p}\ \mathrm{He}^{++}) + e \qquad (1)$$

These states are characterized by high angular momentum of $\bar{p}$ ($N, L \sim 40$) and zero angular momentum of electron. It can be supposed that main de-excitation mechanisms (internal and exterior Auger transitions, Stark-mixing) from these highly excited states to lower states, in which annihilation takes place ($N, L \sim 33$) are very much suppressed, while the antiproton lifetime is determined by cascade of slow radiative transitions. The qualitative arguments in favor of this model were discussed in literature [4, 5, 6].

The main goal of present study is the understanding of the details of the internal Auger de-excitation mechanism in the isolated cluster $(e^- \bar{p}\ \mathrm{He}^{++})$.

We have taken into account the possibility of the antiproton de-excitation via cascade of virtual internal Auger transitions and found this mechanism to be of significant importance, especially in case of multielectron atoms [7]. We based our treatment on the coupled channels formalism (CCF), which enables the precise calculations of the metastable energy levels as well as the Auger-widths with controlled accuracy.

2 The Coupled Channels Formalism

We would like to develop the formalism, which would enable correct description of the three-body wave-function asymptotic behavior (similar to Faddeev's method [9]), and take into account the small ratio of the electron mass to the alpha-particle mass ($\frac{m_e}{M_\alpha} \sim 10^{-3}$). We will use here mainly the same approach, that was developed by one of the authors for the problem of $\bar{p}$-H interaction [8].

Two asymptotic clusters are important in our case. First is the cluster in which the electron is in the bound state in the field of α, while the $\bar{p}$ is far away. We will denote it as $(\alpha e)\bar{p}$. The three-body states in which the $\bar{p}$ is characterized by high N and L ($N, L > 40$), while electron is in 1s state, so that the mean radius of the $\bar{p}$ is greater than that of the electron, may be associated with this cluster. The second cluster is one in which the $\bar{p}$ is in the bound (excited) state in the field of α, while the electron is far away. We will denote it as $(\alpha\bar{p})e$. The three-body states in which the $\bar{p}$ mean radius is less than that of the electron correspond to this cluster, specially the Auger ionization states.

Two different Jacobi frames are closely related to mentioned above asymptotic clusters. The small ratio $\frac{m_e}{M_\alpha}$ enables us to use one Jacobi frame which is convenient for the description of both clusters. This is the frame, which explicitly contains the α-e coordinate and thus related in natural way to the asymptotic cluster $(\alpha e)\bar{p}$:

$$\boldsymbol{r} = \boldsymbol{r}_e - \boldsymbol{R}_\alpha$$
$$\boldsymbol{R} = \frac{m\boldsymbol{r}_e + M\boldsymbol{R}_\alpha}{m + M} - \boldsymbol{R}_{\bar{p}} \tag{2}$$

Here $\boldsymbol{r}_e$ is the coordinate of the electron, $\boldsymbol{R}_\alpha$ is the coordinate of α, $\boldsymbol{R}_{\bar{p}}$ is the coordinate of $\bar{p}$, m is the mass of the electron, M is the mass of α. This coordinate frame is convenient because the α-$\bar{p}$ coordinate, which appears in the wave-functions of the states, corresponding to the cluster $(\alpha\bar{p}\,)e$, agrees within the ratio $m/M \sim 10^{-3}$ with the coordinate $\boldsymbol{R}$ from the equation (2), thus we can substitute the α-$\bar{p}$ coordinate by $\boldsymbol{R}$ and take into account the difference between them in terms of the perturbation treatment. It is the coordinate system that we will be using below.

We would like to represent the three-body wave-function in the form, which explicitly contains information about the two asymptotic clusters, mentioned above. Thus we write the wave-function of the three-body system as a sum of two components:

$$\Phi(\boldsymbol{r},\boldsymbol{R}) = \Phi_1(\boldsymbol{r},\boldsymbol{R}) + \Phi_2(\boldsymbol{r},\boldsymbol{R}) \tag{3}$$

The component $\Phi_1(\boldsymbol{r},\boldsymbol{R})$ should be connected with the cluster $(\alpha e)\bar{p}$. We may express it in the form:

$$\Phi_1(\boldsymbol{r},\boldsymbol{R}) = \phi_{1s}(\boldsymbol{r})\chi(\boldsymbol{R}) \tag{4}$$

Here $\phi_{1s}(\boldsymbol{r})$ is the Coulomb wave-function of the 1s state of the electron in the field of α, $\chi(\boldsymbol{R})$ is the unknown antiproton function.

In order to explicitly take into account the asymptotic behavior of the cluster $(\alpha\bar{p})e$, it is convenient to expand the Φ_2 component in the complete set of the $(\alpha\bar{p})$ eigenfunctions:

$$\Phi_2(\boldsymbol{r},\boldsymbol{R}) = \sum_n g_n(\boldsymbol{r}) f_n(\boldsymbol{R}) \tag{5}$$

Here $f_n(\boldsymbol{R})$ are the Coulomb-type eigenfunctions of the pair $(\alpha\bar{p})$, $g_n(\boldsymbol{r})$ are unknown expansion coefficients, which represent electron wave-functions in the channels characterized by certain $(\alpha\bar{p})$ antiproton quantum numbers set n (the principal quantum number and the angular momentum).

The detailed description of the coupled channels equation system can be found in [7].

3 Virtual Cascade of Auger Transitions

Now we would like to perform the qualitative analysis to understand the nature of the suppression of the internal Auger de-excitation mechanism and find out which channels dominate the Auger-widths of metastable states and are the most important.

With idea of qualitative estimations, we treated the nondiagonal coupling potentials as perturbation.

In terms of this perturbation treatment one can easily see that the expression for metastable Auger-width $\Gamma(N,L)$ contains the product of $2\Delta L$ terms, where $\Delta L = L - \tilde{L}$, and thus can be expressed in the form [7]:

$$\Gamma(N,L) = \beta^{2\Delta L}(N,L)\Gamma_{\mathrm{atom}} \tag{6}$$

Here $\Gamma_{\mathrm{atom}} = 10^{12}\mathrm{s}^{-1}$ is some characteristic value of Auger-rate, which usually appears in atomic processes. The Auger-width for the state with $N = 44, L = 43$ was calculated and found to be of order of $10^1\mathrm{s}^{-1}$, while the dimensionless suppression factor $\beta^{2\Delta L}(N,L) \sim 10^{-11}$ ($\beta \sim 0.3$). Parameter β is some dimensionless mean factor, characterizing the suppression of Auger de-excitation per one virtual transition. This shows that the large number of virtual transitions ($\Delta L = 10$ in case of metastable state $N = 44, L = 43$) which the antiproton has to perform before the electron would be emitted, determines the extremely small Auger-widths of the highly excited metastable states.

While in the cluster ($e^-\bar{p}\,\mathrm{He}^{++}$) the Auger-widths of the highly excited states are much smaller than those of the radiative transitions ($\Gamma_{\mathrm{rad}} \sim 10^6\mathrm{s}^{-1}$), the virtual cascade de-excitation mechanism seems to play much more important role in multielectron atoms with $Z > 2$. In fact, using very crude model of noninteracting electrons and perturbation treatment, one can show that the antiproton interacts independently with $Z-1$ electrons of the atom (one electron is ionized during the cluster formation) and thus each virtual transition becomes $Z-1$ times more intensive, while the rate of the whole cascade increases $(Z-1)^{2\Delta L}$ times. In case of Li we will obtain million times enhancement of virtual cascade Auger de-excitation rate, while in atoms with $Z > 3$ this effect should absolutely cancel the delayed annihilation.

We should mention that the direct Auger transition from initial metastable state to the lower states in which the electron is emitted is much less probable than the virtual cascade mechanism at least for highly excited states $N > 39, L > 38$ (two orders of magnitude for $N = 40, L = 39$) in case of He and is negligible in case of multielectron atoms. In fact the enhancement of direct transition rates in multielectron atoms does not depend on ΔL and is only $Z-1$ times.

As it's seen from the results of calculations in the frame of CCF for $N > 36$ Auger widths (in He case) are larger than 10^{-6} (radiative lifetime). The antiproton state with N = 36 having time of life of order of 10^{-7} could be considered as candidate for so called "fast component" in the delayed annihilation spectrum ($\tau = 2 \times 10^{-7}$). We found that all states with $N < 36$ have lifetimes shorter than 10^{-7}s.

4 Interaction of $\bar{p}\,He^+$ Atom with Helium Media

We would like to mention that the understanding of the delayed annihilation phenomenon is possible only in connection with understanding of the stability of the longliving clusters due to interaction with media atoms. From this point of view we consider of special interest the following reactions:

$$(\bar{p}\,\mathrm{He}^+)_{N,L} + \mathrm{He} \to (\bar{p}\,\mathrm{He}^+)_{N',L'} + \mathrm{He} \tag{7}$$

$$(\bar{p}\,\mathrm{He}^+)_{N,L} + \mathrm{He} \to \mathrm{He} + (\bar{p}\,\mathrm{He}^+)_{N',L'} \tag{8}$$

Reaction (7) is the external Auger de-excitation, while reaction (8) is recombination process, in which Condo-type atom exchange by antiproton with He atom. Each of these mechanisms (still not examined in details) can yield in fast de-excitation and thus violating the assumption of Condo-type atom stability. Also it seems important to take into account the possibility of the formation of complicated molecules, including antiproton and several He ions [10] or He atoms [11].

5 Conclusion

In conclusion we would like to note that the virtual cascade de-excitation mechanism is of significant importance for understanding of the nature of long-living antiprotonic states in He. Taking into account this mechanism it is possible to understand why one docs not see any delayed annihilation in multielectron atoms ($Z > 3$) media. For Li we estimated that the fraction of antiprotons, stopped in Li, should live no longer than 10^{-7} s. The coupled channels approach formulated here, seems to be a convenient tool both for precise calculations as well as for qualitative analysis. The main benefit of this formalism is the possibility to describe the asymptotic behavior of the three-body wavefunction using one coordinate frame. This enables us to obtain simultaneously Auger widths and energy levels taking into account reasonably limited number of channels.

The authors would like to thank I. S. Shapiro for numerous useful discussions, T. Yamazaki and E. Zavattiini for experimental information and fruitful discussions.

References

1. M.Iwasaki et al.: Phys. Rev. Lett. **67** (1991) 1246
2. T.Yamazaki et al.: Nature **361**, 238 (1993)
3. G.T. Condo: Phys. Lett. **9**, 65 (1964)
4. J.E. Russel: Phys. Rev. Lett. **23**, 63 (1969); Phys. Rev. **188**, 187 (1969); Phys. Rev. **A 1**, 721 (1970); Phys. Rev. **A1**, 735 (1970); Phys. Rev. **A1**, 742 (1970)
5. T.Yamazaki, K.Ohtsuki: Phys. Rev. **A45**, 6202 (1992)
6. T.Yamazaki et al.: Phys. Rev. Lett. **72**, 1180 (1994)
7. O.D.Dalkarov, A.Yu.Voronin: in *Proceedings of the International School of Physics of Exotic Atoms*, INFN/AE-94/24; Preprint FIAN-32, 1994; Sov.Pisma JETP, **60**, 158 (1994)
8. A.Yu. Voronin: Sov. Phys. JETP **75**, 416 (1992)
9. L.D. Fadeev: Tr. Mat. Inst. Akad. Nauk SSSR **1**, 69 (1963)
10. L. Bracci, E. Zavattini: Phys. Rev. **A41**, 2532 (1990)
11. J.Carbonell, F.Ciesielski, C.Gignoux, A.Voronin: Preprint ISN-95.23 (1995)

Few-Body Systems Suppl. 8, 438–447 (1995)

Few-Body Systems

The Nijmegen Potentials

J.J. de Swart*, R.A.M.M. Klomp, M.C.M. Rentmeester, Th.A. Rijken

Institute for Theoretical Physics, University of Nijmegen, Nijmegen, The Netherlands

Abstract. A review is given of the various Nijmegen potentials. Special attention is given to some of the newest developments, such as the extended soft-core model, the high-quality potentials, and the Nijmegen optical potentials for NN.

1 Introduction

A large part of the efforts of the Nijmegen group in the last decennia has been concentrated on the study of the baryon-baryon [1, 2, 3], as well as the antibaryon-baryon interaction [4, 5]. In first instance this has been the construction of potentials [6, 7, 8, 9, 10], but later also partial-wave analyses (PWA) of the experimental scattering data [11, 12] were performed. The knowledge obtained in these PWA's was then applied again in the construction of new, improved potentials [13, 14, 5, 15]. This interplay between potential construction and PWA has has turned out to be very fruitful.

We considered extensively the baryon-baryon channels with strangeness S = 0, 1, and 2. The potentials we constructed for the (non-strange) NN-channels were in the beginning all "NN-potentials". We say here explicitly "NN-potentials". We mean with this something else then when we say "*pp*- or *np*-potentials". In NN-potentials charge-independence is assumed for the nuclear part of the potential. For the exchanged mesons and for the nucleons averaged iso-multiplet masses are used, such as the average pion mass $m = (2m_+ + m_0)/3 = 138.4$ MeV, the nucleon mass $M = (M_p + M_n)/2 = 938.93$ MeV, etc. The $I = 1$ part of the NN-potentials was always obtained by fitting the *pp*-data. Lately we have been constructing *pp*-, as well as *np*-potentials. In such potentials the mass differences are properly taken into account. In *pp*-scattering there is only π^0-exchange, while in *np*-scattering one must introduce π^+- as well as π^0-exchange.

In our efforts to describe and understand the inelasticity in *pp*-scattering above the various pion production thresholds, we have studied also the coupled

* *e-mail address:* swart@sci.kun.nl

$I = 1$ NN- and NΔ-channels.

In order to describe the elastic Λp scattering below, as well as above the Σ-production threshold, and the elastic and inelastic Σp scattering, we have been constructing hyperon-nucleon (YN-) potentials [2, 7, 10]. Charge independence breaking was partially taken into account by using the correct Σ-thresholds, and by introducing explicit $\pi\Lambda\Lambda$-couplings via the mechanism of $\Lambda\Sigma^0$-mixing. At various times also the $Y = 0$ potentials [16, 17, 18] such as $\Lambda\Lambda$, ΞN, $\Sigma\Sigma$, etc. were considered.

The baryon-antibaryon ($\overline{\mathrm{B}}$B) potentials [4, 15, 5] were constructed to describe the large amount of elastic $\bar{p}p$-scattering data, and the quasi-elastic $\bar{p}p \rightarrow \bar{n}n$ charge exchange data. Potentials [19, 20] were also constructed to describe the various strangeness exchange reactions, such as $\bar{p}p \rightarrow \bar{\Lambda}\Lambda$, etc.

All Nijmegen NN-potentials (except the HQ-potentials) were developed with in the back of our minds the extension to the YN-channels. This required treatments which could be generalized to the other channels with the help of SU(3). One needs therefore to include exchanges of all mesons of the same meson nonet. Next to the π-meson one needs to include the exchanges of η and η'-mesons. Next to the ρ- and ω-meson one needs to take account of the ϕ-meson.

The various Nijmegen potentials can be grouped into several classes. These are

- **Hard core potentials.**
  Important examples are the potentials Nijm D [6, 7] and Nijm F [8].

- **Soft core potentials.**
  We think here of Nijm78 [9] and its update Nijm93 [13].

- **Extended soft core potentials.**

- **High-Quality *pp*- and *np*-potentials.**
  We would like to call a potential a HQ-potential, when compared with the experimental data it has a $\chi^2/N_d < 1.05$. The only examples [13, 14] of this class are Nijm I, Nijm II, and Reid93.

- **Optical potentials.**

We would like to point out here the existence of the NN-OnLine facility of the Nijmegen group [21], which is accessible via the World-Wide Web. There one can obtain the various Nijmegen e-prints, the fortran codes for some of the Nijmegen potentials, the deuteron parameters and the deuteron wave functions, the phases obtained from the Nijmegen PWA and from the Nijmegen potentials, and predictions for many of the experimental quantities. A direct comparison of these predictions with the Nijmegen NN-data base is also possible.

2 Hard Core Potentials

The Nijmegen D potential [6] was one of our first hard core potentials that had an acceptable χ^2 with respect to the experimental NN-data. The extension [7] to the YN-channels did describe the YN-data well. Because of the succes of this potential in hypernuclear physics [22] we used this potential also for the construction of $\overline{\text{B}}$B-potentials [4, 19]. The elastic and charge exchange $\overline{p}p$-scattering data are well described in the Nijmegen coupled-channel model [4]. Our wish to study also the strangeness exchange reaction $\overline{p}p \longrightarrow \overline{\Lambda}\Lambda$ was the main reason for using the Nijmegen D model in these anti-particle reactions, because this Model D had been tested in the YN-channels. The Nijmegen soft-core potential [9] was then already available, but was not tested yet in YN.

It is very interesting to look at some of parameters of this, now 20 years old, Nijm D potential. The πNN-coupling constant was determined by fitting to the experimental NN-scattering data from before 1969, using the PWA of the Livermore group [23]. We obtained $f^2/4\pi = 0.074$. This must be compared with the present best value $f^2/4\pi = 0.0748$. This 20 year old result shows, that the recently obtained low value for the πNN-coupling constant is not due to recent experimental data, but results from our more sophisticated handling of the data. It was unfortunate that at that time in 1975, when we found this low value, we were so brainwashed by the πN-community in thinking that $f^2/4\pi \simeq 0.080$, that we did not take this result very seriously. It lasted till 1984 before we were convinced that the πNN-coupling constant was indeed so low [24].
This low value for the πNN-coupling constant resulted in a very good deuteron in Nijm D. For the d/s ratio we then found $\eta = 0.0251$. This must be compared with the present best value $\eta = 0.0252(1)$.
The value $\rho(-\varepsilon, -\varepsilon) = 1.776$ fm for the effective range at the deuteron pole must be compared with the present value $\rho(-\varepsilon, -\varepsilon) = 1.764(3)$ fm.
Other consequences of this all are: the pretty good values for the d-state probability $p_d = 5.9\%$, and the uncorrected value for the electric quadrupole moment $Q_0 = 0.272$ fm^2. These are in good agreement with the most recent guesses $p_d = 5.7(1)\%$ and $Q_0 = 0.271(1)$ fm^2.
Another Nijmegen hard core potential that is worth looking at, is the Nijm F model. This model has an NN as well as an YN-version. In fact even an YY-version [17] was constructed, but unfortunately never published. Recent calculations [25] show, that the YN-version of Model F reproduces many of the features in hypernuclear physics.

3 Soft Core Potentials

The Nijmegen soft-core OBE-potential Nijm78 [9] and its updated version Nijm93 [13] are based upon Regge pole theory. The corresponding YN-version [10] was published in 1989. This potential has also been applied to antibaryon-baryon scattering [12, 15, 20], where very good descriptions of the

various reactions, such as elastic scattering, charge exchange scattering, and strangeness exchange scattering, have been obtained.
One of the attractive features of this potential is that the coordinate space version and the momentum space version are exactly equivalent [26]. This at the cost of having only a minimal form of non-locality. In the triton this minimal non-locality has a 100 keV effect on the binding energy [14].

4 The Extended Soft Core Model

We next mention an important improvement on the soft-core OBE-model. Inspired by the chiral quark model, see for example [27], and duality [1, 28, 29], recently there has been constructed the extended soft-core model (ESC model) for the NN interaction. The first results with this ESC model were reported in Refs. [30, 31]. The ESC-model contains, besides soft OBE potentials of [9], also contributions from two-meson exchange diagrams ($\pi\pi$, $\pi\rho$, $\pi\varepsilon$, etc.) [32, 33], and from one-pair and two-pair diagrams. The latter are generated through pair-vertices ($\pi\pi$, $\pi\rho$, $\pi\varepsilon$ etc.) [34]. These meson-pair vertices are, except for a few, all fixed by heavy meson saturation. This way an excellent fit to the NN single energy PWA is achieved with a restricted set of free parameters. This model is still under construction. A preliminary fit is reached with $\chi^2_{p.d.p.} = 1.08$.

5 High Quality Potentials

In the various Nijmegen partial wave analyses [11] of the NN-scattering data we can describe these data with a $\chi^2/N_d \simeq 1.0$. This means that with a potential model description this will also be about the limit. A measure for the quality of potentials is therefore the difference with 1.0 for the value of χ^2/N_d. Let us define high quality (HQ) potentials as having $\chi^2/N_d < 1.05$ when compared directly with the experimental data. We constructed in Nijmegen three potentials of this HQ-type [13, 14]. These are the potentials Nijm I, Nijm II, and Reid93. They have all the excellent value $\chi^2/N_d = 1.03$.

Nijm I
The Nijmegen soft-core Nijmegen potential Nijm78 has been the starting point in the construction of the high quality NN-potentials. This soft-core potential gives already a reasonable good description of all partial waves. In each partial wave separately the description can be improved and made excellently, when we allow the parameters of the potential to be adjusted in each partial wave separately. This leads to the Nijm I potential. This potential has a minimal form of non-locality in the central potential only. There

$$V_c(r) = V(r) - \frac{1}{2m}(\Delta V' + V'\Delta)$$

Nijm II and Reid93
The Nijm II potential is similar to the Nijm I potential, but with all non-locality in each partial wave removed, i.e. $V' \equiv 0$. The Reid93 potential is an update

of the old Reid potential (see ref [13]). The singularities in this potential at the origin are removed by introducing formfactors. The main difference between the Nijm I and the Reid93 potential is just these formfactors. In the Nijmegen potentials an exponential form factor has been used

$$F(k^2) = \exp(-k^2/\Lambda^2),$$

while in the Reid93 potential we used

$$F(k^2) = (\Lambda^2/(\Lambda^2 + k^2))^2 .$$

The presence of form factors is due to the spatial extension of the nucleons and the mesons. Quarkmodels using harmonic oscillator interactions between the quarks will lead quite naturally to exponential form factors [35].

6 The Deuteron

It is instructive to compare the tensor potentials of the 3 HQ-potentials. In doing this we must keep in mind that the quality of the description of the NN-data is in all three cases essentially the same. In Fig. 1 we plot the tensor potential connecting the $^3S_1$ and $^3D_1$ partial waves. We see that in the inner region, $r < 1$ fm, these potentials are quite different. In Fig. 2 we plot the deuteron d-state wave function $w(r)$. It is remarkable that these wave functions are essentially the same, despite the fact that the tensor potentials are different. In Fig. 3 we plot the deuteron s-state wave function $u(r)$. We see again the agreement between these wave functions for large values of r but for values of $r < 0.6$ fm we spot some interesting differences.

There is a fantastic agreement between the values of the deuteron parameters as determined in our PWA's and the values for the same parameters as given by the HQ-potentials. These deuteron parameters are [36]:
the d/s ratio $\eta = 0.0252(1)$ and
the effective range at the deuteron pole $\rho(-\varepsilon, -\varepsilon) = 1.764(3)$ fm.

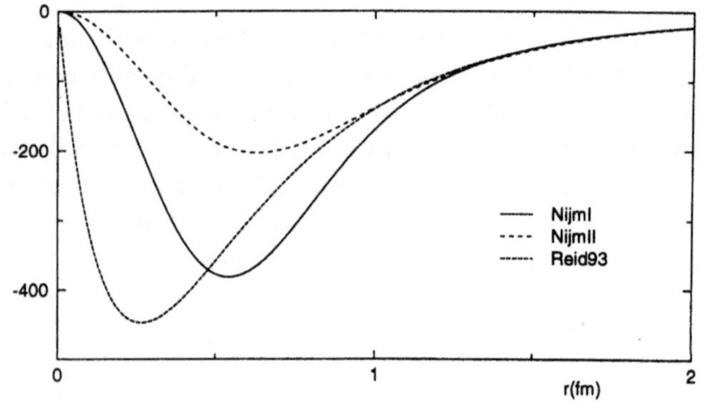

Figure 1. The tensor potential connecting the $^3S_1$ and $^3D_1$ partial waves.

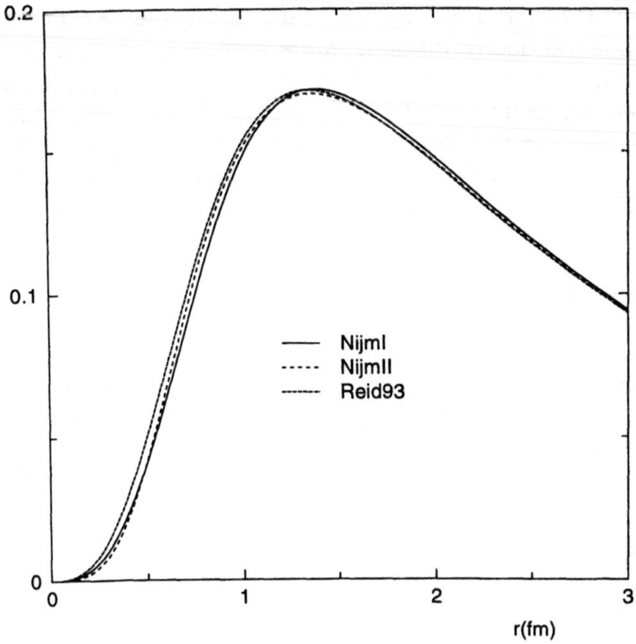

Figure 2. The deuteron d-state wave function $w(r)$.

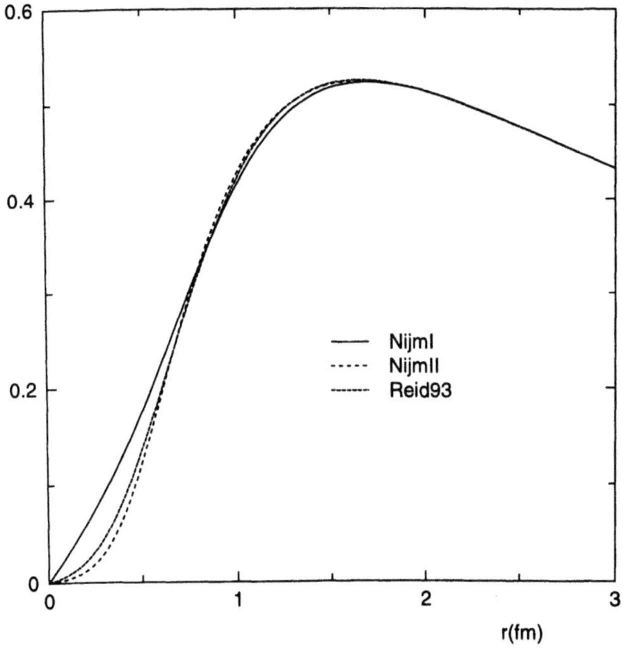

Figure 3. The deuteron s-state wave function $u(r)$.

These values follow directly from the scattering data, and can be considered as the best experimental determinations. More surprising is, that the value of the d-state probability $p_d = 5.7(1)\%$ and the value of the uncorrected electric quadrupole moment $Q_0 = 0.271(1)$ fm^2 are more or less unique. This value of Q_0 does not agree at all with the experimental value [37] $Q = 0.2859(3)$ fm^2. The difference $Q - Q_0 = 0.015$ fm^2 needs to be explained in terms of meson exchange currents, relativistic effects, contributions of Q^6-states, etc.

7 Optical Potentials

Below the threshold(s) for pion-production the NN-potentials are real. When one wants to describe the inelasticities in the scattering above these thresholds, then one has to go to either a complicated coupled channel description or one has to introduce an optical NN-potential: $V = V_R - iV_I$.

The influence and the importance of the imaginary part of the optical NN-potential can clearly be seen in our PWA [38] of the np-scattering data below $T_L = 500$ MeV. In Table 1 we give for various energy intervals the value of χ^2 obtained in our PWA, and the increase in the value of χ^2 when we omit the imaginary part of the potential.

Table 1. The values of χ^2 obtained in our PWA of the np data for various energy intervals. Also given are the number of data in that interval and the increase $\Delta\chi^2$ in the value of χ^2, when we omit the imaginary part of the potential.

Interval	# data	χ^2	$\Delta\chi^2$
0-350	2549	2519	11
350-400	254	306	13
400-450	319	337	105
450-500	866	905	651
0-500	3988	4067	880

Looking at this Table 1 we come to the conclusion that the use of a purely **real** potential works very well up to $T_L = 400$ MeV, it works well up to $T_L = 450$ MeV, and one can say that it works reasonably up to $T_L = 500$ MeV.

Let us next make from the real HQ-potentials optical potentials by adding to them the same imaginary part as was used in our PWA of the np-data below $T_L = 500$ MeV. When we compare now with the experimental np-data below 500 MeV we find for the Nijm I and Nijm II optical potentials $\chi^2/\#$ data $\simeq 2.4$ to 2.45. This is quite a lot larger than the minimum χ^2-values obtained in our PWA of these data. It turns out that especially the $^1F_3$ and $^1D_2$ partial waves are responsible for the big rise in χ^2.

We are looking into the possibility to construct optical potentials that are good up to 1 GeV. In Fig. 4 we give the $^1S_0$ and the $^1D_2$-phase shifts as

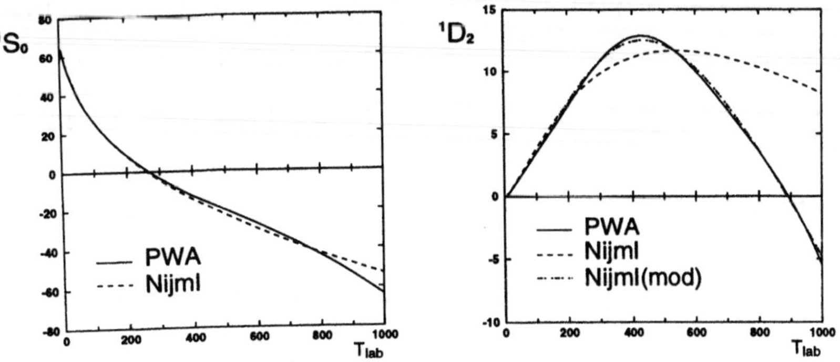

Figure 4. $^1S_0$ and $^1D_2$ phase shifts for the optical Nijm I potential and for the modified version.

determined in a preliminary PWA of all np-data below 1 GeV. In the same figures are also plotted the prediction of the optical Nijm I potential. For the $^1S_0$-phase the description is pretty good up to 1 GeV. For the $^1D_2$-wave one notices quite large differences. However, after refitting we get the modified Nijm I optical potential (Nijm I(mod)), which gives a very good fit to the $^1D_2$-phase shift up to 1 GeV. We think that it will be not too difficult to produce an optical potential which fit the np-data up to 1 GeV.

Acknowledgement. Part of this work was included in the research program of the Stichting voor Fundamenteel Onderzoek der Materie (F.O.M.) with financial support from the Nederlandse Organisatie voor Zuiver-Wetenschappelijk Onderzoek (N.W.O.).

References

1. J.J. de Swart, M.M. Nagels: Fortschritte der Physik **28**, 215 (1978)
2. J.J. de Swart, M.M. Nagels, T.A. Rijken, P.A. Verhoeven: Springer Tracts in Modern Physics **60**, 138 (1971)
3. J.J. de Swart, W.A. van der Sanden, W. Derks: Nucl. Phys. **A 416**, 2992 (1984)
4. P.H. Timmers, W.A. van der Sanden, J.J. de Swart: Phys. Rev. **D 29**, 1928 (1984); Erratum: **D 30**, 1995 (1984)
5. J.J. de Swart, R. Timmermans: *The antibaryon-baryon interactions*, Proceedings LEAP94, Bled, Slovenia
6. M.M. Nagels, T.A. Rijken, J.J. de Swart: Phys. Rev. **D 12**, 744 (1975)
7. M.M. Nagels, T.A. Rijken, J.J. de Swart: Phys. Rev. **D 15**, 2547 (1977)

8. M.M. Nagels, T.A. Rijken, J.J. de Swart: Phys. Rev. **D 20**, 1633 (1979)

9. M.M. Nagels, T.A. Rijken, J.J. de Swart: Phys. Rev. **D 17**, 768 (1978)

10. P.M.M. Maessen, Th.A. Rijken, J.J. de Swart: Phys. Rev. **C 40**, 2226 (1989)

11. V.G.J. Stoks, R.A.M. Klomp, M.C.M. Rentmeester, J.J. de Swart: Phys. Rev. **C 48**, 792 (1993)

12. R. Timmermans, Th.A. Rijken, J.J. de Swart: Phys. Rev. **C 50**, 48 (1994)

13. V.G.J. Stoks, R.A.M. Klomp, C.P.F. Terheggen, J.J. de Swart: Phys. Rev. **C 49**, 2950 (1994)

14. J.L. Friar, G.L. Payne, V.G.J. Stoks, J.J. de Swart: Phys. Lett. **B 311**, 4 (1993)

15. R. Timmermans, *Antiproton-proton scattering at LEAR energies*, Thesis University of Nijmegen, 1991

16. J.J. de Swart, Phys. Lett. **5**, 58 (1963)

17. W.M. Macek, M.M. Nagels, J.J. de Swart (1978) unpublished

18. P.M.M. Maessen, Th.A. Rijken, J.J. de Swart (1990) unpublished

19. P.H.A. Timmers, *Nucleon-antinucleon interaction*, Thesis University of Nijmegen, 1985

20. R.G.E. Timmermans, Th.A. Rijken, J.J. de Swart: Phys. Rev. **D 45**, 2288 (1992)

21. The URL is: `http://NN-OnLine.sci.kun.nl`

22. J. Rozynek, J. Dabrowski: Phys. Rev. **C 20**, 1612 (1979);
C.B. Dover, A. Gal: in *Progress in Particle and Nuclear Physics*, (ed. D. Wilkinson; Pergamon, Oxford, 1984), Vol. 12, pp. 17;
H. Bandõ, S. Nagata: Prog. Theor. Phys. **67**, 522 (1982)

23. M.H. MacGregor, R.A. Arndt, R.M. Wright: Phys. Rev. **182**, 1714 (1969)

24. J.J. de Swart, W.A. van der Sanden, W. Derks: Nucl. Phys. **A 416**, 2992 (1984)

25. Y. Yamamoto, T. Motoba, H. Himeno, K. Ikeda, S. Nagata: Progr. Theor. Phys. Suppl. **117**, 361 (1994)

26. Th.A. Rijken, R.A.M. Klomp, J.J. de Swart: *A Gift of Prophecy (R.E. Marshak Memorial Volume)*, World Scientific, 1995

27. J. Schwinger: Phys. Rev. **167**, 1432 (1968);
S. Weinberg: *ibid.* **166**, 1568 (1968) and **177**, 2604 (1969)

28. R. Dolen, D. Horn, C. Schmid: Phys. Rev. **166**, 1768 (1968);
H. Harari: Phys. Rev. Lett. **20**, 1395 (1968);
F.J. Gilman, H. Harari, Y. Zarmi: Phys. Rev. Lett. **21**, 323 (1968)

29. J.J. de Swart, Th.A. Rijken, P.M. Maessen, R.G. Timmermans: Nuovo Cim. **102A**, 203 (1989)

30. Th.A. Rijken, in *Proceedings of the XIVth European Conference on Few-Body Problems in Physics*, Amsterdam, 1993, edited by B. Bakker and R. von Dantzig, Few-Body Systems, Suppl. 7, 1 (1994)

31. V.G.J. Stoks, Th.A. Rijken: contributed paper to the *14th International Conference on Few-Body Problems, Williamsburg, 1994*, edited by F. Gross, p. 193

32. Th.A. Rijken: Ann. Phys. (N.Y.) **208**, 253 (1990).

33. Th.A. Rijken, V.G.J. Stoks: (submitted for publication)

34. Th.A. Rijken, V.G.J. Stoks: (submitted for publication)

35. F. Fernandez, E. Oset: Nucl. Phys. **A 455**, 720 (1986)

36. C.P.F. Terheggen, R.A.M.M. Klomp, M.C.M. Rentmeester, V.G.J. Stoks, J.J. de Swart (in preparation)

37. D.M. Bischop, L.M. Cheung: Phys. Rev. **A 20**, 381 (1979)

38. R.A.M.M. Klomp, J.J. de Swart (in preparation)

Few-Body Systems Suppl. 8, 448–459 (1995)

Few-Body Systems

The Bonn Potential

K. Holinde *

Institut für Kernphysik, Forschungszentrum Jülich GmbH, D-52425 Jülich, Germany

Abstract. The basic features of the Bonn meson exchange nucleon-nucleon interaction are outlined, and the implications of its nonlocal structure for the predicted triton binding energy are discussed.

1 Introduction

The potential V_{NN} between two nucleons is a quantity of central importance in nuclear physics. It not only serves to conveniently parametrize our empirical knowledge about low energy strong interaction dynamics in the NN sector but it also provides the basic input for calculations determining the properties of finite nuclei. Last but not least it is a useful concept in terms of which we can discuss the decisive question about the relevant degrees of freedom in this area of physics, i.e. baryons and mesons or quarks and gluons, and their respective region of validity.

In this talk I shall first outline the main features of the Bonn meson exchange nucleon-nucleon (NN) interaction, as described in detail in our Physics Reports article [1]. I will then discuss results for the triton binding energy obtained with various versions of the Bonn potential, and point out the importance of doing a consistent and sufficiently complete calculation and of keeping all predicted nonlocalities. Finally, I want to mention two recent extensions of the model: First, the inclusion of correlated $\pi\rho$ exchange so far missing; second, an explicit dynamical treatment of correlated 2π exchange, which should replace the sharp mass σ' and ρ exchange used before [1].

2 The Bonn NN Potential

The development of the Bonn potential started more than 30 years ago. It was initiated by Konrad Bleuler and, in its first simplified one-boson-exchange

* *E-mail address:* kph124@djukfa11.bitnet

(OBE) representation, guided by Klaus Erkelenz. He died much too early in 1973, but was well known in the European Few-Body Physics community from his progress reports (e.g. in Schladming) about our efforts to construct a realistic momentum space meson exchange NN interaction.

The Bonn potential as it should be understood nowadays goes beyond the OBE approximation. Its construction took about 10 years starting from the mid- seventies, which is documented by a series of papers, see ref. [1]. While it was originally designed to be a neutron-proton potential the extension to the proton-proton case has been done shortly afterwards [2].

Starting point for the formal development are baryon-meson vertex functions $W_{B'B\alpha}$ (see Fig. 1) involving not only nucleons but also Δ-isobars, and various mesons below 1 GeV. (Note that the fictitious σ meson, with a mass around 500 MeV, is not included). The analytic structure of these couplings is (essentially) determined by the quantum numbers of particles involved at the vertex. The strength is parametrized by coupling constants $g_{B'B\alpha}$; in addition form factors are included, which in principle depend on all four-momenta of the particles involved at the vertex, i.e. $F_{B'B\alpha} = F(q_{B'}^2, q_B^2, q_\alpha^2)$, and are normalized to 1 when all parameters are on the mass shell. I stress that these form factors represent a truly physical concept since they are meant to parametrize the extended structure of the core and meson cloud contributions at the vertex.

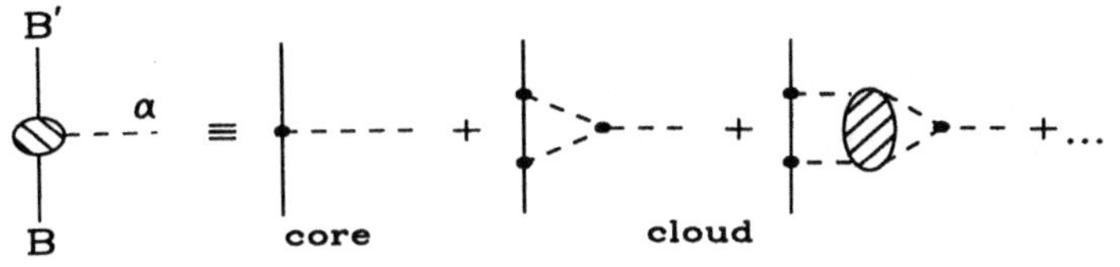

Figure 1. Baryon-meson vertex functions

These vertex functions should be considered as "elementary" building blocks for an effective theory of nuclear reactions. As such they represent a link between QCD and nuclear physics. Ultimately therefore, QCD should determine the values of the coupling constants and the structure of the various form factors. Corresponding calculations are just beginning to emerge. For the πNN vertex, with nucleons restricted to their mass shells, a recent QCD lattice calculation by Liu et al. [3] yields indeed a reasonable value for $g_{NN\pi} = 12.7 \pm 2.4$ and a monopole structure of the formfactor with the cutoff mass $\Lambda_{NN\pi} = 0.75 \pm 0.14$ GeV.

In the Bonn potential [1] all form factors $F_{B'B\alpha}$ are parametrized in a monopole or dipole form, $F_{B'B\alpha}(q_\alpha^2) = \left(\frac{\Lambda_{B'B\alpha}^2 - m_\alpha^2}{\Lambda_{B'B\alpha}^2 + q_\alpha^2}\right)^n$, n=1,2, with adjustable cutoff masses $\Lambda_{B'B\alpha}$. This procedure provides a total of about 10 free parameters (coupling constants and cutoff masses) which have been fixed by a fit to all NN data below pion production. These vertex functions should then be used consistently in other hadronic reactions in order to explore the limits of the meson exchange concept.

In a Hilbert space containing baryons as well as mesons a suitable Hamiltonian can be constructed which consists, apart from the free part, just of these vertex functions. In the (full) Bonn potential [1] it was treated in time-ordered perturbation theory yielding a two-nucleon scattering amplitude T_{NN}, which is obtained by means of a relativistic Schroedinger-equation with a suitably defined kernel (quasi-potential) (see Fig. 2) which is energy-dependent and due to its momentum space representation can keep all predicted nonlocalities. Apart from single meson exchanges the quasipotential contains higher order processes involving crossed two-meson exchanges and the Δ isobar in intermediate states. Correlated two-pion exchange (Fig. 2d) is approximated by sharp mass σ' and ρ exchange.

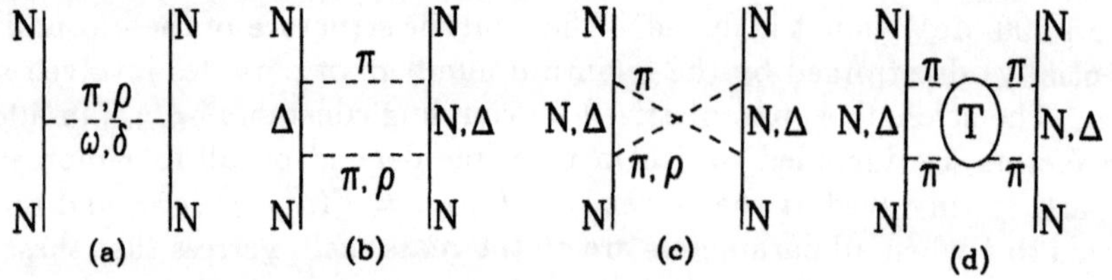

Figure 2. Processes included in the Bonn potential

Due to the suppression of the Δ width for example, the Bonn potential can only serve as a model below pion production. The resulting description of the NN data is good; the achieved χ^2 value per observable of about 2 [2] certainly proves that the meson exchange picture essentially works in the low energy NN system. Furthermore the Bonn potential appears to be a realistic starting point for finite nuclear calculations. Of course questions remain, e.g. about the true nature of ω exchange: The corresponding coupling constant used in the Bonn potential and needed to describe the empirical NN repulsion is about a factor of 2 larger than suggested from SU(3) estimates. Therefore it is well conceivable that ω exchange as used in the Bonn potential consists only partly of true ω exchange but parametrizes in addition different dynamics like e.g. explicit gluon exchange processes as we will hear about in the next talk by A. Faessler.

3 Bonn Potential and Triton Binding

In this section I would like to look at the implications of the Bonn potential for the triton binding energy, naturally an observable of great interest at this conference. In order to put things in a wider perspective I will however start with quite general remarks. Let us assume we have a Hamiltonian of the same type as leading to the Bonn potential, and we want to evaluate a three-body observable as e.g. the triton binding energy. Applying a specific expansion scheme to this Hamiltonian auxiliary quantities namely two- (V_2) and three (V_3)-body forces, can be defined. An incomplete calculation based on V_2 only yields a certain result for the observable, which, in a complete calculation involving also

V_3, will be modified to acquire a certain final value.

The point now is that starting from the same Hamiltonian another expansion scheme can be chosen, which leads to alternative potentials, whose amount and structure, especially their off-shell behaviour, can differ drastically from those obtained before. Likewise an incomplete calculation based on V_2 only can lead to quite a different observable value. Only when a complete calculation is done the same value is obtained as in the first expansion scheme, as it must be since in both cases we start from the same Hamiltonian. Therefore there is no point in testing different off-shell behaviour arising in potentials based on the same Hamiltonian since this is an artifact of using different expansion schemes. Of course the situation is different with potentials generated from different Hamiltonians.

So the message is that potentials depend on the calculational scheme whereas results for observables are scheme-independent provided two- and three- body forces are taken into account up to sufficiently high order. Then a choice of a specific scheme might be more advisable in case it leads to a faster convergence to the final result. In any case, a consistent calculation is required; it would be desastrous to mix two- and three-body potentials derived in different schemes.

Let me give a specific example for illustration. Suppose we start from a Hamiltonian involving vertex functions with nucleons and mesons only. In time-ordered perturbation theory and restricting ourselves to OBE-terms in the two-body potential we obtain an energy-dependent quasi-potential with an exceptionally low deuteron D-state probability of 4.27% (version OBEPT in ref. [1]). Despite the small D-state probability a Faddeev-type calculation [4] based on this potential yields only 6.73 MeV binding, almost 2 MeV below the empirical value of 8.48 MeV. (The reason is simply that due to its energy dependence the potential gets suppressed for the energies relevant in the triton). Alternatively, in another expansion of essentially the same Hamiltonian based on so-called Folded Diagrams we arrive at an energy-independent (instantaneous) OBE-potential (called OBEPF [5]) with both the D-state probability (5.66%) and triton binding (7.83 MeV) in a more conventional region. I want to stress that these large differences obtained with these potentials have nothing to do with slight differences in the description of the empirical NN data and differences in the parameter values but merely reflect the use of completely different calculational schemes. Also, the rule that potentials with a low deuteron D-state probability provide more binding is not general; it only holds when the comparison is made for potentials derived in (essentially) the same scheme. This fact once more shows that the D-state probability is not an observable as has been stressed many times in the past.

If the observed difference in the binding energy is due to the use of different schemes it should become reduced by a more complete calculation. Indeed the inclusion of three-body forces involving nucleons only fulfils this goal: In the time-ordered scheme (Fig. 3a) this contribution is known to provide large additional binding (of about 1 MeV) [6] leading then to almost 8 MeV binding. On the other hand, in the instantaneous scheme of Folded Diagrams (Fig.

3b), strong cancellations occur within this type of three-body forces [7], which leave the former result essentially unaltered. Thus, by doing a more complete calculation i.e. by taking a certain class of three-body forces into account, the results derived in different expansion schemes strongly approach each other.

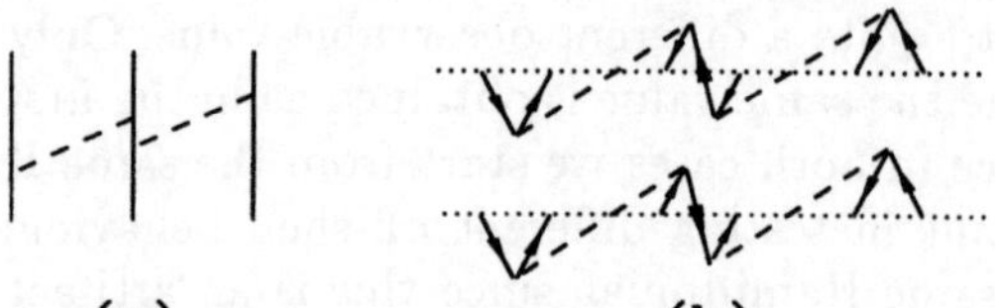

Figure 3. Three-body forces arising in a time-ordered expansion scheme (a) and in an instantaneous approach based on folded diagrams (b).

There are two basic lessons to be learned from this example: First, it is absolutely essential to treat 2- and 3-body forces consistently in the scheme chosen. The amount of 3-body forces evaluated in one scheme can be completely different from that derived in another scheme. Second, the instantaneous scheme is obviously more efficient, i.e. converges much faster, a conclusion which not only holds for this specific example but seems to be generally true for bound nuclear systems [7].

There remains still a discrepancy of about half an MeV compared to the empirical triton binding, which is hopefully removed by the inclusion of further three-body forces involving e.g. the Δ-isobar. This however cannot be done with the present simple Hamiltonian, in which the Δ does not occur. Therefore we have to go back to the Hamiltonian of the (full) Bonn potential treating it however in the instantaneous Folded Diagram representation in order to improve the efficiency. Such a version ("FULLF") has been constructed by Haidenbauer et al [5]. The corresponding interaction has a deuteron D-state probability of 5.22% and yields, in a calculation comparable to that done with OBEPF, a similar triton binding of 7.86 MeV. Surely, the next step is to include 3-body forces involving the nucleon as well as the Δ-isobar, which can and should be done consistently in this expansion scheme using the Hamiltonian of the (full) Bonn potential.

The triton binding energy results for OBEPF and FULLF appear to be somewhat larger (by about 0.2 MeV or so) compared to essentially local potentials constructed e.g. by the Nijmegen group [8]. Since almost all versions of the Bonn potential stay in the original momentum space representation and are thus able to keep the prescribed nonlocalities the additional binding energy is generally been attributed to the extended nonlocal structure. The effect seems to be especially drastic for the so-called Bonn-B potential [9] which provides a sufficiently realistic description of the NN system, with a D-state probability of 4.99%, and a triton binding of 8.14 MeV [10]. This interaction is again a simple OBE-potential based on a still different calculational scheme, namely a 3-dimensional reduction of the Bethe-Salpeter equation [11] proposed by Blankenbecler and Sugar [12]. In this OBE-approximation, the interaction is likewise instantaneous. (It becomes however energy-dependent when one goes beyond the OBE-approximation, like in the (full) Bonn potential).

It turns out that part of the difference between the triton binding (about 0.3 MeV) provided by Bonn-B on one hand and the Folded Diagram potentials FULLF, OBEPF on the other hand arises from differences in the NN phase shift fits (with Bonn-B being superior in the partial waves relevant for the triton). At least 0.1 MeV remains however, which can be traced to the different nonlocal structure of the corresponding potentials.

4 Structure and Role of Nonlocalities

In this section we want to discuss the structure of the nonlocalities present in momentum space potentials and their role for the resulting triton binding energy. We will do this mainly for the Bonn-B potential, which, as said before, is based on the Blankenbecler-Sugar (BbS) [12] reduction scheme. Therefore, the starting point is the 4-dimensional Bethe-Salpeter equation (BS) [11]

$$T = K + \int d^4k\ KGT \ . \tag{1}$$

with an (energy-dependent) kernel consisting of all irreducible Feynman diagrams. In order to simplify this equation various reduction schemes introduce modified propagators g which lead to a 3-dimensional equation. Surely, such a procedure modifies the scattering amplitude, which however can be restored by adding suitable correction terms to the BS-kernel, i.e. we have

$$\begin{aligned} T &= V + VgT \\ V &= K + K(G-g)K + \dots \end{aligned}$$

$\simeq$

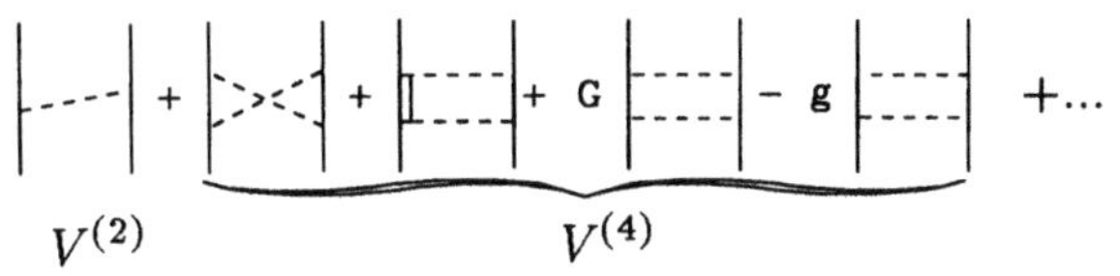

(2)

In case of the BbS-reduction, energy components of (relative) four-momenta are restricted to be zero (whereas they are integrated over in the original BS-equation). This leads to the 3-dimensional equation

$$T(\boldsymbol{k}', \boldsymbol{k}; E) = V(\boldsymbol{k}', \boldsymbol{k}; E) + \int d^3 k''\, V(\boldsymbol{k}', \boldsymbol{k}''; E) \frac{M}{E_{k''}} \frac{1}{E^2 - E^2_{k''}} T(\boldsymbol{k}'', \boldsymbol{k}; E) \ , \tag{3}$$

which after a suitable transformation of amplitudes

$$\tilde{V}(\boldsymbol{k}', \boldsymbol{k}; E) \equiv \sqrt{\frac{M}{E_{k'}}} V(\boldsymbol{k}', \boldsymbol{k}; E) \sqrt{\frac{M}{E_k}} \tag{4}$$

and the same for T acquires the usual Schroedinger form, with $E_k = \sqrt{M^2 + \boldsymbol{k}^2}$ and M the nucleon mass.

From the above it is immediately clear that the nonlocality structure of the original Bethe-Salpeter kernel is modified by such reductions; in addition, the modifications should depend on the kind of reduction chosen. Certainly, in order to remove such ambiguities, systematic corrections have to be done by going beyond the OBE-approximation, as indicated in Eqs. (1) and (2).

In the following I would like to divide the nonlocalities into standard ones which are essentially scheme-independent, and those called non-standard which depend on the reduction scheme. In the first group we have e.g. the m/E factors, which always arise in order to guarantee relativistic unitarity. They disappear when we do a nonrelativistic approximation neglecting 3-momenta versus the nucleon mass. More important is the nonlocal structure of the πNN vertex. Choosing the $i\gamma^5$-coupling in the Hamiltonian, we obtain for the nucleon matrix element

$$\overline{u}(\boldsymbol{k}')\gamma^5 u(\boldsymbol{k}) = \sqrt{\frac{(E_{k'} + M)(E_k + M)}{4M^2}}\,\boldsymbol{\sigma}\cdot\Big(\frac{\boldsymbol{k}'}{E_{k'} + M} - \frac{\boldsymbol{k}}{E_k + M}\Big)\,, \qquad (5)$$

which only after an on-shell approximation $E_{k'} = E_k$ boils down to the well-known local expression $\frac{1}{2M}\boldsymbol{\sigma}\cdot(\boldsymbol{k}' - \boldsymbol{k})$. Similar standard nonlocalities due to the structure of the Dirac spinors occur at the other ρNN, ωNN,... vertices.

Non-standard nonlocalities arise in potential terms which depend on the energy components of four-momenta. The best example is the meson propagator, which in the original Feynman amplitude can be written as

$$\frac{1}{(k_0' - k_0)^2 - (\boldsymbol{k}' - \boldsymbol{k})^2 - m^2}\,, \qquad (6)$$

with energy components completely unrelated to corresponding 3-momenta. The BbS-reduction yields the well known local expression whereas in another reduction scheme advocated by Franz Gross [13] a nonlocal correction term appears. Another example is the ρNN tensor coupling involving $\sigma^{\mu\nu}(k' - k)_\nu$ which after a BbS reduction goes into $\sigma^{\mu i}(\boldsymbol{k}' - \boldsymbol{k})_i$ whereas the Gross-scheme implies an additional nonlocal term $\sigma^{\mu 0}(E_{k'} - E_k)$.

These examples show that there are numerous nonlocalities with prescribed structure which can (and should) be kept in momentum space representations. Clearly, the aforementioned nonrelativistic and on-shell approximations are a priori not justified even for extremely low-energy NN scattering since the 3-momenta are integrated over in the scattering equation. The question then arises how important are such nonlocalities for the evaluation of observables.

We first want to look at the actual role of these nonlocalities in NN scattering. (The following calculations have been done by J. Haidenbauer at our institute). Fig. 4 shows the $^3S_1$ phase shift and ϵ_1 mixing parameter, which have strong impact on the triton binding. Obviously the removement of the nonlocal structure at the πNN vertex together with the neglection of m/E factors in the BONN-B potential provides a drastic change. Of course, by readjusting suitable parameters a description with roughly the same quality can be again obtained within this local approximation. Thus, if the interaction is considered to be purely phenomenological, this approximation does not seem to matter at all, at least in the NN system. However, the parameters in meson exchange NN interactions (coupling constants and cutoff masses) are physical and will be ultimately fixed from QCD. Therefore it indeed matters already at this stage whether an approximation is done which forces you to change parameters.

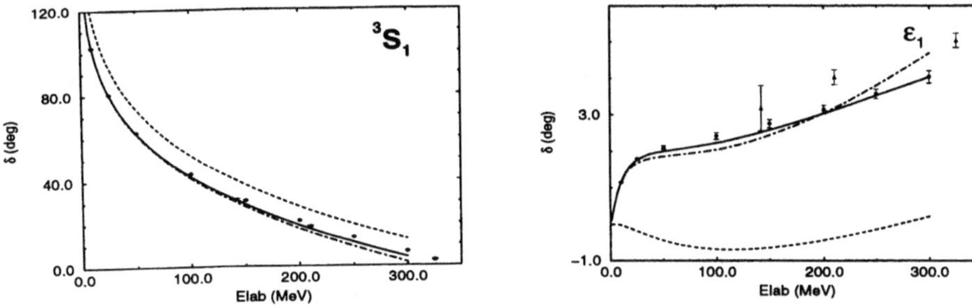

Figure 4. $^3S_1$ NN phase shift and ϵ_1 mixing parameter. The solid curve denotes the predictions of the Bonn-B potential. For the dashed curve, the local approximation (5) has been applied at the πNN vertex, together with the suppression of all m/E factors. With the same approximation a readjustment of parameters leads to the dash-dot curves. The error bars are from the empirical analyses of Nijmegen [14] and Bugg and Bryan [15].

Even after a refit of the S-wave phase shifts, the ϵ_1 mixing parameter and the deuteron observables like binding energy, quadrupole moment etc. consequences of the local approximation remain in the deuteron D-wave as shown in Fig. 5. Compared to the original Bonn-B wave function a sizeable increase occurs which ranges out well above 2 fm leading to a D-state probability of 5.45 % and a corresponding decrease in the triton binding of 0.36 MeV. Further nonlocalities occur; their combined effect on the triton binding seems to be small since Gibson et al. [16] found a total increase of 0.30 MeV obtained by comparing the original Bonn-B potential result with a strictly local but phase-equivalent version obtained by inversion methods.

However one should realize that part of these effects are scheme-dependent (i.e. are generated by non-standard nonlocalities), and thus might change when systematic corrections beyond the OBE-approximation will be applied. Indeed, if in the original Bonn-B potential the tensor coupling of ρ exchange is modified according to the Gross prescription [13] as outlined before the resulting different nonlocal structure leads, after a suitable refit of NN observables, to a sizable modification of the deuteron D-wave shown in Fig. 5, with a new value for p_D of 5.36 % and a reduction of the triton binding by 0.18 MeV.

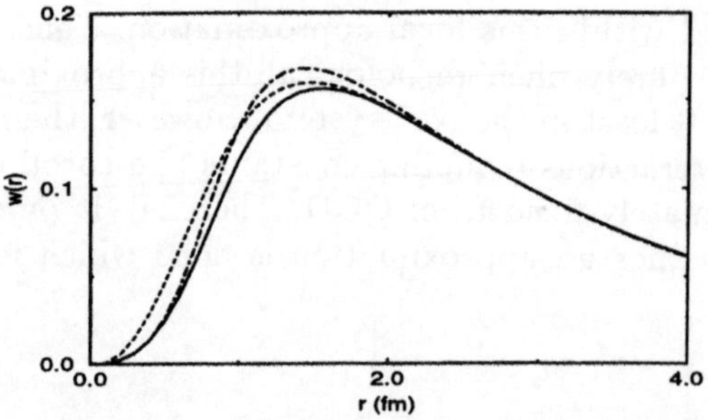

Figure 5. Deuteron D-wave. The solid curve denotes the result of the Bonn-B potential. The dash-dot curve corresponds to those of Fig. 4. For the short-dashed curve, the ρ-meson tensor vertex coupling is evaluated according to the Gross reduction scheme [13] (see text), again with a suitable readjustment of parameters.

5 Recent Improvements in the Dynamical Input

Recently our group in Juelich has performed several extensions in the dynamical input of the (full) Bonn potential, which due to limited space and time however I cannot discuss in detail.

5.1 *Correlated $\pi\rho$ exchange*

There is a strong cancellation, over a relatively broad range of energies and distances, between π and ρ exchange in the tensor channel. Therefore it appears mandatory to always group π and ρ together in order to reach sufficient convergence in the expansion of the irreducible kernel (quasi-potential). This procedure has been an essential guideline when constructing the Bonn potential. However, in ref. [1], it was not followed to a sufficient degree: Whereas, in second order diagrams, $\pi\pi$ as well as $\pi\rho$ exchange have been included for uncorrelated processes (with N and Δ intermediate states) this has not been done for correlated processes; correlated 2π exchange processes have been included (in terms of sharp mass σ' and ρ exchange) but correlated $\pi\rho$ processes have been left out. Recently [18] we have evaluated the contribution to the NN interaction due to correlated $\pi\rho$ exchange by means of dispersion-theoretic methods (quite analogously to the $\pi\pi$ case) based likewise on a realistic meson exchange model for the interaction between a π and a ρ meson [19]. This process has a substantial effect; it provides sizable tensor force which is able to counterbalance the substantial suppression induced in the one-pion exchange potential when going from a πNN monopole cutoff mass of 1.3 GeV (phenomenologically required in the (full) Bonn potential) to a smaller value of about 1.0 GeV, which is more in line with evidence from other sources [17] that this form factor should be soft. I should mention that, with this contribution included, there is absolutely no problem to obtain the correct value for the deuteron quadrupole moment using the lower value of the πNN coupling constant advocated by the

Nijmegen group.

This correlated $\pi\rho$ exchange piece contributes also in the ω- and A_1-meson channel; its inclusion leads to a strong reduction of the required ωNN coupling (by about 30 %), to a value more in line with SU(3) estimates.

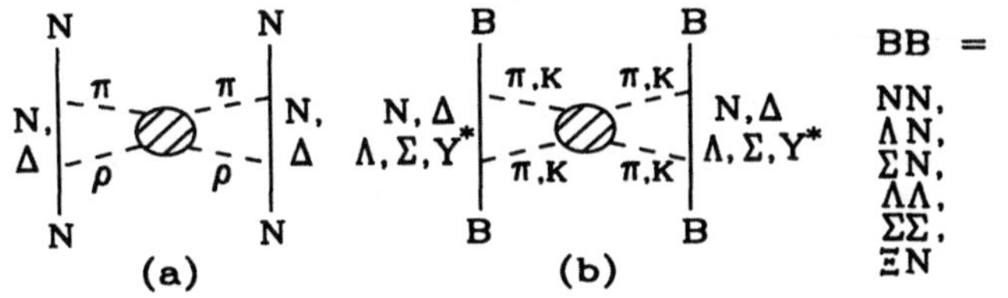

Figure 6. Correlated $\pi\rho$ exchange (a) and correlated 2π plus $K\overline{K}$ exchange (b).

5.2 *Correlated $\pi\pi$ exchange*

So far, in the (full) Bonn potential [1], correlated 2π exchange processes (for $J^P = 0^+$ and 1^- in the t-channel) are included in a simplified way, parametrizing them by σ' and ρ exchange with a sharp mass of 550 and 769 MeV, respectively. While possibly adequate for free NN scattering medium modifications (e.g. change of selfenergies of the intermediate pions) of such processes cannot be considered then.

Therefore we [20] have developed an explicit dynamical model for the correlated 2π as well as $K\overline{K}$ processes of Fig. 6b. Corresponding meson-meson correlations have been taken into account using a meson exchange coupled channel model of $\pi\pi$ and $K\overline{K}$ exchange [21].

6 Summary and Outlook

Basic ingredients of the Bonn potential are vertex functions involving nucleons and Δ-isobars, together with mesons of masses below 1 GeV. These vertex functions represent a link between the fundamental theory of strong interactions, QCD, and nuclear physics, with about 10 physical parameters (coupling constants and cutoff masses), which are adjusted to empirical NN data below pion production. A good description of those data is obtained, which demonstrates that the meson exchange concept essentially works in low-energy NN scattering, and that the Bonn potential seems to be a suitable starting point for realistic nuclear calculations.

While originally the Bonn potential has been derived from the vertex functions using time-ordered perturbation theory this scheme has proven to be less efficient for the evaluation of few- and many-body observables. Therefore an alternative instantaneous representation of the (full) Bonn potential in terms of Folded Diagrams has been constructed, which yields a triton binding energy of about 7.9 MeV, in the conventional region though somewhat larger than obtained with essentially local potentials.

We have demonstrated (using the Bonn-B OBE-potential) that such an increase is due to additional nonlocalities, which are prescribed by the input dynamics and, due to the momentum space representation, can be kept completely. Part of the nonlocalities however depend on the calculational scheme chosen. In order to reduce the resulting ambiguities the OBE-approximation should be avoided by keeping the higher dynamical level reached already in the (full) Bonn potential. This is anyhow required for a consistent calculation of 2- and important 3-body forces involving the Δ-isobar.

Let me end by making some remarks about the various strategies of dealing with the two- and 3-body problem, which are currently followed in our community: The first one is to avoid nonlocalities altogether by starting from a simple local NN potential. With sufficiently many parameters the fit of the NN data is done as best as possible. Such potentials underbind the triton by almost 1 MeV, and a suitable phenomenological 3-body force is added by hand to account for the empirical value of the binding energy. Such a procedure works by definition, and within this strategy the best potential is that which provides the lowest χ^2-value to the NN data.

The second strategy puts the emphasis on the nonlocal structure implied by the underlying dynamics and tries to avoid 3-body forces completely. This requires to exploit scheme-dependent ambiguities (unavoidably present in an OBE restriction) by choosing a specific scheme wich implies a nonlocal structure yielding as much binding as possible. In case additional binding is needed further nonlocalities generated e.g. by additional quark-gluon degrees of freedom have then to be identified and incorporated. Here the best potential is that which provides most binding and in this way comes nearest to the empirical value without the introduction of 3-body forces.

In view of my presentation of the whole subject it should come with no surprise that I advocate a third strategy, which is in a certain sense inbetween the former two and concentrates not so much on reproducing the empirical triton binding but on testing the assumed dynamical input. This means one should take both the nonlocalities and the involved 3-body forces implied by the dynamical input seriously. Ambiguities due to the choice of a specific calculational scheme should be avoided by doing a consistent and sufficiently complete calculation, which surely requires to go beyond the one-boson-exchange approximation. An efficient scheme should be chosen (which probably requires instantaneous interactions at all orders) in order to reach sufficient convergence. A comparison of the resulting "theoretical" value for the triton binding (and any other 3-body observable) with the empirical value will then provide a test of the input dynamics. Only in this way we can hope to learn something about the limits of the conventional meson exchange concept and possible indications of the relevance of other (quark-gluon) degrees of freedom, which should be our ultimate goal.

References

1. R. Machleidt, K. Holinde, Ch. Elster: Phys. Rep. **149**, 1 (1987)
2. J. Haidenbauer, K. Holinde: Phys. Rev. C **40**, 2465 (1989)
3. K.F. Liu, S.J. Dong, T. Draper, W. Wilcox: Phys. Rev. Lett. **74**, 2172 (1995)
4. R.A. Brandenburg, G.S. Chulick, R. Machleidt, A. Picklesimer, R. Thaler: Phys. Rev. C **38**, 1397 (1988)
5. J. Haidenbauer, K. Holinde, M.B. Johnson: Phys. Rev. C **45**, 2055 (1992)
6. C. Pask: Phys. Lett **25 B**, 78 (1967)
7. M.B. Johnson, J. Haidenbauer, K. Holinde: Phys. Rev. C **42**, 1878 (1990)
8. V.G.J. Stoks, R.A.M. Klomp, C.P.F. Terheggen, J.J. de Swart: Phys. Rev. C **49**, 2950 (1994)
9. R. Machleidt: Adv. Nucl. Phys. **19**, 189 (1989)
10. H. Witała, W. Glöckle, H. Kamada: Phys. Rev. C **43**, 1619 (1991)
11. H.A. Bethe, E.E. Salpeter: Phys. Rev. **84**, 1232 (1951)
12. R. Blankenbecler, R. Sugar: Phys. Rev. **142**, 1051 (1966)
13. F. Gross: Phys. Rev. **186**, 1448 (1969)
14. V.G.J. Stoks, R.A.M. Klomp, M.C.M. Rentmeester, J.J. de Swart: Phys. Rev. C **48**, 792 (1993)
15. D.V. Bugg, R.A. Bryan: Nucl. Phys. **A540**, 449 (1992)
16. B.F. Gibson, H. Kohlhoff, H.V. von Geramb: Phys. Rev. C **51**, R465 (1995)
17. S.A. Coon, M.D. Scadron: Phys. Rev. C **23**, 1150 (1981); Phys. Rev. D **42**, 2256 (1990)
18. G. Janssen, K. Holinde, J. Speth: Phys. Rev. Lett. **73**, 1332 (1994)
19. G. Janssen, K. Holinde, J. Speth: Phys. Rev. C **49**, 2763 (1994)
20. H.-C. Kim, J.W. Durso, K. Holinde: Phys. Rev. C **49**, 2355 (1994); A. Reuber: PhD Thesis, University of Bonn, unpublished; A. Reuber, H.-C. Kim, K. Holinde, J. Speth: contribution 159 to this conference and in preparation
21. D. Lohse, J.W. Durso, K. Holinde, J. Speth: Nucl. Phys. **A516**, 513 (1990)

Few-Body Systems Suppl. 8, 460–470 (1995)

Few-Body Systems

The Nature of the Baryon-Baryon Interaction in the Quark-Model

Amand Faessler

Institut für Theoretische Physik, Auf der Morgenstelle 14, Universität Tübingen, D-72076 Tübingen, Germany

Abstract. Deuteron properties and nuclear magnetic moments are studied in the non-relativistic quark cluster model. The quark cluster model is modified to include chiral symmetry. This reduces the number of parameters. The σ meson is exchanged between quarks and not as in earlier versions between nucleons. The charge monopole, quadrupole and magnetic-dipole form factors and the tensor polarization of the deuteron in this microscopic meson-quark cluster model are calculated. The deuteron wave function is derived from a microscopic 6-quark Hamiltonian which, in addition to a quadratic confinement potential, includes the one-pion and the one-gluon exchange potentials between quarks. The electromagnetic current operators are constructed on the quark level, i.e., the photon is coupled directly to the quarks. Aside from the one-body impulse current, pionic and gluonic exchange current corrections are included. Due to the Pauli principle on the quark level, new quark interchange terms arise in the one-body and two-body current matrix elements, that are not present on the nucleon level. While these additional quark exchange currents are small for low momentum transfers, we find that they appreciably influence the electromagnetic structure of the deuteron beyond a momentum transfer of $q = 5\text{fm}^{-1}$.

1 Introduction

The first idea about the nature of the nucleon-nucleon interaction came from Yukawa in 1935. He assumed that the strong interaction between two nucleons is carried by an interaction quantum, which is a particle of a medium heavy mass of about 200 MeV, the meson. After finding the π meson one thought one has found the carrier of the strong nuclear force. But the fifties saw a time where more and more mesons were found which contribute to the nucleon-nucleon interaction. One of the high points of this development was the suggestion of Gregory Breit in 1958 that the short range repulsion should be due to a vector, isoscalar meson, the omega meson (ω) with a mass of about 800 MeV. It was a big success of the meson exchange theory of the nucleon-nucleon interaction when this ω meson was found also experimentally.

But just the ω meson shows that this cannot be the whole story. Flavor symmetry $SU(3)$ predicts from the ρ-nucleon coupling the ω-nucleon coupling squared $g^2_{\omega NN}/(4\pi) = 4.5$. In reality one needs to reproduce the short range repulsion according to the Bonn-potential values between 12 and 24, which corresponds to 10 at momentum transfer zero, which is relevant for the data. Normally flavor $SU(3)$ is only violated within 30 % or less. The need to blow up the $\omega - NN$ coupling constant by a factor 2 indicates that the ω meson must carry a load for which it is not prepared. After we learned that the nucleon is composed out of three valence quarks, gluons and sea quarks it is natural to look on the quark level for the nature of the short range repulsion. Indeed we will show in the next chapter that the short range repulsion of the NN interaction can be understood in the quark model by the symmetry of the 6-valence quarks [1, 2, 3, 4, 5].

The main purpose of this work is to search for quark degrees of freedom in the deuteron properties and NN phase shifts.

In the second chapter we extend [6, 7, 8, 9, 10] the quark cluster model to include chiral symmetry, which is broken afterwards by quark condensates. This connects the sigma meson quark coupling constant to the pion-nucleon coupling. PCAC gives a relation [11] between the sigma meson mass and the pion and quark masses. In this way the number of parameters can be drastically reduced. Furthermore one does not need to have two different values for the quark-sigma meson (or nucleon-sigma meson) coupling for S and higher partial waves [8, 9].

In the third chapter we calculate the deuteron wave function in the quark cluster model. For the electromagnetic deuteron form factors of the elastic electron deuteron scattering we include the impulse term where the photon is directly interacting with the quarks and exchange currents [12, 13, 14, 15].

2 The Quark Model and the Nucleon-Nucleon Interaction

At large distances the nucleon-nucleon interaction can be represented quite successfully by exchanging mesons between the centre of mass of nucleon 1 and the centre of mass of nucleon 2. At smaller distances it makes no sense to exchange mesons between the centre of mass of the two nucleons, since due to antisymmetrization it is not even known where the centre of masses of the two nucleons are, since the antisymmetrizer attaches the 6-valence quarks in each of the ten terms of the antisymmetrization to other nucleons. The difficulty of the meson exchange model at short distances might be indicated by the too large $\omega - NN$ coupling constant needed to reproduce the data as indicated in the introduction. At short distances (below 1 fm), where the quark contents of the two nucleons overlap, the quark degrees of freedom should play a major role in describing the nucleon-nucleon (NN) interaction and especially the short range repulsion.

In principle we should describe the NN interaction by quantum chromodynamics (QCD). But QCD is highly nonperturbative. This shows up in the color confinement phenomenon at large distances and the appearance of chiral condensates $\langle\bar{\psi}\ \psi\rangle \neq 0$ of quarks which break chiral symmetry. As a consequence of chiral symmetry breaking current quarks acquire a constituent mass $m_q = m_N/3 \approx 330\ MeV$. This symmetry breaking only occurs at small momenta $|q| < \Lambda_{\mathrm{CSB}}$. These ideas have been included by Shuryak [6] into an instanton liquid model of QCD. Using this picture Diakonov et al [7] have derived an effective low-momentum Lagrangian,

$$L_{\mathrm{QCD}} = \bar{\psi}^{a}_{f'}(i\gamma^{\beta}\partial_{\beta} - m(q^2)e^{i\gamma_5}(\vec{\pi}\cdot\vec{\tau})\psi^{\alpha}_{f} \tag{1}$$

Here ψ^{α}_{f} is the quark spinor with color α and flavor f, f'. $m(q^2)$ is the dynamical mass of the quarks. From Eq. (1) one can derive the effective interaction between quarks and pions and its chiral partner the σ meson:

$$H'_{\mathrm{chiral}} = m(q^2)\bar{\psi}\ \exp(i\gamma_5\vec{\tau}\cdot\vec{\phi}/f_{\phi})\psi \tag{2}$$

If one linearizes this expression using

$$\begin{aligned} \pi &= \widehat{\vec{\phi}}\, f_\pi \sin(\phi/f_\pi) \\ \sigma &= f_\pi\left[\cos(\phi/f_\pi) - 1\right] \end{aligned} \tag{3}$$

one obtains:

$$H'_{\mathrm{chiral}} = g_{\mathrm{ch}}F(q^2)\bar{\psi}(\sigma + i\gamma_5\vec{\tau}\cdot\vec{\pi})\psi \tag{4}$$

where $g_{\mathrm{ch}} = m_q(0)/f_\pi, F(q^2) = m_q(q^2)/m_q(0)$. This expression differs from the Hamiltonian in the linear σ-model in the modification of the coupling constant g_{ch} by the form factor $F(q^2)$,

$$F(q^2) = \left[\frac{\Lambda^2_{\mathrm{CSB}}}{\Lambda^2_{\mathrm{CSB}} + q^2}\right]^{1/2} \tag{5}$$

This allows us now to write down the potential generated between quark i and j by the exchange of π and σ mesons [8, 10]

$$\begin{aligned} V_{\mathrm{OPE}}(\boldsymbol{r}_{ij}) \;=\; & \frac{1}{3}\alpha_{\mathrm{ch}}\frac{\Lambda^2}{\Lambda^2 - m_\pi^2}m_\pi\left\{\left[Y(m_\pi r_{ij}) - \frac{\Lambda^3}{m_\pi^3}Y(\Lambda r_{ij})\right]\boldsymbol{\sigma}_i\cdot\boldsymbol{\sigma}_j + \right. \\ & \left.\left[H(m_\pi r_{ij}) - \frac{\Lambda^3}{m_\pi^3}H(\Lambda r_{ij})\right]S_{ij}\right\}\boldsymbol{\tau}_i\cdot\boldsymbol{\tau}_j\,, \end{aligned} \tag{6}$$

$$V_{\mathrm{OSE}}(\boldsymbol{r}_{ij}) = -\alpha_{\mathrm{ch}}\frac{4\,m_q^2}{m_\pi^2}\frac{\Lambda^2}{\Lambda^2 - m_\sigma^2}m_\sigma\left[Y(m_\sigma r_{ij}) - \frac{\Lambda}{m_\sigma}Y(\Lambda r_{ij})\right]. \tag{7}$$

Here α_{ch} is the chiral coupling constant, r_{ij} is the interquark distance and S_{ij} is the quark tensor operator,

$$S_{ij} = 3(\boldsymbol{\sigma}_i \, . \, \hat{r}_{ij})(\boldsymbol{\sigma}_j \cdot \hat{r}_{ij}) - \boldsymbol{\sigma}_i \cdot \boldsymbol{\sigma}_j \tag{8}$$

Y(x) and H(x) being the Yukawa functions defined as

$$Y(x) = \frac{e^{-x}}{x} \; , \; H(x) = \Big(1 + \frac{3}{x} + \frac{3}{x^2}\Big) Y(x) \, . \tag{9}$$

The scale for chiral symmetry breaking must lie between 1 GeV and 600 MeV to reproduce the adequate cut-off for the instanton fluctuations (we chose 829 MeV.) [6]

In addition to the π and σ exchange between quarks one introduces in the quark cluster model a confinement

$$V_{\text{con}}(\boldsymbol{r}_{ij}) = -a_c \, (\boldsymbol{\lambda}_i \cdot \boldsymbol{\lambda}_j) \, r_{ij}^2 \, , \tag{10}$$

and the one gluon exchange potential,

$$\begin{aligned} V_{\text{OGE}}(\boldsymbol{r}_{ij}) \;&=\; \frac{1}{4} \alpha_s \, \boldsymbol{\lambda}_i \cdot \boldsymbol{\lambda}_j \left\{ \frac{1}{r_{ij}} - \frac{\pi}{m_q^2} \left[1 + \frac{2}{3} \boldsymbol{\sigma}_i \cdot \boldsymbol{\sigma}_j \right] \delta(\mathbf{r}_{ij}) \right. \\ &\left. - \; \frac{3}{4 m_q^2 \, r_{ij}^3} \, S_{ij} \; + \text{two} - \text{body} \;\; \text{spin} - \text{orbit} \right\} . \end{aligned} \tag{11}$$

Here λ_i are the eight color octet matrices acting on quark i. The two body spin-orbit contribution is not written down. It plays an important role in higher $(L \neq 0)$ partial waves.

The total Hamiltonian of the quark cluster model [1, 2, 3, 4, 5, 6, 7] has with chiraly invariant quark-pion and quark-sigma interaction for the six quarks of two baryons the form:

$$\begin{aligned} \hat{H}_{\text{quark}} = \sum_{i=1}^{6} [m_i + \frac{\mathbf{p}_i^2}{2m_i}] - \hat{T}_{\text{CM}} \;\; + \;\; & \sum_{i<j=1}^{6} [V_{OGE}(\boldsymbol{r}_{ij}) + V_{OPE}(\boldsymbol{r}_{ij}) \\ & + V_{OSE}(\boldsymbol{r}_{ij}) + V_{\text{con}}(\boldsymbol{r}_{ij})] \end{aligned} \tag{12}$$

The quark mass is taken as in our earlier works [1, 4, 17, 18, 19] to $m_q = 313$ MeV. It is essentially $m_N/3 = m_q$. The oscillator length is determined to fit, with the pion cloud around, the nucleon the electric proton root mean square radius $\langle r^2 \rangle^{1/2} = 0.83$ fm.

α_{ch} can be connected with the experimental pion-nucleon coupling [8, 9, 10].

$$\alpha_{\text{ch}} = \left(\frac{3}{5}\right)^2 \frac{g_{\pi NN}^2}{4\pi} \, \frac{m_\pi^2}{4M_N^2} \tag{13}$$

For the mass of the σ meson PCAC provides a relation [11]

Table 1. The parameters

m_q	b	α_s	$g^2_{\pi NN}/4\pi$	a_c	m_σ	Λ_{CSB}
MeV	fm			$\mathrm{MeV \cdot fm^{-2}}$	$\mathrm{fm^{-1}}$	$\mathrm{fm^{-1}}$
313	0.518	0.485	13.7	46.94	3.42	4.2

$$m_\sigma^2 = (2m_q)^2 + m_\pi^2 \tag{14}$$

If one fixes the pion mass by the physical value, one obtains a value for m_σ between 650 and 700 MeV. We chose 675 MeV. The cut off mass for chiral symmetry breaking can be related to the cut off mass in the pion-nucleon form factor [8].

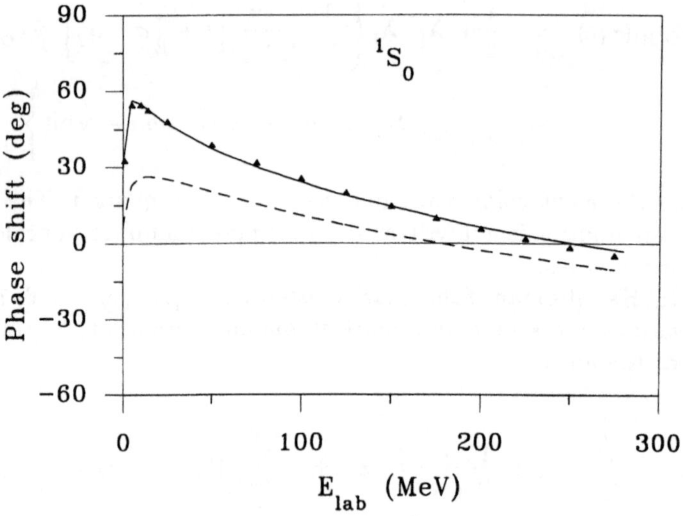

Figure 1. Singlet $^1S_0$ NN phase shift including chiral symmetry and coupling to the $^5D_0$ $N\Delta$ channel (solid line). The dots are the experimental phase shifts by Arndt et al [21]. The dashed line is the result without the inclusion of the $^5D_0$ $N\Delta$ channel.

Figures 1 and 2 show the results for the $^1S_0$ and the $^3S_1$ NN phase shifts. For the $^1S_0$ phase shift the coupling to the $^5D_0$ $N\Delta$ channel is very essential [20].

The quark cluster model explains also nicely the short range repulsion of the nucleon-nucleon interaction [1]. Figure 3 shows the spatial symmetries of

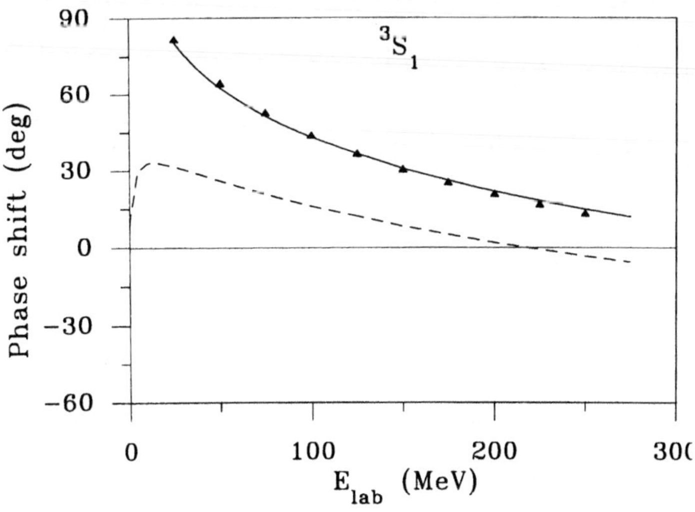

Figure 2. $^3S_1$ NN phase shift. The dots are the experimental values of Arndt et al [21], while the solid line is the result of the present model calculation including chiral symmetry. The dashed curve is the result without coupling to the $^3D_1$ NN partial wave.

the 6-quarks at small distances.

From Fig. 3 one sees that it is more probable by the weight 8/9 to have at short distance the [42] symmetry compared with the completely symmetric orbital wave function [6] which has only the weight 1/9. Fig. 3 indicates also the lowest energy realizations of the orbital symmetries [6] and the mixed orbital symmetry [42]. The last configuration requests at zero distance of the two nucleons ($r = 0$) that two quarks are in the $1p$ state for the lowest energy realization of this configuration. The usual way of representing the two nucleon wave functions by 6 quarks in the $1s$ state is only contained with the probability 1/9. It is obvious from that the [42] orbital symmetry cannot be neglected. If for a moment we neglect 1/9 compared to 8/9 we have at small distances $r \approx 0$ at least two harmonic oscillator quanta excited. Or in other words at least two quarks have to be not in the $1s$ state. For the lowest configuration they are in the $1p$ state. That means at short distances this configuration with the probability 8/9 has at least two harmonic oscillator quanta excited. If one moves again the two nucleons apart at distance r as on the left hand side of Fig. 3, one sees that inside the two nucleons one has no harmonic oscillator quanta excited. Since one has to conserve the number of harmonic oscillator quanta, the two quanta must be contained in the relative motion. If one expands the relative S wave function of the two nucleons in harmonic oscillators

$$u(r_{12}) = \alpha_1|1s\rangle + \alpha_2|2s\rangle + \alpha_3|3s\rangle + \ldots$$

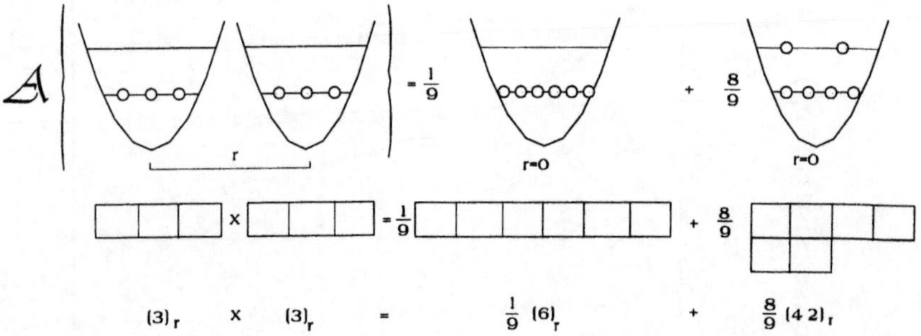

Figure 3. The left hand side shows two nucleons in the quark model at distance r. In each of the two nucleons all three valence quarks are in the $1s$ state. Group theory tells us that if the two nucleons are in a relative orbital S-state the permutation symmetry of all 6 valence quarks can either be only completely symmetric [6] or have the [42] symmetry. The square of the Clebsch-Gordan coefficients of the permutation group of 6 objects gives also the probability to find the completely symmetric spatial representation [6] or the [42] symmetry.

$$\text{with} : \alpha_1 \quad = \quad 0 \text{ (Pauli forbidden)} \tag{15}$$

one finds that the $1s$ amplitude must be zero since all parts of the wave function have to contain at least two harmonic oscillator quanta if one considers the orbital [42] symmetry. Thus the relative wave function $u(r_{12})$ is dominated at small distances by $|2s\rangle$ and therefore has a node near the so-called hardcore radius ($r = 0.4$fm). This node is seen in the asymptotic phase shift measured by the differential cross section. To explain the node one requests that the nucleon-nucleon interaction potential has a hard or soft core at this radius. In reality the node in the wave function is enforced by the orbital [42] symmetry.

3 Deuteron Form Factors

To describe deuteron properties one has first to reproduce the deuteron binding energy of 2.2 MeV. With the σ-nucleon coupling constant $g_\sigma^2/(4\pi) = 5.0$ from chiral symmetry [8, 9] one obtains a binding energy of 2.22 MeV. The results using chiral symmetry are shown in Table 2. The wave function for the proton-neutron relative motion in the deuteron is show in Fig. 4.
The elastic electron deuteron cross section is described by a longitudinal and a transversal form factor

$$\begin{aligned}
\left(\frac{d\sigma}{d\Omega}\right)_{ed} &= \left(\frac{d\sigma}{d\Omega}\right)_{\text{Mott}}[A(q^2) + B(q^2)\tan^2\frac{\theta}{2}] \\
A(q^2) &= F_C^2(q^2) + \alpha_1 q^2 F_Q^2(q^2) + \alpha_2 q^2 F_M^2(q^2) \qquad (16) \\
B(q^2) &= \beta\, q^2(1+q^2)F_M^2(q^2)
\end{aligned}$$

Table 2. Deuteron properties for the chiral σ-nucleon coupling constant $g_\sigma^2/(4\pi) = 5.0$. The D state admixture is given in % while the normalization of the s-wave A_s is indicated in units of $fm^{-1/2}$ and the asymptotic normalization η is dimensionless. The first line gives the NRQM theoretical values, while the second line lists the experimental data. The last line is calculated with the Bonn Potential.

Model	P_D[%]	A_s[$fm^{-1/2}$]	η	r_d[fm]
NRQM	4.91	0.877	0.0261	1.96
Exp.	-	0.884	0.0256	1.96
OBEP	4.99	0.886	0.0264	1.97

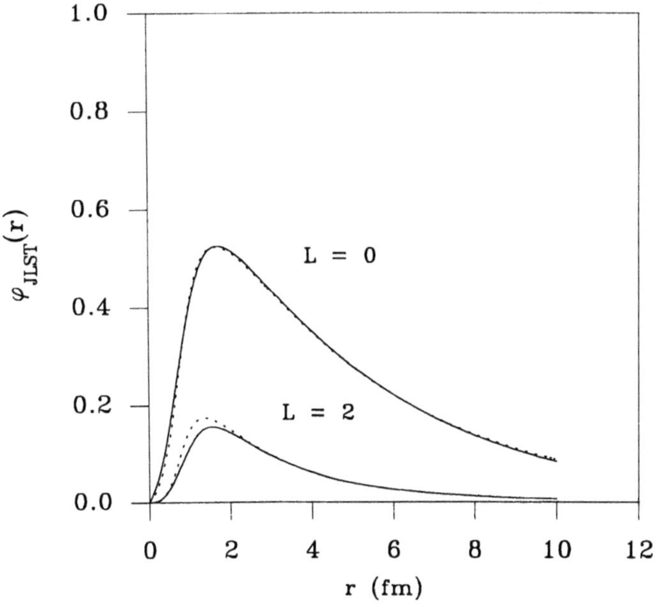

Figure 4. The deuteron S and D wave functions. Full curves: relative deuteron wave function in the quark cluster model; dashed curves: two-nucleon model wave functions calculated with the Paris potential [22].

q is the momentum transfer and F_C, F_Q and F_M indicate the Coulomb, the quadrupole and the magnetic dipole form factors. The transversal form factor $B(q^2)$ is only given by the magnetic dipole form factor $F_M(q^2)$. The longitudinal $A(q^2)$ and transversal form factors are compared in Fig. 5 with the experimental data. Quark exchange diagrams, which are included in Fig. 5, cannot be included phenomenologically on the nucleon level. They represent genuine quark degrees of freedom.

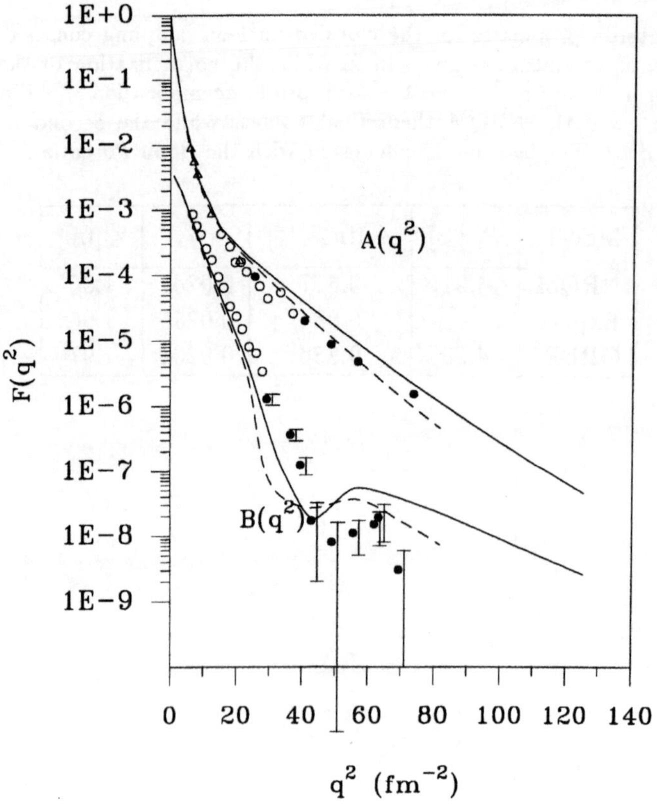

Figure 5. Longitudinal $A(q^2)$ and transversal $B(q^2)$ form-factor of the deuteron including chiral symmetry as a function of q^2. The dashed curve is one of our previous results without chiral symmetry. The data are the circles and dots [23].

4 Conclusions

In this contribution I have essentially communicated three messages:
(i) A short range repulsion of the nucleon-nucleon interaction is not due to the hard or soft core in a static nucleon-nucleon potential but is due to many body symmetries of the 6-valence-quarks of the two interacting nucleons. At short distances the spatial symmetry is of [42] nature with the probability 8/9. This requests that one has at least two harmonic oscillator quanta in the relative wave function. That means one has a node in the interaction region which produces a hard core phase shift in the differential cross section.
(ii) In the elastic electron deuteron scattering genuine quark degrees of freedom can be seen in the transversal form factor which is proportional to the

magnetic dipole form factor $F_M(q)$. These quark degrees of freedom show up at momentum transfers above 5 fm^{-1} or 1 GeV/c. They are essentially due to spin magnetization currents.

(iii) Inclusion of chiral symmetry allows to calculate the NN data without fitting any six quark data. For the $^1S_0$ phase shift the $^5D_0$ $N\Delta$ admixture plays an important role. The two body spin-orbit force from gluon and σ exchange may add or substract in different partial waves. This improves the description of the ^3L_J (with J=L-1, L and L+1) phase shifts without fitting the two-body spin-orbit strength.

Acknowledgement.
I would like to thank Profs. F. Fernández, Zhang Zong-ye and Drs. A. Valcarce, A. Buchmann, L. Glozman, U. Straub and Y. Yamauchi with whom this work has been performed.

References

1. A. Faessler, F. Fernández, G. Lübeck, K. Shimizu: Phys. Lett. **112B**, 201 (1982); Nucl. Phys. **A402**, 555 (1983)
2. M. Oka, K. Yazaki: Progr. Theor. Phys. **66**, 556, 572 (1981)
3. M. Oka, K. Yazaki: *Quarks and Nuclei*. Singapore: World Scientific 1985
4. K. Bräuer, A. Faessler, F. Fernández, K. Shimizu: Z. Phys. **A320**, 609 (1985)
5. K. Shimizu: Rep. Progr. Phys. **52**, 1 (1989)
6. E.V. Shuryak: Phys. Rep. **115** 151 (1984)
7. D.I. Diakonov, V.Yu. Petrov, M. Prassalowicz: Nucl. Phys. **B323**, 53 (1989)
8. F. Fernández, A. Valcarce, U. Straub, A. Faessler: J. Phys. **G19**, 2013 (1993)
   A. Valcarce, A. Buchmann, F. Fernández, A. Faessler: Phys. Rev. **C50**, 2246 (1994)
9. Z. Zhang, A. Faessler, U. Straub, L. Glozman: Nucl. Phys. **A578**, 573 (1994)
   A. Valcarce, A. Buchmann, F. Fernández, A. Faessler: Phys. Rev. **C51**, 1480 (1995)
10. I.T. Obukhosky, A.M. Kusainov: Phys. Lett. **B238**, 142 (1990)
    A.M. Kusainov, V.G. Neudatschin, I.T. Obukhosky: Phys. Rev. **C44**, 2343 (1991)
11. M.D. Scadron: Phys. Rev. **D26**, 239 (1982)
12. A. Buchmann, H. Ito, Y. Yamauchi, A. Faessler: J. Phys. **G14**, 1037 (1988)

13. Y. Yamauchi, A. Buchmann, A. Faessler: Nucl. Phys. **A494**, 401 (1989)

14. A. Buchmann, Y. Yamauchi, A. Faessler: Nucl. Phys. **A496**, 621 (1989)

15. Y. Yamauchi, A. Buchmann, A. Faessler, A. Arima: Nucl. Phys. **A526**, 495 (1991)
L. Glozman, U. Straub, A. Faessler: J. Phys. **G20**, 93 (1994); J. Phys. **G20**, L49 (1994)

16. A. Buchmann, E. Hernández, K. Yazaki: Phys. Lett. **B269**, 35 (1991)
E.M. Henley, G.A. Miller: Phys. Lett. **B251**, 453 (1991)

17. K. Brääuer, A. Faessler, F. Fernández, K. Shimizu: Nucl. Phys. **A507**, 599 (1990)

18. U. Straub: thesis University of Tübingen

19. A. Faessler, U. Straub: Ann. Physik **47**, 439 (1990); Nucl. Phys. **A483**, 686 (1988)

20. A. Valcarce, A. Faessler, F. Fernández: Phys. Lett. **B345**, 367 (1995)

21. R. Arndt et al.: Phys. Rev. **D45**, 3995 (1992)

22. M. Lacombe et al.: Phys. Lett. **B101**, 139 (1981)

23. M.E. Elias et al.: Phys. Rev. **177**, 2075 (1969)
S. Galster et al.: Nucl. Phys. **B32**, 221 (1972)
R.G. Arnold et al.: Phys. Rev. Lett. **35** 776 (1975); **58**, 1723 (1987)

Few-Body Systems Suppl. 8, 471–482 (1995)

Few-Body Systems

Nuclear Forces and Chiral Theories

J. L. Friar[1,2*]

[1] Theoretical Division, Los Alamos National Laboratory, MS B283, Los Alamos, NM 87545, USA

[2] Institute for Nuclear Theory, University of Washington, Physics-Astronomy Building, Box 351550, Seattle, WA 98195-1550, USA

Abstract. Recent successes in *ab initio* calculations of light nuclei (A=2-6) will be reviewed and correlated with the dynamical consequences of chiral symmetry. The tractability of nuclear physics evinced by these results is evidence for that symmetry. The relative importance of three-nucleon forces, four-nucleon forces, multi-pion exchanges, and relativistic corrections will be discussed in the context of effective field theories and dimensional power counting. Isospin violation in the nuclear force will also be discussed in this context.

1 Introduction

The purview of my talk is chiral (symmetry) aspects of nuclear forces. In order to treat this expanding topic, I will ask and answer three questions. The first question is: What is chiral symmetry (CS) and where does it come from? The second question is: What influence does chiral symmetry have on nuclear physics? Finally, my last question is: What are effective field theories, and why should we be interested in them? The answer to the first question will illustrate how QCD plays a significant role even in the low-energy regime appropriate to nuclear ground and low-lying exited states, where quarks and gluons are not the most appropriate degrees of freedom to perform dynamical calculations. The second answer will discuss how the **symmetries** of QCD persist when one uses pions and nucleons as the relevant degrees of freedom in a nucleus, and what those symmetries imply in nuclei (power counting). The last question will outline an approach to nuclear physics (chiral perturbation theory or χPT) that is very recent and does not yet approach the quantitative sophistication of traditional nuclear physics. I will outline a recent calculation of isospin violation in the nuclear force and try to illustrate why this approach is superior in many ways to the traditional approach. Finally, I will conclude and summarize.

**E-mail address:* friar@sue.lanl.gov

2 What is Chiral Symmetry and Why is it Important?

We are all aware of QCD. This theory has made a profound impact on nuclear physics, where it has nevertheless produced very few concrete results. Everyone talks QCD, but in the low-energy regime for more than one nucleon (traditional nuclear physics) there are few successes at relating QCD to nuclear structure. The reasons for this are clear. The success of QCD was founded at first on its theoretical structure. It is a structurally simple theory when written in terms of quark and gluon fields. Indeed, the theory manifests its symmetries in terms of these variables, particularly in the limit of vanishing quark masses.

One such symmetry[1] is chiral $SU(2)_R \times SU(2)_L$ symmetry, where "R" and "L" refer to right-handed and left-handed (helicity) quarks [1] that separately transform in a particular way in the QCD Lagrangian. In other words, those types of quarks don't talk to each other. The vacuum of our world doesn't share this symmetry and spontaneous symmetry breaking is the result, which leads to massless pions (for massless quarks). If the quarks are given a small mass, these Goldstone bosons derive a small mass, as well. The resulting $SU(2)_V \times SU(2)_A$ symmetry has a conserved vector current and partially conserved axial current (PCAC), which would be conserved if pions were massless.

The "simple" symmetries of QCD pose a problem for nuclear physics, however. We don't ordinarily describe nuclei in terms of quarks. Although there is nothing improper with viewing a nucleus as a large "container" filled with quarks and gluons, our basic description is in terms of "physical" degrees of freedom. If one bombards any nucleus with low-energy photons (for example), nucleons are ejected. At somewhat higher energies pions are emitted. These ejecta are clearly the appropriate degrees of freedom for low-energy ($\lesssim$ a few hundred MeV) nuclear physics, as our gedanken experiment shows.

Chiral symmetry can be expressed in terms of the physical degrees of freedom, but the description is more complicated and not particularly obvious. Unfortunately this also means that determining the constraints of CS for a nucleus is nontrivial and is buried in details of the nuclear force. Chiral models (or theories) exist for the various building blocks required for constructing a nuclear force. We will discuss these details later. We can summarize this section by stating:

- Chiral symmetry arises naturally in QCD at the quark level and persists in nuclear physics dynamics expressed in terms of nucleons and pions.

3 What Influence does Chiral Symmetry Have on Nuclear Physics?

The influence of CS (or more generally, QCD) on nuclear physics can be stated on several levels. Most are obvious, but they should be stated nonetheless.

[1] A similar chiral symmetry arises in QED if the electron mass is set to zero. This symmetry accounts for differences between Mott and Rutherford back-scattering of electrons from nuclei. This difference provides a mechanism for the separation of charge and transverse multipoles.

- The pion has a very small mass.
- The pion is a pseudoscalar particle.
- The pion is an isovector particle.
- Chiral symmetry forbids extremely large πN interactions.
- The large-mass scale, Λ, associated with QCD is $\sim$ 1 GeV.

The first four items involve the pion. They and their consequences have been well known for decades (even before QCD). The last item is newer and its consequences for nuclear physics are much more subtle.

The first and most obvious of these properties leads to OPEP being the longest-range part of the nuclear force and plausibly the most important component. The second follows from the Goldstone theorem and the properties of the axial current; it leads to a spin-dependent pion-nucleon interaction. Since the strong-interaction Hamiltonian must be an overall scalar, the pion's negative parity must be balanced by another negative parity, which can only come from a vector (e.g., the pion momentum). This vector must then be contracted with a pseudovector, of which only the nucleon spin suffices. Thus we arise at the usual $\boldsymbol{\sigma} \cdot \boldsymbol{q}_\pi$ form of the πN interaction. When we form OPEP, the resulting spin dependence becomes a tensor force, an important distinction from the much-more-tractable central forces.

These two features account for the OPEP dominance seen in light nuclei. Most of the potential energy derives from the tensor force, while OPEP dominates[2] the potential energy: $\langle V_\pi \rangle / \langle V \rangle \sim 75\%$. This complete dominance is due in part to cancellations, but the overall importance of OPEP is a consequence of chiral symmetry.

The isovector nature of the pion is of the utmost importance in meson-exchange currents. This area of nuclear physics provided the first unambiguous evidence for pion degrees of freedom in nuclei. The motion of **charged** particles in any system produces a current. This fact and the long range of OPEP guarantee the pion a dominant role. Recent work in this area[3] is based on χPT and power counting, but we have no space to pursue this very interesting topic.

Another unambiguous demonstration of pion degrees of freedom in nuclei has an added cachet: it comes with error bars. Beginning approximately fifteen years ago, the Nijmegen group have implemented a sophisticated and successful program of Phase Shift Analysis (PSA) of the NN interactions. Their methodology includes treating all known long-range components of the electromagnetic interaction, such as Coulomb, magnetic moment, vacuum polarization, etc., as well as the tail of the NN interaction beyond 1.4 fm, which includes OPEP. The inner interaction region is treated in a phenomenological fashion. This allows an accurate determination of the πN coupling constants[4]. In order to check for systematic errors in the analysis, they also fitted the masses of the exchanged pions (both charged and neutral) and found:

$$m_{\pi^\pm} = 139.4(10)\,\text{MeV}, \qquad (1a)$$

$$m_{\pi^0} = 135.6(13)\,\text{MeV}. \tag{1b}$$

The small error bars ($\lesssim 1\%$) further demonstrate the importance of OPEP in the nuclear force. They are currently investigating the tail of the rest of the NN interaction.

A valuable byproduct of this work is the ability to construct potentials by directly fitting to the data, rather than to phase shifts, and to utilize the entire NN data base. Several potential models, such as the Argonne V_{18} model, have been constructed in this way and fit the NN data base far better than any previous attempts. One useful corollary of this work[5] is that a baseline has been set for the triton binding energy ($\sim$ 7.62 MeV) using local NN potentials. Nonlocal potential components are mandated by relativity and these are currently under intensive investigation.

3.1 An Opinionated Symmetry

The constraints of chiral symmetry are most easily stated in the form of "opinions", which are detailed below. Because this symmetry is not realized in the usual way (viz., comparing matrix elements of operators that should be (nearly) equal), our tests of CS in nuclei are somewhat indirect. We choose to express the results of such tests as: chiral symmetry and dimensional power counting have an **opinion** about

- the relative size of various components of the two-nucleon force (2Nf);
- the relative sizes of three-nucleon forces (3Nf) and 2Nf;
- the relative sizes of four-nucleon forces (4Nf) and 2Nf, $\cdots$;
- the relative sizes of various components of the nuclear electromagnetic currents, such as the impulse approximation, pion-exchange currents, heavy-meson-exchange currents, etc.;
- the size of relativistic corrections to various parts of the dynamics.

These "opinions" are the result of dimensional power counting – expressing the results of dynamical calculations in terms of powers of the ratio of an average momentum (or energy) component, $\bar{p}$, to Λ, the large-mass QCD scale. Thus, $(\bar{p}/\Lambda)^N$ represents the progression in the sizes of various operators in the nuclear medium. There are rules[6] for determining N for a given case, allowing comparisons to be made. Moreover, in light nuclei we expect $\bar{p} \sim m_\pi c^2$. This estimate is useful but should not be taken too literally.

Power counting is not merely based on our wishes (however strong they may be), but rather on an analysis[6] of the structure of the (generic) effective Lagrangian (based on QCD) underlying nuclear physics. That analysis suggests that the various (dimensional) couplings are given by powers of f_π, the pion decay constant (~93 MeV), Λ (discussed above), and dimensionless constants ($\sim$ 1). The latter assumption is what makes the whole scheme quantitative. Thus, the sizes of individual nuclear operators are expected to be expressible as $\sim (\bar{p}/\Lambda)^N$. One last ingredient is needed in order to make the scheme useful. If N were negative, higher-order terms in perturbation theory (such as loops) could become quite large. Chiral symmetry prevents that and mandates[6] $N \geq 0$. This non-obvious and nontrivial requirement allows a convergent perturbation theory for low-energy strong-interaction dynamics. Also not obvious is the fact that this condition is equivalent to "pair suppression", an *ad hoc* (but phenomenologically necessary) procedure that historically was used to eliminate large and unphysical two-pion-exchange forces.

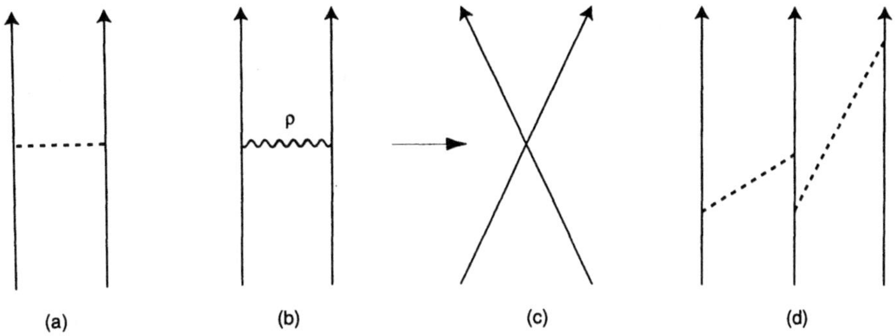

Figure 1. Time-ordered perturbation theory diagrams for nuclear potentials in χPT, with OPEP shown in (a), ρ-exchange in (b) becomes a contact interaction in (c), while overlapping pion exchanges contribute to the 3Nf in (d). Pions are depicted by dashed lines, while nucleons are shown as solid lines.

We can now discuss the various "opinions" in turn. The smallest values of N for the nuclear force correspond to OPEP (shown in Fig. (1a)) and a generic short-range interaction, illustrated in Fig. (1c). Typically the latter might arise from ρ-exchange, as indicated in Fig. (1b), whose range is shrunk to a point. Massive, unstable particles, resonances, etc., cannot propagate very far in the low-energy regime appropriate to effective field theories, and their interaction range is therefore shrunk to a point. Finite-range effects are introduced as derivatives of zero-range interactions. Higher values of N correspond to n-body forces with $N \sim 2n$, and 2-body forces arising from loops. A typical (time-ordered) contribution to a three-nucleon force is shown in Fig. (1d). Thus we expect 4Nf to be smaller than 3Nf, and the latter to be smaller than 2Nf. Indeed, the rule of thumb is that adding nucleons irreducibly to a process increases N by two (each) and adding a loop adds two also. Examples of the latter are two-pion-exchange two-nucleon forces and vertex corrections to one-pion exchange. These forces are therefore considerably weaker than OPEP.

We can also ask the obvious question: do models exist that violate the constraints of CS and generate large 2Nf, 3Nf, $\cdots$? The answer is yes and the problem is indicated in Fig. (2). Pure PS coupling (i.e., γ_5) of a pion to a nucleon leads to large "pair" terms as shown in this figure. These terms are each

$\sim (M/m_\pi)$ larger than chiral models generate, where M is the nucleon mass. Two such factors are thus ~ 50 times larger than what is physically allowed. Very large many-body forces, such as the 4Nf in Fig. (2b), would severely limit the tractability of nuclear physics calculations. They are forbidden by the chiral condition, $N \geq 0$. Thus, CS makes many-nucleon forces small and nuclear physics tractable.

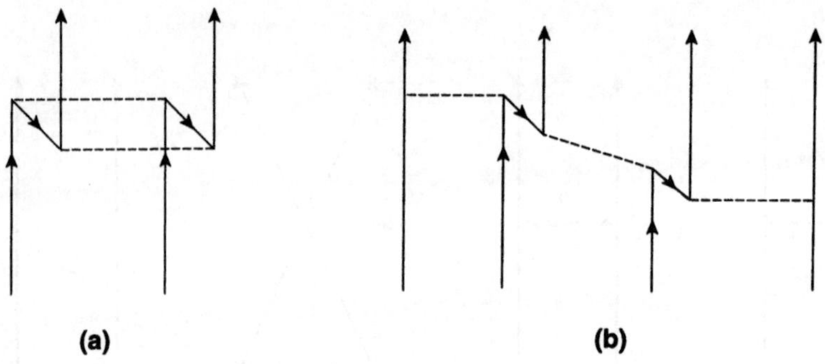

Figure 2. Time-ordered perturbation theory diagrams that emphasize "pair" contributions to the NN force in (a) and to the 4Nf in (b). Pions are depicted by dashed lines, while nucleons are shown as solid lines.

Finally, we note that relativistic corrections fit into this scheme, but have nothing explicitly to do with chiral symmetry. The question of what determines Λ, or even what quantities Λ might subsume, has been deliberately avoided. We simply note that the nucleon mass, M, has a size $\sim \Lambda$, and that an expansion in powers of $1/M$ should behave like an expansion in $1/\Lambda$ (i.e., it should behave like $(\bar{p}/\Lambda)^N$).

3.2 Nuclear Few-body Calculations

Testing these ideas in few-nucleon systems is relatively easy. Few areas of nuclear physics have made such substantive progress in the previous 10 years as few-nucleon physics[7]. We are now solving the Schrödinger equation accurately[8] for the A = 2-6 systems. Such accurate solutions are usually called "exact" or "complete" if their error is $\lesssim 1\%$. The emergence of Green's Function Monte Carlo techniques as the method of choice for nuclear ground states has made accurate calculations possible for systems that were thought to be far out of our reach 10 years ago.

A subset of results from a recent calculation[8] is shown in Table 3.1, including the approximate date when "exact" solutions were first obtained for each nuclear state. The calculations were solutions of the Schrödinger equation for a Hamiltonian containing a recent (accurately fit) NN force and a weak 3Nf, which was adjusted to fit the ^{3}H binding energy. There was no 4Nf. The agreement between theory and experiment is excellent.

Table 3.1. Calculated and experimental ground-state energies (in MeV) of few-nucleon systems, together with (approximate) dates when they were first accurately solved for "realistic" potentials.

$^{A}X(J^{\pi})$	$^{2}H(1^{+})$	$^{3}H(\frac{1}{2}^{+})$	$^{4}He(0^{+})$	$^{5}He(\frac{3}{2}^{-})$	$^{5}He(\frac{1}{2}^{-})$	$^{6}Li(1^{+})$
Solved	~1950	1984	1987	1990	1990	1994
Expt.	-2.22	-8.48	-28.3	-27.2	-25.8	-32.0
Theory	-2.22	-8.47(2)	-28.3(1)	-26.5(2)	-25.7(2)	-32.4(9)

We can also extract from these results the (average) amounts of potential energy accruing from 2Nf and 3Nf, and an upper limit estimate of 4Nf from the error bar on the α-particle energy:

$$\langle V_{NN}\rangle \sim 20\,\text{MeV/pair}\,, \tag{2a}$$

$$\langle V_{3Nf}\rangle \sim 1\,\text{MeV/triplet}\,, \tag{2b}$$

$$\langle V_{4Nf}\rangle \lesssim 0.1\,\text{MeV/quartet}\,. \tag{2c}$$

This geometric progression is in accordance with power-counting predictions. We emphasize again that weak many-nucleon forces are essential for tractability. Although we can use the vast amount of NN scattering data to fit the 2Nf, we have no such options for the 3Nf.

As we stated earlier, the long-range two-pion-exchange 2Nf is weak because of CS. There have been a number of recent papers[9] treating aspects of the problem. Ordóñez et al. was the first to develop a chiral force. Ballot et al. shows how CS arranges cancellations to keep the force weak. Because this potential is relatively weak, no experimental demonstration of its existence yet exists.

4 Effective Field Theories and Isospin Violation in the NN Force

We have already mentioned effective field theories several times. These theories[10] can be viewed as approximations to a known theory (such as QCD) or to an unknown (as yet) theory that is valid to a larger energy scale. Such theories are typically non-renormalizable and can even be nonrelativistic (or semirelativistic). A price is paid for the lack of renormalizability. As the order of the calculation increases, more and more parameters appear that must be fit to data, making higher-order calculations both more difficult and less predictive. What saves the scheme is that for sufficiently low energies the expansion is a series in $\bar{p}/\Lambda$, and should converge fairly rapidly in most circumstances.

Why is this better than simply using a model? Models typically have fewer parameters, for example. The strengths of this scheme are at least threefold:

- The individual terms in the Lagrangian are based on symmetry (e.g., CS).

- It is **not** a model – at low energies it should have all of the content of the original or covering theory (but less predictive power).
- Power counting allows one to estimate with reasonable accuracy the size of various contributions without detailed calculations.

The latter is an extremely useful and powerful technique, as we have seen.

As an illustration we will review the work of van Kolck[11] on isospin violation in the nuclear force, and estimate the size of isospin violation in one-pion-range nuclear forces by comparing to results from the Nijmegen PSA. In the latter work, after accounting for well-established forms of isospin violation such as the NN Coulomb force, magnetic moment interactions between the nucleons, the $n-p$ mass difference in the nucleon's kinetic energy, etc., three πN coupling constants were determined[4]. These correspond to π^0-exchange between two protons, $f^2_{\pi^0 pp}$, or the exchange between a neutron and proton of a π^0, $f_{\pi^0 pp} f_{\pi^0 nn}$, or a charged pion, $f^2_{\pi^c np}$. OPEP depends linearly on f^2, which we define as

$$f^2 = \frac{1}{4\pi}\left(\frac{g_A\, m_{\pi^+}\, d}{2 f_\pi}\right)^2 , \tag{3}$$

where g_A is the axial-vector coupling constant and $d-1$ is the Goldberger-Treiman(GT) discrepancy (> 0) and a measure of chiral-symmetry breaking. In terms of d and $G \sim 13$ (the "pseudoscalar" form of the πN coupling constant) we write the GT relation[12] in the form

$$\frac{G}{M} = \frac{g_A\, d}{f_\pi} . \tag{4}$$

For reference purposes we note that setting d to 1 and using current values of g_A and f_π produces $f^2(d=1) = 0.0718(5)$ in eq. (3).

Table 4.1. Pion-nucleon coupling constants determined by the Nijmegen[4] PSA.

$f^2_{\pi^0 pp}$	$f_{\pi^0 pp} f_{\pi^0 nn}$	$f^2_{\pi^c np}$	Type
0.0751(6)	0.0752(8)	0.0741(5)	np and pp
	0.0745(9)	0.0748(3)	np only

Two sets of coupling constants are available and are shown in Table 4.1. The combined np and pp data sets yield the three values in the first line, while a preliminary solution for np scattering alone gives the two values in the second line. These values are all roughly 4% greater than the value of $f^2(d=1)$, implying that $d-1$ is $\sim 2\%$, much lower than most previous values. We note that systematic effects would enter these results in different ways. The pp data must be carefully corrected for the Coulomb interaction, which plays only a very minor role in np scattering. We also note that charge-symmetry breaking (CSB) in the pion-range force involves a $pp-nn$ comparison and does not contribute in the second (np only) case, but charge dependence (CD) will. In qualitative terms our three experimental numbers from Nijmegen determine the isospin-symmetric d and the CSB and CD πN coupling constants.

The formalism for analyzing these results has been developed by van Kolck[11]. Isospin violation (IV) can be divided into 3 convenient types based on its origin. The u-d quark-mass difference generates one type of IV that has the (tensorial) character of the third component of total isospin, and should be proportional to ϵm_π^2, where $\epsilon = \frac{m_d - m_u}{m_d + m_u} \sim 0.3$ characterizes the quark masses. These contributions are also chiral-symmetry breaking. Hard electromagnetic (EM) processes at the quark level will have a different (tensorial) isospin character and generate the second type, which should be proportional to $\bar{\delta} m_\pi^2$, the (squared) difference of the pion masses. Soft EM processes form the third category and are of order α, the fine structure constant. There are a daunting number of individual diagrams that must be calculated[13] and we will present here only a much simplified overview of where individual mechanisms (most of which are well known) fit into this scheme.

The first difficulty is that we must try to place these categories into an approximate (size) relationship with each other. Van Kolck has argued that since the pion-mass difference is primarily of type II (i.e., EM) and the nucleon-mass difference mostly type I (i.e., quark-mass difference), these terms should be considered on the same level. Although this prescription probably underestimates most EM contributions slightly, it results in the entries in Table 4.2. The order refers to power counting, with OPEP roughly of order 0, while tree and loop refer to the structure of the corresponding diagrams, such as those in Fig. (3). The upper entry in each box describes the fundamental interaction, while the lower (bold-faced) entry describes its effect on the nuclear dynamics.

Some examples can be correlated with entries in Fig. (3). Isospin violations are indicated by a cross, so that Fig. (3a) indicates the CD effect of the pion-mass difference on OPEP. This is listed in the first box (1-tree) together with the pp Coulomb force shown in Fig. (3b). Smaller EM effects (magnetic moment interactions, etc.) are listed in the third box under Breit interaction. The shift of nucleon masses ($\delta M_N \sim M_n - M_p$) has three consequences: (1) an overall energy shift that does not affect physics; (2) a change in the kinetic energies, $p^2/2M$, of the individual nucleons according to whether they are neutrons or protons (indicated in the third box); (3) the effect of the mass difference inside loops. The nucleon-mass splitting is proportional to t_z, the third component of total (nucleon) isospin, which equals (Z-N)/2 and is fixed. On the other hand in loops where pions are circulating it is possible for an np pair to virtually dissociate into $pp\pi^-$. This changes the **nucleons'** isospin and demonstrates that loops with different nucleon masses can in principle play a role in IV.

All of the effects listed above (except loop contributions) are included in the Nijmegen analysis, and we don't need to correct for them. Two additional categories do play a role: the effect of isospin violation in the $\pi^0 N$ coupling constants (indicated by the cross in Fig. (3c)), and $\pi - \gamma$ range NN forces[14] of the type shown in Fig. (3d) and the fourth box in Table 4.2. The latter are not included in analyzing NN scattering data and their effect is unknown, although their order is the same as CD modifications of f^2.

Table 4.2. Contributions to isospin violation in the NN force in orders 1,2, and 3 beyond the usual OPEP, including tree and loop processes, are shown in individual boxes. The first entry of each type corresponds to the elementary vertex, while directly below the single line (in boldface) is the corresponding nuclear contribution. Charge-dependent and charge-symmetry-breaking processes are denoted CD and CSB, respectively.

ORDER	QUARK MASSES	ELECTROMAGNETIC
		Coulomb interaction
1-tree	Nucleon masses	Pion masses
		pp Coulomb force
	Overall energy shift	**OPEP(masses) [CD]**
	πN interaction	Nucleon masses
2-tree	Pion masses	
	OPEP(π_0) [CSB]	**Overall energy shift**
	OPEP(masses) [CD]	
	Nucleon kinetic energies	Other nucleon EM
3-tree		πN interaction
	Nucleus kinetic energy	**Breit interaction**
		OPEP(π_0) [CD + CSB]
	Nucleon masses and KE	Nucleon masses and KE
	Pion masses - OPEP	Pion masses - OPEP
3-loop	OPEP(π_0) [CSB]	**OPEP(π_0) [CD and CSB]**
		$\pi-\gamma$ NN forces [CD]

Ignoring loops and terms that contribute to them, we can write van Kolck's chiral-symmetry-breaking and isospin-violating terms that contribute at tree level in the form:

$$L \sim -\frac{1}{f_\pi}\overline{N}\boldsymbol{\sigma}\cdot\boldsymbol{\nabla}[g_A\, d\, \boldsymbol{t}\cdot\boldsymbol{\pi} - \frac{\beta_1}{2}\pi_0 - \frac{\bar{\beta}_{10}}{2}t_z\pi_0]N\,. \tag{5}$$

The three terms in order are the usual πN coupling (including the GT discrepancy) and the CSB and CD πN vertices. This form can be analyzed with the Nijmegen coupling constants to produce the results in Table 4.3. We had previously estimated $d-1$; the values of β_1 and $\bar{\beta}_{10}$ are consistent with zero and there is no evidence of any systematic disagreement between the two sets of results. We note that the dimensional-power-counting estimate of β_1 is $\sim 6\cdot 10^{-3}$, while that of $\bar{\beta}_{10}$ is $\sim 2\cdot 10^{-3}$. Thus, our results (consistent with zero) are also consistent with expectations of a very small isospin violation.

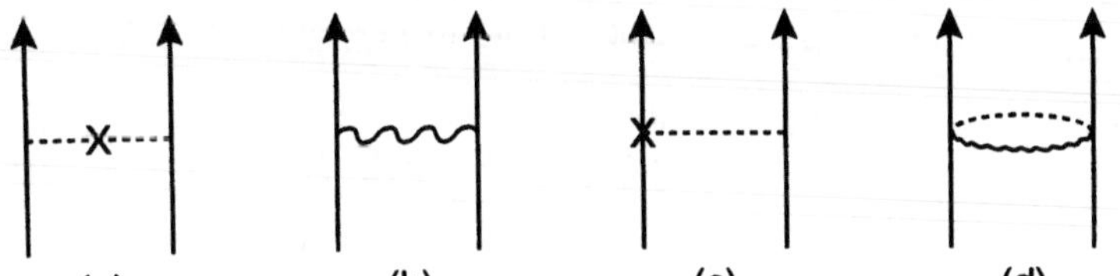

Figure 3. Nuclear isospin-violating interactions, with the pion-mass-difference effect on OPEP shown in (a), the static Coulomb interaction in (b), the isospin-violating pi-nucleon coupling illustrated in (c), and the double-seagull (transverse-)photon-pion exchange sketched in (d). Solid lines are nucleons, dashed lines are pions, small wavy lines are (virtual) transverse photons, while large wavy lines are Coulomb interactions.

Table 4.3. GT discrepancy and isospin-violating pion-nucleon coupling constants determined by the Nijmegen PSA.

$d-1$	β_1	β_{10}	Type
1.6(5)%	$1(9)\cdot 10^{-3}$	$-19(16)\cdot 10^{-3}$	*np* and *pp*
2.1(5)%		$5(16)\cdot 10^{-3}$	*np* only

5 Summary

Chiral symmetry arises at the quark level in QCD and persists in descriptions based on pions and nucleons as effective degrees of freedom. One-pion exchange dominates in the binding of light nuclei and in meson-exchange currents. Chiral symmetry provides order in nuclear forces: without this symmetry nuclear physics would be intractable. Turning the argument around, the tractability of nuclear physics provides strong evidence for chiral symmetry, which weakens N-body forces as N increases and n-pion exchanges compared to OPEP. Isospin-violation upper limits in OPEP obtained from the Nijmegen PSA are compatible with dimensional-power-counting estimates. Finally, few-nucleon systems continue to be the testing ground for new ideas in nuclear physics because of our ability to calculate accurately in those systems.

Acknowledgement. This work was performed under the auspices of the U. S. Department of Energy. J.J. de Swart and M. Rentmeester generously provided preliminary results for our analysis.

References

1. H. Georgi: *Weak Interactions and Modern Particle Theory,* (Benjamin, Menlo Park, 1984)

2. J.L. Friar, B.F. Gibson, G.L. Payne: Phys. Rev. C **30**, 1084 (1984); R.B. Wiringa: Phys. Rev. C **43**, 1585 (1991); S.C. Pieper, R.B. Wiringa, V.R.

Pandharipande: Phys. Rev. C **46**, 1741 (1992)

3. T.-S. Park, D.-P. Min, M. Rho: Phys. Rev. Lett. **74**, 4153 (1995); M. Rho: Phys. Rev. Lett. **66**, 1275 (1991); T.-S. Park, D.-P. Min, M. Rho: Phys. Rep. **233**, 341 (1993)

4. V.G.J. Stoks, R.A.M. Klomp, M.C.M. Rentmeester, J.J. de Swart: Phys. Rev. C **48**, 792 (1993); R.A.M. Klomp, V.G.J. Stoks, J.J. de Swart: Phys. Rev. C **44**, R1258 (1991); V. Stoks, R. Timmermans, J.J. de Swart: Phys. Rev. C **47**, 512 (1993)

5. J.L. Friar, G.L. Payne, V.G.J. Stoks, J.J. de Swart: Phys. Lett. **B 311**, 4 (1993); R.B. Wiringa, V.G. J. Stoks, R. Schiavilla: Phys. Rev. C **51**, 38 (1995)

6. S. Weinberg: Nucl. Phys. **B363**, 3 (1991); Phys. Lett. **B251**, 288 (1990); Phys. Lett. **B295**, 114 (1992); A. Manohar, H. Georgi: Nucl. Phys. **B234**, 189 (1984)

7. J.L. Friar: in *Proceedings of the XIV$^{\underline{th}}$ International Conference on Few-Body Problems in Physics*, ed. by F. Gross, AIP Conference Proceedings **334**, 323 (1995)

8. B.S. Pudliner, V.R. Pandharipande, J. Carlson, R.B. Wiringa: Phys. Rev. Lett. **74**, 4396 (1995)

9. C. Ordóñez and U. van Kolck: Phys. Lett. **B291**, 459 (1992); L.S. Celenza, A. Pantziris, and C.M. Shakin: Phys. Rev. C **46**, 2213 (1992); C. Ordóñez, L. Ray, and U. van Kolck: Phys. Rev. Lett. **72**, 1982 (1994); J.L. Friar, S.A. Coon: Phys. Rev. C **49**, 1272 (1994); C.A. da Rocha and M.R. Robilotta: Phys. Rev. C **49**, 1818 (1994); M.C. Birse: Phys. Rev. C **49**, 2212 (1994); C.A. da Rocha, M.R. Robilotta: Preprint hep-ph/9502371; J.-L. Ballot, M.R. Robilotta, C.A. da Rocha: Preprint hep-ph/9502369

10. H. Georgi: Annu. Rev. Nucl. Part. Sci. **43**, 209 (1993)

11. U. van Kolck: Thesis, University of Texas, (1993); and to be published

12. M.L. Goldberger, S.B. Treiman: Phys. Rev. **110**, 1178 (1958); Y. Nambu: Phys. Rev. Lett. **4**, 380 (1960)

13. U. van Kolck, J.L. Friar, and T. Goldman: Phys. Rev. C (in preparation)

14. J.L. Friar, S.A. Coon: Phys. Rev. C (submitted)

Few-Body Systems Suppl. 8, 483–494 (1995)

Few-Body Systems

Field Theory Approach in Few-Body Systems

J.A. Tjon*

Institute for Theoretical Physics, University of Utrecht, P.O. Box 80.000, 3508 TA Utrecht, The Netherlands

Abstract. The treatment of dynamics and the e.m. interaction in composite systems is discussed within a field theory framework. Canonically this is done using the Bethe-Salpeter equation, where because of technical reasons one has to rely to one-meson-exchange type of force models. A novel approach is also described in which exact nonperturbative boundstate solutions of a field theory can be reconstructed. This may serve as a testing ground of possible few-body equations.

1 Introduction

It is commonly assumed that a nonrelativistic quantum description of the nuclear system based on hadrons as effective degrees of freedom applies. Considerable progress has been achieved in the actual solution of the resulting boundstate and scattering equations for few nucleon systems using realistic spin-isospin dependent nucleon-nucleon interactions and including also effective three-nucleon potentials. In so doing, highly successful agreement has been found with many experimental data at lower energies, though there exist certain cases which need further clarification.

In view of existing and oncoming facilities like CEBAF to probe the behaviour of hadronic systems at higher energies and shorter distances, there is clearly a need for a theoretical description of the dynamical processes in nuclei where relativistic covariance is included in some way. Since the electromagnetic interaction is often used as the probe for such a strongly interacting system, also a proper theoretical treatment of this is needed. In particular, the e.m. current has to be chosen such that it is consistent with the given nuclear dynamics. Moreover, the limits of validity of a hadronic description of the nucleus in the considered energy regime is an important issue and should be studied with some care. Before reaching any definite conclusions at least the importance of effects due to relativity has to be settled.

**E-mail address:* tjon@fys.ruu.nl

There exist two main theoretical frameworks for implementing relativistic covariance and which differ basically in adopting a particle or field formulation. The first one is the relativistic hamiltonian approach [1, 2], which generalizes nonrelativistic quantum mechanics using a fixed particle number description consistent with the Poincaré algebra. One important ingredient is the determination of the mass operator of the composite system, being one of the generators of the Poincaré group. The nuclear interaction appearing in the mass operator is obtained by linking it up to the nonrelativistic NN potential [3]. Clearly such an approach has the virtue of using information from nonrelativistic physics, without having to rely on the specific underlying microscopic dynamics. On the other hand, one has to worry at a later stage how to correct this for contributions arising from meson excitation processes. This obviously happens in the case of the construction of the e.m. current operator. In this talk, I will concentrate on the second line of approach, which is based on field theory models, where implicitly it is assumed that the nuclear interaction is mediated through the meson-nucleon dynamics.

2 Formalism

For definiteness, let us consider for simplicity a ϕ^3-theory, consisting of two types of scalar fields ϕ and χ with the interaction lagrangian

$$\mathcal{L}_{int} = g\chi|\phi|^2. \tag{1}$$

The S-matrix elements are given by the N-point functions

$$\Phi_N = \langle 0|T[\phi(x_1)\phi(x_2)...\phi(x_N)]|0\rangle. \tag{2}$$

Since g is assumed not to be small, we are in general interested in finding the nonperturbative solutions of these N-point functions. Similar as in the nonrelativistic situation we may write down some (multiple) scattering equations. In the 2-particle sector (N=4) this is given by the Bethe-Salpeter equation (BSE). Defining the vertexfunction T through

$$\Phi_4 = G_0 + G_0TG_0, \tag{3}$$

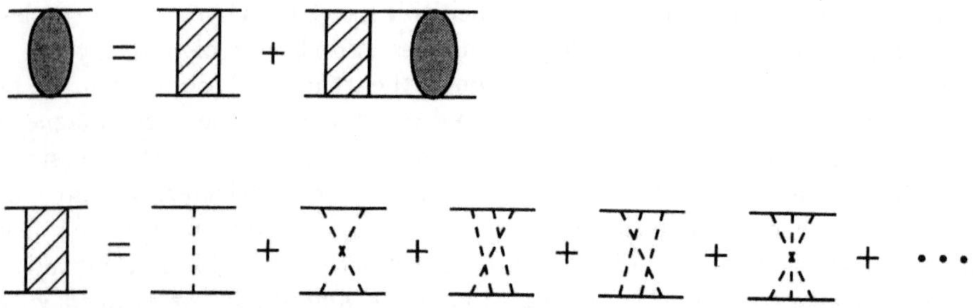

Figure 1. Diagrammatic representation of the Bethe-Salpeter equation with the kernel given by the set of all two-particle irreducible graphs.

with G_0 the free two-particle Green function, the BSE for T has the form shown schematically in Fig. 1, where the driving force is given by the set of all 2-particle irreducible graphs. It is a linear integral equation, where the intermediate state integrations are over the relative four-momentum k.

Although the BSE has formally the same form as the nonrelativistic Lippmann-Schwinger equation an essential difference is that an additional integration occurs, corresponding to the relative energy k_0. It is due to the appearance of this additional integration variable that the singularity structure is much more complicated than in the nonrelativistic case. In particular, because of the presence of the relative energy variable k_0 retardation effects occur and the 4-point function contains time dependences on the individual times of the two particles.

Because of the complexity of the driving force, in practice the BSE has been mostly solved in the socalled ladder approximation, i.e. of the set of 2-particle irreducible graphs only the one-meson exchange contribution is kept. It should be noted that in the case of unequal mass scattering, we donot recover the correct one-body limit from the ladder BSE when we take one of the valence particle masses to ∞. This is because of certain cancelations between the direct and crossed ladder graphs. However, for the case of equal mass scattering as for NN scattering letting $m \to \infty$ in the ladder BSE leads to the proper limit, corresponding essentially to the nonrelativistic case. In this limiting case, we may safely neglect the singularities from the 'potential'. In doing so we can carry out analytically the k_0 loop integral in the BSE. As a result the BSE reduces to a 3-dimensional integral equation, which has a similar structure as the Lippmann-Schwinger equation, but has the nice feature of being relativistic covariant. The resulting relativistic equation is the instantaneous or equal-time (ET) quasi-potential equation. It has been used extensively in the literature [4, 5, 6, 7] and is one of in principle infinite set of possible relativistic quasi-potential equations.

3 A Nuclear Force Model

The BSE and relativistic quasi-potential (QP) approaches have been applied to the study of the two-nucleon system [8, 9]. For the driving force was taken the relativistic one-boson-exchange (OBE) model, consisting of $\pi, \sigma, \omega, \rho, \eta$ and δ meson exchanges. To regulate the high momentum dependence OBE type of meson-nucleon form factors were introduced, depending only on the meson momenta square. Clearly the scalar model has to be extended to also include the full spin complication. Because of covariance one has in general in addition to the known nonrelativistic states also relativistic components in the amplitude, corresponding to negative energy spinor states of the nucleons. The meson-nucleon couplings were fitted to get a reasonable description of the NN phases.

The described model is clearly manifestly covariant. Since it is a field-theory based model, the e.m. interaction can be introduced in a standard way. In particular, Ward-Takahashi identities exist, so that current conservation is satis-

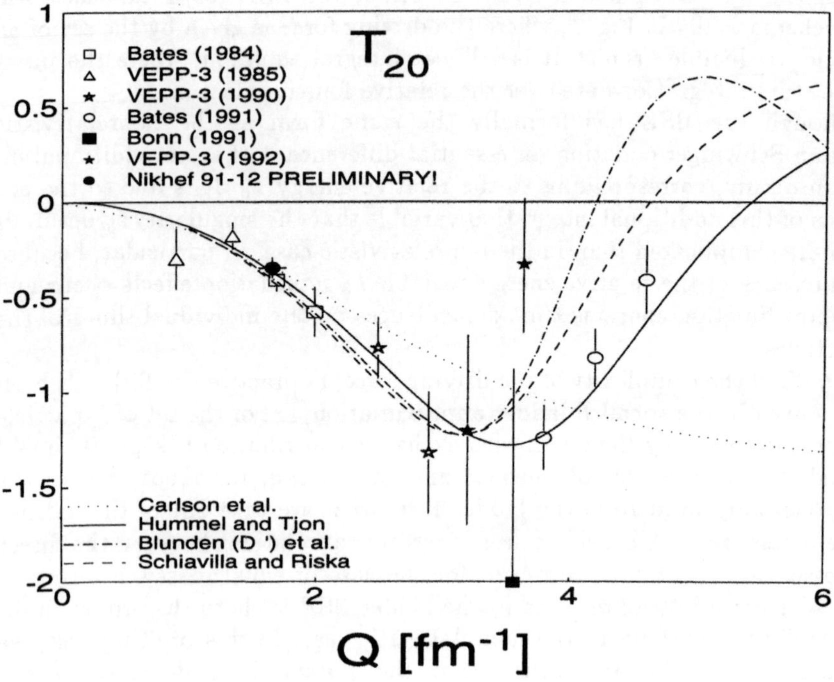

Figure 2. Elastic tensor polarization T_{20} of the deuteron as a function of the momentum transfer Q. References of the experimental data can be found in [10].

fied. For the single nucleon current was taken

$$J_\mu = \gamma_\mu F_1(Q^2) + \frac{\sigma_{\mu\nu}Q^\nu}{2M_N} F_2(Q^2) \tag{4}$$

with on-shell e.m. nucleon form factors F_n.

Meson-photon interactions can clearly also be introduced in the model, leading to meson exchange currents (MEC). In estimating these contributions, e.m. meson form factors have to be adopted, for which among others the vector meson dominance model has been assumed [5]. Calculations with this model has been carried out for both elastic and inelastic electron deuteron scattering. As an example, in Fig. 2 is shown the T_{20} prediction for elastic eD scattering, together with the experimental world data [10]. The single nucleon e.m. current contribution to the deuteron current matrixelements leads to a result which is rather close to the nonrelativistic prediction. This should be contrasted to the conventional MEC calculations where the socalled pair term correction shifts the nonrelativistic result substantially. The prediction of such a calculation [11] is also displayed in the figure (dashed curve). It should be noted that since the nucleons are treated as Dirac particles in the relativistic calculation the pair term contribution is automatically included in that case. This shows that

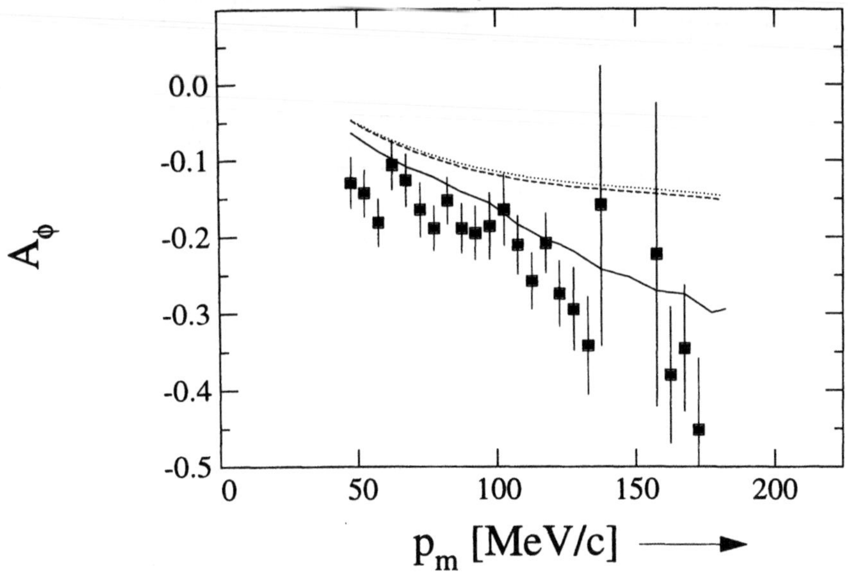

Figure 3. Asymmetry parameter A_ϕ as a function of the missing momentum p_m. Solid curve is the relativistic prediction, while the two other curves are nonrelativsitic calculations.

one should be careful in carrying out simple conventional MEC calculations, where implicitly it is assumed that corrections to the internal dynamics of the composite system need not be considered. It has indeed explicitly been verified in a consistent QP model of the strong and e.m. interaction, that the pair term is cancelled almost entirely by contributions related to the relativistic components in the deuteron [12]. Dropping the pair term in the conventional perturbative analysis yields a close agreement between the two calculations.

An interesting result due to relativity was recently found in structure function studies in inelastic eD scattering. The measurement of the asymmetry parameter A_ϕ, which is directly connected to the interference structure function R_{LT}, was found in better agreement with relativistic predictions than with the nonrelativistic ones. This effect is predominantly due to the use of the full relativistic e.m. current operator. The Nikhef-K data [13] are displayed in Fig. 3, together with various theoretical predictions [14, 15]. Similar effects have been seen at higher momentum transfer [16].

Although reasonable agreement is found with the experimental data, it should be noted that various issues have to be studied in more detail. Here we list some of them. Since we are dealing with bound nucleons, there may be off-shell dependences in the e.m. nucleon form factors. Some model studies [17, 18], based on one-meson-loop calculations indicate that effects of 10 % or more can occur at higher momentum transfer. Also the Q-dependence of photon-meson form factors is found to be very model dependent and sizably

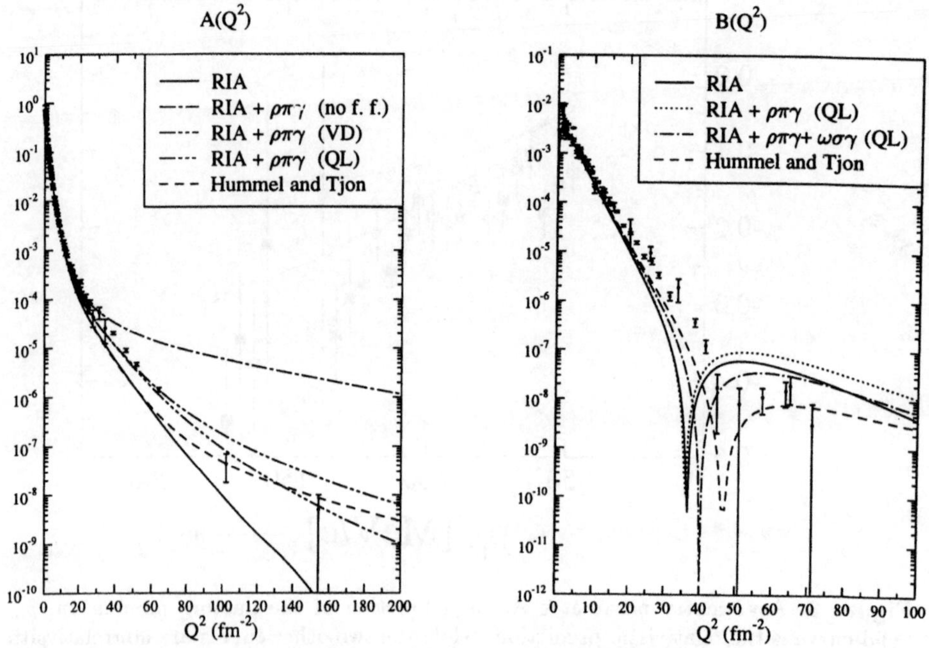

Figure 4. The e.m. deuteron form factors A and B. Sensitivity on choice of e.m. meson form factor in the $\pi\rho\gamma$ MEC current. A softer form factor is found in the quark loop model (QL) as compared to the vector meson dominance model (VD). Effect of the $\omega\sigma\gamma$ MEC contribution is also shown.

different from the vector meson dominance model. In Fig. 4 is shown a calculation of Ito and Gross [19] for the e.m. deuteron form factors A and B using a constituent quark loop model. Similar results have been obtained by Mitchell and Tandy [20] and Tiemeijer [21] in confining models. Finally, different force models can lead to different predictions of the deuteron e.m. form factors at high momentum transfer. It should be important to understand in more detail what these sensitivities are and which one are physically acceptable.

4 The Three-Nucleon System

We now turn to discuss the relativistic correction to the binding energy of the tri-nucleon system. As is well known, extensive nonrelativistic studies with realistic two-nucleon forces have given evidence that the triton is underbound by about 1 MeV in such calculations. Possible reasons may be three-nucleon forces and effects of relativity. To study the three-nucleon system in the relativistic framework one needs the 6-point function. Performing a multiple scattering series resummation one readily derives a set of integral equations of the Watson-Faddeev type. Neglecting the 3-particle irreducible graphs and assuming for simplicity scalar particles, the T-matrix can be written as $T = T^1 + T^2 + T^3$,

Table 1. Calculated three-nucleon binding energies for various realistic two-nucleon interactions using the BSWF equations, nonrelativistically (NR) and two versions of QP prescriptions. The two Paris interactions differ in the choice of form factors in the separable representation.

Interaction	BSWF	QP1	QP2	NR
Paris-1	7.535	7.389	7.435	7.245
Paris-2	7.474	7.363	7.408	7.183
Bonn	8.201	8.083	8.140	7.822

where T^i satisfies for the boundstate problem the coupled set of integral equations

$$T^i = \sum_{j \neq i} K_{ij} T^j \tag{5}$$

with

$$K_{ij} = t_i \Delta_j \Delta_k . \tag{6}$$

where i, j, k $(i \neq j \neq k)$ are cyclic permutations of $1, 2, 3$. In Eq. (6) Δ_n is the propagator of particle n and t_i is the 2-body t-matrix between particle j and k. The 3-particle irreducible graphs correspond in this framework to 3-body forces. It should be noted however that the notion of such forces is framework dependent and that the reduction of Eq. (5) to for example a nonrelativistic equation can also lead in principle to the presence of an effective 3-body force in the reduced equation.

As compared to the nonrelativistic case we have an additional integration over the relative energy. As input for Eq. (5) we need a covariant 2-body t-matrix, depending in general on the invariants P^2, $P.k$ and k^2, P and k being the total and relative momenta of the two-body system. Using a separable representation of realistic NN interactions like Paris and Bonn [22], these potentials were relativized in ref. [23] by replacing $\boldsymbol{k}^2 \rightarrow k^2$ and hence implicitly assuming that the interactions are not dependent on the total energy P. Using these relativized 2-body interactions as input, the three-nucleon boundstate problem has been studied [23], by solving the Bethe-Salpeter-Watson-Faddeev equations (BSWF), treating the spin in a nonrelativistic way. In Table 4, some of the predictions are shown.

Using a QP approach the complexity of these equations are comparable to the nonrelativistic ones, except for possibly additional spin-angular momentum channels due to the presence of negative energy spinor states. Using dispersion relation based quasi-potential prescriptions, similar results are found as for the BSWF calculations, being also in agreement with other calculations [24, 25]. The relativistic corrections to the binding energy as found from these studies are attractive, but small i.e. of the order of 0.3 MeV. These results should be contrasted with other calculations using relativistic hamiltonian dynamics

[26, 27], which predict a repulsive effect (see however ref. [28, 29]). Although the approaches cannot be compared directly, one obvious difference is that the off-mass shell dependences have been neglected in the hamiltonian dynamics.

More recently, calculations have been carried out within a field-theory based framework of the spectator model, where also a repulsive correction to the binding is found [30, 31]. In this approach, the relative energy variable is constrained by setting some of the particles on mass-shell. As compared to the dispersion relation based QP, some additional energy dependence is introduced in the 2-body t-matrix. This may be the underlying reason that the the attraction is changed into a repulsion. In the case of the dispersion relation based prescription the energy-dependence is essentially kept minimal by setting the relative energy equal to zero. From these studies we may infer that results are sensitive to details of the dynamics and in particular the off-energy shell extrapolations may play an important role in the final answer.

5 Feynman-Schwinger Approach

Up to now we have discussed field-theory based models where certain approximations are made to be able to extract the nonperturbative solutions. In particular, all practical calculations are based on the ladder theory. Obviously, it is interesting to be able to find exact nonperturbative solutions beyond the ladder approximation. Such calculations can then serve as a tool to investigate issues like reliabilty of quasi-potential approximation schemes.

Recently a method has been suggested to reconstruct exact boundstate solutions of field theory [32]. It is based on the Feynman-Schwinger (FS) representation. Considering the ϕ^3-theory as discussed in Section 2 we may write in the coordinate representation for the 4-point function the functional form

$$\Phi_4(x\bar{x}|y\bar{y}) = N' \int D\phi^* \int D\phi \int D\chi e^{-S} \Psi_f^* \Psi_i, \qquad (7)$$

where the initial state is given by $\Psi_i = \phi^*(y)\phi(\bar{y})$ and the final state $\Psi_f = \phi^*(x)\phi(\bar{x})$. S is the action expressed in terms of the lagrangian as $S = \int d^4x\mathcal{L}$ and N' is a normalization factor. As demonstrated in ref. [32] we may successively integrate out the ϕ and χ fields. As a consequence the field description is essentially replaced by a quantum particle description. For the class of all direct and crossed ladders we get

$$\begin{aligned}\Phi_4(x\bar{x}|y\bar{y}) &= N' \int_0^\infty ds_1 \int_0^\infty ds_2 \int (Dz)_{xy} \int (D\bar{z})_{\bar{x}\bar{y}} \\ &\times e^{-m^2(s_1+s_2)-\frac{1}{4}\int_0^{s_1} \dot{z}^2{}_\mu d\tau - \frac{1}{4}\int_0^{s_2} \dot{\bar{z}}_\mu^2 d\bar{\tau}} W, \qquad (8)\end{aligned}$$

where W is given by

$$W = \exp\{+g^2 \int_0^{s_1} d\tau \int_0^{s_2} d\bar{\tau} \Delta(z(\tau) - \bar{z}(\bar{\tau}))\} \qquad (9)$$

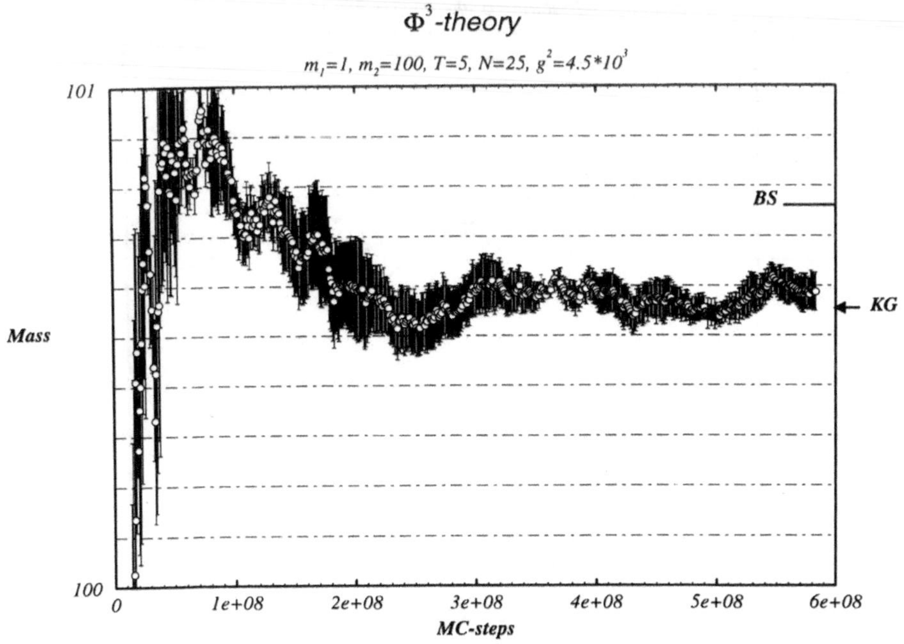

Figure 5. Boundstate mass for the set of all direct and crossed ladders as a function of Monte Carlo steps at $T = 5\bar{m}$. The ϕ^3-theory is studied with $m = 1, \bar{m} = 100$ and $\mu = 0$.

with Δ the free propagator of the χ particle with mass μ in coordinate space. In Eq. (8) the functional integrals are subject to the boundary conditions that $z(0) = y$, $z(s) = x$ and $\bar{z}(0) = \bar{y}$, $\bar{z}(s) = \bar{x}$. They can in principle be evaluated by dividing up the paths in $z(\tau)$ such that $Dz = \prod_{n=1}^{N} \frac{d^4 z(n)}{(2\pi\varepsilon)^2}$ with $N\varepsilon = s$ and similarly for the $\bar{z}$ path.

To study the boundstate problem we may assume that we can carry out the Wick rotation and hence Eq. (8) can be studied in Euclidean space. Since the integrand in Eq. (8) has a positive measure in the Euclidean case, the Metropolis Monte Carlo method can be applied to calculate the functional. The 4-point function Φ_4 clearly depends on the total time $T = x + \bar{x} - y - \bar{y}$. From the spectral decomposition we have

$$\Phi_4 = \sum_n c_n e^{-m_n T}, \tag{10}$$

so that the mass spectrum m_n can in principle be extracted from the T-dependence.

In Fig. 5 is displayed the logarithmic derivative of Φ_4 as a function of Monte Carlo steps at a timescale of T where the groundstate is dominant. The calculations are done for the case of unequal masses with $m = 1, \bar{m} = 100$, while the

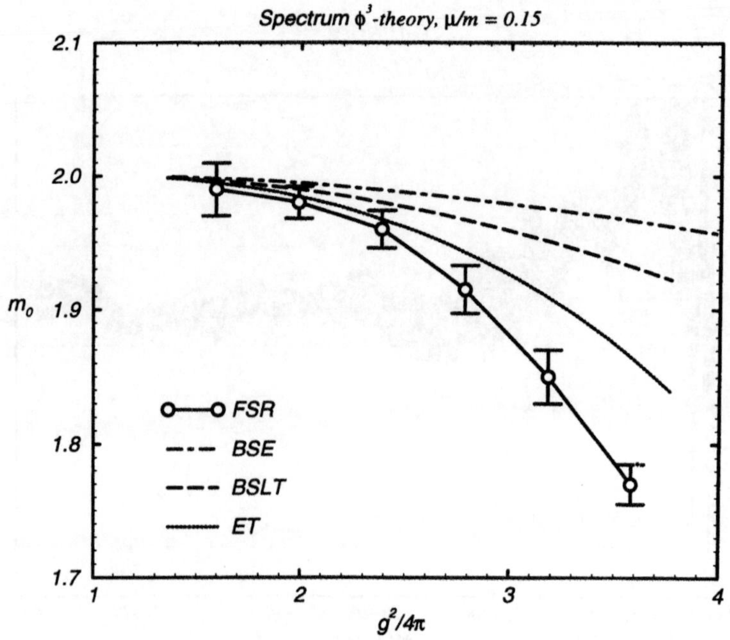

Figure 6. Dependence of the boundstate mass m_0 on the coupling constant g for the set of all direct and crossed ladders. The ϕ^3-theory is studied with $m = \bar{m} = 1$ and $\mu = 0.15$. Also are shown the ladder BSE and various QP results.

mass of the exchanged particle is taken to be zero. The FS calculations predict a groundstate mass of 100.57 for the coupling constant $g^2 = 4.510^3$. In the one-body limit we expect to get the Klein-Gordon (KG) equation for a particle in the presence of an external field as generated by the heavy particle. As can be seen from the figure we do indeed confirm that the FS calculation recovers the KG prediction. On the other hand, the BSE in the ladder approximation yields a result with considerably less binding.

We may also consider the issue how well the ladder BSE and various QP approximations compare with the exact result in the case of equal masses ($m = \bar{m} = 1$) [33]. This is shown in Fig. 6 for the case of an exchanged mass $\mu = 0.15$. We see, that the ladder BSE leads also in this case to considerably less binding, whereas the various QP predictions are more attractive. In particular, the ET prescription is remarkably close to the exact answer. In some way the QP's do seem to include some of the higher order crossed graph contributions in a proper way. At smaller couplings the various predictions are in agreement with each other. This suggests, that for a case like the NN system we may hope that the ladder series is a sound starting point independent of which scheme is used and correcting for this by including the cross-box graph in the interaction may be a good procedure.

6 Concluding Remarks

We have discussed here a particular approach to describe the nonperturbative dynamics of few-body composite systems. It is based on an effective field-theory approach, where Lorentz covariance can be taken into account in a natural way. Moreover, mesonic excitation processes can be included since an underlying meson theory is used and as a result also the e.m. interaction can in principle be treated in a consistent way.

Approximation methods are in general needed in practice. Besides the question of the construction of the driving force, the type of scattering equations to use are important issues. Much work will be needed to find out which approaches will lead to physically reliable and acceptable predictions. Since we may not expect to be able to solve the complete nonperturbative problem it would be clearly of interest to have some model systems, where exact solutions can be constructed and for which approximation schemes can be tested. For a simple scalar ϕ^3-theory exact solutions of the groundstate have been constructed for the direct and crossed ladders. The employed Feynman-Schwinger method is very effective. More recently it has been used to construct the exact solutions of the scalar QED [33, 34]. It will be an essential task to investigate the spin complication in this formalism and furthermore apply the FS method to QCD.

As noted an alternative approach exists for formulating relativistic equations based on the hamiltonian formulation. It is clear that attempts have to be made to make the connection between the two approaches.

References

1. P.A.M. Dirac: Rev. Mod. Phys. **21**, 392 (1949).
2. B. Bakamjian, K.H. Thomas: Phys. Rev. **21**, 1300 (1953).
3. P.L. Chung et al.: Phys. Rev. C **37**, 2000 (1988).
4. E.E. Salpeter: Phys. Rev. **87**, 328 (1952).
5. E. Hummel, J.A. Tjon: Phys. Rev. C **42**, 2354 (1990).
6. P.C. Tiemeijer, J.A. Tjon: Phys. Rev. C **48**, 896 (1993); ibid. C **49**, 494 (1994).
7. N.K. Devine, S.J. Wallace: Phys. Rev. C **48**, 973 (1993); Few-Body Systems Suppl. **7**, 395 (1994).
8. J. Fleischer, J.A. Tjon: Phys. Rev. D **21**, 87 (1980).
9. M.J. Zuilhof, J.A. Tjon: Phys. Rev. C **22**, 2369 (1980).
10. K. de Jager: Trieste Workshop, May 1995.
11. R. Schiavilla, D.O. Riska: Phys. Rev. C **43**, 437 (1991).

12. M.J. Zuilhof, J.A. Tjon: Phys. Lett. **84B**, 31 (1979).

13. M. van de Schaar et al.: Phys. Rev. Lett. **66**, 2855 (1991).

14. W. Leidemann, H. Arenhövel: Nucl. Phys. **A393**, 385 (1983).

15. E. Hummel, J.A. Tjon: Phys. Rev. C **49**, 21 (1994).

16. J.F.J. van de Brand: Few-Body Systems Suppl. **7**, 151 (1994).

17. J. Koch, R. Naus: Phys. Rev. C **36**, 2459 (1987).

18. P.C. Tiemeijer, J.A. Tjon: Phys. Rev. C **42**, 599 (1990).

19. H. Ito, F. Gross: Phys. Rev. Lett. **71**, 2555 (1993).

20. K. Mitchell, P. Tandy: to be published.

21. P. Tiemeijer: private communication.

22. J. Haidenbauer et al.: Phys. Rev. C **30**, 1822 (1984); ibid. C **33**, 439 (1986); ibid. C **33**, 456 (1986);

23. G. Rupp, J.A. Tjon: Phys. Rev. C **37**, 1729 (1988); ibid. C **45**, 2133 (1992).

24. A.D. Jackson, J.A. Tjon: Phys. Lett. **32B**, 9 (1970);

25. F. Sammarruca, D.P. Xu, R. Machleidt: Phys. Rev. C **46**, 1636 (1992).

26. V.S. Bhasin, H. Jacob, A.N. Mitra, Phys. Lett. **32B**, 16 (1970).

27. J. Carlson et al.: Phys. Rev. C **47**, 484 (1993).

28. B.L.G. Bakker, L.A. Kondratyuk, M.V. Terentiev: Nucl. Phys. **B158**, 497 (1979).

29. L.A. Kondratyuk, F.M. Lev, V.V. Soloviev: Few-Body Systems **7**, 55 (1989); F.M. Lev: this conference.

30. F. Gross: Phys. Rev. C **26**, 2203 (1990).

31. F. Gross at. al.: this conference.

32. Y.A. Simonov, J.A. Tjon: Ann. Phys. **228**, 1 (1993); T.E. Nieuwenhuis, Y.A. Simonov, J.A. Tjon: Few-Body Systems Suppl. **7**, 286 (1994).

33. T.E. Nieuwenhuis: Ph-D thesis, 1995.

34. T.E. Nieuwenhuis, J.A. Tjon: Phys. Lett **B** (1995), in press.

Few-Body Systems Suppl. 8, 495–503 (1995)

Few-Body Systems

Application of the Few-Body Theory to Deep Inelastic Lepton-Nucleon Scattering

F. M. Lev*

Laboratory of Nuclear Problems, Joint Institute for Nuclear Research, Dubna, Moscow region 141980, Russia

Abstract. Usually the few-body theory is used in low and intermediate energy physics while high energy processes are described by perturbative QCD. However the perturbation theory does not apply to the bound state problem, and in particular it is not clear whether this theory applies to deep inelastic lepton-nucleon scattering (DILNS). We consider DILNS in a model in which the nucleon is the relativistic N-body bound state (where N can be finite or infinite) and the electromagnetic or weak current operator responsible for DILNS properly commutes with the representation generators of the Poincare group. It is explicitly shown that the internal light cone momentum fraction does not coincide with the Bjorken variable x even in the Bjorken limit, and this gives a natural qualitative explanation of the deficiency of the parton model sum rules.

1 Current Operators for Deep Inelastic Lepton-Nucleon Scattering

To describe some process of deep inelastic lepton-nucleon scattering (DILNS) it is necessary to construct the operator of the electromagnetic or weak current responsible for the transition *nucleon→hadrons* in this process. Let $\hat{J}^\mu(x)$ ($\mu = 0,1,2,3$) be such an operator where x is a point in Minkowski space.

The translational invariance of the current operator implies that

$$\hat{J}^\mu(x) = \exp(\imath\hat{P}x)\hat{J}^\mu(0)\exp(-\imath\hat{P}x) \qquad (1)$$

where $\hat{P}$ is the four-momentum operator of the system under consideration. This relation makes it possible to reduce the problem of seeking $\hat{J}^\mu(x)$ to the problem of seeking $\hat{J}^\mu(0)$. The latter should satisfy Lorentz invariance which implies that

$$[\hat{M}^{\mu\nu}, \hat{J}^\mu(0)] = -\imath(g^{\mu\rho}\hat{J}^\nu(0) - g^{\nu\rho}\hat{J}^\mu(0)) \qquad (2)$$

*E-mail address: lev@nusun.jinr.dubna.su

where $\hat{M}^{\mu\nu}$ are the representation generators of the Lorentz group and $g^{\mu\nu}$ is the metric tensor in Minkowski space ($\mu, \nu = 0, 1, 2, 3$).

Our intuition tells that when the momentum transfer in the electromagnetic or weak process is very large then the virtual photon or W-boson interacts only with one constituent, and the process is so quick that the interaction between this constituent and the remnants of the target can be neglected. Therefore the current operator can be taken in the impulse approximation (IA):

$$\hat{J}^{\mu}(0) = J^{\mu}(0) \tag{3}$$

where $J^{\mu}(0)$ is the free current operator, i.e. the operator $\hat{J}^{\mu}(0)$ is the same as for noninteracting particles.

If Eq. (3) takes place then Eq. (2) will be valid only if $\hat{M}^{\mu\nu} = M^{\mu\nu}$, i.e. the operators $\hat{M}^{\mu\nu}$ are the same as for noninteracting particles. This condition defines the so called point form of relativistic dynamics [1]. Let us note that the parton model is a consequence of the IA in the front form [2]. Therefore the current operator in the parton model is incompatible with relativistic invariance.

If the current operator satisfies the continuity equation $\partial \hat{J}^{\mu}(x)/\partial x^{\mu} = 0$ then, as follows from Eq. (1), this equation can be written in the form

$$[\hat{J}^{\mu}(0), \hat{P}_{\mu}] = 0 \tag{4}$$

Let the initial nucleon state $|N\rangle$ be the eigenstate of the four-velocity operator with the eigenvalue G' and the mass operator with the eigenvalue m. Analogously, let the final state $|f\rangle$ be the eigenstate of the four-velocity and mass operators with the eigenvalues G'' and M'' respectively. It is always possible to consider the process in the reference frame in which $\mathbf{G}'' + \mathbf{G}' = 0$ (see ref. [3] for details). We choose the coordinate axes in such a way that $G'_x = G'_y = 0$. Then, as follows from Eq. (4), the x and y components of the operator $\hat{J}^{\mu}(0)$ are not constrained by the continuity equation and the 0 and z components satisfy the relation

$$G'^{0}(M'' - m)\langle f|\hat{J}^{0}(0)|N\rangle = G''^{z}(M'' + m)\langle f|\hat{J}^{z}(0)|N\rangle \tag{5}$$

In DILNS m is the nucleon mass, and in the Bjorken limit (when $G'^{0} = -G''^{z}$ and $M'' \gg m$) the interaction of the struck quark with the remnants of the target can be neglected [4]. Then it follows from Eq. (5) that the condition (3) for all the components of the operator $\hat{J}^{\mu}(0)$ is compatible with the continuity equation.

2 Explicit Realization of the N-body Representation in the Point Form of Dynamics

Let us first consider a system of N free particles with the 4-momenta p_i and masses m_i ($i = 1, ...N$). It is of no importance for us whether the number N is finite or infinite. If p is the 4-momentum of some particle then $\mathbf{p}_{\perp}$ means the projection of p onto the plane xy, and, instead of the temporal and z components of p, we use the $\pm$ components defined as $p^{\pm} = (p^0 \pm p^z)/\sqrt{2}$.

For simplicity the spin variables will be always suppressed. Then the Hilbert space H for the system under consideration is the space of functions $\varphi(\mathbf{p}_{1\perp}, p_1^+, ...\mathbf{p}_{N\perp}, p_N^+)$ such that

$$\int |\varphi(\mathbf{p}_{1\perp}, p_1^+, ...\mathbf{p}_{N\perp}, p_N^+)|^2 \prod_{i=1}^{N} d\rho(\mathbf{p}_{i\perp}, p_i^+) \quad < \infty \tag{6}$$

where

$$d\rho(\mathbf{p}_{\perp}, p^+) = \frac{d^2\mathbf{p}_{\perp} dp^+}{2(2\pi)^3 p^+} \tag{7}$$

We define $P = p_1 + ... + p_N$, $M_0 = |P| \equiv (P^2)^{1/2}$, and $G = PM_0^{-1}$. Let $\beta(G) \equiv \beta(\mathbf{G}_{\perp}, G^+) \in SL(2, C)$ be the matrix with the components

$$\beta_{11} = \beta_{22}^{-1} = 2^{1/4}(G^+)^{1/2}, \quad \beta_{12} = 0, \quad \beta_{21} = (G^x + \imath G^y)\beta_{22} \tag{8}$$

and $L[\beta(G)]$ be the element of the Lorentz group corresponding to $\beta(G)$. We also define the relative four-momentum of particle i as

$$k_i = L[\beta(G)]^{-1} p_i \quad (i = 1, ...N) \tag{9}$$

The 4-vectors p_i have the components $(\omega_i(\mathbf{p}_i), \mathbf{p}_i)$, and the 4-vectors k_i have the components $(\omega_i(\mathbf{k}_i), \mathbf{k}_i)$ where $\omega_i(\mathbf{k}) = (m_i^2 + \mathbf{k}^2)^{1/2}$. In turn, only $N-1$ vectors $\mathbf{k}_i$ are independent since, as follows from Eqs. (8) and (9), $\mathbf{k}_1 + ... + \mathbf{k}_N = 0$. Therefore $L[\beta(G)]$ has the meaning of the boost, and $\mathbf{k}_i$ are the momenta in the c.m. frame.

Let us define

$$d\rho(\text{int}) = 2(2\pi)^3 M_0^3 \delta^{(3)}(\mathbf{k}_1 + \cdots + \mathbf{k}_N) \prod_{i=1}^{N} d\rho(\mathbf{k}_{i\perp}, k_i^+) \tag{10}$$

We also define the "internal' space H_{int} as the space of functions $\chi(\mathbf{k}_1, ...\mathbf{k}_N)$ such that

$$||\chi||^2 = \int |\chi(\mathbf{k}_1, ...\mathbf{k}_N)|^2 d\rho(\text{int}) \quad < \infty \tag{11}$$

Then the space of functions satisfying Eq. (6) can be realized as the space of functions $\varphi(\mathbf{G}_{\perp}, G^+)$ with the range in H_{int} and such that

$$\int ||\varphi(\mathbf{G}_{\perp}, G^+)||^2 d\rho(\mathbf{G}_{\perp}, G^+) \quad < \infty \tag{12}$$

It is easy to show that $M_0 = \omega_1(\mathbf{k}_1) + ... + \omega_N(\mathbf{k}_N)$. Let us introduce the variables

$$\xi_i = \frac{\omega_i(\mathbf{k}_i) + k_i^z}{M_0(\mathbf{k}_1, ...\mathbf{k}_N)} \in (0, 1) \tag{13}$$

Therefore $(\mathbf{k}_{1\perp}, \xi_1, ...\mathbf{k}_{N\perp}, \xi_N)$ is also a possible choice of the internal momentum variables, and these variables are constrained by the relations $\mathbf{k}_{1\perp} + ... + \mathbf{k}_{N\perp} = 0$, $\xi_1 + ... + \xi_N = 1$.

When the particles interact with each other, the simplest realization of the N-body representation in the point form is $\hat{P} = G\hat{M}$, $\hat{M}^{\mu\nu} = M^{\mu\nu}$, where G is the free four-velocity operator and $\hat{M}$ is the mass operator which commutes with the system spin operator and acts only through the variables of the space H_{int} [5,6]. The internal nucleon wave function χ is the eigenfunction of $\hat{M}$ with the eigenvalue m: $\hat{M}\chi = m\chi$.

We assume that the nucleon is indeed described by such a representation. Roughly speaking this means that we take into account only the N-body confining interaction which ensures binding of the constituents inside the nucleon. Of course, there also exist other interactions but we assume that they can be considered in the framework of perturbative QCD. Otherwise speaking, the difference between our approach and the usual one is that the leading order of the perturbation theory is not the free theory but the theory in which the nucleon is already bound.

3 The Relation Between the Internal Light-Cone Momentum Fraction and the Bjorken Variable x

As noted in Sec. 1, it is convenient to consider DILNS in the reference frame where $G'^z = -G''^z$, $\mathbf{G}'_\perp = \mathbf{G}''_\perp = 0$. For definiteness we choose the case $G'^z > 0$.

Let q be the four-momentum of the virtual photon or W-boson and P' and P'' be the four-momenta of the initial and final systems: $P' = mG'$, $P'' = M''G''$. We consider only the Bjorken limit when $|q^2| \gg m^2$ and $x \equiv |q^2|/2(P'q)$ is not too close to 0 or 1.

Since $P'' = P' + q$, it is obvious that

$$M'' = [\frac{|q^2|(1-x)}{x}]^{1/2} \tag{14}$$

and in the reference frame under consideration

$$q^0 = (M'' - m)G'^0, \quad \mathbf{q} = -(M'' + m)\mathbf{G}' \tag{15}$$

As follows from these expressions

$$(G'^z)^2 = \frac{(M''-m)^2 - q^2}{4M''m} \tag{16}$$

and in the Bjorken limit we get from Eqs. (14) and (16)

$$(G'^z)^2 = \frac{|q^2|^{1/2}}{4m[x(1-x)]^{1/2}} \tag{17}$$

Therefore $G'^z \gg 1$ in the Bjorken limit and in the reference frame under consideration

$$G'^+ = \sqrt{2}G'^z, \quad G''^+ = \frac{1}{2\sqrt{2}G'^z} \tag{18}$$

Suppose that the virtual photon or W-boson is absorbed by particle i. Let M_i be the mass of the free system $1, ...i-1, i+1, ...N$. Then it follows from Eq. (13) that

$$M_0 \equiv M_0(\mathbf{k}_{i\perp}, \xi_i, M_i) = [\frac{m_i^2}{\xi_i} + \frac{M_i^2}{1-\xi_i} + \frac{\mathbf{k}_\perp^2}{\xi_i(1-\xi_i)}]^{1/2} \tag{19}$$

and

$$(M_i^2 + \mathbf{k}_i^2)^{1/2} - k_i^z = M_0(\mathbf{k}_i, M_i)(1-\xi_i) \tag{20}$$

If the current operator satisfies Eq. (3) then the four-momenta of the system $1, ...i-1, i+1, ...N$ in the initial and final states are equal to each other (impulse approximation!). The internal four-momentum of this system in the initial state is obviously equal to $[(M_i^2 + \mathbf{k}_i^2)^{1/2}, -\mathbf{k}_i]$ and the same quantity in the final state is equal to $[(M_i^2 + \mathbf{k}_i''^2)^{1/2}, -\mathbf{k}_i'']$ where $\mathbf{k}_i''$ is the relative momentum $\mathbf{k}_i$ in the final state (and we take into account that the quantities M_i in the initial and final states are the same). Therefore, as follows from Eq. (9)

$$L(G')[(M_i^2 + \mathbf{k}_i^2)^{1/2}, -\mathbf{k}_i] = L(G'')[(M_i^2 + \mathbf{k}_i''^2)^{1/2}, -\mathbf{k}_i''] \tag{21}$$

We assume that the internal nucleon wave function is such that only the contribution of the transversal momenta $|\mathbf{k}_{i\perp}| \leq m_0$ is important where m_0 is a parameter such that $m_0^2 \ll |q^2|$. It follows from the $\perp$ component of Eq. (21) that $\mathbf{k}_{i\perp} = \mathbf{k}_{i\perp}''$ and therefore in the Bjorken limit

$$M'' = (M_i^2 + \mathbf{k}_i''^2)^{1/2} + (m_i^2 + \mathbf{k}_i''^2)^{1/2} \approx 2|k_i''^z| \tag{22}$$

It is also obvious that in the reference frame under consideration $k''^z < 0$. Therefore, as follows from Eqs. (14,17-20,22), the + component of Eq. (21) gives

$$M_0(\mathbf{k}_{i\perp}, \xi_i, M_i)(1-\xi_i) = m(1-x) \tag{23}$$

instead of the well-known relation $\xi_i = x$ in the parton model.

This relation shows that the difference between the parton model and our approach depends on the difference between the free mass M_0 and the real nucleon mass m. In the nonrelativistic approximation $M_0 = m$ since the binding and kinetic energies can be neglected and both quantities are equal to $m_1 + ... + m_N$. However there is no reason to believe that the nucleon is the nonrelativistic system.

As follows from Eqs. (19) and (23), the explicit expression of ξ_i as a function of $\mathbf{k}_i, M_i, x$ is

$$\xi_i = \frac{m_i^2 + \mathbf{k}_{i\perp}^2}{m_i^2 + \mathbf{k}_{i\perp}^2 + M'^2(1-x)^2} \quad \text{if} \quad m_i = M_i,$$
$$\xi_i = \frac{1}{2}\{1 - \alpha_i - \beta_i + [(1-\alpha_i-\beta_i)^2 + 4\alpha_i]^{1/2}\} \quad \text{if} \quad M_i > m_i,$$
$$\xi_i = \frac{1}{2}\{1 - \alpha_i - \beta_i - [(1-\alpha_i-\beta_i)^2 + 4\alpha_i]^{1/2}\} \quad \text{if} \quad M_i < m_i \tag{24}$$

where

$$\alpha_i = \frac{m_i^2 + \mathbf{k}_{i\perp}^2}{M_i^2 - m_i^2}, \quad \beta_i = \frac{M'^2(1-x)^2}{M_i^2 - m_i^2} \tag{25}$$

Since $x \in [0,1]$, it follows from Eq. (24) that $\xi_i \in [\xi_i^{\min}, 1]$ where $\xi_i^{\min} = \xi_i^{\min}(\mathbf{k}_{i\perp}, M_i)$ is a function of $\mathbf{k}_{i\perp}, M_i$ which can be defined from Eq. (24) at $x = 0$. It is easy to see that $0 < \xi_i^{\min} < 1$.

4 Qualitative Explanation of the Deficiency of the Parton Model Sum Rules

The present theory of DILNS is based on the operator product expansion (OPE) [7] and the Altarelli-Parisi equations [8] which are equivalent to the OPE. Strictly speaking the OPE has been proved only in the framework of the perturbation theory while such a theory does not apply to the bound state problem. Therefore applicability of the OPE to DILNS is the assumption which should be verified. Many textbooks consider deep inelastic processes in the framework of the Feynman parton model [9] proposed for explaining the phenomenon of Bjorken Scaling [10]. However there exist only a very few cases when the results obtained in the framework of the OPE agree with the parton model. This occurs when the anomalous dimensions of the Wilson coefficients are equal to zero and the momentum transfer q is so large that all higher twist effects (which are of order $(m^2/|q^2|)^n$) and the perturbative QCD corrections (of order $\alpha_s(q^2)^n$, where $\alpha_s(q^2)$ is the QCD running coupling constant and $n = 1, 2...$) can be neglected.

We shall always assume that q indeed satisfies such properties. Then the investigation of the cases when the OPE agrees with the parton model is of special interest. Indeed, as noted above (see also ref. [11] for details) the current operators in the parton model do not properly commute with the Lorentz group generators. Therefore the validity of the OPE in the given cases is questionable.

The parton model sum rules which can be derived from the general theory are the sum rules derived in refs. [12-15]. The difficulties in the experimental verification of these sum rules are that many data are obtained at not very large $|q^2|$ (a few GeV^2) and the results essentially depend on the ways of extracting the nucleon data from the nucleus ones and extrapolating the data to low x. Nevertheless, the analysis of the recent precise CCFR data [16] in ref. [17] shows that the experimental value of the Gross-Llewellyn Smith sum rule [14] is smaller than the theoretical value, even if the corrections of order $\alpha_s(q^2)$ and $\alpha_s(q^2)^2$ are taken into account.

There also exist sum rules which one way or another are based on the parton model. According to the recent precise results of the NMC group [18] the experimental value of the integral defining the Gottfried sum rule [19] is equal to 0.235 ± 0.026 instead of $1/3$ in the parton model.

The EMC result [20] for the first moment of the proton polarized structure function $g_1(x)$ is $\Gamma_p = 0.126 \pm 0.010 \pm 0.015$ while the Ellis-Jaffe sum rule [21] predicts $\Gamma_p^{EJ} = 0.171 \pm 0.004$.

The most impressive results of the parton model sum rules are those concerning the quark contribution to the nucleon's momentum and spin. The first sum rule (see, for example, the discussions in the textbooks [22]) says that quarks carry only a half of the nucleon momentum, and this fact is usually considered as one of those which demonstrates the existence of gluons. The second sum rule known as "the spin crisis" says that the quark contribution to the nucleon spin is comparable with zero (a detailed discussion of the spin crisis can be found, for example, in papers [23-25]). Of course, these results are not in direct contradiction with constituent quark models since the latter are successful only at low energies. Nevertheless, our experience can be hardly reconciled with the fact that the role of gluons is so high.

The above discussion gives all grounds to conclude that the values given by the parton model sum rules systematically exceed the corresponding experimental quantities.

The results of the preceding section make it possible to give a simple qualitative explanation of the deficiency of the parton model sum rules. First of all we note that the consequence of the fact that $\xi_i \in [\xi_i^{\min}, 1]$ when $x \in [0,1]$ is that *the data on deep inelastic lepton-nucleon scattering do not make it possible to determine the ξ_i distribution of quarks at* $0 < \xi_i < \xi_i^{\min}$. Of course, the quantity $\xi_i^{\min}$ can be determined only if some concrete model of the nucleon is assumed.

Now the fact that the parton model sum rules give values exceeding the corresponding experimental quantities can be qualitatively explained as follows. In the parton model the integration over $x \in [0,1]$ can be replaced by the integration over $\xi_i \in [0,1]$, and the right-hand-side of the parton model sum rules becomes proportional to the normalization integral (11). At the same time, in our approach the integration over $x \in [0,1]$ can be transformed to the integration over $\xi_i \in [\xi_i^{\min}, 1]$. Therefore the right-hand-side of the corresponding sum rules does not contain the contribution of $\xi_i \in [0, \xi_i^{\min}]$, and it is natural to expect that the corresponding integral is smaller than the normalization integral (11).

For the unambiguous verification of the parton model sum rules, and especially the sum rules [12-15] they should be extracted directly from the experimental data at large $|q^2|$, not using the $|q^2|$ evolution determined from the OPE or Altarelli-Parisi equations. The above considerations show that such experimental data are crucial to solve the problem whether perturbative QCD applies to deep inelastic lepton-nucleon scattering.

Acknowledgement. The author is grateful to A. Buchmann, F. Coester, A.E. Dorokhov, S.B. Gerasimov, S. Gevorkyan, I.L. Grach, A.V. Efremov, B.L. Ioffe, A.B. Kaidalov, V.A. Karmanov, N.I. Kochelev, E.A. Kuraev, L.A. Kondratyuk, S. Kulagin, E. Leader, G.I. Lykasov, L.P. Kaptari, A. Makhlin, S.V. Mikhailov, I.M. Narodetskii, N.N. Nikolaev, V.A. Novikov, E. Pace, M.G. Sapozhnikov, G. Salme, S. Simula, N.B. Skachkov, O.V. Teryaev, Y.N. Uzikov and H.J. Weber for valuable discussions and to S.J. Brodsky, S.D. Glazek, L.L. Frankfurt and M.P. Locher for useful remarks. This work was supported by grant No. 93-02-

3754 from the Russian Foundation for Fundamental Research.

References

1. P.A.M. Dirac: Rev. Mod. Phys. **21**, 392 (1949)
2. F. Coester: in *Lectures on quarks, mesons and nuclei*, ed. W.-Y. Pauchy Hwang, vol. 2 (World Scientific, Singapore, 1989); H.J. Weber: Phys. Rev. **D49**, 3160 (1994); Z.Dziembowski, C.J. Martoff, P.Zyla: Phys. Rev. **D50**, 5613 (1994)
3. F.M. Lev: Ann. Phys. (NY) **237**, 355 (1995)
4. R.K. Ellis, W. Furmanski, P. Petronzio: Nucl. Phys. **B212**, 29 (1983)
5. B. Bakamdjian, L.H. Thomas: Phys. Rev. **92**, 1300 (1953)
6. S.N. Sokolov: Teor. Mat. Fiz. **36**, 193 (1978)
7. K.G. Wilson: Phys. Rev. **179**, 1499 (1969); K.G.Wilson, W. Zimmerman: Commun. Math. Phys. **24**, 87 (1972)
8. G. Altarelli, G. Parisi: Nucl. Phys. **B126**, 298 (1977)
9. R.P. Feynman: Phys. Rev. Lett. **23**, 1415 (1969); J.D.Bjorken, E.A. Paschos: Phys. Rev. **185**, 1975 (1969)
10. J.D. Bjorken: Phys. Rev. **179**, 1547 (1969)
11. F.M. Lev: hep-ph 9505373
12. S.L. Adler: Phys. Rev. **142**, 1144 (1966)
13. J.D. Bjorken: Phys. Rev. **163**, 1767 (1967)
14. D.J. Gross and C.H. Llewellyn Smith: Nucl. Phys. **B14**, 337 (1969)
15. J.D. Bjorken: Phys. Rev. **147**, 1467 (1966)
16. CCFR Collab. P.Z. Quintas et. al.: Phys. Rev. Lett. **71**, 1307 (1993)
17. A.L. Kataev, A.V. Sidorov: Report CERN-TH 7235/94 (CERN, Geneva, 1994); Report E2-94-344 (JINR, Dubna, 1994); A.E.Dorokhov: Pisma ZHETF **60**, 80 (1994)
18. New Muon Collaboration. M. Arneodo et al.: Phys. Rev. **D50**, R1 (1994)
19. K. Gottfried: Phys. Rev. Lett. **18**, 1174 (1967)
20. EM Collaboration, J. Ashman et al.: Nucl. Phys. **328**, 1 (1989)
21. J. Ellis, R.L. Jaffe: Phys. Rev. **D9**, 1444 (1974); **D10**, 1669 (1974)

22. F.E. Close: *An Introduction to Quarks and Partons* (Academic Press, London-New York, 1979); F.J. Yndurain: *Quantum Chromodynamics. An Introduction To The Theory of Quarks and Gluons* (Springer Verlag, New York - Berlin - Heidelberg - Tokyo, 1983)

23. J. Ellis, M. Karliner: Report CERN-TH-6898/93 (CERN, Geneva, 1993)

24. B.L. Ioffe: Report ITEP 61-94 (ITEP, Moscow, 1994)

25. A.E. Dorokhov, N.I. Kochelev: Particles and Nuclei **26**, 5 (1995)

Few-Body Systems Suppl. 8, 504–519 (1995)

Few-Body Systems

From the Structure of Nucleons to that of Complex Nuclei: the Role Played by Few-Body Physics

C. Ciofi degli Atti

Department of Physics, University of Perugia, and INFN Sezione di Perugia, Via A. Pascoli, I-06100 Perugia, Italy

Abstract. The major achievements in the field of few-nucleon systems reported at Peniscola are stressed and the important role played by Few-Body physics in the study of the structure of the nucleon and complex nuclei is illustrated by several examples.

1 Introduction

The organisers of the XVth European Few-Body Conference (EFB-XV) asked me to give a Closing speech, leaving me the freedom to choose its format. The only thing that was fixed was the allotted time: 40 generous minutes. Being this too much for a pure formal Closing: "Very good Conference, thanks to our Spanish friends!", and not seeing the real usefulness of a canonical Summary: "x stressed that, y pointed out this, and so on", I decided to use my 40' in order: i) to stress and reiterate some of the results reported at Peniscola in the field of few-nucleon systems; ii) to illustrate the role played by Few-Body physics in the investigation of the structure of nucleons and complex nuclei.

The reason why I decided to discuss only few-nucleon systems can be explained by quoting a nice description of Few-Body Physics given by Amand Faessler in his Summary talk at the XII International Few-Body Conference (FB-XII) held in Vancouver in 1989 [1]:

> In the house of few-body physics there live quite different people. In the lower floors of atomic and molecular physics we are putting in new carpets, more fashionable furniture and discuss on which wall will hang the new paintings. In the next higher floor of nuclear physics we are sometimes still moving a wall, putting in a new staircase, but the ceiling, the floor and the windows are already there. In the next higher floor of quarks and gluons the building is not yet finished, the wind is blowing through the rooms and from the top is raining in.

Even if the construction of the building has much progressed during the last six years, such a picture remains still valid. As a matter of fact, during EFB-XV we were witness of the continuous enrichment of the lower atomic–molecular floor [2], and of the extensive efforts to put the last, quark–gluon floor in a better shape, so as, at least, to recover from rain and wind. But what in my view was most impressive is the relevant qualitative and quantitative works which has been done to put the second, nuclear floor in its final shape. It is for this reason that I think we are at the right time to summarise the progress that occurred in the nuclear sector of the Few-Body house.

To this end, let me start by listing my reasons why few-nucleon systems have to be considered with particular care and attention:

1. they allow a quantitative check of the standard model of nuclear physics, i.e. the description of nuclei as aggregates of non relativistic nucleons interacting through realistic Nucleon-Nucleon (NN) interactions;

2. they teach us about the dynamics of bound hadrons, for by investigating the way they respond to different probes, basic quantities related to the nuclear wave function, e.g. different kinds of density matrices, can be studied in a more reliable way than in complex nuclei;

3. they can provide information on several relevant aspects of the structure of complex nuclei;

4. they represent a microlaboratory for studying models of the nucleon.

2 The Standard Model of Nuclear Physics

The exact solution of the non relativistic Schroedinger equation is a prerequisite for any significant progress towards the understanding of few-nucleon systems and complex nuclei. In my opinion three major progresses in this field were reported at Peniscola:
1.1 the bound-state calculations for A=3,4,5,6[3, 4, 5]
1.2 the continuum calculations for A=3 [6, 7]
1.3 the report on different relativistic-like calculations for the bound states of A=3,4 [8, 9] (see also [10]).

The results for bound state calculations of A=3,4 are summarised in Tables 1 and 2. It is evident that calculations based upon completely different techniques converge to the same results, which means that the 3 and 4 body problems can now be solved exactly and that the standard model can quantitatively be checked. The continuum calculations for A=3 are presented in Fig.1̃.

A very good agreement between data and theoretical calculations can be noticed and, more important, a comparison between different computational techniques [11] yields consistent results. The possibility to generate continuum wave functions allows one to attack the calculation of nuclear processes on few-nucleon systems leading to various disintegration channels. It is hoped that

Table 1. The binding energy and D-wave probability of ^{4}He (without Coulomb interaction), obtained by the Faddeev–Yakubowsky method with different two-nucleon interactions (after [12]).

Potential	$P_D[\%]$	$j_{max}=3$	$j_{max}=4$
Paris [13]	5.77	24.26	24.32
Nijmegen [14]	5.39	25.03	25.10
AV_8 [15]	6.08	25.11	25.31
AV_{14} [15]	6.08	24.62	24.73
Bonn B [16]	4.99	27.04	27.07
Bonn A [16]	4.38	28.11	28.15
Ruhrpot [17]	5.85	–	24.60
Experiment	–	28.30	–
Without Coulomb	–	–	29.0

Table 2. Results for the ^{4}He binding energy obtained by different methods using various potentials including Coulomb (C) and Three Nucleon Interaction (TNI). FY: Faddeev–Yakubowsky [12]; GFMC: Green–Function–Monte Carlo [18]; CHH: Correlated–Hyperspherical–Harmonics [4].

Potential	B_{FY}	B_{GFMC}	B_{CHH}
AV_8	25.31	25.8±0.1	25.34
$AV_{14}+C$	23.92	24.2±0.2	23.93
$AV_{14}+C+TNI$	–	28.2±0.2	27.48

very soon exact calculations of continuum states for A=3 could be carried out at higher energies and momentum transfers where extensive experimental programs (e.g. the CEBAF ones) are planned.

The ground state of ^{3,4}He is definitely underbound when only two-nucleon interactions are considered. The introduction of three-body forces, either motivated by field theoretical arguments, or modelled by the requirement to produce the correct saturation properties of few-nucleon systems and nuclear matter, can provide an adequate description of the binding energy, but one would like to understand the miscroscopic reason of the underbinding. Relativistic effects do not provide at the moment a clear cut answer to such a question [8], and calculations with QCD motivated interaction [20] cannot yet sytematically be performed in few-nucleon systems. It is clear that in order to understand the reason for the underbinding, one has to look into the structure

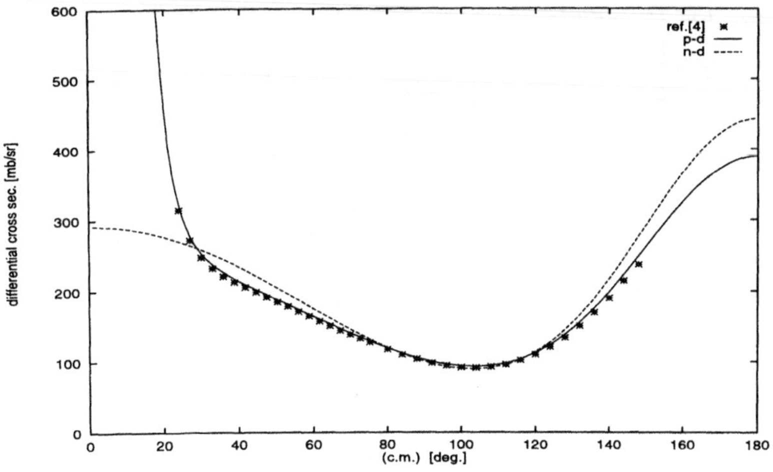

Figure 1. $p-d$ (solid line) and $n-d$ (dashed line) cross sections calculated with the Av18+UR NN potential. Experimental data from [19] (after [7])

of the wave function, which could be done, in principle, by investigating how few-nucleon systems respond to probes of different kinds. The knowledge of the wave function or, better, of some portions of it (viz. various density matrices), would allow one to obtain information on those basic quantities which characterise any many-body system, like e.g. the momentum distributions, the correlation function, etc.

3 Probing Few–Nucleon Systems

Different probes can be used to investigate few-nucleon systems [21], but to illustrate my points I will consider, without any prejudice for hadronic probes [22], electromagnetic processes, in particular inclusive electron scattering, first, because this is the simplest type of process which can be treated theoretically, second, because a wealth of experimental data has been collected on it in a wide domain of kinematics, which includes the quasi elastic scattering from bound nucleons, the excitation of nucleon resonances in the medium, and the Deep Inelastic Scattering (DIS) region where the photon couples with the quarks of a bound nucleon.

In first Born approximation, the hadronic tensor which determines the cross section depends upon two nuclear response functions $W^A_{1(2)}(\nu, Q^2)$ (ν and Q are the energy and four momentum transfers, respectively), whose calculation requires the knowledge of the non-diagonal one-body spectral function $P(\boldsymbol{k}, \boldsymbol{k}', E)$ where $\boldsymbol{k}$ and $\boldsymbol{k}'$ are nucleon momenta and E their removal energy. Due to the difficulty in obtaining $P(\boldsymbol{k}, \boldsymbol{k}', E)$, a simple, and frequently used approach, is the Plane Wave Impulse Approximation (PWIA), which corresponds to using

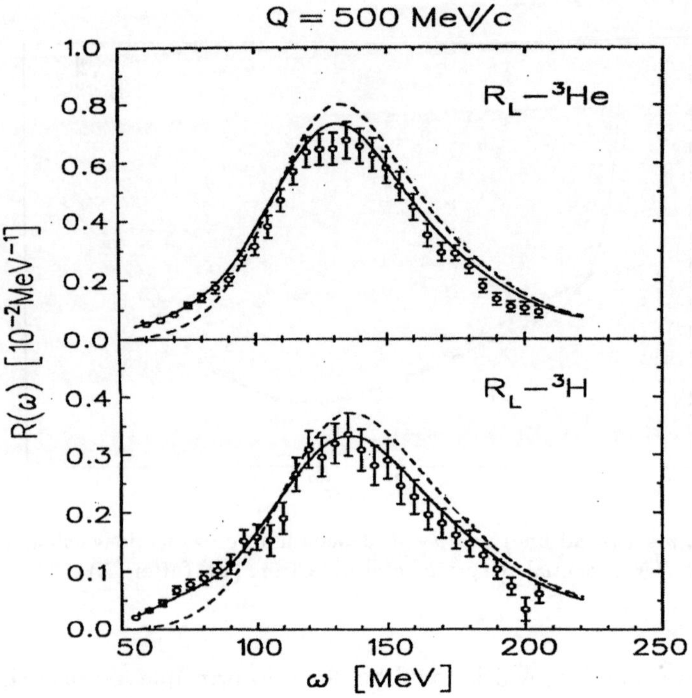

Figure 2. The longitudinal cross section for inclusive electron scattering off ^{3}He and ^{3}H. Dashed line: PWIA. Full line: PWIA+FSI. Experimental data from [27] (after [26])

the diagonal part of $P(\boldsymbol{k}, \boldsymbol{k}', E)$, defined by

$$P^N(k, E) = \sum_{f_{A-1}} \left|\langle \boldsymbol{k}, N; \psi^f_{A-1} | \psi_{J\mathcal{M}} \rangle\right|^2 \delta(E - E^f_{A-1} + E_A) \qquad (1)$$

where $|\psi_{J\mathcal{M}}\rangle$ is the ground state of the target nucleus, $|\psi^f_{A-1}\rangle$ the eigenstate of the $(A-1)$ nucleon system, $|\boldsymbol{k}, N\rangle$ the plane wave for the nucleon $N \equiv p(n)$.

Since Eq. (1) represents the joint probability to find in the nucleus a nucleon with momentum k and removal energy E, the PWIA yields for the cross section a convolution integral involving the structure function of an off-shell nucleon and the diagonal spectral function. For an arbitrary value of A, the latter is known only in the case of the mean field scheme. To go beyond such a model, one has to care about correlations, and this makes the derivation of $P(k, E)$ a difficult task. One is therefore forced to make approximations (this mainly concerns complex nuclei, for in the case of Nuclear Matter a many-body spectral function has been obtained both in the BBG and Variational approaches [23, 24, 25]). Once a sound model for the spectral function is obtained, the Final State Interaction (FSI) is introduced by some approximate schemes. Such a procedure might appear a very unsatisfactory one, but it is the best one can do, and since the understanding of the structure of complex nuclei is one of the primary goal of nuclear physics, serious efforts are to be made in order to understand the degree of reliabilty of these model approaches. In this respect calculations in few-body systems could be of great usefulness. Exact calculations for the inclusive cross section for electron scattering off ^{3}He and ^{3}H can at present be performed [26]. An example is shown in Fig. 2 (more recent results [28] were presented at the Conference). It can be seen from the exact calculations that the FSI is very relevant: it decreases the cross section at the peak and increses it at the low energy tail, the net result being a good agreement with the experimental data.

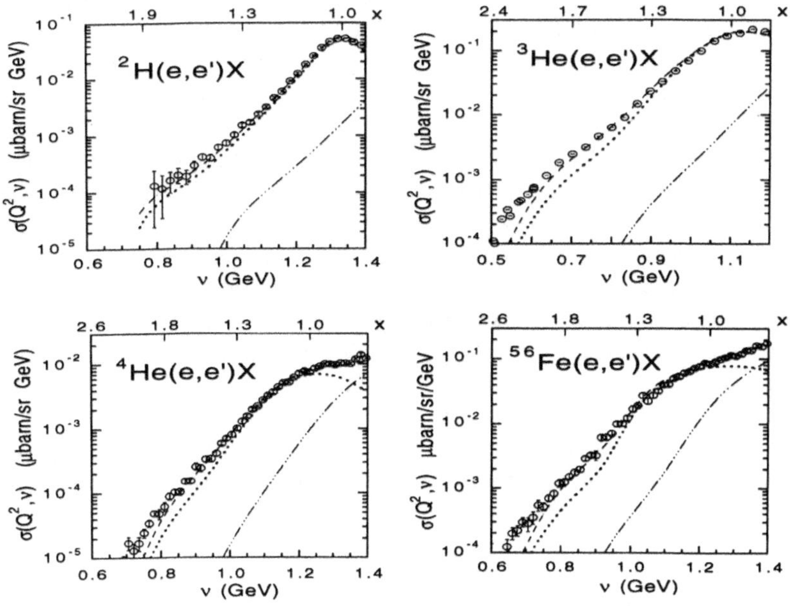

Figure 3. The inclusive cross section from few-body systems and complex nuclei. Dotted line:PWIA. Dashed line: PWIA+FSI. The double-dotted-dashed line represents the contribution from inelastic channels. Experimental data from [32]. $x = Q^2/2M\nu$ is the Bjorken variable. (After [29])

As already pointed out, for heavier nuclei the theoretical description of inclusive scattering has to rely on models. I will briefly present the results of one of these [29], which has interesting relationships with results obtained in few nucleon systems. The model is based upon a spectral function [30, 31] where correlations, which produce high momentum components, are due to ground state configurations characterised by two strongly correlated nucleons having almost equal and opposite momenta; moreover the correlation is similar to the one in a deuteron, the main difference being that the center-of-mass of the correlated pair moves in the mean field created by the other $A-2$ nucleons.

When the photon hits the nucleus, it may knock out either a mean–field, or a correlated nucleon; in the former case, the FSI can be taken into account by the usual multiple scattering theory, whereas in the latter case the FSI is the one acting between the two nucleons interacting in the continuum by the same NN interaction which produced the initial state correlations. The results of such an approach are presented in Fig. 3. It can be seen that the same FSI which is taken into account exactly in the case of A=2, can give a reasonable description in case of complex nuclei as well. More important, by comparing Figs. 2 and 3, it can be seen that the effect of FSI appears to be qualitatively the same in model and exact calculations. It should be pointed out that the experimental data considered in the two cases correspond to very different momentum transfers, but the extension of the model calculations to low momentum transfer confirm the same features. Thus exact calculations of FSI in few-nucleon systems are not only relevant for understanding the structure of the few-nucleon systems themselves but, provided the leading contribution to the FSI can be identified and calculation extended to high momentum transfer, they also help to justify the models which have to be developed for complex nuclei. It has been pointed out previously, that the motivation for probing few-body systems was the investigation of the nuclear wave function, so as to understand the reason for the underbinding of the systems and to obtain information about the nucleon dynamics. It can be seen that inclusive data can reasonably be explained even if the binding energy is not. This simply means that inclusive scattering is not very sensitive to the details of the wave function. More exclusive experiments, e.g. $(e, e'N)$, $(e, e'2N)$, $(e, e'\pi N)$, etc [33] are necessary in order to have the possibility to map out the nuclear wave function. As for the information that these data can provide on nucleon dynamics, it has been shown [34] that by means of a y-scaling analysis, and taking carefully into account the effect of FSI, useful information on the momentum distribution $n(k)$ and the spectral function can be obtained from the low energy side of the cross section.

4 Few–Body Systems vs. Complex Nuclei

It has recently been found that some quantities pertaining to the dynamics of a nucleon in a complex nucleus are similar to the same, properly rescaled, quantities in a few-nucleon system (for original references see [31]). These concern particularly the high momentum and high removal energy components. To illustrate this point,we show in Fig. 4 the nucleon momentum distributions for systems ranging from A=2 to Nuclear Matter.

It can be seen that: i)the mean field picture breaks down at $k > 1.5-2\,\text{fm}^{-1}$ (the value of the "Fermi momentum"), which means that $n(k)$ for higher momenta is governed by correlations; ii) for k larger than the Fermi momentum, $n(k)$ scales to the deuteron momentum distribution and, apart from the scaling factor, becomes A–independent; this means, in turn, that high momentum components are determined by the local properties of the nuclear medium, namely by the NN interaction. The sharp differences of the momentum distributions

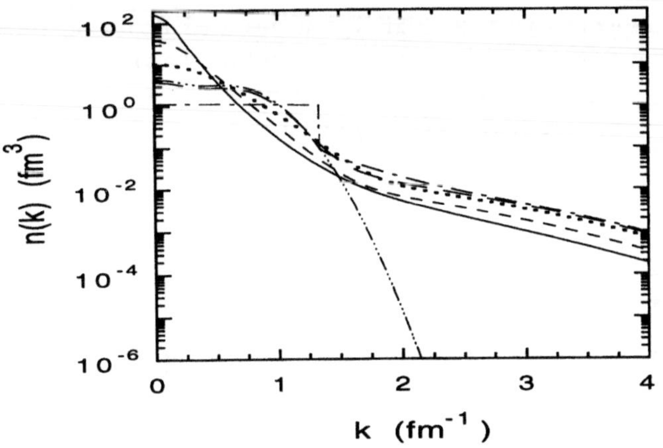

Figure 4. The theoretical many-body nucleon momentum distribution in ^{2}H (full), ^{3}He (dashed), ^{4}He (dots), ^{56}Fe (long dashed), and Nuclear Matter (short-long dashed); the double-dotted-dashed line is the mean-field result for ^{56}Fe (for original references see [31])

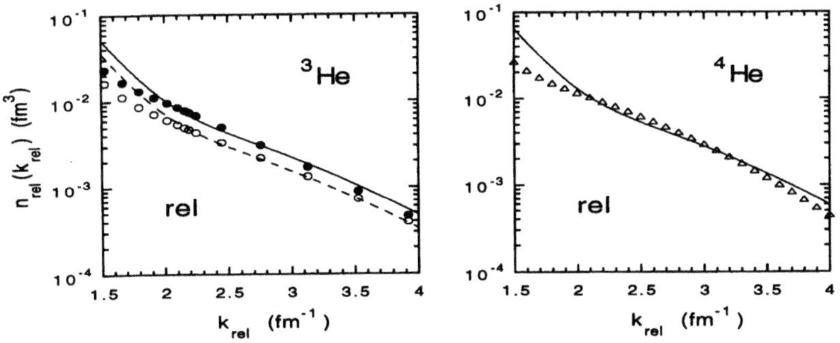

Figure 5. The high momentum part of the relative momentum distribution of a NN pair in ^{3}He and ^{4}He. Open (full) dots: *pp*(*pn*) pair from three-body Faddeev calculations [35]. Triangles: NN pair from the variational calculation of [36]. The full and dashed lines represent the rescaled momentum distribution of the deuteron (After [31])

at low k reflect the mean field properties of the nucleus, whereas their scaling at high momentum transfer reflects the local properties of the NN interaction in the medium.

The similarity between the two-, few- and many-nucleon systems is reiterated in Fig. 5 where the relative momentum distribution in the deuteron is compared with the ones of a NN pair in ^{3}He and ^{4}He.

It can be seen that the two-nucleon relative momentum distribution in few-nucleon systems appears to be, at high k, the rescaled deuteron momentum distribution. In Fig. 6 we show the partial momentum distribution

$$n_f(k) = \int_{E_{min}}^{E_f} P(k, E) dE \tag{2}$$

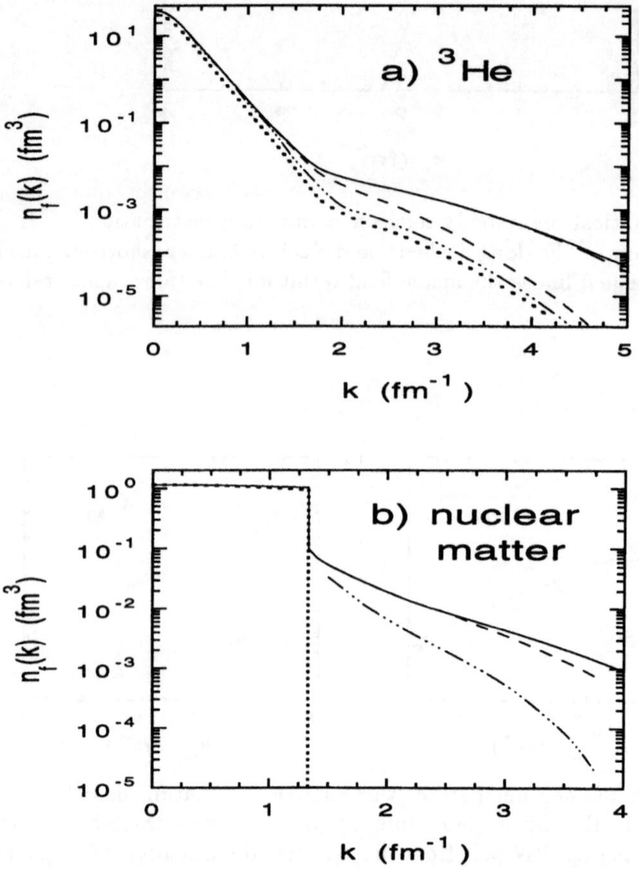

Figure 6. The saturation of the momentum sum rule in ^{3}He [37] (a) and Nuclear Matter [25] (b). The solid line is the full momentum distribution $n(k)$. In case of ^{3}He the dot–dashed, dashed, and long dashed lines correspond to the partial momentum distribution integrated up to E_f=17.75, 55.5, 305.5 MeV, whereas for nuclear matter the dot–dashed and dashed lines correspond to E_f=100 and 300 MeV, respectively.

It can be seen that at high values of k the energy distribution of a nucleon in the three–body system and nuclear matter behave in a similar way. The similarity between the high momentum and removal energy distributions in few- and many-nucleon systems has been used to derive the spectral function employed in the calculation of the inclusive cross sections shown in Fig. 3. As a matter of fact, the main ingredients of this spectral function are the momentum distributions of the relative and CM motions of a pair in the medium. If these are taken from few-body calculations a spectral function for nuclear matter is obtained, which is in very good agreement with the one obtained from many-body calculations. Thus, the knowledge of such basic quantities pertaining the dynamics of a nucleon in the two- and three-nucleon systems (e.g.the momentum and energy distributions), can provide useful information on the structure of complex nuclei. Moreover, the fact that at high values of k and E there is a strong similarity between light and complex nuclei, is evidence that in this region nuclear structure is mainly governed by NN correlations.

5 Polarised Few-Nucleon Systems and Nucleon Spin Structure Functions

With the recent advent of high quality polarised beams and targets, polarised few-nucleon systems started to play a relevant role in the study of spin structure functions of the nucleon [38, 39]. The scattering of polarised electrons off polarised ^3He, viz. the processes $^3\vec{\mathrm{He}}(\vec{e},e')X$, $^3\vec{\mathrm{He}}(\vec{e},e'N)X$, etc. can provide valuable information on the neutron properties, provided nuclear effects can reliably be evaluated. Let us consider inclusive scattering. If we have linearly polarised electrons impinging on a spin 1/2 target A, the hadronic tensor which describes the scattering in Born Approximmation depends, besides the usual unpolarised nuclear structure functions $W^A_{1(2)}(\nu, Q^2)$, upon the polarised or spin structure functions $G^A_{1(2)}(\nu, Q^2)$. These quantities can be obtained from the asymmetries $A_{||}$ and $A_\perp$:

$$A_{||} = (d\sigma_{\uparrow\uparrow} - d\sigma_{\uparrow\downarrow})/(d\sigma_{\uparrow\uparrow} + d\sigma_{\uparrow\downarrow}) = a_1 G_1^A + a_2 G_2^A \tag{3}$$

$$A_\perp = (d\sigma_{\uparrow\rightarrow} - d\sigma_{\uparrow\leftarrow})/(d\sigma_{\uparrow\rightarrow} + d\sigma_{\uparrow\leftarrow}) = b_1 G_1^A + b_2 G_2^A \tag{4}$$

where a and b are proper kinematical factors and $d\sigma_{\uparrow\uparrow(\uparrow\downarrow)}$ is a short hand notation for the inclusive cross section corresponding to the scattering off a target with polarisation vector parallel ($\uparrow\uparrow$) (antiparallel ($\uparrow\downarrow$)) to the electron spin and $\sigma_{\uparrow\rightarrow(\uparrow\leftarrow)}$ the one for a target polarisation vector perpendicular ($\uparrow\rightarrow$) (antiperpendicular ($\uparrow\leftarrow$)) to the electron spin. The spin structure functions $G_{1(2)}(\nu, Q^2)$ have different physical meaning depending on the value of ν and Q^2. Let us consider the nucleon ($A = N$). At the elastic kinematics, $\nu = Q^2/2M$, the spin structure functions are defined in terms of well known quantities, the nucleon elastic Sachs form factors, so that by performing quasi elastic experiments off a polarised ^{3}He target one could obtain information on the neutron elastic form factors. Away from the elastic kinematics, in the resonance ($W_{\mathrm{th}} < W < 2$ GeV, W being the invariant mass) and DIS ($W \gg 2$ GeV) regions, one investigates the internal spin structure of the nucleon. Thus $G^N_{1(2)}(\nu, Q^2)$ provide information on various models for the nucleon structure. Moreover they satisfy several sum rules which are derived from general principles which can therefore be tested in a model independent way. To illustrate the relevance of the experimental knowledge of the spin structure functions, let me consider two examples. It is well known that in the Bjorken limit $\nu, Q^2 \to \infty$, x constant, one has, using the parton model,

$$\nu G_1^N(\nu, Q^2) = g_1^N(x, Q^2) \to g_1^N(x) = \frac{1}{2}\sum_i q_i^2[f_i^\uparrow(x) - f_i^\downarrow(x)] \tag{5}$$

q_i being the charge of the flavor i and $f_i^{\uparrow(\downarrow)}$ the momentum distribution of the quark with spin parallel ($\uparrow$) or antiparallel ($\downarrow$) to the nucleon spin. The asymmetry in this limit becomes $A = g_1^N(x)/F_1^N(x)$, where $F_1^N(x) = MW_1^N(x)$ is the usual unpolarised structure function which is known from unpolarised inclusive scattering. On the other hand, $g_1^N(x)$ satisfies the Bjorken sum rule [40]

$$\Gamma_p(Q^2) - \Gamma_n(Q^2) = (1/6)(g_A/g_V)(1 - \alpha_s(Q^2/\pi) + \ldots) \tag{6}$$

where

$$\Gamma_N(Q^2) = \int_0^1 dx g_1^N(x, Q^2) \tag{7}$$

and $g_{A(V)}$ are the axial (vector) beta–decay weak coupling constants. For real photons, one has the Gerasimov–Drell–Hearn (GDH) sum rule for the nucleon [41]

$$I_{\mathrm{GDH}}^N(Q^2 = 0) = \int_{\nu_{th}}^{\infty} \frac{d\nu}{\nu}[\sigma_{1/2}^N(\nu, 0) - \sigma_{3/2}^N(\nu, 0)] = -\frac{2\alpha\pi^2}{M^2}\kappa^N \tag{8}$$

where $\sigma_{1/2(3/2)}(\nu, Q^2)$ is the total photoabsorption cross section for total helicity 1/2 (3/2) and κ is the nucleon anomalous magnetic moment. For electron scattering one can define a GDH integral $I_{\mathrm{GDH}}(Q^2)$, which, considering the relation between the total photoabsorption cross sections and the spin structure functions, has the following form [42]

$$I_{\mathrm{GDH}}^N(Q^2) = \frac{16\alpha\pi^2}{Q^2}\int_0^1 dx g_1^N(x, Q^2) = \frac{16\alpha\pi^2}{Q^2}\Gamma_N(Q^2) \tag{9}$$

A calculation of the Q^2 evolution of the GDH integral from the real photon limit to DIS shows an enormous variation of the helicity structure of $\gamma^* - N$ coupling (cf Fig. 7), due to the excitation of nucleon resonances.

An experimental investigation of the Q^2 evolution of the GDH integral would therefore provide stringent tests of the spin structure of the nucleon. The investigation of the spin structure function of the proton has been carried out in the DIS region and very few data exist in the resonance region. In case of the neutron, the experimental activity is only beginning through the use of polarised deuterium or ^{3}He targets. The knowledge of the neutron spin structure functions (which have to be extracted from data on nuclear targets) is of great relevance: first, it provides information on the structure of the neutron itself; secondly, it allows the experimental check of the Bjorken and GDH sum rules. To obtain the neutron structure functions from the nuclear ones, one

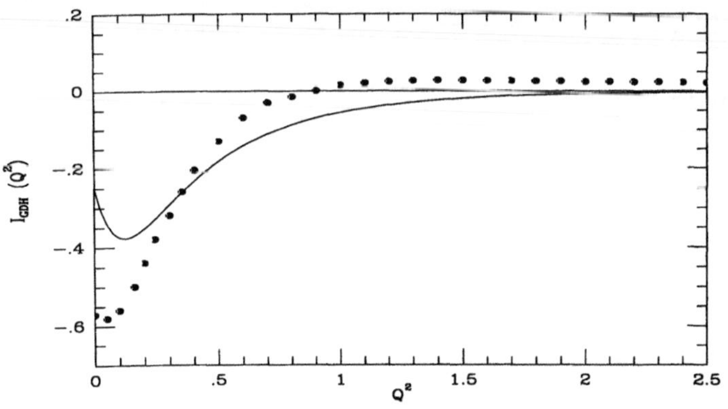

Figure 7. The proton (dotted [43]) and neutron(full [44]) GDH integral (8) vs Q^2 calculated within the constituent quark parton model

has first of all to relate the two quantities. It is here that theoretical few-body physics is playing a crucial role. One has to take into account, first, the Fermi motion and binding of the neutron (EMC effect) and, second, the proton contamination to the asymmetry, generated by the S' and D waves of the three–body wave function.

Recently [45, 46] the neutron and ^{3}He spin structure functions have been related by a convolution model similar to the one used to calculate unpolarised cross sections, the only relevant difference being the replacement of the spin independent spectral function (Eq. (1)) by the spin dependent one which is a 2x2 matrix whose elements are [47]

$$P^N_{\sigma,\sigma',\mathcal{M}}(\boldsymbol{k},E) = \sum_{f_{A-1}} \langle \boldsymbol{k},N,\sigma;\psi^f_{A-1}|\psi_{J\mathcal{M}}\rangle\, \langle\psi_{J\mathcal{M}}|\psi^f_{A-1};\boldsymbol{k},N,\sigma'\rangle \, \delta(E-E^f_{A-1}+E_A) \tag{10}$$

where $|\psi_{J\mathcal{M}}\rangle$ is the ground state of the target nucleus polarized along $\boldsymbol{S}_A$, and $|\boldsymbol{k},N,\sigma\rangle$ is the plane wave for the nucleon. Let us briefly summarise the experimental situation and theoretical interpretations in this field. In case of quasi elastic kinematics experimental data have been obtained at BATES [38]. The interpretation show a relevant role played by nuclear effects which however are under control. An example is given in Fig. 8.

The theoretical results shown there do not include FSI, which have to be considered for a quantitative analysis of the data. The proton contribution to the asymmetry is a relevant one, which makes the extraction of the neutron form factor a difficult task (a proper kinematical set up which reduces the proton contribution has been suggested in [45]). Nuclear effects have been shown [48] to play a minor role in the DIS region, where the ^{3}He structure function can be reliably approximated by

$$g_1^3(x) = [2p_p g_1^p(x) + p_n g_1^n(x)] \tag{11}$$

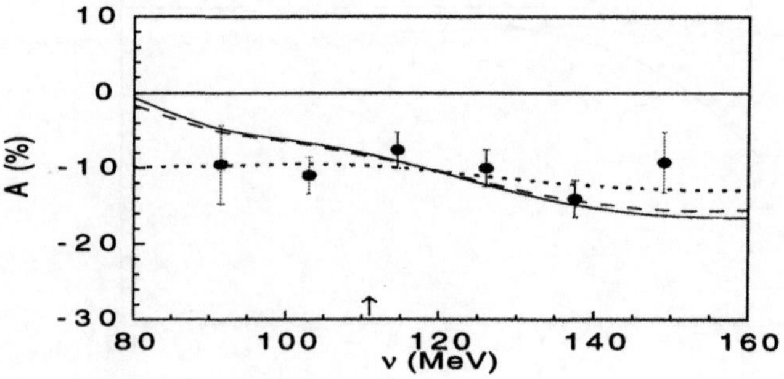

Figure 8. The $A_{T'}$ asymmetry vs the energy transfer ν in the process $^3\vec{\mathrm{He}}(\vec{e},e')X$. Experimental points from [50]. Theoretical curves are from [45] (full) and [46] (dashed). The dotted line represents the proton contribution and the arrow denotes the position of the quasi elastic peak

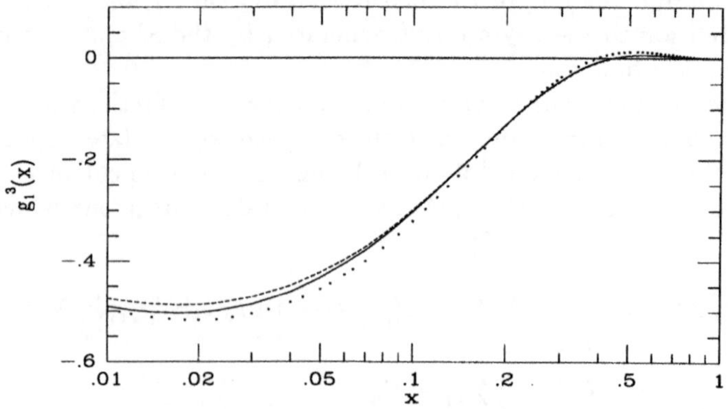

Figure 9. The spin structure function of ^{3}He (full) compared with the free neutron one (dotted) and with the Eq. (11) (dashed). (After [48])

where $g_1^{p(n)}$ are the free proton (neutron) spin structure function and $p_{p(n)}$ the effective proton (neutron) polarization which are very well known from three-body calculations [49]; thus, as shown in Fig. 9, for $x < 0.7$ nuclear effects can safely be taken care of by (11).

In the resonance region only preliminary theoretical results are available which seem to show that nuclear effects are very relevant and that the simple formula (11) cannot be applied [44]. Theoretical work in this field is called for, since extensive experimental programs are planned at MAMI and CEBAF to investigate the GDH sum rule and its Q^2 evolution, both on the proton, deuteron and ^{3}He.

6 Conclusions

The theoretical and experimental investigations of the few-nucleon systems have reached a high degree of sophistication and accuracy; such an activity has important impacts on the knowledge of the structure of nucleons and nuclei. What I presented in my talk is only a small fraction of the large number of fundamental topics discussed during the Conference, viz. the refinement of our knowledge of the two- and three-body forces [14, 16, 51, 52], the developments of proper approaches to implement relativistic effects [8], the application of few-body techniques to other fields (e.g. quark, atomic, and molecular physics and astrophysics [53]), the physics of antiprotonic and strange systems [54], the muon catalyzed fusion [55], and others.

It is evident that Few-Body physics is a vital and rich field, and while the merits are due to our whole Community, they are certainly due to those people who periodically undertake the heavy job of organising that important forum represented by the EFB Conference. It is therefore with pleasure and acknowledgment that I thank, on behalf of all of you, *Rafael Guardiola, Jesús Navarro, Eulogio Oset, Josep Ros* and *Vicente Vento*, for having organised such a useful and stimulating Conference.

References

1. A. Faessler: Nucl. Phys. **A508**, 617c 12(1990)
2. E. Krotscheck: this Conference; O. Kartavcev: this Conference; I. Lindgren: this Conference
3. W. Gloeckle: this Conference
4. S. Rosati: this Conference
5. J. Carlson: this Conference
6. H. Witała, D. Hüber, W. Gloeckle: Phys. Rev. **C49**, R14 (1994)
7. A. Kievsky, M. Viviani, S. Rosati: Phys. Rev. **C52**, R15 (1995)
8. J.A. Tjon, F.M. Lev: Panel on *Relativistic effects*, this Conference
9. F. Gross: this Conference
10. F. Coester: Panel on *Relativistic approaches to hadrons and nuclei at medium energies*, 7th ICTP Workshop, May 1995, Trieste, World Scientific, to be published
11. D. Hüber et al.: Phys. Rev. **C51**, 1100 (1995)
12. H. Kamada, W. Gloeckle: this Conference; W. Gloeckle and H. Kamada: Phys. Rev. Lett. **71**, 971 (1993)

13. M. Lacombe et al.: Phys. Rev. **C21**, 861 (1980)

14. J.J. de Swart: this Conference

15. R.B. Wiringa, R.A. Smith, P.L. Ainsworth: Phys. Rev. **C29**, 1207 (1981)

16. K. Holinde: this Conference

17. D. Plümper, J. Flender, M.F. Gari: Phys. Rev. **C49**, 2370 (1991)

18. J.Carlson, R. Schiavilla: Few-Body Systems Suppl. **7**, 349 (1994)

19. K. Sagara et al.: Phys. Rev. **C50**, 576 (1994)

20. A. Faessler: this Conference

21. J.Arvieux and J. van den Brand: Panel on *Few-Body Systems probed in several ways*, this Conference

22. G.Smith: this Conference; E. Oset: this Conference

23. A. Ramos, A. Polls, W.H. Dickoff: Nucl. Phys. **A503**, 1 (1989)

24. M. Baldo et al.: Nucl. Phys. **A545**, 741 (1992)

25. O. Benhar, A. Fabrocini, S. Fantoni: Nucl. Phys. **A505**, 267 (1989)

26. E. van Meijgaard, J.A. Tjon: Phys. Rev. **C45**, 1463 (1992)

27. K. Dow et al.: Phys. Rev. Lett. **61**, 1706 (1988)

28. J. Golak et al.: this Conference and to appear in Phys. Rev. **C**

29. C. Ciofi degli Atti, S. Simula: Phys. Lett. **325B**, 276 (1994)

30. C. Ciofi degli Atti, S. Simula, L. Frankfurt, M. Strikman: Phys. Rev. **C44**, R7 (1991)

31. C. Ciofi degli Atti, S. Simula: to appear

32. B. D. Day et al.: Phys. Rev. **C48**, 1849 (1993)

33. T. Walcher: this Conference

34. C.Ciofi degli Atti, E. Pace, G. Salmé: Phys. Rev. **C43**,1155 (1991)

35. H. Meier-Hajduk, P.U. Sauer, W. Thies: Nucl. Phys. **A395**, 332 (1983)

36. H. Morita,Y. Akaishi, H. Tanaka: Progr. Theor. Phys. **79**, 863 (1988)

37. C.Ciofi degli Atti, E. Pace, G. Salmé: Phys. Lett. **147B**, 14 (1984)

38. S. Kowalski: this Conference

39. M. Litmaath: this Conference

40. J.D. Bjorken: Phys. Rev. **148**, 1467 (1966)

41. S. B. Gerasimov: Yad. Fiz. **2**, 839 (1965); S.D. Drell, A.C. Hearn: Phys. Rev. Lett. **162**, 1520 (1966)

42. M. Anselmino, B. L. Ioffe, E. Leader: Sov. J. Nucl. Phys. **49**, 136 (1989)

43. V. Burkert, Z. Li: Phys. Rev. **D47**, 46 (1993)

44. S. Scopetta, C. Ciofi degli Atti: 7th ICTP Workshop, May 1995, Trieste, World Scientific, to appear

45. C. Ciofi degli Atti, E. Pace, G. Salmé: Phys. Rev. **C51**, 1108 (1995)

46. R.W. Schultze, P.U. Sauer: Phys. Rev. **C48**, 38 (1993)

47. C. Ciofi degli Atti, E. Pace, G. Salmé: Phys. Rev. **C46**, R1591 (1992)

48. C. Ciofi degli Atti, S. Scopetta, E. Pace, G. Salmé: Phys. Rev. **C48**, R968 (1993)

49. J. Friar et al.: Phys. Rev. **C42**, 2310 (1990)

50. H. Gao et al.: Phys. Rev. **C51**, R546 (1994)

51. J. Friar: this Conference

52. T. E. O. Ericson: this Conference

53. L. Kiseleva: this Conference; O. Oberhummer: this Conference

54. E. Klempt: this Conference; A. Boudard: this Conference; G. Anton: this Conference; V. Belayev: this Conference

55. C. Petitjean: this Conference

List of Participants

Adam, R.
Department of Physics
University of Natal
Private Bag X01
Scottsville 3209
South Africa

Alt, E.O.
Institut für Physik (THEP)
Universität Mainz
Staudingerweg 7, D-55099 Mainz
Germany

Anklin, H.
Department of Physics
University of Basel
Basel
Switzerland

Anton, G.
Physikalisches Institut
Universität Bonn
Nussallee 12
D-53115 Bonn 1
Germany

Arvieux, J.
Laboratoire National Saturne
Centre d'Etudes de Saclay
F-91191 Gif-sur-Yvette Cedex
France

Bakker, B.L.G.
Free University
Dept. of Physics and Astronomy
De Boelelaan 1081
Amsterdam NL-1081 HV
The Netherlands

Barnea, N.
Racah Institute of Physics
Hebrew University
Jerusalem 91904
Israel

Belyaev, V.
Laboratory of Theoretical Physics
Joint Institut for Nuclear Research
141980 Dubna
Russia

Blokhintsev, L.
Institute of Nuclear Physics
Moscow State University
Vorob'jovy Gory
Moscow 119899
Russia

Borromeo, M.
Dipartimento di Fisica
Universitá di Perugia
Via A.Pascoli
06100 Perugia
Italy

Boudard, A.
SPHN-DAPNIA
CEA-Saclay
F-91191 Gif-Sur-Yvette
France

Braghieri, A.
INFN - Sezione di Pavia
Via BASSI 6
Pavia I-27100
Italy

Cano, F.
Departamento de Física Teórica
Universidad de Valencia
Avda. Dr. Moliner, 50
46100-Burjasot (Valencia)
Spain

Carbonell, J.
Institut des Sciences Nucléaires
53, Av. des Martyrs
Grenoble 38026 Cedex
France

Carlson, J.A.
Los Alamos National Laboratory
M.S. B283
Los Alamos, NM 87545
USA

Chandler, C.
Dept. of Physics and Astronomy
University of New Mexico
Albuquerque, NM 87131
USA

Chmielewski, K.
Institute of Theoretical Physics
University of Hannover
Appelstrasse2
D-30167 Hannover
Germany

Ciesielski, F.
Institut des Sciences Nucléaires
53 Avenue des Martyrs
Grenoble 38026 Cedex
France

Ciofi degli Atti, C.
Dipartimento di Fisica
Universita di Perugia
Via A. Pascoli
I-06100 Perugia
Italy

Cotanch, S.R.
North Carolina State University
Raleigh NC 27695-8202
USA

Dalkarov, O.
Lebedev Physical Institute
Leninsky pr. 53
Moscow 117924
Russia

Damjanović, M.
Faculty of Pharmacy
Department of Physics
Vojvode Stepe 450, P.O.B. 146
Beograd YU-11000
Jugoslavija

de Swart, J.J.
Institute for Theoretical Physics
University of Nijmegen
P.O. Box 9010
NL-6500 GL Nijmegen
The Netherlands

Dmitrieva, L.
Department of Mathematical
and Computational Physics
Faculty of Physics
St.Petersburg State University
St.Petersburg 198904
Russia

Desplanques, B.
Institut des Sciences Nucléaires
53 avenue des Martyrs
F-38026 Grenoble Cedex
France

Dortmans, P.J.
Department of Physics
University of Melbourne
Parkville 3052
Australia

Eden, J.
Institut für Theoretische Physik II
Ruhr Universität Bochum
Universitaetstrasse 150
D-44780 Bochum
Germany

Efimov, G.V.
Bogolubov Laboratory of
Theoretical Physics
Joint Institute for Nuclear Research
141980 Dubna, Moscow region
Russia

Efros, V.
Russian Research Centre
Kurchatov Institute
Kurchatov Square, 1
Moscow 123182
Russia

Eiró, A.M.
Centro de Fisica Nuclear
Universidad de Lisboa
Avda. Profesor Gama Pinto, 2
P 1699 LISBOA CODEX
Portugal

Ellerkmann, G.
Physikalisches Institut
Universität Bonn
Endenicher Allee 11-13
Bonn, D-53115
Germany

Ericson, T.E.O.
Theory Division
CERN, CH-1211 GENEVE 23
Switzerland

Fabre de la Ripelle, M.
Institut de Physique Nucléaire
Université Paris-Sud
P.O. Box 1
F-91406 Orsay Cedex
France

Fabrocini, A.
INFN
Universitá di Pisa
Piazza Torricelli 2
I-56100 Pisa
Italy

Faessler, A.
Institut für Theoretische Physik
Universität Tübingen
Auf der Morgenstelle 14
D-72076 Tübingen
Germany

Fernández, F.
Grupo de Física Nuclear
Universidad de Salamanca
E-37008 Salamanca
Spain

Friar, J.L.
Los Alamos National Laboratory
M.S. B283
Los Alamos, NM 87545
USA

Gibbs, W.R.
Department of Physics
New Mexico State University
Las Cruces NM 88003
USA

Glöckle, W.
Institut für Theoretische Physik
Ruhr-Universität Bochum
Universitaetsstr. 150
P.O. Box 102148
D-44780 Bochum
Germany

Gómez, J.M.G.
Departamento de Física Nuclear
Universidad Complutense
MADRID
Spain

Gómez-Tejedor, J.A.
Departamento de Física Teórica
Universidad Valencia
Avda. Dr. Moliner, 50
46100-Burjasot (Valencia)
Spain

González, P.
Departamento de Física Teórica
Universidad Valencia
Avda. Dr. Moliner, 50
46100-Burjasot (Valencia)
Spain

Gross, F.
Physics Division MS 12 H
CEBAF
Newport News, VA 23606
and
Department of Physics
College of William and Mary
Williamsburg, VA 23185 USA

Guardiola, R.
Departamento de Física Nuclear
Universidad Valencia
Avda. Dr. Moliner, 50
46100-Burjasot (Valencia)
Spain

Haberzettl, H.
Department of Physics
The George Washington University
Washington, D.C. 20052
USA

Haidenbauer, J.
Institut für Kernphysik
Forschungszentrum Jülich
Postfach 1913
D-52425 Jülich
Germany

Hernández, E.
Grupo de Física Nuclear
Universidad de Salamanca
E-37008 Salamanca
Spain

Hirenzaki, S.
Departamento de Física Teórica
Universidad Valencia
Avda. Dr. Moliner, 50
46100-Burjasot (Valencia)
Spain

Holinde, K.
Institut für Kernphysik
Forschungszentrum Jülich
Postfach 1913
D-52425 Jülich
Germany

Hüber, D.
Institut für Theoretische Physik 2
Ruhr-Universität Bochum
D-44780 Bochum
Germany

Jans, E.
NIKHEF-K
P.O. Box 4395
NL-1009 AJ Amsterdam
The Netherlands

Kalantar-Nayestanaki, N.
Kernfysisch Versneller Instituut
Zernikelaan 25
NL-9747 AA Groningen
The Netherlands

Kaldor, U.
School of Chemistry
Tel Aviv University
69978 Tel Aviv
Israel

Kamada, H.
Institut für Theoretische Physik II
Ruhr-Universität
NB6 / 125, Universitätstr. 150
Bochum D44780
Germany

Kartavtsev, O.I.
Bogolubov Laboratory of
Theoretical Physics
Joint Institute for Nuclear Research
Dubna, Moscow region, 141980
Russia

Kievsky, A.
Istituto Nazionale di Fisica Nucleare
Dipartimento di Fisica
Piazza Torricelli 2
56100 Pisa
Italy

Kim, Y.E.
Department of Physics
Purdue University
West Lafayette, IN 47907–1396
USA

Kiseleva, L.
University of Cambridge
Institute of Astronomy
Madingley Road
Cambridge CB3 0HA
UK

Kistryn, St.
Institute of Particle Physics
ETH Hoenggerberg HPK
CH-8093 Zürich
Switzerland

Klempt, E.
Institut fur Strahlen- und Kernphysik
Universität Bonn
Nussallee 14 - 16
D53115 BONN
Germany

Klomp, R.A.M.M.
University of Nijmegen
P.O. Box 9010
NL-6500 GL Nijmegen
The Netherlands

Kobushkin, A.
Bogoliubov Institute of
Theoretical Physics
National Academy of Sciences
of Ukraine
Kiev 252143
Ukraine

Kochelev, N.
Joint Institute for Nuclear Research
Laboratory of Theoretical Physics
Dubna 141980
Russia

Kopaleishvili, T.
High Energy Physics Institute
Tbilisi State University
9, University st.,
380 086 Tbilisi
Georgia

Kowalski, S.
Lab. for Nuclear Science
MIT 26-427
Cambridge, MA 02139
USA

Krivec, R.
J. Stefan Institute
Department of Theoretical Physics
Jamova 39, PO Box 100
Ljubljana SI-61111
Slovenia

Krotscheck, E.
Institut für Theoretische Physik
Johannes Kepler Universität
A4040 LINZ
Austria

Krutenkova, A.
Institute of Theoretical and
Experimental Physics
B.Cheremushkinskaya,25
Moscow 117259
Russia

Kuperin, Yu.
Department of Mathematical
and Computational Physics
Faculty of Physics
St.Petersburg State University
St.Petersburg 198904
Russia

Kvitsinsky, A.A.
Physics Department
California State University
Long Beach, CA 90840
USA

Larsen, S.
Department of Physics
Temple University
Philadelphia, PA 19122
USA

Leidemann, W.
Dipartimento di Fisica
Università di Trento
Povo (Trento) I-38050
Italy

Lengyel, V.
Uzhgorod State University
Department of Theoretical Physics
Voloshyna st. 32/7
Uzhgorod 294000
Ukraine

Lev, F.M.
Laboratory of Nuclear Problems
JINR, DUBNA
Moscow region 141980
Russia

Lindgren, I.
Department of Physics
Chalmers University of Technology
University of Gothenburg
S-412 96 Gothenburg
Sweden

Litmaath, M.
NIKHEF-K
P.O. Box 4395
NL-1009 AJ Amsterdam
The Netherlands

Loiseau, B.
Institut de Physique Nucléaire
Université Paris-Sud
P.O. Box 1
F-91406 Orsay Cedex
France

Marco-Rubio, E.
Departamento de Física Teórica
Universidad Valencia
Avda. Dr. Moliner, 50
46100-Burjasot (Valencia)
Spain

Martinus, G.H.
Kernfysisch Versneller Instituut (KVI)
Zernikelaan 25
9747 AA Groningen
The Netherlands

Maschuw, R.
Institut für Strahlen- und Kernphysik
Universität Bonn
Nussallee 14-16
53115 Bonn
Germany

Matveenko, A.
Bogoliuvov Laboratory of
Theoretical Physics
Joint Institute for Nuclear Research
Moscow Region DUBNA
Russia

Moliner, P.I.
Departamento de Física Nuclear
Universidad Valencia
Avda. Dr. Moliner, 50
46100-Burjasot (Valencia)
Spain

Motovilov, A.K.
Bogoliubov Laboratory of
Theoretical Physics
Joint Institute for Nuclear Research
141980 Dubna
Russia

Mukhamedzhanov, A.
Cyclotron Institute
Texas A&M University
College Station, Texas 77843
USA

Nagarajan, M.A.
Laboratori Nazionali di Legnaro
Via Romea, 4
Legnaro (Padova)
I-35020 ITALY

Najmeddine, M.
Center for Nuclear Studies
The George Washington University
1520 N Pierce st, apt 8
Arlington, VA 22209
USA

Narodetski, I.
Institute of Theoretical and
Experimental Physics
ITEP, B.Cheremushkinskaya 25
Moscow 117259
Russia

Navarro, J.
IFIC
CSIC-Universidad de Valencia
Avda. Dr. Moliner, 50
46100-Burjasot (Valencia)
Spain

Nemoto, S.
Institute of Theoretical Physics
University of Hannover
Appelstrasse2, D-30167 Hannover
Germany

Nieuwenhuis, T.
Institute for Theoretical Physics
University of Utrecht
Princetonplein 5, P.O. Box 80.006
Utrecht 3508 TA
The Netherlands

Niskanen, J.A.
University of Helsinki
Department of Physics
Box 9
FIN-00014 Helsinki
Finland

Oberhummer, H.
Institut für Kernphysik
Techn. Universität Wien
Wiedner Hauptstr. 8-10
A-1040 Wien
Austria

Orlandini, G.
Dipartimento di Fisica
Università di Trento
Povo (Trento) I-38050
Italy

Oryu, S.
Department of Physics
Science University of Tokyo
2641 Yamazaki
Noda Chiba 278
Japan

Oset, E.
Departamento de Física Teórica
Universidad Valencia
Avda. Dr. Moliner, 50
46100-Burjasot (Valencia)
Spain

Ostrovsky, V.N.
Institute of Physics
The University of St Petersburg
198904 St Petersburg
Russia

Pace, E.
Dipartimento di Fisica
Università di Roma 'Tor Vergata'
Via della Ricerca Scientifica, 1
I-00133 Roma
Italy

Pascalutsa, V.
Institute for Theoretical Physics
University of Utrecht
Princetonplein 5, P.O. Box 80006
Utrecht 3508 TA
The Netherlands

Petitjean, C.
Paul Scherrer Institut
CH-5232 Villigen PSI
Switzerland

Pinto, J.P.
Centro de Fisica Nuclear
INIC
Av. Gama Pinto 2
P-1699 Lisboa Codex
Portugal

Plessas, W.
Institute for Theoretical Physics
University of Graz
Universitätsplatz 5
Graz A-8010
Austria

Polls, A.
Departament d'Estructura
i Constituents de la Matèria
Universitat de Barcelona
Diagonal 647
08028 Barcelona
Spain

Porras, I.
Departamento de Física Moderna
Universidad de Granada
18008 Granada
Spain

Protasov, K.
Institut des Sciences Nucléaires
53, Avenue des Martyrs
38026 Grenoble Cedex
France

Prymas, W.
Institute for Medium Energy Physics
Austrian Academy of Sciences
Boltzmanngasse 3
A-1090 Wien
Austria

Rahimov, A.
Theoretical Physics Department
Tashkent State University
Tashkent-700095
Uzbekistan

Ramos, A.
Departament d'Estructura
i Constituents de la Matèria
University of Barcelona
Diagonal 647
08028 Barcelona
Spain

Rentmeester, M.
University of Nijmegen
Institute for Theoretical Physics
Toernooiveld 1
Nijmegen 6525 ED
The Netherlands

Ros, J.
Departamento de Física Teórica
Universidad Valencia
Avda. Dr. Moliner, 50
46100-Burjasot (Valencia)
Spain

Rosati, S.
INFN
Università di Pisa
Piazza Torricelli 2
I-56100 Pisa
Italy

Rosina, M.
Department of Physics
University of Ljubljana
Jadranska 19, POB 64
Ljubljana SI-61111
Slovenia

Rubio-Melon, F.J.
University of Nijmegen
Institute for Theoretical Physics
Faculty of Science
Nijmegen 6525 ED
The Netherlands

Saghai, B.
Commisariat a L'Energie Atomique
SPHN, Bat. 703
Centre d'Etudes de Saclay
Gif-sur Yvette
91191 CEDEX
France

Salmè, G.
Istituto Nazionale di Fisica Nucleare
Sezione Sanitá
Viale Regina Elena 299
Rome I-00161
Italy

Sakuler, W.
Institut für Kernphysik, TU Wien
Wiedner Hauptstraße 8-10
Wien A-1040
Austria

Sandhas, W.
Physikalisches Institut
Universität Bonn, D-53115 Bonn
Germany

Santos, F.D.
Centro de Fisica Nuclear
Universidad de Lisboa
Avda. Profesor Gama Pinto, 2
P 1699 LISBOA CODEX
Portugal

Sauer, P.U.
Institute of Theoretical Physics
University of Hannover
Appelstrasse2, D-30167 Hannover
Germany

Schadow, W.
Physikalisches Institut
Universität Bonn
Endenichev Allee 11-13
D-53115 Bonn 1
Germany

Schellingerhout, N.W.
Institute of Theoretical Physics
University of Hannover
Appelstrasse2, D-30167 Hannover
Germany

Scopetta, S.
Dipartimento di Fisica
Università di Perugia
Via A. Pascoli
I-06100 Perugia
Italy

Shebeko, A.
Bogoliubov Laboratory of
Theoretical Physics
Joint Institute for Nuclear Research
Dubna 141980
Russia

Shmatikov, M.
Russian Research Center
Kurchatov Institute
123182 Moscow
Russia

Simula, S.
Istituto Nazionale di Fisica Nucleare
Sezione Sanitá
Viale Regina Elena 299
Rome I-00161
Italy

Smith, G.
TRIUMF
4004 Wesbrook Mall
Vancouver, B.C., V6T 2A3
Canada

Sofianos, S.A.
Physics Department
University of South Africa
Pretoria 0001
South Africa

Spaltro, C.
NIKHEF-K
P.O. Box 41882
1009 DB Amsterdam
The Netherlands

Sprung, D.W.
Dept. of Physics and Astronomy
McMaster University
Hamilton, Ont. L8S 4M1
Canada

Strokovsky, E.A.
Laboratory for High Energies
Joint Institute for Nuclear Research
141980, Dubna, Moscow region
Russia

Szczurek, A.
Institute of Nuclear Physics
ul.Radzikowskiego 152
Kraków 31-342
Poland

Tjon, J.A.
Institute for Theoretical Physics
University of Utrecht
Princetonplein 5, De Uithof
P.O. Box 80006
NL-3508 TA Utrecht
The Netherlands

Tornow, W.
Department of Physics
Duke University and
Triangle Universities
Nuclear Laboratory
Durham, NC 27708
USA

Traini, M.
Dipartimento di Fisica
Università di Trento
Povo (TN) I-38050
Italy

Truhlík, E.
Institute of Nuclear Physics
Czech Academy of Sciences
Rez n. Prague CZ 250 68
Czech Republic

Valcarce, A.
Grupo de Física Nuclear
Universidad de Salamanca
E-37008 Salamanca
Spain

van Dantzig, R.
NIKHEF-K
P.O. Box 4395
NL-1009 AJ Amsterdam
The Netherlands

van den Brand, J.F.J.
Nikhef
P.O.Box 41882
NL–1009 DB AMSTERDAM
The Netherlands

Van de Vyver, R.
Dept. Subatomic and Radiation Phys.
RUG
Proeftuinstraat 86
B-9000 Gent
Belgium

van Leeuwe, J.J.
NIKHEF-K
P.O. Box 4395
NL-1009 AJ Amsterdam
The Netherlands

Vasilevsky, V.
Dep. Mathematics and Informatics
University of Antwerp (RUCA)
Groenenborgerlaan, 171
Antwerpen B2020
Belgium

Vento, V.
Departamento de Física Teórica
Universidad Valencia
Avda. Dr. Moliner, 50
46100-Burjasot (Valencia)
Spain

Viviani, M.
INFN
Universitá di Pisa
Piazza Torricelli 2
I-56100 Pisa
Italy

von Witsch, W.
University of Bonn
Institut fur Strahlen- und Kernphysik
Nussallee 14-16
D-53115 Bonn

Vorobiov, A.
High Energy Division
Petersburg Nuclear Physics Institute
Gatchina 188350
Russia

Wagenbrunn, R.F.
Institute for Theoretical Physics
University of Graz
Universitätsplatz 5
Graz A-8010
Austria

Walcher, T.
Institut für Kernphysik
Universität Mainz
J.J. Becher Weg 45
D-55099 Mainz
Germany

Walter, R.
Department of Physics
Duke University
DURHAM NC 27708
USA

Weber, S.
Institut für Theoretische Physik III
Univ. Erlangen-Nuernberg
Staudtstrasse 7
D-91058 Erlangen
Germany

Yang, S.N.
Department of Physics
National Taiwan Univ.
Taipei 10764
Taiwan

Yarevsky, E.
Department of Mathematical
and Computational Physics
Faculty of Physics
St.Petersburg State University
St.Petersburg 198904
Russia

Zejma, J.
Institute of Physics
Jagellonian University
Kraków
Poland

List of Authors

Springer-Verlag
and the Environment

WE AT SPRINGER-VERLAG FIRMLY BELIEVE THAT AN international science publisher has a special obligation to the environment, and our corporate policies consistently reflect this conviction.

WE ALSO EXPECT OUR BUSINESS PARTNERS – PRINTERS, paper mills, packaging manufacturers, etc. – to commit themselves to using environmentally friendly materials and production processes.

THE PAPER IN THIS BOOK IS MADE FROM NO-CHLORINE pulp and is acid free, in conformance with international standards for paper permanency.

GPSR Compliance

The European Union's (EU) General Product Safety Regulation (GPSR) is a set of rules that requires consumer products to be safe and our obligations to ensure this.

If you have any concerns about our products, you can contact us on ProductSafety@springernature.com

In case Publisher is established outside the EU, the EU authorized representative is:

Springer Nature Customer Service Center GmbH
Europaplatz 3
69115 Heidelberg, Germany

Batch number: 10726298

Printed by Printforce, the Netherlands